Contributions from Science Education Research

Contributions from Science Education Research

Edited by

Roser Pintó
Universitat Autònoma de Barcelona, Spain

Digna Couso
Universitat Autònoma de Barcelona, Spain

Springer

A C.I.P. Catalogue record for this book is available from the Library of Congress.

ISBN-13 978-1-4020-5031-2 (HB)
ISBN-13 978-1-4020-5032-9 (e-book)

Published by Springer,
P.O. Box 17, 3300 AA Dordrecht, The Netherlands.

www.springer.com

Printed on acid-free paper

TABLE OF CONTENTS

PART 3 SCIENCE TEACHERS' KNOWLEDGE, PRACTICE AND EDUCATION

PART 4 LEARNING AND UNDERSTANDING SCIENCE

PART 5 TEACHING AND LEARNING SCIENTIFIC CONCEPTS

PART 6 INNOVATIVE TEACHING–LEARNING ENVIRONMENTS IN SCIENCE EDUCATION

LIST OF CONTRIBUTORS

Ana S. Afonso University of Minho, Portugal

Virginie Albe Didactique des questions scientifiques, économiques et sociales émergentes, Ecole Nationale de Formation Agronomique, France

José M. Almudí Department of Applied Physics, University of the Basque Country, Spain

Annemarie M. Andersen The Danish University of Education, Denmark

Claudia von Aufschnaiter University of Hannover, Germany

Milena Bandiera Biology Department, "Roma Tre" University, Italy

Thorsten Bell Leibniz – Institute for Science Education (IPN), University of Kiel, Germany

J. Bellou Department of Primary Education, University of Ioannina, Greece

H. B. Berg Department of Teacher Education and School Development, University of Oslo, Norway

Amanda Berry Monash University, Australia

Kerst Boersma Centre for Science and Mathematics Education, Utrecht University, The Netherlands

Maja Brückmann IPN – Institute for Science Education, Kiel, Germany

Marion Budde Universum Science Center Bremen, Germany

Christian Buty UMR ICAR, COAST group, Université Lyon 2, France

Modesto Cabellos Dept. DCEM, Universitat de Barcelona, Spain

Ivo Čáp University of Žilina, Slovakia

Marina Castells Dept. DCEM, Universitat de Barcelona, Spain

Mikel Ceberio Department of Applied Physics, University of the Basque Country, Spain

Josep M. Cerveró Dept. DCEM, Universitat de Barcelona, Spain

Constantinos P. Constantinou Learning in Science Group, University of Cyprus, Turkey

M. Consuelo Domínguez-Sales Dept. de Didáctica de las Ciencias Experimentales y Sociales, Universidad de Valencia, Spain

Zoubeida R. Dagher University of Delaware, USA

Onno de Jong Centre for Science and Mathematics Education, Utrecht University, The Netherlands

Brigid Downing Manchester Metropolitan University, UK

Søren Dragsted The Danish University of Education, Denmark

Reinders Duit IPN – Institute for Science Education, Kiel, Germany

Juan Enciso Dept. DCEM, Universitat de Barcelona, Spain

Sibel Erduran University of Bristol, UK

Robert H. Evans Graduate School of Arts and Sciences, Wake Forest University, Winston-Salem, North Carolina, USA

Peter J. Fensham Monash University/Queensland University of Technology, Australia

Hans Fischer Research Group and Graduate School: Teaching and Learning of Science, University of Essen, Germany

Carles Furió-Más Dept. de Didáctica de las Ciencias Experimentales y Sociales, Universidad de Valencia, Spain

Bernhard Gerber University of Applied Sciences, Berne, Switzerland

John K. Gilbert University of Reading, UK

Damien Givry CNRS – Université Lyon, France

Wolfgang Gräber Leibniz – Institute for Science Education (IPN), University of Kiel, Germany

Jenaro Guisasola Department of Applied Physics, University of the Basque Country, Spain

Christina Hart University of Melbourne, Australia

Avi Hofstein Department of Science Teaching, The Weizmann Institute of Science, Israel

Monika Hüther University of Duisburg-Essen, Germany

María P. Jiménez-Aleixandre Universidade de Santiago de Compostela, Spain

Doris Jorde Department of Teacher Education and School Development, University of Oslo, Norway

Yasmin B. Kafai University of California, Los Angeles, USA

Alexander Kauertz Research Group and Graduate School: Teaching and Learning of Science, University of Essen, Germany

Brenda Keogh Millgate House Publishing & Consultancy Ltd, UK

Mira Kipnis Department of Science Teaching, The Weizmann Institute of Science, Israel

Birte Knierim University of Applied Sciences, Berne, Switzerland

A. Kontogeorgiou Department of Primary Education, University of Ioannina, Greece

Peter Labudde University of Applied Sciences, Berne, Switzerland

John Leach Centre for Studies in Science and Mathematics Education, The University of Leeds, UK

Jean-François Le Maréchal UMR ICAR, COAST group, Université Lyon 2, France

Tamar Levin School of Education, Tel Aviv University, Israel

Pere López Dept. DCEM, Universitat de Barcelona, Spain

John Loughran Monash University, Australia

Jane Maloney Institute of Education, University of London, UK

Linda McGuigan Centre for Research in Primary Science and Technology (CRIPSAT), University of Liverpool, UK

Helmut F. Mikelskis University of Potsdam, Germany

Silke Mikelskis-Seifert IPN – Institute for Science Education, Kiel, Germany

Tassos A. Mikropoulos Department of Primary Education, University of Ioannina, Greece

Katherine A. Muir University of Wyoming, USA

Pamela Mulhall Monash University, Australia

Patricia Murphy Centre for Education and Educational Technology, The Open University, UK

Stuart Naylor Millgate House Publishing & Consultancy Ltd, UK

Anja Neumann Leibniz – Institute for Science Education (IPN), University of Kiel, Germany

Hans Niedderer Mälardalens Högskola, Sweden

Feral Ogan-Bekiroglu Marmara University, Turkey

Jonathan Osborne King's College, University of London, UK

Marios Papaevripidou Learning in Science Group, University of Cyprus, Turkey

Bülent Pekdağ Faculty of Education, University of Balıkesir, Turkey

Vaughan Prain La Trobe University, Australia

Dimitri Psillos Aristotle University of Thessaloniki, Greece

Thorid Rabe University of Potsdam, Germany

Carlos Reigosa High School Fuentesnuevas, León, Spain

Thomas Reyer Research Group and Graduate School: Teaching and Learning of Science, University of Essen, Germany

Terry Russell Centre for Research in Primary Science and Technology (CRIPSAT), University of Liverpool, UK

Elwin Savelsbergh Centre for Science and Mathematics Education, Utrecht University, The Netherlands

Sascha Schanze Institute for the Didactics of Chemistry, University of Hannover, Germany

Shirley Simon Institute of Education, University of London, UK

James D. Slotta Ontario Institute for Studies in Education (OISE), University of Toronto, Canada

Helene Sørensen The Danish University of Education, Denmark

T. Strømme Department of Teacher Education and School Development, University of Oslo, Norway

Sigmar-Olaf Tergan Knowledge Media Research Center (KMRC), Tübingen, Germany

Maike Tesch IPN – Institute for Science Education, Kiel, Germany

Heike Theyßen University of Dortmund, Germany

Andrée Tiberghien UMR ICAR, COAST group, Université Lyon 2, France

David F. Treagust Science and Mathematics Education Centre, Curtin University of Technology, Australia

Russell Tytler Deakin University, Australia

A.E. van der Valk Centre for Science and Mathematics Education, Utrecht University, The Netherlands

Jan H. van Driel ICLON – Leiden University Graduate School of Teaching, The Netherlands

Arend J. Waarlo Centre for Science and Mathematics Education, Utrecht University, The Netherlands

Tili Wagner Department of Science Teaching, The Weizmann Institute of Science, Israel

Manuela Welzel University of Education Heidelberg, Germany

René Westra Centre for Science and Mathematics Education, Utrecht University, The Netherlands

Elizabeth Whitelegg Centre for Education and Educational Technology, The Open University, UK

Christoph T. Wodzinski IPN – Institute for Science Education, Kiel, Germany

Zacharias C. Zacharia Learning in Science Group, University of Cyprus, Turkey

Albert Zeyer University of Zurich, Switzerland

José L. Zubimendi Department of Applied Physics, University of the Basque Country, Spain

PREFACE

In August 2005, over 540 researchers in the field of Science Education Research met at the International European Science Education Research Association (ESERA) Conference in Barcelona, Spain. This was the fifth biannual ESERA conference since the foundation of the Association in 1995. As an organisation, ESERA has not stopped growing in the number of researchers becoming its members. The ESERA Conferences and Summerschools have been increasingly chosen as the international arena where to discuss and make public the new ideas and findings in the Science Education field. In this last ESERA conference in Barcelona, researchers came from all over the world from more than 40 different countries, mainly from Europe, but also from Asia, North and South America, Australia and Africa. Together with long experienced scholars, more than 150 young researchers discussed their work. As a result, 400 papers[1] were presented in the conference. Interestingly, more than 130 of them were presented as Simposia among different researchers from diverse research groups, providing enriched scenarios where specific themes of Science Education could be more widely discussed.

The increase in the number of researchers presenting their works in the ESERA conference is in part the result of the matureness of the Association and the Science Education Research field, but also responds to a global concern about Science Education over the last decade. Today there is an evidenced decrease in the number of students interested in school science in many countries, together with worrying results on students' scientific literacy worldwide. These findings from international comparisons among students' outcomes and interest in school science have reached the public arena, permeating the political, social and economical agenda of countries and institutions. A particular concern for this Conference was, then, the role and scope of Science Education Research in this problematic context. The contributions from Science Education Researchers to this public debate are extremely important, both in terms of adequate interpretation of findings and research-based proposals to guide future action. The name of the Conference, Contributions of Research to Enhancing Students' Interest in Learning Science (CRESILS), intended to show the importance given to the present problematic scenario by the Science Education Research community, together with the acknowledgment of its own crucial role in producing knowledge about how to address it.

The idea of Contributions of Research to address the present problems of Science Education was stressed in the Conference themes, and so, the researchers' contributions offered theory, findings and methods to guide future decisions and practices in an informed way. They intended to have a positive influence on the teaching and learning that takes place both in formal and informal contexts, in real or virtual scenarios, and everywhere science is to reach and be reached by citizens.

The present book wants to offer the reader some of these interesting Contributions from Science Education Research. This volume includes edited versions of the 37 most outstanding papers presented during the conference, including the lectures of the invited keynote speakers. All contributions have been selected for their quality, variety and interest, with the aim to offer the present panorama of the Science Education Research field worldwide. The review of these papers was done following a double-blind process which relied on the expertise of a group of distinguished scholars in the field.

The book is divided in 9 parts, covering a wide range of relevant topics for science education research nowadays, from more general to more specific ones. part 1 deals with the issue of students' interest in learning science, discussing the results of international studies such as TIMMS and PISA; presenting science education research results for enhancing students' interest and studying gender' issues and its relation with interest in science. Part 2 presents new approaches to Science Education research and discusses the impact of research in actual practice. Part 3 deals with Science Teachers Education, Knowledge and Practices, showing the importance given by the Science Education Research field to teachers, being this the topic with more papers presented both in the conference and the book. Parts 4,5 and 6 are devoted to research on teaching and learning. We have divided them according to their focus when studying the teaching and learning situation. Part 4 presents papers with a stronger focus in the general process of learning science. Part 5 includes studies focused on the actual teaching and learning of specific scientific concepts. Part 6 presents studies where innovative teaching and learning environments, either formal or informal, are analysed. The remaining three chapters of the book are devoted to specific themes in Science Education which are acquiring a great interest in the last years. These are Part 7 about Models and Modelling, Part 8 about Discourse and Argumentation in Science Education and Part 8 about Multimedia and Computer tools for teaching and learning science. We hope this classification would help the reader, even though the interrelation among different fields of educational research makes possible to classify many of the contributions in more than one of the mentioned sections

We want to thank the following people for their readiness to collaborate in the review process of this book and the quality of their suggestions:

Jaume Ametller
Claudia von Aufschnaiter
Paul Black
Ivo Čàp
Costas Constantinou
Justin Dillon
Reinders Duit
Mercè Izquierdo
María Pilar Jiménez-Aleixandre

Doris Jorde
John Leach
Martine Meheut
Robin Millar
Dimitris Psillos
Elena Sassi
Rosa Maria Sperandeo-Mineo
Andrée Tiberghein
David Treagust
Matilde Vicentini
Laurance Viennot

Finally, we want to acknowledge the support of the Spanish and Catalan educational authorities (Ministerio de Educación y Ciencia; Departament d'Universitats Recerca i Societat de la Informació; Departament d'Educació) and the Centre de Recerca per a l'Educació Científica i Matemàtica (CRECIM) of the Universitat Autònoma de Barcelona (UAB) in the publication of the present book.

The editors:
Roser Pintó
Digna Couso

NOTE

[1] The whole of the contributions to the conference are published in the R. Pintó and D. Couso (Eds) *Proceedings of the Fifth International ESERA Conference on Contributions of Research to Enhancing Students' Interest in Learning Science*, Barcelona, Spain. ISBN 689-1129-1

PART 1

STUDENTS' INTEREST IN LEARNING SCIENCE

PETER J. FENSHAM

1. INTEREST IN SCIENCE: LESSONS AND NON-LESSONS FROM TIMSS AND PISA

Abstract: The contemporary issue of students' interest in science has caught the attention of both the current large-scale, international science assessment projects, Trends in Mathematics and Science Study (TIMSS) and the Programme for International Study Assessment (Reading, Mathematics and Science (PISA). The manner in which these student attributes have been measured in TIMSS has, however, suffered from a number of methodological inadequacies. As just one small part of very large data collections the instruments used do not measure up to those that are now expected for these affective attributes by main stream researchers. Furthermore, the project's findings lack specificity in relation to both the target of the affect and in terms of the students' experiences with science with which their affect can be associated. Accordingly, the findings hitherto provide little insight into what curriculum authorities could do to counteract the now alarming lack of interest in science among students in the more developed countries. The PISA Science project for 2006 is committed to correct some of these weaknesses and its approach, in which affect is now explicitly an expected outcome of learning science, is described. PISA Science has already confirmed the role of *Science as A Story* as an approach to teaching/learning that may have positive affective outcomes

Keywords: Affective Assessment, Experience of Science Classrooms, Interest in Science TIMSS, PISA Science, Science as a Story

1. INTRODUCTION

In many developed countries there is a worrying lack of interest in science among secondary students. This lack of interest in the lower secondary years means fewer students choose to study the physical science in the later secondary years. In turn, fewer higher achieving students continue science studies at university. In Japan, where science study is required in senior high school and the comparative international student achievement is very high, the same lack of interest occurs in junior high school and university science study. This feature of science education in the more developed world stands in marked contrast to the much more positive affect that has been found among students in more developing countries (Sjøberg and Schreiner, 2006). In the complexities of the more developed world, the issue of interest in science has to be seen in wider terms than those of schooling. It involves

---

P. J. Fensham, Faculty of Education, Building 6, Monash University, Victoria 3800, Australia

R. Pintó and D. Couso (eds.), Contributions from Science Education Research, 3–10.

the comparative attraction of alternatives in out-of-school contexts, wider societal attitudes, and employment prospects.

In 2002, independent qualitative studies in three more developed countries reported similar findings about students' experience of studying science in the compulsory secondary years (England: Osborne and Collins, 2001; Australia: Lyons, 2006: Sweden: Lindahl, 2003).

Science classes were transmissive in character, Science content was abstract and not relevant to the lives of students, and Science was at the difficult end of the subjects being studied (even for the successful).
Do not continue to study the sciences (unless there are compelling extrinsic reasons).

These qualitative studies are limited in their representativeness, but they have identified features of school science education that may be linked to the issue of interest in science and which, in some of the newer science curricula, are being addressed (Aikenhead, 2005; Millar, 2006).

TIMSS, the Third International Mathematics and Science Study (after 1999, renamed Trends in Mathematics and Science Study) and PISA, the Programme for International Student Assessment, are current, large scale projects concerned with science achievement among school students within and across many countries. Both give some attention to affective aspects of students' response to science. TIMSS collected nationally representative data from Year 4 and Year 8 students in 1994/5, 1998/9 (Year 8 only), and 2002/3, while PISA has collected them in 2000, 2003 and currently in 2006. Since many more developed countries have participated, it is reasonable to ask, *Do these projects contribute to the understanding of lack of interest in science, and can they suggest ways to remedy this situation?*

2. FINDINGS

In the 20002/3 TIMSS study, the percentages of students with high *Self-confidence* ranged from 20% to 69%, and students in a number of the highest achieving countries had some of the lowest average values (see Ogura (2003) below for Japan). Within countries there is a low but significant correlation between *Self-confidence* and *Achievement.* The association between *Enjoyment of learning science* (one item in the *Valuing* variable) and *Achievement* is even lower – a not very helpful finding. The repetition of the TIMSS study enabled significant change in the percentage *Enjoying learning science* to be observed in a few countries from 1994/5 to 2002/3. In the three reportings thus far from TIMSS, the measures of student affect about science and of teachers' views of students' affect have all been in the high inference category, from which little can be learnt about how to change the levels of student interest.

The PISA project has more explicitly recognised a distinction between motivation and engagement with school and interest in a specific subject. The former are seen as the "energy bases" for learning in school and beyond school (OECD, 2001). Interest in a particular subject, it is argued, can affect the degree and continuity of engagement in a student's learning of this subject and hence, the depth of understanding that is achieved, but this effect can be largely independent of the student's general learning. Girls in Thailand provided an example of this, when educationally, upwardly mobile girls were found to be high achievers in physics (a compulsory subject in the senior secondary years), but had very low interest in this subject (Klainin and Fensham, 1989). Similarly, Ogura (2003) has reported that the role of science in university entrance examinations in Japan relates to the low interest/high achievement results among students in that country. The PISA data for mathematics provides examples of countries in which almost every possible combination of these affective variables are operating, but does not include other in-country specific data to suggest bases for these relations.

Because *Science* was a minor domain in PISA in 2000 and 2003, the only affective data collected in these studies related to Reading and to Mathematics and to schooling as a whole. The data and the findings in the Mathematics study are discussed at length in the international report of the 2003 testing, but little emerges of a definitive nature about how *Value* and *Interest* relate to cognitive achievement, or how interest is so different in some countries (OECD, 2004).

3. METHODOLOGICAL SHORTCOMINGS

3.1 Proviso

In defence of the shortcomings in their methodology and findings, it should be said that neither TIMSS nor PISA (prior to 2006), has been primarily been a study about students' interest in science or science learning. Both projects are primarily concerned with comparative measures of student science learning, albeit with very different foci for this learning.

3.2 Validity of Measures

Both projects, in addition to their test of science learning achievement, have collected a great deal of data via questionnaires from the students and their schools (TIMSS and PISA), their teachers (TIMSS) and their parents (PISA) These data are to explore factors that may be associated with the students achievement, nationally and internationally. Among the many items for response in the TIMSS student questionnaires, only a small number of items were used to construct measures of affective variables. These measures are thus higher inference measures than would now usually be acceptable in studies specifically concerned with affect and science (see Hoffmann, 1998). For example, the variables, *Liking of science* and *Attitude to science* in TIMSS 1994/5 (Year 4) were based on one item and three items respectively; while *Self-confidence in learning science* and *Valuing science* had 4 and 7 items in TIMSS 2002/3. The wording of these items refer to the students' personal sense of *Valuing science* in their education, now and for their personal futures, and not to science as a research or applied enterprise in society. Again, the teachers' estimate in TIMSS, 2002/2003 of *Disinterest among their students* was just one item of a six item potpourri scale that also included *Range of backgrounds* and S*tudents with special needs!*

By using the generic term, *"science"* in the questionnaire items, it is quite an assumption that the 14 and 15 year old students across, and within the diverse countries, will be responding to the same meaning of this word. Gardner (1975) pointed out many years ago that *"science"* is an umbrella word, and like *"a table"*, is multidimensional. To use it in an undifferentiated way is like asking, *How large is a table?* without defining that *large refers to*" *the table's length* or *leg height*, etc. In the same way, the study of science at school, let alone science in society, has a number of different components; and it is at this differentiated level that affective information would be useful. The projects' constructs have not yet been designed to be measured at such differentiated levels.

In PISA, for both the 2000 and 2003 studies, affective measures were constructed only for the major domain – Reading and Mathematics respectively. These were more subtly constructed in a manner that holds some promise for Science in 2006, when it is the major domain (see below).

3.3 Location of Student Affect in the Projects' Instruments

In TIMSS, in 1994/5 and its repeated studies in 1999 and 2003, and in PISA Reading and Mathematics in 2000 and 2003, the placement of the items for the affective variables were located, as indicated above, in the Student and Teacher Questionnaires that were separate from the Achievement Test. By placing these items in instruments that were designed to measure the student's personal context, the project managers, in effect, treat the affective constructs as personal attributes, like sex, parents' education, books in the home, etc. Their divorce from the cognitive Achievement Test means they are not presented as expectations of learning science, nor can they be related to the specific science topics in the Test. The measures of the affective constructs are treated as independent variables that may be correlatable with cognitive achievement, rather than as outcomes of learning that are directly associable with the science at issue in the test unit.

3.4 Comparative Aspect

The projects make no use of the comparative nature of affective constructs. Many of our attitudes and interests are determined comparatively, and studies have shown that comparing student interest in science as a rank order among other subjects being studied can give different information from simply asking about science. The students' responses in the qualitative studies above were very often couched in comparative terms, for example, *the absence of discussion, or place for my opinion, in science compared with other subjects*. Ogura (2003) in his large study in Japan asked students about their interest in each of the school curriculum's main subjects. Science was not alone in losing interest from Year 6 onwards, but following a question about the intrinsic worth of the content of these school subjects, science and mathematics remained low, compared with the positive responses that were now expressed for social sciences and humanities.

3.5 Interpretation of Affective Responses

Ogura's study is an example of how linked items can greatly assist the interpretation of primary responses. In its cognitive achievement test PISA Science, in contrast with the isolated items in TIMSS, the format is of a number of presenting science contexts with several associated items. This immediately has a linkage format that has enabled secondary analyses with interesting results to be made (Kjærnsli and Lie, 1999). Nevertheless, prior to PISA's 2006 study, there has been no deliberate design of linking items to elucidate how to interpret a student's response to the primary affective items. For example, items about the experiences of learning science in school (like those in the qualitative studies above) could enable connections to be explored between affect and classroom experiences. It is these links that can assist authorities to take curriculum initiatives in content, pedagogy and assessment that may improve interest in science.

4. PISA SCIENCE 2006

For the PISA Science test in 2006, steps have been taken to improve on several of these shortcomings. The development of the **Framework for Science** as the major domain in 2006 has benefited from the publications that have so well documented the lack of interest in science and the recent literature on its possible links to students' classroom experiences.

The definition of scientific literacy has been extended and now includes *"an individual's willingness to engage in science-related issues and with ideas of science, as a reflective citizen"* (OECD, 2006, p. 23). Consequently, affective items will be included in the sets of items associated with a number of the contextual science units that will constitute the student Achievement Test. These affective items are constructed to represent two variables – *Interest in science* and *Support for scientific enquiry* – which will be scored separately from the cognitive items that, in 2006, will lead to scores in three scientific competencies. The presence of these items in such a prestigious test of science learning will be a clear message that affect about science, as well as cognition of it, is intended to be an outcome of learning for all students. This step will also recognize the differentiated and comparative nature of affect, since the items will now refer to the specific science in each contextual unit. Students will be able to react positively to some, and negatively to others. This will offer students something like the range of response to particular science issues that has marked the more recent studies of public understanding of science (Bauer and Schon, 1993).

The constructs, *Interest in science* and *Support for scientific enquiry* are elaborated in the Framework for 2006 as follows.

4.1 Interest in Science

- Indicate curiosity in science and science related issues and endeavours
- Demonstrate willingness to acquire additional scientific knowledge and skills, using a variety of resources and methods
- Demonstrate willingness to seek information and have an ongoing interest in science, including consideration of science careers

4.2 Support for Scientific Enquiry

- Acknowledge the importance of considering different scientific perspectives and arguments
- Support the use of factual information and rational explanation
- Express the need for logical and careful processes in drawing conclusions (OECD, 2006, p. 37)

The student's questionnaire will also include items that will give measures of these two affective constructs in a non-contextualised form along with a third construct, *Responsibility towards resources and environments.* This will enable an assessment of whether students' attitudes vary in and out of context and whether they vary between contexts.

The other Sections of the student questionnaire have been substantially revised, compared with its form in 2000 and 2003. The important socio-economic and family support constructs will still be measured, but five sets of items are now included that relate to science and science education. They will probe the students' views about science, their awareness (and its source) of specific scientific and environmental topics, their interests in science learning in- and out-of-school, their awareness and interest in science careers, and their experience of the teaching and learning in science classrooms. The last of these aspects is especially detailed, covering the occurrence of 18 different activities – many relating directly to the qualitative studies' findings listed at the beginning of this paper.

These sets of items have been deliberately included to provide the opportunity to explore their links with the responses about interest in science. That is, they will enable secondary analyses to derive which features of school science in a given country do relate to interest and which do not. Each of them represents a science education condition that is amenable to change by school authorities and teachers who wish to remedy the interest issue.

The parents of their students are the other group, that schools and teachers have some hope of engaging in initiatives to improve interest in science. Accordingly, the Parent's Questionnaire for 2006 is gathering data on the parents' views on science as a possible career, and on their awareness of, and concern about the same science and environmental issues as on the Student Questionnaire. These can be related to the student's responses to see where reinforcement of attitudes is occurring – positively or negatively.

In these ways the PISA Science testing in 2006 offers much greater promise to address the issue of lack of interest in science than has been the case in the earlier international studies.

5. SCIENCE AS A STORY: POSITIVE EVIDENCE FROM PISA

Teaching *Science as a Story* was strongly advocated in the mid-1990s as an approach that may have positive affective outcomes as well as providing content learning in more authentic scientific and applied contexts (Campbell et al., 1994; Malcolm, 1996). Hitherto, there have been few research reports on what is achieved when this approach is implemented. PISA Science in 2000 and 2003 has indirectly provided very positive evidence for the worth of this approach, when the storyline is embedded as part of the assessment of science learning. *Reading* was the dominant domain in PISA 2000, and in all 32 participating country girls substantially outperformed boys. In the limited *Science* testing, the decision was made to concentrate on the students' capacity to understand and analyse actual media reports involving science – as one important aspect of scientific literacy. This led to a science test that involved more reading than usual, and might have favoured girls in the light of the *Reading* Test in which girls performed significantly better in all countries. The remarkable findings in both the 2000 and 2003 *Science* Tests of 26 countries showing no significant gender difference can only be explained, I believe, in terms of the engaging nature of the media reports that introduced the test items in the test units. Many of these were little stories (e.g. the unit about Dr. Semmelweis and fatal maternity disease, OECD, 2001), a style journalists clearly know is of interest to readers and which contrasts with traditional textbook presentations of science.

REFERENCES

Aikenhead, G.S. (2005) *Science for Everyday Life: Evidence – based practice*. New York: Teachers College Press.

Bauer, M. and Schon, I. (1993) Mapping variety in public understanding of science, *Public Understanding of Science*, 2(2), 141–155.

Campbell, B., Lazonby, J., Millar, R., Nicholson, P., Ramsden, J. and Waddington, D. (1994) Science the salters' approach – A case study of the process of large scale curriculum development, *Science Education*, 78(5), 415–447.

Gardner, P.L. (1975) Attitudes to Science: A review, *Studies in Science Education*, 2, 1–41.

Hoffmann, L., Krapp, A., Renniger, K.A. and Baumert, J. (eds.) *Interest and Learning*. Kiel, Germany: IPN.

Kjærnsli, M. and Lie, S. (1999) PISA and Scientific Literacy: Similarities and differences between the Nordic countries, *Scandinavian Journal of Educational Research*, 48(3), 271–286.

Klainin, S. and Fensham, P.J. (1989) Successful achievements by girls in physics learning, *International Journal of Science Education*, 11(1), 101–112.

Lindahl, B. (2003) *Pupils' Responses to School Science and Technology?* (Published dissertation), Göteborg: Acta Universitatis Gothobugensis.

Lyons, T. (2006) Different Countries, Same Science Classes: Students' experiences of school science in their own words, *International Journal of Science Education*, 28(6), 591–614.

Malcolm, C. (ed.) (1996) *Could We? Should We? Science for Year 10*. Carlton, Vic., Australia: Curriculum Corporation.

Millar, R. (2006) Twenty first century science: Insights from the design and implementation of a scientific literacy approach in school science, *International Journal of Science Education*, 28(13) 1523–1544.

OECD (2001) *Knowledge and Skills for Life: First results from PISA 2000*. Paris: OECD.

OECD (2004) *Learning for Tomorrow's World: First results from PISA 2003*. Paris: OECD.

OECD (2006) *Assessing Scientific, Reading and Mathematical literacy: A Framework for PISA 2006.* Paris: OECD.

Ogura, Y. (2003) *Informal Science Education for Promoting Children's Science Learning in Japan.* Paper presented at 2003 International Seminar on Improvement of Students' Science Achievement and Attitude through Informal Science Education, Seoul, Korea, December 5–6.

Osborne, J. and Collins, S. (2001) Pupils' views of the role and value of the science curriculum: A focus group study, *International Journal of Science Education*, 23(5), 448–467.

Sjøberg, S. and Schreiner, C. (2006) How do learners in different cultures relate toscience and technology? Results and perspectives from the project ROSE (the relevance of Science Education), *APFSLT: Asia-Pacific Forum on Science Learning and Teaching*, 7(1) Foreword.

DAVID F. TREAGUST

2. RESEARCH-BASED INNOVATIVE UNITS FOR ENHANCING STUDENT COGNITIVE OUTCOMES AND INTEREST IN SCIENCE

Abstract: This paper is organised around two issues, firstly locating interests in science within science education research and secondly providing examples of four types of research-based approaches to curriculum. The first issue, from the extant literature, is concerned with present trends in science enrolments and interest in science, primarily in developed nations. Three questions organise this Section: do current science curricula achieve the intended affective (interest) and cognitive (achievement) goals; how are interests in science best measured; and what is the contribution of research on conceptual change for science achieving affective and cognitive goals? The second issue provides examples of the evaluation of innovative 'units' – interpreted as topics within a curriculum – in science for enhancing affective and cognitive goals, both from other colleagues and from my own work. The relative incidence and viability of the four research types are discussed. The paper concludes with recommended research needed to move the field of interests in science forward, claiming that without research to determine whether or not innovative research-based curricula or topics within a curricula are perceived by students as being interesting, and that successful learning outcomes ensue, there is little likelihood of arresting this decline of enrolments in science

Keywords: Affective goals, Cognitive gaols, Conceptual change, Enrolment trends in science, Innovative curricula, Interest

1. LOCATING INTERESTS IN SCIENCE WITHIN SCIENCE EDUCATION RESEARCH

My research agenda tends to be oriented towards cognitive outcomes of students' learning of science so it was a challenge to be asked to talk on issues related to the theme of this conference, namely, interests in science in the context of research-based units at all levels of science education. This seemed like a worthwhile challenge and the first part of this paper is a review of some relevant literature, with some examples of types of research-based approaches to curriculum 'units' in science for enhancing affective and cognitive goals, particularly from my own work.

Science education has a long tradition of ensuring the improvement of cognitive outcomes and there is an assumption that attitudes and interest will naturally follow. Certainly research into the relationship between interest in science and

David F. Treagust, Science and Mathematics Education Centre, Curtin University of Technology, GPO Box U1987, Perth WA 6845 Australia

R. Pintó and D. Couso (eds.), Contributions from Science Education Research, 11–26.

achievement has been conducted over four decades. Recent research, however, has not definitively identified a causal link between interest and achievement (see Osborne et al., 2003).

2. WHAT ARE PRESENT TRENDS IN SCIENCE ENROLMENTS AND INTEREST IN SCIENCE?

Recent reports from the findings of the Programme for International Student Assessment (PISA) (OECD, 2004) and Trends in Mathematics and Science Studies (TIMSS) (2006) show a wide range of scientific literacy among the nations that participated among school aged youth. Results were often less than expected in some nations and even where results were acceptable, one only has to read the local, national and international press to realise that all is not well with science education in our communities. Whether or not students performed well on these international tests, there are numerous reports about the declining student enrolments and interests in science at both high school and university levels (Lyons, 2006). Such data have been reported in Australia (Dekkers & De Laeter, 2001; Goodrun et al., 2001; Thomas, 2002), in the UK (Murphy & Beggs, 2003), in Germany (Haas, 2005), and in France (Nature, 2002). That this topic is of international significance was illustrated by the need for a conference entitled 'Declining Student Enrolment in Science & Technology: Is It Real? What Are The Causes? What Can Be Done?' (OECD Global Science Forum, 2005) that involved over 200 participants from 26 countries debating ways to attract young people to science.

Although more students are studying at a higher education level in most western nations, the proportion taking up science and engineering and especially physical sciences is declining (Hassan & Treagust, 2003; OECD Global Science Forum, 2005). Certainly, the physical sciences no longer attract students as they once did and these reports are not restricted to science education journals. In the UK, chemistry student enrolments are down by 25% and physics students in France have fallen by 46% in recent years. Australia, Germany, the USA and Japan all face similar declining enrolments in the physical sciences (Lyons, 2006). In the UK, Onlooker (2005) in a pharmaceutical journal writes "Is science on the way out of education?" describing the closure of the chemistry department at the University of Exeter and an eminent chemist returning his honorary degree. An article in *Science* on 4 Feb 2005 described how similar closures of science departments have occurred to many universities during the past decade. That this is a concern becomes most evident when the government, in this case in Ireland, begins to redress the falling enrolments in chemistry (Childs, 2002) after science that has been neglected over the many years.

2.1 What Does the Literature say about Trends in Science Achieving Intended Affective and Cognitive Goals and How are Interests in Science Best Measured?

Extensive literature reviews of students' attitudes to science have been conducted approximately each decade (Gardner, 1975; Osborne et al., 2003; Schibeci, 1984;

Simpson et al., 1994;) and provide a useful measure of the state of this construct amongst school aged youth – at least in English speaking countries. These data are consistent with Ramsden's (1998) view about the need to gather data to provide an up-to-date baseline of information about students' attitudes to science and interests.

In Australia Goodrun et al. (2001) reported that almost 40% of secondary school students indicated that they never got excited about what they were doing in science and 22% indicated that they were bored in science. Of course the difficulty level or at least perceived difficulty level and science perceived as a foreign language might negatively influence students' interest in science.

The Programme for International Student assessment (PISA) addresses a scientific literacy framework and is interested in providing information about the declining interests in studying science. In 2006, students' attitudes are being evaluated by PISA in three ways – interest in science, support for scientific enquiry, and responsibility for sustainable development – to provide an international portrait of students' general appreciation of science, their specific scientific attitudes and values, and their responsibility toward selected science-related issues that have national and international ramifications. Earlier, PISA 2000 examined interest in mathematics and reported that about half of 15 year olds considered mathematics to be important but rather fewer thought this a reason for pursuing it further. Across all OECD countries, an average of only 14% agreed that mathematics is fun and would not want to give it up. In an attempt to find out why Swedish students loose interest in science in secondary school, Lindahl (2003, 2005) documented the effects of school science experiences on attitudes and intentions to enrol in senior science courses showing convincingly that students were discouraged by school science which they found it boring.

For the affective measure of attitude, most studies tend to use questionnaires comprising several scales that have been developed with high levels of reliability and validity. Examples include, the *Test of Science Related Attitudes* (TOSRA) (Fraser, 1981) for use with students from grades 4 to 10 that has seven scales such as Enjoyment of Science, Inquiry of Science and Future Careers. More recently, Dalgety et al. (2003) developed and reported on a Chemistry Attitudes and Experiences Questionnaire used with first year chemistry students and Jarvis and Pell (2002) constructed a reliable instrument Pupil's Attitudes to Science for primary school students aged 5–10 years that includes scales such as Interest in Science.

However, measurement of interests by use of questionnaires limits students' responses to pre-defined categories and facilitates statistical analysis though there is little opportunity to establish the meanings behind student responses. So while we have data on such questionnaires that are valid and reliable, what do the analyses tell us about interest and enjoyment in science? This is a key issue in the recent debate about the need for research-based studies in education because the data are generally recommended to be gathered in a way that allows quasi-experimental studies (Maxwell, 2004).

Research often goes in cycles and I am reminded of the symposium held at the IPN in Kiel Germany in1984 on Interests in Science and Technology Education that resulted in an edited volume by Lehrke et al. (1985). Papers dealt with issues relevant today, declining interest in science, reducing disadvantages for girls in science, programs to improve motivation and development and implementation of questionnaires to measure interests. Studies continued to be conducted in Germany by Häussler and Hoffmann (2000) about students' interests in science providing a framework for future action. A more recent conference on interest and learning was held in Kiel in 1998 (Hoffmann et al., 1998).

A growing number of researchers are focusing their work upon the emotional and motivational aspects of learning science though I believe that this research is in its infancy. Alsop and Watts (2003) partly filled this void when they edited a special issue of the IJSE looking at the effect of affect on learning science stating that this is an "often overlooked domain" (p. 1044). Recently Alsop (2005) has edited a book looking at affect in the teaching of science. In their special issue, Alsop and Watts provided articles with different methodologies from the familiar statistical analyses of attitudes to science to the personalised, emotionally charged account of teaching and learning.

Briefly, Osborne et al. (2003) from the UK conducted an excellent review and analysis of the literature on attitudes towards science; among many issues, the authors showed a decline in students pursuing post-compulsory science education in the UK. The authors concluded that "research continues to record declining attitudes but has little to say about how this problem might be remediated but suggest that the answer might evolve from the growing body of literature exploring motivation, interest and task value" (p. 1045). A second paper by Zusho et al. (2003) from the US looked at how the level of motivation and use of specific cognitive and self-regulatory strategies changed over time and how these motivational and cognitive components in turn predicted students' course performance in chemistry. Motivation and strategy use were assessed at three time points over the course of one semester using self-report instruments completed by 458 students. The results of the study showed a decline in students' motivational levels and use of rehearsal and elaboration strategies over time but an increase is students' use of organisational and self-regulatory strategies. The motivational components of self-efficacy and task value were the best predictors of final performance. A third paper by Teixiera Dos Santos and Mortimer (2003) from Brazil in two different classes using ethnographic research in a way that sought to include raw emotions such as pleasure pain, fear surprise, sadness and annoyance. The findings from the study suggested that science teachers might benefit from scholarly empirical work exploring aspects of facial expressions of the psychology of emotion. The final paper by Perrier and Nsengiyumva (2003) from France/Rwanda showed how hands-on science with child victims of violence and war developed from isolated, silent and sad to being enthusiastic and displaying positive attitudes. What I gleaned from these papers is the need for more studies that meet the goals of being legitimate as educational research but which take into account the context and meanings that students make of the answers they give to questions.

More recently, Zembylas (2005) from the University of Cyprus has examined in some considerable detail the relation between the cognitive and emotional aspects of learning science by examining three perspectives. The first perspective is embedded in and draws upon conceptual change theory, the second perspective is informed by the idea of the social and cultural construction of learning, the third perspective is informed by post structural insights involving language and discursive practices. Zembylas contends that emotion is more integrated with cognition in the last perspective and least in the first perspective.

2.2 *What is the Contribution of Research on Conceptual Change for Science Achieving Affective (and Cognitive) Goals?*

There is ample evidence in research on learning and instruction that cognitive and affective issues are closely linked. However, the number of studies on the interaction of cognitive and affective factors in the learning process is limited. There are, for instance, many studies on the relations between interests and acquisition of science concepts. However, these studies are usually restricted to correlations between interests and cognitive results of learning. The interplay of changes of interests and conceptual change is investigated only in a small number of studies.

The conceptual change perspective asserts that a concept has to be built upon students' prior ideas about that concept and that the learning process has to be embedded in supporting conditions including "motivation, interests and beliefs of learners and teachers as well as classroom climate and power structure" (Duit & Treagust, 1998, p. 15). For more than 20 years, researchers in this area have examined teaching and learning of science concepts from different perspectives alongside the purely cognitive one; one such perspective is the role of motivation and interest in conceptual learning (Pintrich et al., 1993). Motivation can be either extrinsic (i.e., reward for an activity) or intrinsic (i.e., interest in an activity). Recently, there has been a shift in focus from a personal motivation perspective to one that takes into account classroom contextual factors. Such a shift is in line with social constructivist approaches in science education and as discussed by Zembylas.

To construct a holistic picture of learning, it is both possible and beneficial to consider a learning situation from differing theoretical perspectives of conceptual change. For example, rather than only considering conceptual changes in knowledge that a student constructs in moving from, say a prescientific notion to a scientific view of a concept, a more complete and informative picture would be painted if these changes were viewed from a multidimensional perspective. The way the student views a concept in terms of its status (Posner et al., 1982) or its ontological category (Chi et al., 1994) or the motivational and contextual factors (Pintrich et al., 1993) necessary to precipitate the conceptual change also should be considered. From this point, it becomes clear that a multi-dimensional framework utilising differing perspectives of conceptual change to view a learning situation has merit.

The multi-dimensional framework for interpreting conceptual change by Tyson et al. (1997) includes, for instance, an epistemological, ontological and an affective domain, though the affective domain needs to be more fully elaborated. It appears that it is fruitful to merge ideas of conceptual change and theories on the significance of affective factors. It also seems to be most valuable to view the issue of interests in science and science teaching from the perspective of conceptual change. Clearly, it is an important aim of science instruction to develop interest in much the same way as to develop students' pre-instructional conceptions towards the intended science concepts.

Within this conceptual framework is the question of whether analogies can contribute to conceptual change. Dagher (1994) suggested that analogies contributed to students' level of comfort and security, raising the students' interest and enhancing their imaginative potential. Venville and Treagust (1996) analysed a teaching situation where an analogical model was used to describe the function of the heart. The analysis was performed and the data interpreted from Pintrich et al.'s motivational/affective perspective; the authors found that the simplicity of the heart model and the way the teacher overtly used it for motivation raised the students' self-efficacy, convincing them that learning this difficult topic was something they could achieve. The authors concluded that the analogical model of the heart contributed to the process of conceptual change by acting as a "motivator".

3. TYPES OF RESEARCH-BASED APPROACHES FOR DEVELOPMENT OF CURRICULUM 'UNITS' IN SCIENCE FOR ENHANCING AFFECTIVE AND COGNITIVE GOALS?

To address the goals of this paper, I have used four research-based approaches to curriculum development and implementation, adapted from the typology described in Gilbert, de Jong, Justi, Treagust and van Driel (2004), and comment about their relative incidence and viability.

3.1 Development and Evaluation of Research-based Curriculum Units Intended to Enhance Affective and Cognitive Goals

Many curriculum developments have been undertaken but never researched and evaluated in terms of the influence on affective and cognitive goals. Fortunately, however, there are excellent examples of projects where curriculum materials were produced and implemented, after which their effectiveness in the classroom was evaluated with the aim of improving their capacity to support the learning of scientific ideas. One example is that described by Barker and Millar (2000) who conducted a longitudinal study of 400 upper secondary level students at 36 schools in England following post-compulsory conventional A-level chemistry courses and a context-based course, *Salters Advanced Chemistry*. The study employed a series of diagnostic questions on key areas of chemical understanding, administered at three points over an 18-month period. Understanding of the following chemical ideas was probed: elements, mixtures and compounds, conservation of mass and reacting masses, characteristics of chemical reactions, chemical bonding, energy changes in chemical reactions, rates of reaction, and equilibrium reactions. The study indicated that there were no significant differences in understanding between the two groups, though the context-based approach appeared to offer a slight advantage in developing ideas about chemical bonding and thermodynamics. Common areas of difficulty across all courses also emerged, in particular with ideas about ionic bonding, inter-molecular forces and open system chemical reactions. However, the students who experienced the gradual introduction and revisiting of ideas (such as chemical bonding and thermodynamics) in different contexts at several points during the course appeared to develop better understanding of these ideas than students following more conventional courses.

Developing science curricula that are located in context is a growing area of interest in science education resulting most recently in a special issue of the *International Journal of Science Education* (Pilot & Bulte, 2006) and an edited book (Nentwig & Waddington, 2005). Most of the authors contributing to these volumes have presented various types of evaluation in terms of meeting both affective – interest or motivation – or achievement goals and are well worth reviewing.

In my own work, I have been interested in the effect of analogical-based instruction to engender interest as measured by a conceptual change approach. Research has shown that analogical teaching approaches can enhance student learning although analogies for teaching and learning can be a friend or a foe depending on the approach taken by the teacher (Harrison & Treagust, 2006). An example of analogies being friends is from the findings of a study involving the teacher's use of analogies in a topic/unit on optics in a 10th-grade physics class in an all-girls school (Treagust et al., 1996). The curriculum design involved an approach for teaching with analogies in an effective manner (Treagust et al., 1998). The teacher's use of a cart with wheels moving obliquely over different surfaces as an analogy for refraction of light successfully engendered conceptual change in student learning about the refraction of light. One class was taught without this analogy and another with the analogy. On a standard test at the end of the unit, comparison of the means and the patterns for the two classes indicated that the non-analogy class scored as well as the analogy class. However on responses to an interview protocol involving ray tracing tasks then scored as a conventional test, the analogy group produced explanations of refraction which were of a higher status than those in the non-analogy group.

It is asserted that the analogy enabled these higher status explanations to be generated; alternatively, these data can be interpreted from the perspective that the analogy was acting simply as a tool of explanation. It is certainly likely that the concrete nature of the analogy enabled the students to remember what they had learned about refraction. Students have difficulty articulating the refraction phenomenon; subsequently, it would seem reasonable that students who possess a familiar analogy will produce the dichotomy seen in the class results. That is, the analogy provided students with language they could comfortably utilise to transform an abstract idea into an articulate explanation. The non-analogy group lacking a familiar means for describing their conceptions could explain the difference in status between the two classes. The point at issue in this paper is the need to measure and record both the cognitive and the affective outcomes. The cognitive outcomes are discussed in terms of results on tests. The affective domain is much more difficult to report in any conventional way – there was no questionnaire involved.

In this study, we reported case studies of students to illustrate where the analogy's affective dimension shows how crucial it was to learning. One such case involved Dana who was vague and unenthusiastic during the initial phase of the interview. When asked "what do you think will happen to the ray of light when it hits the surface of the water?" she responded, "It will probably be stopped and spread." Her sketch depicting this and her subsequent responses when asked "Why does that happen?" were "I don't know", and a similar reply was given for other questions. Dana did not volunteer any explanation, however simple, to account for refraction. However, after Dana had completed three sketches in the interview, the interviewer asked her if she could think of any simple analogy that would help you explain to a friend why those pencils appear to be bent? After a little hesitation Dana recalled "a car type of thing with wheels when it was changing from a piece of carpet to paper". In the subsequent interview, Dana confidently answered the interviewer's questions about the optical phenomena under investigation, acknowledging that when she thought of the wheels, she could work out which way the light rays bend. Dana's answers to the subsequent ray tracing problems (glass block and prism) were among the best we saw of the 39 students. Her physics score on these tasks became 10/10 and this from a student the teacher placed in the bottom 25% of the class and from a student who failed the Optics unit! The tape recording of the interview adds another facet to this story. Until the advent of the analogy, Dana was quietly spoken and disinterested. When the analogy was recalled, she became enthusiastic and the interview produced another four pages of transcript. Dana's initial answer to the first question was inaccurate and showed little evidence of understanding. The introduction of the analogy led to her becoming dissatisfied with her initial answer to the first question and she confidently changed her vague sketch into a correct response.

3.2 Action Research Involving Research-based Curriculum Units Intended to Enhance Affective and Cognitive Goals

Reports of action research are not widely found in the science education research literature, probably because teachers and instructors at all levels too often lack the skills, incentive, and time, to conduct such studies. Even if the teachers and instructors do conduct such studies, the formal education system rarely provides teachers with incentives to write-up their work for publication. In this paper, I report on action research studies conducted by two of my doctoral students – one in university chemistry and a second in high school science. Working with first year university chemistry students in Thailand, Chaimaneewong (2005) developed POEs to teach key concepts for the rate of reaction using three optimized POE models based on inconsistent or discrepant events in science education. The reason for introducing these POEs was because the university's chemistry curriculum was not ensuring understanding or interest, having been developed from English language textbooks for more than 30 years. What is stated in the books tends to be translated, rearranged and used by instructors to explain concepts to Thai students in the classroom. The foreign textbooks contain a tremendous amount of facts, principles, rules, laws, theories and exercises in almost the same fashion, such that only the most intelligent readers can grasp the concept. Thai teachers inevitably use the same sequence of course structure for describing the content to students, while students passively listen and take notes.

To address this concern, Chaimaneewong designed and developed an action research cycle of plan, act, observe, and reflection, to produce three optimized Predict-Observe-Explain (POE) models based on inconsistent or discrepant events in science education. Discrepant events play a key role in the POE model, but it is the intended inconsistency of the model rather than unrecognized inconsistency that is important (Fensham & Kass, 1988). The three core questions of using anomalous data suggested by Chinn and Brewer (1993) have also been taken into consideration for the three optimized POE models: What are different responses a student can make from anomalous data? What are the different conditions that lead to anomalous data? What is observable? Three main tasks were conducted in Chaimaneewong's study. The first task involved the optimization on the three POE models. The models, which are concerned with basic chemical kinetics, were designed and trialed with university students in an action research study. The three types of chemical reactions – acid-base neutralization, a clock reaction, and a decomposition reaction – were rearranged to meet the POE model's criteria. Students' preinstructional conceptions were the most important clues used to guide this development. The results showed that students who encountered anomalous data were able to construct their own knowledge after responding to the POE models. Students' responses to an attitude test produced a mean average of 4.17 (out of 5) indicating students had a highly positive attitude towards the POE model.

In another action research study, Liew (2004) sought to explore the effectiveness of the Predict-Observe-Observe-Explain (POE) teaching/learning technique to diagnose grades 9, 10, 11, and 12 students' understanding of science and identify students' level of achievement. Data collected were interpreted using three theoretical perspectives, namely, Chi et al.'s theory of ontological categories, Hewson and Hennessey's conceptual change theory to determine the epistemological status of students' understanding of science, and Chinn and Brewer's model to classify types of students' responses to contradictory observations. This purpose of using this methodology was to obtain an in-depth, plausible and credible account of students' understanding and their level of achievement.

POE tasks were concerned with heat and the expansion of water, solubility of salt, and power and resistance of light globes. The data revealed common ideas amongst students that are contrary to scientists' science; furthermore, students showed that they were able to articulate their own ideas based on the POE tasks. The findings of this research revealed that these POEs were effective in capturing a range of possible student observations and prediction outcomes when worded in an open-ended format. Quality information on students' understanding and on the way they responded to contradictory data was obtained when POEs were administered by teacher demonstrations and were designed to produce phenomena that were clear, immediate and had only one aspect to observe. Furthermore, the results of this research suggest that POEs are effective in diagnosing students' understanding of science and their level of achievement. Students engaged in this kind of science teaching show positive attitudes towards science classes.

3.3 Research-based Units Incorporating Multiple External Representations Intended to Enhance Affective and Cognitive Goals

Recently, new perspectives on computer-based multiple representations used in instruction (Ainsworth, 1999) have provided a robust framework to interpret analogical models and their relatives such as metaphors which are used for reasoning. According to Ainsworth's (1999) conceptual analysis of existing computer-based multi-representational learning environments (Ainsworth et al., 1997), there are three major functions that multiple external representations (MERs) serve in learning situations – to complement, to constrain and to construct. The first function of MERs in Ainsworth's (1999) functional taxonomy is to use representations that provide complementary information or support complementary cognitive processes so that learners can reap the benefits of the combined advantages such as using both diagrams and verbal-textual representations. The second function is to use a familiar representation to constrain the interpretation (or misinterpretation) of a less familiar representation so as to help learners develop a better understanding of the domain. The third function of MERs is to encourage learners to construct deeper understanding of a phenomenon through abstraction of, extension from and relations between the representations. In school learning, multiple representations provide new opportunities to engender student motivation, interest and understanding.

Biology teachers and biology educators are increasingly using technology to supplement their biology teaching. Indeed, a variety of computer programs have been used over the past two decades to enhance student learning of genetics. Multiple representations can be verbal-textual, mathematical, visual, or real-life observations and have long been used by biology teachers in their teaching. What is different is that these multiple external representations are now dynamically linked in interactive multimedia programs such as *BioLogica* (Concord Consortium, 2006) which is an exemplar of such programs for learning introductory genetics. The multiple representations and the tools for manipulating these representations in *BioLogica* are designed to provide students with a challenging and interactive environment for learning genetics across the multiple levels of organisation. For example, in the Monohybrid activity a student first predicts the offspring phenotypes, does a simulation of a cross, visualises the process and results, and then explains his/her reasoning on the screen. The student is then presented with challenges and some embedded assessment questions and a real-world human genetics problem to solve.

However, for students to benefit from the interactions with the multiple representations, they need to be engaged in mindful learning which depends on students' willingness and interest in learning.

Tsui and Treagust (2003) reported on a case study in an Australian school to examine student motivation when the teacher incorporated *BioLogica* during his teaching of genetics with a Grade 10 class. The genetics topic lasted six weeks and the experienced teacher engaged his Grade 10 students in *BioLogica* activities involving fictitious dragons in three separate sessions, each of about one hour's duration. Data were collected from student interviews, online tests, computer data logs files, classroom observations, field-notes and lesson transcripts, and other documents, such as teacher's teaching plans, hand-outs for students and test papers and the sessions were videotaped and transcribed.

Student motivation was examined by analysing seven student interviewees' responses to the interview questions and online questionnaires about their experience using *BioLogica*. Three salient features of the multiple representations in *BioLogica* related to motivation and identified from multiple data sources were: *instant feedback*, *flexibility*, and *visualization*. Student motivation was interpreted using constructs of *curiosity*, *control*, *fantasy*, and *challenge* which were partly based on Malone and Lepper's (1987) intrinsic motivations.

Interviewees' perceptions about learning with BioLogica included appreciation of: Instant feedback while solving problems: "[on paper] you're not sure if you're right or not [but] on the computer you can actually see if you are right"; Flexibility, allowing students to work at their own pace and go back to check their answers during the activities. "With *BioLogica* you can go back if you forget"; Visualization which involves linked visual-graphical representations in *BioLogica* may deepen students' understanding of the connection between representations and concepts that require multilevel thinking. In particular, visualization can help students to "make meaningful connection between processes and their observable manifestations (e.g., the connection between meiosis/fertilisation and Mendelian genetics) "It's good with the Dragons because it's a total change, so you understand eventually why they have different traits.

Student motivation was measured by Control in that *BioLogica* promoted feelings of control on the part of the learner using computers instead of reading a textbook or listening to the teacher. "Instead of just doing it [genetic crosses] out of a book or just being told it, you get to actually see it and decide yourself and see the outcome of it." *BioLogica* evoked in students mental images of situations not actually existing. In particular, the fictitious Dragons are different from the humans that the students see every day. When asked why they liked the Dragons, two students fantasised: "Yeah, because it [a Dragon] is made up; it's not real; it makes it more fun; like if you had humans on the computer, it's a bit boring because you see them every day ... with animals, it's more interesting". *BioLogica* provided some interactive challenging problems with the Dragons. Some students tried to talk genetics like scientists during their interviews: "Well the *BioLogica* work [activity] I don't know what it's called, exactly, like the way that you had to pick which cells were chosen to make the new offspring. The dominant and recessive genes showed whether they actually had that characteristics or not." A second source of evidence about *BioLogica* engendering motivation came from the online responses of 14 participants to an open-ended question in the posttest about their learning experience using *BioLogica.* Overall, students enjoyed learning genetics with interactive multimedia and displayed conceptual change along the social/affective dimension related to different motivating features of the multimedia.

With regards to epistemological conceptual change, most students improved their genetics reasoning but only in the easier types and in a pattern similar to the results of Grade 10 students in another non-laptop school (see Tsui & Treagust, in press). The pretest-posttest improvement in genetics reasoning was similar to the non-laptop school despite the students in this laptop school having more interactions with computer-based multiple representations. With regards to ontological conceptual change, most students exhibited ontological conceptual change within the category of matter, i.e., gene as an inactive to an active particle. According to Venville and Treagust's (1998) framework, most students' conceptions were only intelligible-plausible (IP). Analyses of the interview transcripts using Thorley's (1990) framework (see also Hewson & Lemberger, 2000) indicated that some high-achieving students' postinstructional conceptions were intelligible-plausible-fruitful (IPF). One high-achieving student displayed three-dimensional conceptual change (Tyson et al., 1997): After completing six *BioLogica* activities, she indicated that she enjoyed learning and believed that her understanding of genetics was enhanced by these activities (Tsui & Treagust, 2003).

3.4 Research-based Science Curricula from a Particular Psychological Perspective Intended to Enhance Affective and Cognitive Goals

There is a very low proportion of this type of research, probably because it can only take place effectively within partnerships between psychologists and science educators. Indeed, the politics and funding of universities make such collaborations difficult to produce and sustain. For example, Adey and Shayer (1994) looked at the implications of Piagetian psychology as a way of understanding the learning of the sciences at secondary school level.

The Model of Educational Reconstruction described by Kattmann et al. (1995; 1997) closely links analysis of science content structures, analyses of the educational significance of the content, research on teaching and learning processes and development of instruction. The model has three components. The content structure is analysed from the scientific point and from aims, the educational analysis includes studies in which way the content to be taught is linked to the other topics of instruction, to what extent it is exemplary for other content and the meaning the content has for students and for society. The third component involves how the instruction is to be delivered.

4. WHAT RESEARCH IS NEEDED TO MOVE THE FIELD OF INTERESTS IN SCIENCE FORWARD?

A great deal of research has been conducted examining the affective and cognitive domains in relation to science classrooms. Indeed, there are well developed and well argued theoretical positions upon which to base further more detailed research. However, there is need for further research of the type described in Section 3 that evaluates the effectiveness and efficacy and enjoyment by students of science curriculum topics, based on sound research. Several ways forward with this research agenda can be taken from the studies discussed in Section 1 that have demonstrated the use of several approaches, each with a well developed theoretical basis, to investigate the affective domain in science education. Furthermore, an approach that takes into account the multi-dimensional interpretive framework discussed in Section 2 can be used to broaden and enhance this research agenda so that the ontological, epistemological and social/affective aspects of a learning situation can be best investigated together to determine how they facilitate or hinder learning viewed as conceptual change. That such research is vitally important cannot be underestimated because there is a real decline in enrolments of students in science at both the secondary and tertiary levels. Without research to determine whether or not innovative research-based curricula or topics within a curricula are perceived by students as being interesting, and that successful learning outcomes ensue, there is little likelihood of arresting this decline of enrolments in science.

REFERENCES

Adey, P. & Shayer, M. (1994). *Really Raising Standards.* London: Routledge.

Ainsworth, S. E. (1999). The functions of multiple representations. *Computers & Education*, 33(2/3), 131–152.

Ainsworth, S. E., Bibby, P. A. & Wood, D. J. (1997). Information technology and multiple representations: New opportunities – new problems. *Journal of Information Technology for Teacher Education*, 6(1), 93–104.

Alsop, S. (2005). *Beyond Cartesian Dualism. Encountering Affect in the Teaching and Learning of Science.* Dordrecht, The Netherlands: Springer.

Alsop, S. & Watts, M. (Eds.) (2003). Science education and affect. Special Issue, *International Journal of Science Education*, 25(9).

Barker, V. & Millar, R. (2000). Students' reasoning about basic chemical thermodynamics and chemical bonding: what changes occur during a context-based post-16 chemistry course. *International Journal of Science Education*, 22(11), 1171–1200.

Chaimaneewong, K. (2005). *Developing a Constructivist Learning Approach in Chemical Kinetics in Rajabhat University of Thailand: An Action Research Study.* Unpublished Doctor of Science Education thesis, Curtin University of Technology, Perth.

Chi, M. T. H., Slotta, J. D. & de Leeuw, N. (1994). From things to processes: A theory of conceptual change for learning science concepts. *Learning and Instruction*, 4, 27–43.

Childs, P. E. (2002). Securing the future of chemistry: A case study of developments in chemical education in Ireland. *Chemistry Education: Research and Practice in Europe*, 3(3), 361–369.

Chinn, C. A. & Brewer, W. F. (1993). The role of anomalous data in knowledge acquisition: A theoretical framework and implication for science instruction. *Review of Educational Research*, 63(1), 1–49.

Concord Consortium. (2006). *BioLogica.* Retrieved 6 November 2006 http://biologica.concord.org/webtest1/web_labs.htm

Dagher, Z. (1994). Does the use of analogies contribute to conceptual change? *Science Education*, 76(6), 601–614.

Dalgety, J., Coll, R. K. & Jones, A. (2003). Development of chemistry attitudes and experiences questionnaire (CAEQ). *International Journal of Science Education*, 40(7), 649–668.

Dekkers, J. & de Laeter, J. (2001). Enrolment trends in school science education in Australia. *International Journal of Science Education*, 23(5), 487–500.

Duit, R. & Treagust, D. F. (1998). Learning in science – from behaviourism towards social constructivism and beyond. In B. J. Fraser and K. Tobin (Eds.), *International handbook of science education, Part 1.* (pp. 3–25). Dordrecht, The Netherlands: Kluwer.

Fensham, P. J. & Kass, H. (1988). Inconsistent or discrepant events in science instruction. *Studies in Science Education*, 15, 1–16.

Fraser, B. J. (1981). *Test of Science Related Attitudes* (TOSRA). Hawthorn, Victoria. Australian Council of Educational Research.

Gardner, P. L. (1975). Attitudes to science: A review. *Studies in Science Education*, 2, 1–42.

Gilbert, J. K., Justi, R., Van Driel, J. H., De Jong, O. & Treagust, D. F. (2004). Securing a future for chemical education. *Chemical Education: Research and Practice*, 5(1), 5–14.

Goodrun, D., Hackling, M. & Rennie, L. (2001). *The status and quality of teaching and learning science in Australian schools.* A research report prepared for the Department of Education, Training and Youth Affairs. Canberra.

Harrison, A. G. & Treagust, D. F. (2006). Teaching and learning with analogies: Friend or Foe? In P. J. Aubusson, A. G Harrison and S. M. Ritchie (Eds.), *Metaphor and Analogy in Science Education.* (pp. 11–24). Dordrecht, The Netherlands: Springer.

Haas, J. (2005). The situation in industry and the loss of interest in science education. *European Journal of Education*, 40(4), 405–416.

Hassan, G. & Treagust, D. F. (2003). What is the future of science education in Australia? *The Australian Science Teachers Journal*, 49(3), 6–15.

Haussler, P. & Hoffmann, L. (2000). A curricular frame for physics education: Development, comparison with students' interests, and impact on students' achievement and self-concept. *Science Education*, 84, 689–705.

Hewson, P. & Lemberger, J. (2000). Status as the hallmark of conceptual learning. In R. Millar, J. Leach & J. Osborne (Eds.), *Improving Science Education: The Contribution of Research* (pp. 110–125). Buckingham, UK; Philadelphia, PA: Open University Press.

Hoffmann, L., Krapp, A., Renninger, K. A. & Baumert, J. (Eds.) (1998). *Interest and Learning.* Kiel, Germany: Institute for Science Education at the University of Kiel.

Jarvis, T. & Pell, A. (2002). Effect of the challenger experience on elementary children's attitudes to science. *Journal of Research in Science Teaching*, 39, 979–1000.

Kattmann, U., Duit, R., Gopengiesser, H. & Komorek, M. (1995, April). *Educational Reconstruction – Bringing together Issues of Scientific Clarification and Students' Conceptions.* Paper presented at the annual meeting of the National Association for Research in Science Teaching, Chicago.

Kattmann, U., Duit, R., Gopengiesser, H. & Komorek, M. (1997). Das Modell der didaktischen Rekonstruktion – Ein Rahmen für naturwissenschaftenschaftsdidaktische Forschung und Entwicklung [The model of educational reconstruction. A model for science education research and development]. *Zeitschrift fuer Didaktik der Naturwissenschaften*, 3(3), 3–18.

Lehrke, M., Hoffmann, L. & Gardner, P. L. (Eds.) (1985). *Interests in Science and Technology Education.* Institute for Science Education, University of Kiel, West Germany.

Liew, C.-W. (2004). *The Effectiveness of Predict-observe-explain Technique in Diagnosing Students' Understanding of Science and Identifying their Level of Achievement.* Unpublished Doctor of Science Edcuation thesis, Curtin University Perth Australia.

Lindahl, B. (2003). Pupil's responses to school science and technology? A longitudinal study of pathways of upper secondary school. English summary of a thesis, Kristianstad University, Sweden. Retrieved November 3 2006 from http:// www.mna.hkr.se/~ll/summary.pdf

Lindahl, B. (2005). *How Does School Homework Contribute to Pupils' Understanding and Attitudes to Science?* Paper presented at the biennial meeting of the European Science Education Research Association, Barcelona, September.

Lyons, T. (2006). Different countries, same science classrooms: Students' experiences of school science in their own words. *International Journal of Science Education*, 28(6), 591–613.

Malone, T. W. & Lepper, M. R. (1987). Making learning fun: A taxonomy of intrinsic motivations for learning. In R. Snow & M. Farr (Eds.), *Aptitude, Learning, and Instruction: Vol. 3 Cognitive and Affective Process Analyses* (pp. 223–253). Hillsdale, NJ: Erlbaum.

Maxwell, J. A. (2004). Causal explanation, qualitative research, and scientific inquiry in education. *Educational Researcher*, 33(2), 3–11.

Murphy, C. & Beggs, J. (2003). Children's perceptions of school science. *School Science Review*, 84(308), 109–116.

Nature. (2002). Selling science to the young. *Nature*, 417(6884), 1.

Nentwig, P. & Waddington, D. (Eds.) (2005). *Making it Relevant: Context Based Learning of Science*. Muenster, Germany: Waxmann Verlag Gmbh.

OECD/PISA (2004). Learning for tomorrow's world: First results from PISA 2003. Paris: OECD.

OECD Global Science Forum. (2005). Declining Student Enrolment in Science & Technology: Is It Real? What Are The Causes? What Can Be Done? Conference held in Amsterdam The Netherlands, 14–15 November 2005. http://www.caos.nl/ocw/programme.html

Onlooker. (2005). Is science on the way out of education? *The Pharmaceutical Journal*, 274, 280.

Organisation for Economic Co-operation and Development. (2005). *Learning for Tomorrow's World: First Results from PISA 2003*. Paris: Author.

Osborne, J., Simon, S. & Collins, S. (2003). Attitudes toward science: A review of the literature and its implications. *International Journal of Science Education*, 25(9), 1049–1080.

Perrier, F. & Nsengiyumva, J.-B. (2003). Active science as contribution to the trauma recovery process: preliminary indications with orphans from 1994 genocide in Rwanda. *International Journal of Science Education*, 25(9), 1111–1128.

Pilot, A. & Bulte, A. M. (Eds.) (2006). Context-based chemistry education. Special issue. *International Journal of Science Education*, 28(9).

Pintrich, P. R., Marx, R. W. & Boyle, R. A. (1993). Beyond cold conceptual change: The role of motivational beliefs and classroom contextual factors in the process of conceptual change. *Review of Educational Research*, 6, 167–199.

Posner, G. J., Strike, K. A., Hewson, P. W. & Gertzog, W. A. (1982). Accommodation of a scientific conception: Toward a theory of conceptual change. *Science Education*, 66, 211–227.

Ramsden, J. (1998). Mission impossible? Can anything be done about attitudes to science? *International Journal of Science Education*, 20(2), 125–137.

Schibeci, R. A. (1984). Attitudes to science: An update. *Studies in Science Education*, 11, 26–59.

Simpson, R. D., Koballa, T. R. Jr., Oliver, J. S. & Crawley, F. E. Jr. (1994). Research on the affective dimension of science learning. In D. Gabel (Ed.), *Handbook of research on science teaching*. New York: McMillan.

Teixeira dos Santos, F. M. T. & Mortimer, E. F. (2003). How emotions shape the relationship between a chemistry teacher and her high school students. *International Journal of Science Education*, 25(9), 1095–1110.

Thomas, J. (2002). Maths is still looking for a future in Australia. *Australasian Science*, 23(4), 17.

Thorley, N. R. (1990). *The role of the conceptual change model in the interpretation of classroom interactions*. Unpublished doctoral dissertation, University of Wisconsin-Madison, Wisconsin, USA.

TIMSS. (2006). International test scores: Poor U.S. test results tied to weak curriculum. Retrieved from http://4brevard.com/choice/international-test-scores.htm

Treagust, D. F., Harrison, A. G. & Venville, G. J. (1998). Teaching science effectively with analogies: An approach for pre-service and in-service teacher education. *Journal of Science Teacher Education*, 9(2), 85–101.

Treagust, D. F., Harrison, A. G., Venville, G. J. & Dagher, Z. (1996). Using an analogical teaching approach to engender conceptual change. *International Journal of Science Education*, 18, 213–229.

Tsui, C.-Y. & Treagust, D. F. (2003). Genetics reasoning with multiple external representations. *Research in Science Education*, 33, 111–135.

Tsui, C.-Y. & Treagust, D. F. (2007). Understanding genetics: Analysis of secondary students' conceptual status. *Journal of Research in Science Teaching*, 44(2), 205–235.

Tyson, L. M., Venville, G. J., Harrison, A. G. & Treagust, D. F. (1997). A multi-dimensional framework for interpreting conceptual change in the classroom. *Science Education*, 81, 387–404.

Venville, G. J. & Treagust, D. F. (1996). The role of analogies in promoting conceptual change in biology. *Instructional Science*, 24, 295–320.

Venville, G. J. & Treagust, D. F. (1998). Exploring conceptual change in genetics using a multidimensional interpretive framework. *Journal of Research in Science Teaching*, 35, 1031–1055.

Zembylas, M. (2005). Three perspectives on linking the cognitive and the emotional in science learning: Conceptual change, socio-constructivism and poststructuralism. *Studies in Science Education*, 41, 91–116.

Zusho, A., Pintrich, P. R. & Coppola, B. (2003). Skill and will: the role of motivation and cognition in the learning of college chemistry. *International Journal of Science Education*, 25(9), 1081–1094.

ELIZABETH WHITELEGG, PATRICIA MURPHY
AND CHRISTINA HART

3. GIRLS AND PHYSICS: DILEMMAS AND TENSIONS

Abstract: This paper reports on some of the findings from a review of research commissioned by the Institute of Physics into the participation of girls in physics. The review was commissioned in response to concern about the continuing decline in the numbers of girls studying physics post-16 in England. The review includes 177 sources of national and international research literature on the participation of girls in science and in physics and is a narrative review covering 161 pages. The review findings reveal a complex picture of the reasons for girls continuing decline in participation related to their lack of meaningful access to physics which is constrained by a complex web of interactions in girls' curriculum and assessment experience. When this is combined with perceptions of the representation of physics it results in a reduction in girls' self-efficacy and self-concept in the subject as they progress through schooling. The review recommends that purposes for studying physics need to be made explicit for girls in particular, and that this should happen within their curriculum experience rather than outside it. Relevance of the subject to girls' lives outside the classroom is as important as prior knowledge so curriculum interventions and teachers should take this into account. Staff development is needed to help teachers develop strategies to increase the participation of girls and this is particularly important where single sex teaching is used. Long term evaluation of different approaches, further research into the difficulty of physics and access to achievement data is needed

Keywords: Difficulty, Gender, Influences and course choices, Pedagogy, Physics, Physics self-concept, Relevance and attitudes, Representations of physics, Single-sex

1. INTRODUCTION

Over the past decade recruitment to physics A-levels have been continuing to decline, particularly for girls and, coupled with the closure of some university physics departments, this has positioned physics as a 'vulnerable strategic subject'. The Institute of Physics (IOP) has been concerned to uncover sources of this problem and develop strategies to alleviate it, so commissioned a review of research on the participation of girls in physics in order to provide evidence to inform policy-related decision making. This chapter outlines some of the messages from the review and the recommendations that have emerged from it, but space does not allow full consideration of all the aspects considered in the full review (Murphy and Whitelegg, 2006a).

Elizabeth Whitelegg, Centre for Research in Education and Educational Technology, The Open University, Walton Hall, Milton Keynes, UK

R. Pintó and D. Couso (eds.), Contributions from Science Education Research, 27–36.

1.1 Historical Framework

The problem of girls' participation in physics needs to be seen in a social and historical framework. Throughout the nineteenth century and for most of the twentieth, education opportunities for different social classes and genders within those differed not only in the opportunities to attend school, and the type of school available, but also on the subjects taught and curriculum offered within those subjects. Girls and boys were educated to fit their future roles based on their social class and on their gender – girls for the domestic sphere (as wives and mothers within their own homes or as domestic servants in the homes of others) and boys for work in the public sphere outside the home. The sciences, in particular, were taught in the gender-segregated ways which continued to some extent and in some schools until well into the mid- to late-twentieth century. The introduction of the National Curriculum in 1988 in England, Wales and Northern Ireland made all the sciences – physics, chemistry and biology, compulsory for the majority of students. However, compulsory study of physics in this new curriculum has not resulted in an increase in uptake of the subject post-16, particularly for girls. As evidence from the review suggests, the historical and culturally driven representation of physics that developed in earlier times continues to affect some teachers', parents' and students' perceptions of the subject. We argue that physics continues to reflect its historical tradition and can be said to be "inheriting a gender culture". (Murphy and Whitelegg, 2006b, p. 285)

2. METHODS

This is a narrative review, detailed and through, drawing on the practice of systematic reviews by identifying criteria for inclusion, and developing a systematic approach using key words and mapping the field. Hundred and seventy-seven sources of national and international research literature were selected for inclusion. Criteria for selection of the studies examined were research that showed:

- Clarity of purpose and data collection tools and methods used.
- Studies that showed attention to validity and reliability of the analysis and its interpretation.
- Large scale empirical studies using
  - descriptive analyses
  - analytical techniques (e.g., factor analysis using multivariate and univariate analyses).
- Small scale in depth follow-ups of large scale studies.
- Theoretical papers offering well-grounded insights.

Initially, the brief from the IOP was to cover research over last 15 years but in the process of doing the review we found it necessary to include earlier seminal work that was unique and had not been repeated. The Institute also initially only wanted us to consider UK research but this was extended to include research in other countries where work had transferable messages for UK. The research also focussed on the period of secondary education (11–18 years), only examining research from the primary phase were it was shown to have an important influence on what happened in secondary. Significant studies of post-16 physics were also included as many of the pre-16 studies were concerned with science rather than with physics. Similarly research at Higher Education level was included when it produced messages transferable to the secondary phase.

The literature review is about gender and physics, so how gender is interpreted by the literature is important. The literature recognises gender as socially constructed and not as a fixed attribute of an individual determined by their genes. Gender is therefore not the same as biological sex, but is concerned with how an individual relates to a situation or in this case to a subject. So in the review and this paper, when referring to groups of boys or girls, it should be born in mind that within these groups there will be individuals who are not reacting as a stereotypical 'boy' or 'girl' but who may be behaving or identifying with the opposite gender group. So the differences referred to are trends for a group and not attributable to all members of that group.

A second concern related to the field itself, as there is limited recent research into gender and physics particularly in England. It was therefore felt that the absence of research evidence was equally important as its presence. Consequently a wide range of research literature was explored to point up:

- where understandings about the nature of the problem of girls' participation in physics were emerging in a consistent way;
- how understanding about the issues and their impact had evolved;
- where there was limited research evidence and where there was an absence of evidence.

3. ANALYSIS AND FINDINGS

The review is divided into seven Sections and what follows in this paper is a very short summary of some, but not all of the findings from the review. The evidence cited and the sources referenced here form a small part of the research examined and discussed in the full report.

The nature of the problem of girls' lack of participation in physics is multifaceted and so no single factor can be studied in isolation and be said to account for the problem. The way the outcomes of the review and this paper have been organised is just one way, amongst many, of organising the messages from the literature.

3.1 Interests, Motivation, Course Choices and Career Aspirations

The research indicates that interest and enjoyment of science declines throughout secondary school and that the decline accelerates from age 14 particularly for many girls. This decline in interest linked to girls' self-efficacy in physics and this leads to girls' experiencing physics as difficult (Häussler, 1987; Hoffmann, 1997; Reid and Skryabina, 2002; Lindhal, 2003; Osborne et al., 2003). However even those who retain an interest in science and in physics, may not choose to continue with the subject when they experience a number of other negative factors, such as those outlined below. So interest and enjoyment alone are not sufficient reasons to continue studying physics.

Career intentions have an important influence on continuing to study the subject. Students, particularly boys, will continue to study if they see it as essential for a career that is appropriate for them (Stokking, 2000). Many girls are not aware of careers that may be suitable for them and where physics plays an important part. Prior achievement also has an important influence on course choices (Sharp et al., 1996; Reid and Skryabina, 2002). Gender is not found to be a direct influence on students' attitudes to physics (Stokking, 2000). More recent research points to the significance of students' evolving physics self-concept, that is their perception of their present and future possible selves in relation to the subject, and this in turn is influenced by how they experience physics in schools. It is in this respect that gender effects might influence attitude development (Krogh and Thomsen, 2005). Because of the greater decline in self concept amongst girls, compared to boys, even girls who are successful in science, and in physics, will ascribe their success to factors other than ability and so still not believe they are good enough to continue with a subject that is generally recognised as difficult (see Section 5 for a discussion of difficulty of physics). Males rate themselves as more successful learners and are more willing to consider maths and science irrespective of their success.

"The type of course studied is a significant influence on both boys' and girls' enjoyment and motivation to learn physics, and this is linked, particularly for some girls, to the match between their goals for their learning and the goals of the course. More girls than boys report that they value social applications and want more social relevance in their physics courses, which can be linked to the higher recruitment and retention of girls to physics courses that emphasise real-life applications." (Murphy and Whitelegg, 2006a)

3.2 Relevance and Curriculum Interventions

Relevance is reported in the review in terms of usefulness either to daily life or to students' goals. Both boys and girls (but more girls than boys) consider that physical sciences are not personally relevant (Stokking, 2000), and personal relevance plays a key role in many girls' motivation for learning (Murphy and Elwood, 1998; Osborne and Collins, 2000). Gender mediates learning from an early age and affects what students become familiar with. It is differences in what is familiar that lie behind what students report they consider personally relevant. Gender differences in what students consider personally relevant influence their perceptions of their competency, so if students, particularly girls, do not perceive a subject as personally relevant, they may also not perceive themselves as competent in it, particularly if they lack self belief in their ability in the subject.

What boys, more than girls, pay attention to and engage with is generally valued and judged relevant in physics and this is evidenced by the sort of applications that have often been used to illustrate physics concepts in the past. More recent approaches have tried to widen participation in physics by adopting a context-led approach, based in contexts that are socially relevant. Advocates of this approach claim that this type of curriculum better meets the needs of all students and there is evidence to support this (Boaler, 1997; Bennett et al., 2003). The characteristics of a context-based/humanistic curriculum are those where:

- Social situations are used to organise and determine the content studied and assessed.
- The social situation and the problems within it provide the purpose for learning.
- The social situations vary between those of relevance to students' daily lives and concerns and wider social issues of concern to societies generally.
- Physics is represented as a social practice, physics knowledge as a social construction open to change and influenced by social, political, historical and cultural factors.
- The values implicit in physics practices and knowledge are matters for discussion and critique between students and their teachers.

Differences between what girls and boys have learned is relevant and of personal value affects the problems they perceive and whether and how they engage with these problems. So a curriculum that uses this approach must be careful to use context that are relevant to a wide range of learners. Some boys, particularly high achieving ones are more content with abstract nature of physics or its strategic usefulness and some reject context-based approaches as they have learned not to pay attention to social contextual features and feel disadvantaged if asked to do so (Whitelegg and Edwards, 2001).

3.3 Teacher Effects

Teachers' behaviours and attitudes are a key influence on student attitude, motivation, achievement and continuing participation (Labudde, 2000) so an examination of the effects teachers have on their students is an important part of the review. However, the review of the literature revealed that there were no recent *UK-based* empirical studies in science, consequently the Section examines evidence from studies from abroad and some earlier UK-based studies where they still have relevance today. Teachers' expectations have significant effects on students' self-concepts in physics and teachers of physics hold lower expectations for girls in physics than they do for boys (Kenway and Gough, 1998). Supportive teacher-student relationships are more important for girls than boys (Sharp, 2004), particularly in physics where girls' self-concept is less positive than boys'. Personal teacher support was found to be a key predictor of attitudes to physics (Krogh and Thomsen, 2005).

Boys' as a group receive more teacher attention than girls (Kelly, 1988) and the nature of the feedback differs such that girls are more likely to receive negative feedback on quality of work and boys on behaviour. This allows boys to retain confidence in their ability. A study that looked at boys' confidence when they received the sort of feedback more usually given to a girl resulted in boys loosing confidence in their academic abilities. (Dweck in Kelly, 1988) A key finding from another study in Israel (Zohar and Bronshtein, 2005) was that girls with average grades were not encouraged to study physics whereas boys with similar grades were. If teachers hold different perceptions of girls' and boys' abilities in physics based on their gender and if they also see physics as primarily a boys' subject, these subtle and not so subtle messages are likely to be transmitted to the students.

Maintenance of students' autonomy and responsibility for learning are key factors in encouraging participation in science. Teachers can introduce strategies to develop these qualities in their students. Such strategies (typically found in context-based curriculum) include:

- investigative laboratory work;
- group and class discussion where alternative views are considered and valued;
- problem-solving and project-based activities where students are the decision-makers;
- creative writing involving a wide range of genres in which science understanding is communicated to the public.

3.4 Single Sex School and Groupings

There is very little research into the impact and effectiveness of single-sex groupings in co-ed schools in physics, particularly in England. What research there is suggests that any increase in achievement of girls in single-sex schools is due more to social, environmental and economic factors than the single-sex nature of the school (Sammons et al., 1994; Smithers and Robinson, 1995; Smithers and Robinson, 1997; Elwood and Gipps, 1999). However, some recent studies in England suggest that girls' achievement can be increased by single-sex schooling, particularly for girls of lower ability, but these studies in common with most others in this area was unable to take the social, environmental and economic factors into account (Spielhofer et al., 2002; GSA, 2004). Research does show that girls' subject preferences at younger ages are less polarised in single-sex schools than in co-education schools, but that polarisation occurs later on at around age 15 (Ormerod and Duckworth, 1975; Stables, 1990; Colley et al., 1994).

Research from countries outside the UK show that only when pedagogy and curriculum are effective and inclusive and teachers are gender sensitive do single-sex groupings enhance girls' achievement and self-concept (Parker and Rennie, 2002).

For a significant proportion of girls in single sex schools the research suggests that decline in interest and enjoyment is attributed by the research to curriculum experience (e.g. content overload and fast-pace of study, particularly in top groups), rather than single/mixed sex schooling. (Murphy and Whitelegg, 2006a)

3.5 Measures and Perceptions of Difficulty

Research (Fitz-Gibbon and Vincent, 1994) has found a measured difference of up to one grade lower for physics compared to other subjects. Other unpublished research shows that the difference is between a half to one whole grade (Conway personal communication, 2004). However, regardless of the size of the difference, or the technical concerns about the validity of such measures, the difficulty of physics is accepted as common knowledge amongst teachers and affects who they consider suitable to study physics. Ability in mathematics also affects students' achievement and confidence in physics, and teachers' views on the suitability of a student for A-level physics study (Sharp et al., 1996). Students' perception of physics as difficult increases with age. This is related to an increase in mathematical demand and increased sense of inadequacy, particularly for girls.

3.6 Performance Patterns in Physics

In this Section, the review considers results of research that focus on the English examination and assessment system and considers the way the specific factors in this system affect girls' participation in physics in England. Some of these factors are unique to the English system and so messages cannot be given that also sit within international contexts. This Section of this paper is therefore limited to a discussion of the factors that have relevance to an international audience. Readers are advised to consult the full review for further discussion.

International survey findings indicate that boys' are demonstrating higher performance in physics examinations although the trend shows a decreasing gap between boys' and girls' performance. The situation in England is similar to this. The majority of students in England and Wales take the Double Award Science GCSE at age 16. In this examination girls overall outperform boys. However, examination of one examination board's results by gender for the physics, chemistry and biology components of the Double Award separately show that girls' overall higher performance is associated with their performance on the biology and chemistry components of the examination. (A similar trend is found on national science tests at age 14.) For the Triple Award examination at 16, where students are entered for the three science subjects separately, far fewer girls than boys are entered for physics and their performance relative to boys is lower at the higher and lower grades in England and Wales but not in Scotland and Eire.

3.7 Impact of Assessment Techniques

The review found little recent research into the effect of situating a physics assessment problem within a real-life context. Social contexts are underrepresented in physics assessments, but those that are found are of more relevance to boys' interests than to girls'.

However it is likely that social contexts and human dilemmas are more likely to be judged relevant and of value by girls more than boys. Some boys are disinterested in relating physics to a context and prefer an abstract approach.

The format of the assessment also affects performance. Research (Murphy, 1982; Powney, 1996) suggests that girls do less well than boys on multiple choice formats, compared with short free response formats. Boys do equally well on both formats. There is also evidence that girls do less well relative to boys on items with graphical and figural data that require this type of response too. The evidence for item effects is equivocal as the effect depends on the interaction with other aspects of the assessment items (Murphy, 1982; Gipps and Murphy, 1994; Willingham and Cole, 1997; Ruddock et al., 2003). The format matters if it alters the construct being assessed.

4. KEY FINDINGS

The review findings reveal a complex problem that limits students', particularly girls', access to physics. Access is constrained via a web of interactions within the curriculum and assessment experience. However, changing access alone will not change girls' participation. More fundamental reconsideration of the contribution of physics to students' future lives is needed.

The limitation on access leads to girls' increased sense of inadequacy and the growing belief of the difficulty in the subject. The Review has revealed that there are many factors in physics curriculum, teaching and assessment that undermine or deny girls' sense of competency. But perceptions of competence alone are not sufficient to influence girls' choices; girls need to perceive a future in physics that will help girls achieve their goals.

Recommendations emerging out of the review suggest that:

- Interventions need to come early in the physics curriculum and its assessment in order to make explicit the value and purposes for studying physics and the range of social issues and careers that it informs.
- Formative strategies to elicit and address differences between students' views need to be developed and made available to teachers. Views of relevance are as important in learning physics as is prior knowledge.
- Long term evaluation of physics curriculum interventions are needed that consider the impact of the curriculum on subgroups of students and on teachers.
- Pedagogic changes are needed that alter the teaching role and the teacher-student relationship so initial teacher training and continuing professional development programmes for physics teachers need to emphasise:
  - *how to develop good relationships with students, particularly with girls;*
  - *how to exhibit leadership and understanding in the classroom;*
  - *the use of a variety of teaching strategies, particularly in open-ended lab work.*
- There is a need for a coherent programme of staff development before any wider implementation of strategies used with single sex teaching groups.
- Further research into the basis for the belief in the difficulty of physics is needed.
- Access to achievement data will help to challenge physics teachers' beliefs about girls' achievement and potential. These beliefs are powerful in shaping girls' self-concept.

ACKNOWLEDGEMENT

Grateful acknowledgement is given to the Institute of Physics for funding this work.

REFERENCES

Bennett, J., et al. (2003). A systematic review of the effects of context-based and Science-Technology-Society (STS) approaches in the teaching of secondary science. *Research Evidence in Education Library*. (London: EPPI-Centre, Social Science Research Unit, Institute of Education.)

Boaler, J. (1997). *Experiencing School Mathematics: Teaching Styles, Sex And Setting*. (Buckingham: Open University Press.)

Colley, A., et al. (1994). School subject preferences of pupils in single-sex and co-educational secondary schools. *Educational Studies 20*(3): 379–385.

Elwood, J. and C. Gipps (1999). *Review of Recent Research on the Achievement of Girls in Single-Sex Schools.* (London: Institute of Education University of London.)

Fitz-Gibbon, C. T. and L. Vincent (1994). *Candidates' Performance in Public Examinations in Mathematics and Science*, CEM Centre Report for the School Curriculum and Assessment Authority: *30.*

Gipps, C. and P. Murphy (1994). *A Fair Test?: Assessment, Achievement and Equity.* (Buckingham, UK: Open University Press.)

GSA. (2004). GSA Survey of maths, science and modern languages at A level Summer 2004. Leicester, Girls Schools Association.

Häussler, P. (1987). Measuring students' interest in physics – design and results of a cross-sectional study in the Federal Republic of Germany. *International Journal of Science Education 9*(1): 79–92.

Hoffmann, L. (1997, September). *An Intervention Project Promoting Girls' and Boys' Interest in Physics: Opening the Door to Physics for Girls.* ESERA Symposium: Gender Issues in Physics. (Paper presented at the first Conference of the ESERA, Rome.)

Kelly, A. (1988). Gender differences in teacher-pupil interactions: A meta-analytic review. *Research in Education 39*: 1–24.

Kenway, J. and A. Gough (1998). Gender and science education in schools: a review with attitude. *Studies in Science Education 31*: 1–30.

Krogh, L. B. and P. Thomsen (2005). Studying students' attitudes towards science from a cultural perspective but with a quantitative methodology: Border crossing into the physics classroom. *International Journal of Science Education 27*(3): 281–302.

Labudde, P. (2000). Girls and physics: teaching and learning strategies tested by classroom interventions in grade 11. *International Journal of Science Education 22*(2): 143–157.

Lindhal, B. (2003, July). *Changing the Subject and Teaching to Get More Students to Science and Technology?* (Paper presented at the 11th Gender and Science and Technology Conference, Mauritius.)

Murphy, P. and J. Elwood (1998). Gendered experiences, choices and achievement – exploring the links. *International Journal of Inclusive Education 2*(2): 95–118.

Murphy, P. and E. Whitelegg (2006a). *Girls in the Physics Classroom: A Review of Research on the Participation of Girls in Physics.* (London: Institute of Physics: 1–61. An electronic copy can be downloaded from: http://www.iop.org/Our_Activities/Schools_and_Colleges/Making_a_Difference/Policy/page_5534.html (last accessed on 15th September 2006.)

Murphy, P. and E. Whitelegg (2006b). Girls and Physics: continuing barriers to 'belonging'. *The Curriculum Journal 17*(3): 281–305.

Murphy, R. J. L. (1982). Sex differences in objective test performance. *British Journal of Educational Psychology 52*: 213–219.

Ormerod, M. B. and D. Duckworth (1975).*Pupils' Attitudes to Science: A Review of Research.* (Windsor, Berks: National Federation for Educational Research.)

Osborne, J. and S. Collins (2000). *Pupils' and Parents Views of the School science Curriculum.* (London: Wellcome Trust.)

Osborne, J., et al. (2003). Attitudes towards science: a review of the literature and its implications. *International Journal of Science Education 25*(9), 1049–1079.

Parker, L. H. and L. J. Rennie (2002). Teachers' implementation of gender-inclusive instructional strategies in single-sex and mixed-sex science classrooms. *International Journal of Science Education 24*(9), 881–897.

Powney, J. (1996). *Gender and Attainment: A Review.* (Edinburgh: Scottish Council for Research in Education (SCRE))

Reid, N. and E. A. Skryabina (2002). Attitudes towards physics. *Research in Science and Technological Education 20*(1), 67–80.

Ruddock, G., et al. (2003). *Where England stands in the Trends in International Mathematics and Science Study (TIMSS) 2003: Summary of national report for England.* (Slough: NFER Department for Education and Skills, 12.)

Sammons, P., et al. (1994). *Continuity of School Effects.* (London: Institute of Education.)

Sharp, C., et al. (1996). *The Take-Up of Advanced Mathematics and Science Courses – Summary Report.* (Slough: National Foundation for Educational Research report for SCAA, 19.)

Sharp, G. (2004). *A Longitudinal Study Investigating Pupil Attitudes Towards their Science Learning Experiences from a Gender Perspective.* Unpublished PhD thesis. (Milton Keynes, UK: The Open University.)

Smithers, A. and P. Robinson (1995). *Co-educational and Single-sex Schooling.* (Manchester: Centre for Education and Employment Research, University of Manchester.)

Smithers, A. and P. Robinson (1997). *Co-education and Single-sex – Revisited.* (Uxbridge: Brunel University.)

Spielhofer, T., et al. (2002). *The Impact of School Size and Single-sex Education on Performance.* (Slough: NFER for the Local Government Association.)

Stables, A. (1990). Differences between pupils from mixed and single-sex schools in their enjoyment of school subjects and in their attitudes to science and to school. *Educational Review 42*(3): 221–230.

Stokking, K. (2000). Predicting the choice of physics in secondary education. *International Journal of Science Education 22*(12): 1261–1283.

Whitelegg, E. and C. Edwards (2001). Beyond the laboratory: Learning physics in real life contexts. (in R. Duit (Ed.), *Research in Science Education: Past, Present and Future.* (pp. 337–342) Dordrecht, Netherlands: Kluwer Academic Publishers.)

Willingham, W. W. and N. S. Cole (1997). *Gender and Fair Assessment.* (Mahwah, New Jersey/London: Lawrence Erlbaum Associates.)

Zohar, A. and Bronshtein B. (2005). Physics teachers' knowledge and beliefs regarding girls' low participation rates in advanced physics classes. *International Journal of Science Educatiion* 27(1): 61–77.

PART 2

SCIENCE EDUCATION RESEARCH: NEW APPROACHES AND LINKS TO PRACTICE

JOHN LEACH

4. CONTESTED TERRITORY: THE ACTUAL AND POTENTIAL IMPACT OF RESEARCH ON TEACHING AND LEARNING SCIENCE ON STUDENTS' LEARNING

Abstract: The actual and potential impact of research on the practice of teaching science has been discussed within both academic and policy communities. Radically different conclusions have been advanced

This paper presents evidence from one study to demonstrate the potential positive impact of research on teaching and learning science on students' understanding of science. Three short teaching sequences (around 6 hours) were designed by a group of researchers working with a group of teachers, drawing explicitly upon insights from research. Tests of the students' conceptual understanding were applied both before and after teaching. Furthermore, identical test data were collected from classes of similar students in the same schools, who were following the school's usual approach to teaching. In cases where students who followed the designed teaching sequences achieved measurably better results than their peers following the school's usual approach to teaching, other teachers in different schools (who had not been involved in the design of the teaching sequences) implemented the teaching sequences with their students. Students following the designed teaching achieved significantly better scores on tests of conceptual understanding after teaching than their peers who followed their school's usual approach, irrespective of whether their teacher was involved in the design of the teaching or not

The terms *research evidence-informed* and *research evidence-based* practice are introduced (Millar et al., 2006), and used to discuss the implications of findings from research such as that reported in this paper for practice

Keywords: Impact of research, Research evidence-informed, Research evidence-based, Students' understanding of science, Teaching sequence

1. RESEARCH IN SCIENCE EDUCATION: A SUCCESS STORY?

The main English language journals[1] devoted to science education research are publishing more and more issues per year. New, regionally-focused journals are being established. The number of papers being submitted to international conferences is higher than ever, and the diversity of nations represented at such conferences is growing. There is probably more science education research going on now than ever before. Although few would claim that this research enterprise has had a major

John Leach, Centre for Studies in Science and Mathematics Education, The University of Leeds, Leeds LS2 9JT, UK

R. Pintó and D. Couso (eds.), Contributions from Science Education Research, 39–57.

impact on the practice of science education, within the research community it is not hard to find optimism about the *potential* of science education research to make a positive impact on policy and practice (see, for example, John R. Staver's Presidential Address, given at the (US) National Association for Research in Science Teaching this year; Staver, 2005).

However, less optimistic judgements have been voiced, within both the research and the policy communities. Critiques within the policy community tend to be focused upon educational research in general, rather than science education research in particular. The critique presented by David Hargreaves in the UK is typical:

> *"Given the huge amounts of educational research conducted over the past fifty years or more, there are few areas which have yielded a corpus of research evidence regarded as scientifically sound and as a worthwhile resource to guide professional action. "* (Hargreaves, 1996: 2)

The same point has been made by others in the UK, including the then Secretary of State for Education (Blunkett, 2000), academics reporting on behalf of the inspectorate of schools (Tooley and Darby, 1998), and on behalf of the government Department for Education and Skills (Hillage et al., 1998). In order to address this perceived difficulty, the US government Department of Education's strategic plan (2002) aims to transform education into an evidence-based field, based upon the use of randomised controlled trials (RCTs) as the preferred methodological approach.

Within the research community, however, more specific critiques have been directed at research on teaching and learning science.[2] The key point that has come out of philosophical critiques of research on teaching and learning science centres upon the distinction between *knowledge* – which is warranted by empirical evidence – and *belief* – which is not. The constructivist perspectives used by researchers are portrayed as being based on a flawed empiricist view of scientific knowledge (and, by extrapolation, the learning of scientific knowledge) (Matthews, 1992; 1997; Nola, 1997). According to the critique, constructivist research in science education assumes that individuals – whether they are scientists generating new knowledge or students 'constructing' personal beliefs – base their intellectual activities upon *sensory information*. Knowledge is therefore judged according to whether it fits with sense experience, rather than whether it can be warranted as a true account of the material world, and learning is seen as an *individual* process of sense-making. However, this portrayal of the epistemological underpinnings of research on teaching and learning science has been rejected by some researchers, who portray science learning as a process of developing personal understanding of knowledge that is located in communities of practice (e.g. Leach and Scott, 2003; Tiberghien, 1996).

I believe that there are some cases where research on teaching and learning science has had a positive impact upon the practice of science teaching. One example is research which aims to clarify, in detail, the aims of teaching in a particular area of content, and to embody this knowledge into artefacts for use in teaching. Halloun and Hestanes (1985) used research on university students' conceptual difficulties in the area of mechanics to develop a test instrument for use by teachers in formative assessment (the Force Concept Inventory, FCI). Since the development of the FCI, there has been a significant amount of curriculum development which aims to focus teaching more precisely on students' conceptual difficulties, and to use the FCI to evaluate the success of the teaching (e.g. Savinainen et al., 2005). In the UK, Millar and Haines (2005) have developed formative assessment instruments with a similar purpose, for use in the lower secondary school.

However, the difficulties of conducting research with a view to improving the practice of teaching science should not be underestimated. Piet Lijnse has argued that research on the didactics of science has the potential of generating knowledge about how to improve the teaching of scientific content (in this case, physics content) – while noting that the research enterprise to date has substantially failed to achieve this potential:

"I still remember my disappointment when, as a newly appointed didactician, I had to develop an innovative series of lessons to introduce quantum mechanics at secondary school. I turned to theories of education and educational psychology for help. However, hardly any such help appeared to be available, a frustrating result which unfortunately was (and is) in line with the 'traditional' scepticism of physicists concerning the 'soft' sciences." (Lijnse, 2000: 309)

Lijnse goes on to illustrate developmental research, a cyclical process of generating knowledge about teaching specific content which is of high 'didactical quality', and others have reported cyclical processes of design and evaluation research which aims to develop knowledge about teaching and learning that is relevant to practice (e.g. Brown, 1992; Cobb et al., 2003; Duit et al., 1997; Méheut, 2004).

There are, however, significant methodological difficulties in conducting research which aims to generate reliable knowledge about teaching scientific subject matter. Firstly, in order to claim that one approach to teaching is 'better' than another, it is necessary to have agreement about the outcomes that will be used to define quality. These might include students' performance in official test scores (which may or may not correlate with the same students' scores on tests of conceptual understanding). Furthermore, quality might be judged in terms of the mean score, giving a measurement of the performance of all students, or alternatively the performance of an elite who are likely to go on to further study in science. Alternatively, judgements might be made about students' and teachers' enjoyment of the teaching, or the cost of implementation of an approach relevant to its benefit on some criterion. Deciding upon the nature of quality in science teaching is a question of values.

If different criteria are used to judge the quality of science teaching in different situations, it is certainly not possible to identify unique 'best practice'. Furthermore, there is ample evidence that individual teachers make changes as they implement designed science teaching approaches (Pintó, 2005), with the result that the practice *as intended by the planners* can be significantly different from the practice *as implemented by teachers*. The proposed approach may not fit the teachers' (or their students') preferred ways of working, or may be judged to require an investment of effort in change that is disproportionate to the likely gains. Teachers may judge key features of planned teaching approaches as flawed, and not implement them. Teachers usually have good reasons for doing the things that they do, even though these reasons may not always stand up to closer scrutiny. The challenge for transferring research insights from one site to another lies in enabling teachers to recognise which features of a design are *central* to its rationale, and therefore should be modified with extreme caution, and which features are less critical.

In a recent study in the UK, we investigated practitioners' stated views about the interface between research and practice (Ratcliffe et al., 2005). An important insight to come out of this work is that although science teachers in the sample were positive about the potential of science education research to make a positive impact on practice, generally speaking they said that they were unlikely to change their practices based on evidence from research if that evidence was not consistent with their professional experience. Furthermore, they were rather sceptical about research which did not adhere to methodological standards characteristic of some areas of the natural sciences (controlled tests, large sample sizes, quantitative data, etc.). The science teachers said that their colleagues and curriculum materials had a very strong influence on their practice. This raises an important issue for science education researchers working on teaching and learning science. In the UK, the traditional metaphor for the influence of research on practice is *dissemination*: researchers generate knowledge and this is then *given* to teachers for implementation. Funding has been provided for small-scale research projects to produce 4-page leaflets summarising key findings for teachers, and professional journals such as the *School Science Review* often carry short reports of research for teachers. However, this *dissemination* approach does not fit comfortably with what science teachers say about the factors likely to influence their practice. Rather than research findings, they say that their practice is most likely to be affected by things that can be used in the classroom, such as teaching sequences.

Over the years, many science education researchers have developed teaching sequences for use by practitioners. However, producing teaching sequences is not a process of *applying* research findings about teaching and learning in a logical way, so that any two people would take the same findings and come up with the same teaching sequence. Rather, those writing teaching sequences are involved in a process of *creating* teaching sequences that are informed by their knowledge of the culture of classrooms, students, teachers and resources – as well as research findings about teaching and learning.

We have introduced the phrase *research evidence-informed practice* to describe educational practices that draw systematically upon research findings in their design (Millar et al., 2006). We use the phrase *research evidence-based practice* to refer to practices that have been evaluated, where there is evidence that they are better (in some sense) than other possible approaches.

In the next section, I will describe findings from a project in which we drew upon insights from research on teaching and learning to design teaching sequences, and then evaluated those sequences in terms of students' content learning. I then go on to discuss the extent to which the designed teaching sequences represent research evidence-informed practice and research evidence-based practice.

2. DESIGNING AND EVALUATING RESEARCH EVIDENCE-INFORMED TEACHING SEQUENCES

A group of researchers and teachers worked collaboratively to design three short teaching sequences, aimed at lower secondary school students. The teaching sequences addressed introductory ideas about simple electric circuits, particles and change, and plant nutrition. In designing the teaching sequences, we drew upon two fundamentally different groups of insights from research. The first group includes findings from empirical research studies, which are commonly referred to as 'evidence'. By contrast, the second group of insights are theoretical, and include insights into the nature of knowledge, thinking, learning and teaching. These insights have been developed through research and scholarship in science education and other fields, and are typically referred to as 'theory' rather than 'evidence'. Nonetheless, however, they influence how findings from empirical studies are interpreted and used as *evidence* about something. We refer to such insights about the nature of knowledge, thinking, learning and teaching as our *theoretical framing perspective.*

We refer to our theoretical framing perspective as *social constructivism*, and have described it in detail elsewhere (Leach and Scott, 2003). We see teaching and learning science as involving an introduction to the social language of school science against a 'backdrop' of everyday reasoning. This theoretical perspective frames our approach to planning and implementing science teaching. Although the perspective is based upon scholarship, and has been subjected to scrutiny from other researchers, it can not really be viewed as *evidence*. Indeed, although there are studies in the literature which present evidence that is most plausibly explained by a social constructivist perspective on learning, we find it hard to imagine how this (or any) perspective on learning might be 'proved' through empirical evaluation. Rather, the perspective provides a lens though which research evidence can be interpreted and drawn upon in the design and evaluation of science teaching. In the following paragraphs, I describe how we used the social constructivist perspective on learning to interpret research evidence, in designing the three teaching sequences.

The pioneering work of Jean Piaget, in the early part of the 20th Century, established that there are commonalities in children's everyday accounts of the natural world. For most school science topic areas, the research literature on teaching and learning science now contains several studies (and some review articles) on learners' likely reasoning about the topic prior to teaching – and on how that reasoning changes as a result of teaching. We shall refer to this as the learners' *everyday reasoning*. The literature may also highlight common difficulties and misunderstandings that learners develop *during* teaching within the topic domain.

In relation to the three topic areas focused on in this study, the following examples illustrate the kind of evidence about students' everyday reasoning that was found in the research literature:

- *Simple electrical circuits*: 'Electricity' (which is not differentiated into clear and distinct ideas of charge, current, energy) is thought to leave a battery (the source) and move around the circuit in a temporal sequence, with the result that there is a pause between connecting a circuit and observing behaviour such as a lamp lighting, and that its effect is 'used up' by each component in turn as it passes through. (e.g. Shipstone, 1988).

- *Plant nutrition*: Plants are thought to ingest their food 'ready-made' in a similar way to animals. When students encounter teaching about photosynthesis, they often retain a model of food as ingested through the roots, and find it implausible that biomass in plants is synthesised mainly from water and carbon dioxide (e.g. Bell, 1984).
- *Particles and change:* Matter is thought of as being continuous rather than particulate. When students encounter simple particulate models of matter through teaching, they often assume that particles are tiny bits of the substance itself, with the same properties (such as melting or expanding) (e.g. Driver et al., 1994).

Such insights into learners' everyday reasoning provide a starting point for thinking about the design of teaching sequences. Building on this starting point, we have developed the notion of *learning demand* (Leach and Scott, 2002) to characterise more precisely the intellectual task faced by learners in coming to understand the scientific account of a given topic. The learning demand characterises the ways in which the scientific account (or more specifically the 'school science' view) of a particular natural phenomenon, differs from everyday views of that phenomenon. From this point of view, learning science involves coming to terms with the conceptual tools, and associated epistemology and ontology, of school science. If the differences between school science and everyday ways of reasoning are great, because there is little overlap between the concepts and associated epistemology and ontology of school science and everyday views, then the topic in question appears difficult to learn and to teach. Conversely, if there is considerable overlap between school science and everyday views, the learning demand is small and students may think that the school science account is 'easy' or 'obvious'.

At this point it is worth reflecting for a moment on the nature of the evidence which we are using in this part of the planning of the teaching sequence. The concept of learning demand itself follows from the social constructivist perspective on learning in setting up the comparison between everyday and school science views. Actually identifying learning demands involves drawing on research evidence about students' domain specific reasoning about natural phenomena. This evidence is empirical, based on students' written or oral responses to research questions, and much of it has been replicated through multiple studies around the world. The concept of learning demand therefore provides a bridge between findings of empirical research on students' reasoning, and the design of teaching approaches.

Having identified the learning demands for each of the three teaching sequences, we then considered how these demands might be addressed through teaching. We started by identifying a set of *Teaching Goals* for each sequence. The teaching goals make explicit the ways in which the students' ideas and understandings are to be worked on through the intervention and guidance of the teacher, in order to overcome the identified learning demand. Once the teaching goals have been clarified, attention can be given to considering what teaching activities might be used to address these goals. Our overall approach to planning is therefore based on the following overall scheme:

1. Identify the *school science* knowledge to be taught;
2. Consider how this area of science is conceptualised in the *everyday* reasoning of students;
3. Identify the *learning demand* by appraising the nature of any differences (conceptual, epistemological, ontological) between 1 and 2;
4. Design a *teaching sequence* to address each aspect of this learning demand:
   a) identifying the *teaching goals* for each phase of the sequence.
   b) planning a sequence of *activities* to address the specific teaching goals.
   c) specifying how these teaching activities might be linked to appropriate *forms of classroom communication.*

2.1 Planning a Teaching Sequence

I will illustrate the process by which the teaching sequences were designed through the case of the physics teaching sequence, which presented introductory ideas about electric circuits. In the context of the National Curriculum for Science in England and Wales (DfEE, 1999), the school science knowledge to be taught involves developing a model of energy transfer via an electric current where:

- The current in a series circuit depends on the number of batteries and the number and nature of other components;
- Current is not 'used up' by components;
- Energy is transferred from batteries and other sources to other components in electrical circuits.

This model is developed in subsequent phases of the National Curriculum by introducing the concept of voltage.

A review of the literature on teaching and learning about simple electric circuits was conducted, and the following characteristic patterns in students' reasoning were identified (see, for example: Psillos, 1998; Shipstone, 1988).

- The circuit is not viewed as a *whole system*, with changes occurring virtually simultaneously in all parts (for example, when a switch is closed charges are set into motion in all parts of the circuit together). Instead, students often explain effects in terms of *sequential* models, where any disturbance travels in one direction, affecting circuit components in succession. Thus, when an extra resistive component (such as a bulb) is added in series to a circuit, students often predict that the 'first component' after the battery gets most, or all, of the energy.
- Students often think about electric circuits in terms of a *source* (the battery) and a *consumer* (for example, a bulb). This can lead to problems:
  - the charge which constitutes an electric current is considered to origincate in the battery (the *source*).
  - the battery is considered to provide a fixed electric current.
  - when an extra battery is added to a circuit the extra current is thought to come from the additional battery (the *source*).

- electric current and energy are not differentiated, with students suggesting that the current is used up in a bulb (the *consumer*).

- The size of the electric current is estimated to be less in high resistance parts of a circuit (such as a bulb filament) than in other parts (such as the connecting leads).

In relation to broader epistemological issues, it is likely that the students will have little experience of using a scientific model which involves moving between the 'theoretical world' of the model (based on the abstract concepts of charge, current and energy) and the 'real world' of observations and measurements (Tiberghien, 1996). They are also likely to have little appreciation of the fact that scientific models are intended to be applied generally to a wide range of contexts (Driver et al., 1996).

By comparing the school science and everyday views identified above, the following learning demands were developed. Learning in this area involves the students in coming to

- Develop abstract scientific concepts of charge, current, resistance and energy in the context of explaining the behaviour of simple electric circuits.
- Understand that the battery is the source of energy for the circuit.
- Understand that energy is transferred in the circuit and not the current.
- Understand that the charges originate in the circuit and not in the battery.
- Understand that an electric circuits behaves as whole system such that a change in one part of the circuit affects all other parts of the circuit simultaneously.
- Understand that the electric circuit model based on concepts of charge, current, resistance, energy can be used to predict and explain the behaviour of a wide range of simple circuits.

The first five elements of the learning demand involve conceptual issues whilst the final element relates to more general epistemological matters.

The first step in designing the teaching sequence itself involved formulating a set of teaching goals, which were based on the learning demands and specified more directly the nature of the pedagogical interventions to be taken by the teacher. The first set of teaching goals is conceptual:

To *build on* the ideas that:

- batteries make things work.
- electricity/current flows.

To *introduce, and support the development of*, the ideas that:

- an electric current consists of a flow of charge.
- the electric current provides the means for transferring energy.
- components such as bulbs introduce resistance to the circuit. The resistance restricts the flow of charge, reducing the current flowing around the whole circuit and causing energy transfer as the charge passes through the resistance.

To *draw attention to, and to emphasise*. the idea that:

- the battery is the energy source for the circuit.
- the charges originate in the circuit and not solely in the battery.
- the electric current does not get used up.
- it is energy which is transferred in resistances.

To *emphasise throughout* that:

- electric circuits behave as whole systems such that a change in one part of the circuit affects all other parts of the circuit simultaneously.

Progressively to *differentiate between*:

- the theoretical concepts of charge, current, energy: emphasising that the electric current does not get used up, rather it is energy which is transferred in resistances to the surroundings.

The second set of teaching goals is epistemological:

To *introduce, and support the development of*, the idea that:

- the electric circuit model based on concepts of charge, current, resistance, energy can be used to predict and explain the behaviour of a wide range of simple circuits.

It is instructive to compare this list of detailed teaching goals with what is typically presented to teachers in curriculum specifications. Curriculum specifications typically provide information about 'what is to be taught' at a macro level. In contrast, the teaching goals derived from research evidence (through the learning demand analysis) provide a much more fine-grained analysis of the learning points which need to be addressed by the teacher.

The teaching sequence was designed to address the teaching goals listed above and consists of four lessons. In Lesson 1, two diagnostic questions were set for the students to work through individually, with a view to the teacher being able to probe their initial understandings of a simple electric circuit. The teacher then introduces a starter activity involving a 'BIG Circuit'. This is a simple circuit including a battery and bulb, which stretches all of the way around the laboratory. The students are asked to predict what will happen when the Big Circuit is completed. This activity was designed to challenge student thinking about the 'battery as source' (as identified through the learning demand analysis) and debates typically ensue in class on whether the bulb will light 'straight away' or 'after a short time'. Completing the circuit shows that the bulb lights 'immediately' and attention is then turned to developing a model to account for this finding: 'What is going on inside the circuit to allow this to happen?'

The basic elements of the electric circuit model are introduced via a teaching analogy. The approach taken in the teaching sequence emphasises the importance of introducing the teaching analogy *systematically*, making explicit links between the home and target domains (Glynn et al., 1995). Given the importance of making clear links to the analogy, it was decided that rather than include a number of different teaching analogies in the sequence (each of which would demand careful introduction and development), the focus would be on just one.

The particular teaching analogy selected addresses the key teaching goal of differentiating between charge and energy. In the analogy the electric circuit is represented by a continuous line of vans (charges), collecting loaves of bread (energy) at a bakery, and carrying them round to a supermarket (bulb) where they are delivered and dissipated. There has been considerable debate in the literature about the advantages and disadvantages of specific teaching analogies, in the context of electric circuits (see, for example, Schwedes and Dudeck, 1996). We recognise that charges-carrying-energy analogies have their drawbacks, relating in particular to the detail of what happens during the transient phases after a circuit is completed or broken, to how well they can account for the behaviour of circuits with several loads, and whether they introduce distracting ideas about variations in the speed of the carriers. We came to the conclusion, however, that such weaknesses are unlikely to be detrimental to student progress during the first steps of developing an electric circuit model.

Although many of the activities used in the teaching sequence look similar to activities used in many science classrooms, it is worth underlining how the teaching sequence differed from the approaches typically used in British classrooms. Perhaps the most significant difference is the proportion of time devoted to important issues in students' learning through activities such as The BIG Circuit, in comparison to typical teaching approaches. Furthermore, the overall sequence was designed so that the same teaching analogy was used throughout the four lessons, as the teacher supported the students in coming to understand the effects of changing the numbers of cells and bulbs on the electric current and energy transfer. Time was spent systematically developing the analogy and showing how it relates to actual circuits and to a scientific model of the behaviour of circuits, and students were provided with the opportunity to use the scientific model and analogy to generate explanations. Although analogies are often used by British science teachers when introducing simple explanations of electric circuits, they are rarely built up and used systematically to bridge between a scientific model and real circuits. The teaching activities also address the epistemological teaching goal of coming to appreciate the generalisability of scientific models by demonstrating the consistent application of one model to a range of contexts. It is not typical for British science teachers to address the way in which models are used to generate explanations of electric circuits.

Planning the sequence of activities involved using a number of different sources of research evidence. Some of this evidence relates to general pedagogical perspectives, such as the use of analogies and formative assessment (Black and Wiliam, 1998) whilst other evidence is taken from studies which focus directly on teaching and learning about electrical circuits. What precisely do these sources of evidence tell us? In relation to use of analogies, the general perspectives provide helpful advice about issues which need to be addressed whatever the context (such as the systematic introduction of the analogy and careful linkage to the target case). The studies relating specifically to electrical circuits offer discussions of the pros and cons of particular analogies (in relation to the physics subject matter), details about the use of particular analogies in classroom settings and in some cases sets of data relating to student learning. What these teaching studies do *not* provide is any kind of appraisal of the *relative* effectiveness of different analogies in supporting student learning.

The teaching sequences also contained guidance for teachers about different approaches to classroom communication at different points in the teaching. This guidance was based upon the theoretical framing perspective (see Leach and Scott, 2002).

This approach to identifying teaching goals is strongly informed by research evidence. However, it is important to recognise that the process of developing teaching activities inevitably involves judgements about selecting one activity rather than another, and that this step is less well informed by research evidence to evaluate different approaches. The influences upon our planning are summarised in Figure 1.

In developing teaching activities, it is also clear that effective lessons must motivate both students and teachers and conform to their expectations of what 'good' science lessons look like. No matter how 'research evidence-informed' a teaching activity is, if it is perceived as being dull, it is unlikely to prompt engagement of students and meaningful learning.

In order to move from research evidence-informed teaching goals to sequences of lessons ready for the classroom, we drew upon the professional expertise of a group of 9 teachers. These teachers attended planning meetings at the university where they contributed their professional knowledge to discussions about teaching activities to address specific teaching goals. In some cases the teachers suggested possible activities and at other times offered feedback on ideas proposed by the research team. The contribution of the teachers involved creative insight and professional knowledge about 'what would work', rather than knowledge of research and scholarship on teaching and learning science, which came chiefly from the researchers. The researchers involved had all also had extensive experience in the past as classroom teachers, which they also brought to bear on the design.

Looking back over this account, it is instructive to return to the question of what different insights, including research evidence, were employed in designing the teaching sequences, and what was the source of each kind of insight. The range of insights used are summarised in Figure 2:

As previously stated, the three teaching sequences are not unique solutions to addressing the learning demands identified for the three content areas. We see

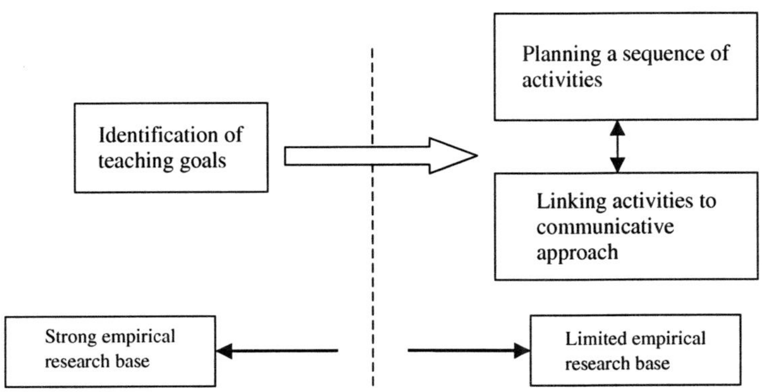

Figure 1. Using research evidence to inform the design of teaching

Insight	**Source of insight**
1. Theoretical framing perspective: a social constructivist perspective on teaching and learning science	– Scholarly writing in the academic literature on teaching and learning
2. Planning tools	
– Learning demand	– Developed from a social constructivist perspective on learning and teaching, and knowledge of the existence of research evidence about students' reasoning.
– Communicative approach	– Scholarly writing in the academic literature – empirically validated as an *analytical* tool (but not as a *planning* tool leading to enhanced learning).
3. Students' reasoning	– multiple (and frequently replicated) empirical studies
4. Design of teaching sequences	
– general pedagogical perspectives	– reflective articles, position papers, analytical accounts in science education and psychological research
– teaching and learning specific topics	– classroom-based empirical studies, but with limited comparative evaluation of learning data.
– teachers' professional knowledge	

Figure 2. Insights informing the design of the teaching sequences

them as *worked examples,* which present one way in which research evidence-informed design principles for teaching can be met. This allows for the possibility of improving and adapting the worked examples for use in different settings, while remaining faithful to the underlying design principles.

2.2 Evaluating the Teaching Sequences

The teaching sequences were evaluated in three ways. Firstly we were interested to know whether students attained conceptual understanding consistent with the aims of the teaching. The teaching sequences were taught to at least 1 group of students by each of the 9 teachers who helped to develop them (the Development Phase teachers). We used a number of short diagnostic questions focusing upon key conceptual content to assess the students' understanding. The majority of these questions were administered both before and after teaching, although in some cases we did not judge it appropriate to ask students certain questions prior to teaching. (For example, it did not seem sensible to ask questions about the readings on ammeters prior to teaching students what an ammeter is.) Some of the questions were set in contexts that were covered in the teaching sequences, whilst others were designed to test students' ability in applying concepts in novel situations. The questions followed a structure whereby students were asked to make a prediction about behaviour (such as whether a lamp would light in an electric circuit), followed by an explanation of that behaviour based upon a taught model.

Secondly, we were interested to know something about the effectiveness of the teaching sequences compared to other teaching approaches. We therefore compared the conceptual understanding of the students who followed the designed sequences (as measured with the diagnostic questions) with the performance of similar students *in the same school*, who had followed the school's usual teaching programme. This evidence was used to inform a judgement about the effectiveness of the designed teaching approach in promoting students' understanding, compared with the teaching approaches normally used in their schools. We have discussed the methodological difficulties in making such comparisons elsewhere (Leach and Scott, 2002).

Thirdly, we were interested to know whether teachers who were not involved in the design of the teaching sequences could use them with similar results as the Development Phase teachers, in cases where students following the teaching sequences had significantly higher test scores on conceptual understanding.

In addition to the nine Development Phase teachers, we worked with nine Transfer Phase teachers, who were not involved in developing the teaching sequences. The Transfer Phase teachers taught in schools of similar profile to those of the Development teachers (ranging from affluent outer suburban areas to the inner city), and implemented the written teaching sequences with at least one class of students. Pre- and post-test data were collected from students in Transfer Phase classes, and similar data were collected from students following the school's usual teaching approach. The chemistry teaching sequence was significantly different in structure from the two other sequences. It aimed to revise and extend students' understanding of the use of a simple particulate model of matter to account for physical and chemical change processes, rather than introduce ideas about physical and chemical change for the first time. It was not possible, however, to identify baseline classes covering similar content to the designed teaching sequence, and we did not therefore use the chemistry teaching sequence in the Transfer Phase of the study.

Data from the diagnostic questions completed by students who followed the biology and physics teaching sequences in the Development and Transfer Phases were analysed as follows. First, students' predictions were coded as correct or not. Their explanations were then categorised into three groups:

- Responses broadly consistent with the scientific model introduced in the teaching ('Consistent');
- Responses that are consistent with the taught model but incomplete in some respect ('Incomplete');
- Other responses ('Other').

These coded responses were averaged across all of the diagnostic questions used.

Using data from both the Development and Transfer phases (physics and biology), there are 15 pairs of 'experimental' and 'baseline' classes, in the same school, who had pre-test scores that are not significantly different[3] and where we therefore concluded that it is valid to make comparisons between the effectiveness of the designed teaching and the school's usual approach.

To what extent did students who followed the research evidence-informed biology and physics teaching sequences develop conceptual understanding consistent with the teaching goals of the sequences? Ideally all students, after teaching, would be able to make correct predictions for the diagnostic questions. Looking across both sets of data, in 14 of the 15 experimental classes, at least 79.9% of student predictions were correct after following the designed teaching sequences. In addition, all students' explanations, after teaching, would ideally be consistent with the conceptual models taught. Realistically, however, it would be surprising if all students wrote explanations, for *all* questions, that drew upon *all* features of the taught conceptual models, particularly given that the diagnostic questions have an open format which provides minimal cueing about the level of detail required. Our more realistic aim, therefore, is that all students' explanations would be coded as either consistent or incomplete, with as many as possible being consistent. Students' actual results fall some considerable way short of this ideal. There are only two classes for which more than 85% of students' explanations were coded as either consistent or incomplete, and in 7 out of 15 classes less than 50% of students' explanations were coded as either consistent or incomplete.

Nevertheless, looking across our evaluations of the biology and physics teaching sequences, students who followed the designed teaching sequences in all 15 cases were significantly more likely to draw upon the conceptual models introduced in teaching, when compared with students in associated baseline groups. Although mean scores for students in the Development Phase are higher than those for students in the Transfer Phase, this difference is not statistically significant.

3. RESEARCH EVIDENCE-BASED PRACTICE?

This study is relatively small in scale, involving only 15 cases where there is no evidence to challenge the validity of making comparisons between the outcomes of experimental and baseline teaching. Conclusions and implications need to be advanced with this in mind. Nonetheless, our findings lead us to conclude that it is possible to develop research evidence-informed teaching materials in such a way that teachers act upon many key design principles in their teaching, and that there is a significantly enhanced gain in students' learning. We noted modest but significant improvements in students' use of conceptual models after teaching, compared with those achieved by students following the schools' usual teaching programmes. This study was not designed to provide conclusive evidence about the cause of any enhanced student learning, and students' enhanced performance might in fact be caused by features of the teaching other than the design principles that informed the writing of the teaching sequences. Nonetheless, we believe that the existence of a link between the design principles of the teaching and students' performance is the most plausible explanation.

Our findings pose a challenge to Matthews' (1997) claim that there is no evidence that science teaching informed by insights from 'constructivist' research on science learning is better at promoting conceptual understanding amongst students. On the contrary, they lead us to suggest that teaching arising from curriculum development focussing on the teaching of key scientific concepts and informed by research, can indeed result in improvements to students' conceptual understanding.

Is it appropriate to use the evidence presented in this paper to support a claim that science teachers who use the physics and biology teaching sequences in the future are undertaking *research evidence-based practice*? Indeed, based on the evidence presented here should all science teachers be encouraged to use the teaching approaches set out in the two experimental schemes?

Our response to the first question is a qualified 'yes'. The evidence presented in this study shows that, for whatever reason, students who followed the designed teaching sequences as implemented by their science teachers were significantly more likely to achieve the learning aims of those teaching sequences than their peers following the school's usual approach. To this extent, in terms of the definition of research evidence-based practice presented earlier in this paper, the teaching sequences could be used in future with the justification that there is research evidence to suggest that their use is likely to lead to improvements in students' understanding of the target content.

However, we raise several qualifications to this claim. The evidence produced in our study compares students' learning following the use of the teaching sequences, *in a very small sample of schools*, with what might have been achieved if they had followed the school's usual teaching. Teachers in other schools not involved in this study might therefore argue that their existing approach has features which, in principle, appear at least as well conceptualised as the designed teaching sequences. Coupled with such teachers' experience and confidence in using their existing approaches, there is insufficient evidence to refute the claim that using the existing teaching approach might be at least as effective as using one of the teaching schemes. A much larger trial, involving a much wider range of schools and 'normal' practices would be needed to establish this more convincingly, or to refute it.

Another limitation in the evidence base collected in this study is the sampling strategy. Ideally, both students and teachers would be randomly allocated between 'experimental' and 'baseline' groups, in order to avoid the risk that measured effects are attributable to the kind of teachers who opted into the study, rather than the pedagogical approach used. However, it is not possible in the UK for researchers to design studies that have this level of impact on the running of schools.

Furthermore, in deciding whether future implementations of the teaching sequences constitute research evidence-based practice, it is necessary to specify which practice is being judged as research evidence-based. The following claims about teaching practices might be made, and evaluated in terms of the findings of this study:

- Teachers should use the physics and biology teaching sequences in a manner consistent with their design principles.
- Teachers should use the physics and biology teaching sequences in whatever way they judge to be appropriate. (There was, in fact, considerable diversity in

the way the teachers in the study used he teaching sequences, yet significant improvements in students' performance were noted).

- Teachers should use a teaching sequence which has the critical features of the designed ones, namely that it systematically addresses a set of research evidence-informed teaching goals, using a specified communicative approach. (The designed teaching sequences are clearly not the only way to address these research evidence-informed teaching goals, using this communicative approach).

The study does not provide an evidence base which could refute any of the above claims. It is not, however, possible to support the first claim, because no teacher in the sample implemented all aspects of the design principles about communicative approach. Although the data from this study supports the second and third claims, the sample size and range of 'worked examples' used in this study are not sufficient to establish either claim conclusively. This leaves a very considerable diversity in the range of practices that might be viewed as research evidence-based.

We think it is helpful to draw a distinction between *research evidence-based practice*, and *appropriate practice*. As we have shown, it is entirely legitimate to justify a decision to use the teaching sequences on the grounds that there is research evidence that they 'work'. However, we also believe it is legitimate for a teacher to argue that another approach is more appropriate, possibly on the grounds that it is better matched to the needs of teachers and/or students in a particular setting, or that the improvements in understanding demonstrated by students in our study are not sufficient to justify the effort of changing an existing approach. The first of these arguments could be rendered untenable by data from a larger study. But even a larger study could not completely undermine the second argument. In other words, a judgement remains to be made about the quality and relevance of the evidence base, and the size of any effect noted, before it is sensible to make decisions about investing time and energy into changing practice. In order to claim that using one particular teaching sequence is better than a range of other practices, it is necessary to have data from a random sample of schools, as well as a range of other teaching approaches that are commonly used. The evidence provided by our study falls considerably short of this. Therefore, our answer to the question of whether science teachers should be required to use the designed teaching sequences is 'no'. However, we nonetheless believe that this study provides evidence to suggest that using the teaching sequences might well result in improvements in students' understanding of the target content. Furthermore, the fact that the design principles of the teaching sequences are explicit means that teachers are in a position to make decisions about the implementation of the teaching sequences, and possible modifications, in a principled way.

Rather than asking whether the implementation of the teaching sequences – or other educational practices – is research evidence-based or not, we think it is more helpful to ask about the nature of the evidence base on which claims rest, and the legitimate use of such evidence in informing practice.

4. POSTSCRIPT: RESEARCH EVIDENCE-INFORMED TEACHING AND STUDENTS' MOTIVATION TO LEARN SCIENCE

Evidence about the number of students opting to study science, and interview studies with young people themselves (Osborne and Collins, 2001), suggests that British students in secondary education do not find their science lessons particularly motivating. The most commonly-suggested solutions to this problem include increasing the amount of practical work in science classes, and making science education 'more relevant' by including examples familiar to students from their everyday lives or industry (e.g. Bell, 2004).

The group responsible for designing the teaching sequences presented in this paper was mindful of the need to develop an approach that would be motivating to students and teachers. This aspect of the design, however, was informed by professional judgement rather than research evidence. The teaching sequences did not include more practical work than would be typical in British science classrooms, and they made no attempt to motivate students by selecting examples thought to be familiar or interesting. Rather, our intention was to motivate students through the satisfaction of developing conceptual understanding.

Many people will no doubt view this aspiration as naïve. Although our study was not designed to look at the influence of the designed teaching on students' (or teachers') motivation, some insights came to light through the interviews that we conducted with teachers in the Development and Transfer phases of the project after they had implemented the teaching. Without exception, the biology and physics teachers in the Development and Transfer phases said that the designed teaching sequences were well-received by their students, and several said that they had noted a significant improvement in their students' motivation. Furthermore, several of the Development phase teachers made positive comments about the process of designing research evidence-based teaching:

"To me it seemed a much better way of going about it and I felt quite excited about the approach. (...) It's not the only way but it's better than what I do now (...) and I just sort of got excited about it."

"I don't think in any other lesson that I've done, have I ever gone into using the analogy in that much depth. I might have mentioned it in passing, but not really probed the children for their understanding of the analogy. That's the big difference here and I think it's really valuable."

These quotations hardly form an appropriate body of evidence to use in making judgements about the effect of the teaching sequences upon students' (and teachers') motivation. However, I think that it lends support to the idea that motivation for science learning can come from teaching that succeeds in introducing traditional conceptual subject matter to students in such a way that they understand it.

ACKNOWLEDGEMENTS

This work was conducted through the research network *Towards evidence-based practice in science education*, which involved the Universities of York, Leeds, Southampton and King's College London, and was funded through the Economic and Social Research Council's Teaching and Learning Research Programme. The work was conducted in collaboration with a number of colleagues. The team who conducted the empirical study involved Jaume Ametller, Andy Hind, John Leach, Jenny Lewis and Phil Scott. Phil Scott led the group working on the physics teaching sequence whose development is illustrated in this article. The notion of research evidence-informed and research evidence-based practice was developed in collaboration with Robin Millar, Jonathan Osborne and Mary Ratcliffe. I acknowledge the contribution of these colleagues, while taking responsibility for the content of this paper myself.

NOTES

[1] Due to my linguistic (in)competence, I focus upon literature published in the English language in this paper.
[2] The phrases 'research on teaching and learning science' and 'research on the didactics of science' have both been used in this paper, because both have appeared in English language writing. I have used both phrases to refer to research which has a substantial focus on the teaching and/or learning of specific subject matter.
[3] In fact, two of the experimental groups were significantly better at making predictions about the behaviour of phenomena, though their explanations were not significantly different from students in the baseline groups. On that basis, we did not exclude them from the sample.

REFERENCES

Bell, B. (1984). *Aspects of secondary students' understanding of plant nutrition: Summary report.* (Leeds: Centre for Studies in Science and Mathematics Education)

Bell, D. (2004). Editorial: Engaging with science. *Education in Science, 212*, 4

Black, P. and Wiliam, D. (1998). *Inside the black box: Raising standards through classroom assessment.* (London: King's College)

Blunkett, D. (2000). Influence or irrelevance: Can social science improve government? *Research Intelligence, 71*, 12–21

Brown, A. (1992). Design experiments: theoretical and methodological challenges in creating complex interventions in classroom settings. *Journal of the Learning Sciences*, 2(2), 141–178

Cobb, P., Confrey, J., diSessa, A., Lehrer, R. and Schauble, L. (2003). Design experiments in educational research. *Educational Researcher, 32*(1), 9–13

DfEE. (1999). *Science: The national curriculum for England.* (London: DfEE)

Driver, R., Leach, J., Millar, R. and Scott, P. (1996). *Young people's images of science.* (Buckingham: Open University Press)

Driver, R., Squires, A., Rushworth, P. and Wood-Robinson, V. (1994). *Making sense of secondary science: Research into children's ideas.* (London: Routledge)

Duit, R., Kormorek, M. and Wilbers, J. (1997). Studies on educational reconstruction of chaos theory. *Research in Science Education, 27*(3), 339–357

Glynn, S., Duit, R. and Thiele, R. (1995). Teaching science with analogies: A strategy for constructing knowledge. (In S. M. Glynn and R. Duit (Eds.), *Learning science in the schools: Research reforming practice* (pp. 247–273). Mahwah, NJ: Erlbaum.)

Halloun, I. and Hestanes, D. (1985). Common sense conceptions about motion. *American Journal of Physics, 53*, 1056–1065

Hargreaves, D. H. (1996) *Teaching as a research-based profession: Possibilities and prospects* (Annual Lecture). (London: Teacher Training Agency)

Hillage, J., Pearson, R., Anderson, A. and Tamkin, P. (1998). *Excellence in research on schools.* (London: Department for Education and Employment)

Leach, J. and Scott, P. (2002). Designing and evaluating science teaching sequences: an approach drawing upon the concept of learning demand and a social constructivist perspective on learning. *Studies in Science Education*, 38, 115–142

Leach, J. and Scott, P. (2003). Individual and sociocultural perspectives on learning in science education. *Science and Education*, 12(1), 91–113

Lijnse, P. L. (2000). *Didactics of science: The forgotten dimension in science education research?* (In R. Millar, J. Leach and J. Osborne (Eds.), *Improving science education: The contribution of research* (pp 308–326). Buckingham, UK: Open University Press.)

Matthews, M. R. (1992). Constructivism and empiricism: an incomplete divorce. *Research in Science Education*, 22, 299–307

Matthews, M. R. (1997). Introductory comments on philosophy and constructivism in science education. *Science and Education*, 6(1), 5–14.

Méheut, M. and Psillos, D. (2004). Teaching learning sequences: aims and tools for science education research. *International Journal of Science Education, 26*(5), 515–535

Millar, R. and Haines, V. (2005). EPSE teaching resources. Retrieved July 2005 from http://www.york.ac.uk/depts/educ/projs/DiagQuestionsIntro_Dec02.htm.

Millar, R., Leach, J., Osborne, J. and Ratcliffe, M. (2006). *Improving subject teaching: Lessons from research in science education.* (London: RoutledgeFalmer)

Nola, R. (1997). Constructivism in science and science education: A philosophical critique. *Science and Education, 6*, 55–83

Osborne, J. and Collins, S. (2001). Pupils' views of the role and value of the science curriculum: a focus group study. *International Journal of Science Education, 23*(5), 441–467

Pintó, R. (2005). Introducing curriculum innovations in science: Identifying teachers' transformations and the design of related teacher education. *Science Education, 89*(1), 1–12

Psillos, D. (1998). Teaching introductory electricity. (In A. Tiberghien, E. L. Jossem and J. Barojas (Eds.), *Connecting research in physics education with teacher education: An ICPE Book*, International Commission on Physics Education. Retrieved January 2005 from http://www.physics.ohio-state.edu/~jossem/ICPE/TOC.html)

Ratcliffe, M., Bartholomew, H., Hames, V., Hind, A., Leach, J., Millar, R. and Osborne, J. (2005). Evidence-based practice in science education: the researcher-user interface. *Research Papers in Education*, 22(2), 69–86

Savinainen, A., Scott, P. and Viiri, J. (2005). Using a bridging representation and social interactions to foster conceptual change: Designing and evaluating an instructional sequence for Newton's third law. *Science Education, 89*(2), 175–195.

Schwedes, H. and Dudeck, W.-G. (1996). Teaching electricity by help of a water analogy (how to cope with the need for conceptual change). (In G. Welford, J. Osborne, P. Scott (Eds.), *Research in science education in Europe* (pp. 50–63), London: Falmer Press.)

Shipstone, D. M. (1988). Students' understanding of simple electrical circuits. *Physics Education*, 23 (2), 92–96

Staver, J. R. (2005). *Let's change before it's too late: A speech given at the awards banquet of the Annual Meeting of NARST, Dallas, TX, 7 April 2005.* Retrieved July 2005 from http://www.educ.sfu.ca/narstsite/news/e-narstnews7-12-05.pdf.

Tiberghien, A. (1996). Construction of prototypical situations in teaching the concept of energy. (In A. G. Welford, J. Osborne and P. Scott (Eds.), *Science education research in Europe: Current issues and themes* (pp 100–114). London: Falmer.)

Tooley, J. and Darby, D. (1998). *Educational research: A critique.* (London: Office for Standards in Education)

US Department of Education. (2002). *Strategic plan 2002–2007.* (Washington, DC: US Department of Education). Online. Retrieved February 10, 2005, from http://www.ed.gov/about/reports/strat/plan2002-07/plan.pdf

ANDRÉE TIBERGHIEN AND CHRISTIAN BUTY

5. STUDYING SCIENCE TEACHING PRACTICES IN RELATION TO LEARNING: TIME SCALES OF TEACHING PHENOMENA

Abstract: This chapter presents an analysis of the teaching practices in the perspective of relating teaching and learning. This analysis involves three time scales: macro (months), meso (about ten minutes) and micro (seconds). Different concepts are involved in the meso and micro analyses. The theoretical approaches at the meso scale are mainly based on the notion of *chronogenesis* that is included in a larger theoretical framework. This notion recovers the phenomena of evolution in time of the production of knowledge of the class: the taught knowledge. This taught knowledge is considered as staged in the classroom by a joint action of the teacher and the students. Moreover, the taught knowledge has been decomposed into smaller elements of knowledge than the themes. Two physics classes (grade 10) have been analysed. The current results show that this theoretical framework leads to differentiate the two classes. Even if the institutionalized knowledge is similar, the order of presentation, the duration of similar themes, the relations between elements of knowledge, their rhythm of introduction, their "re-use" after a first introduction can be different. These differences can create a variety of learning possibilities in the classes. These several dimensions of the taught knowledge at meso and micro levels are candidates to be characteristics of a class relevant to relate teaching and learning

Keywords: Classroom practice, Comparison of physics classrooms, Decomposition into elements of knowledge, Taught knowledge, Teaching, Time scale

1. INTRODUCTION

Relations between teaching and learning are involved in a variety of research studies. These relations are complex and are still to be elucidated. In science education research, several types of studies deal with these relations, in particular: the evolution of students' conceptions with a teaching sequence and the design of teaching sequences, teacher training, teaching practices and the relation between teaching practices and students' performances. These studies can be differentiated into two sets: (1) studies dealing with teaching innovation or improvement and those dealing with "ordinary" teaching in the sense that the researcher does not intervene in the teaching situations except to collect data; (2) studies using a methodology of case study and those using more quantitative methods for large-scale investigations.

Andrée Tiberghien, UMR ICAR, 5, av. Pierre Mendès-France, CP 11 69676 BRON Cedex, University Lyon 2

R. Pintó and D. Couso (eds.), Contributions from Science Education Research, 59–75.

Whatever their perspectives all these studies show the complexity of the teaching activity itself and also the complexity of the relationships between teaching and learning.

In this chapter, we propose an analysis rather typical of complex phenomena involving an approach in terms of time scales. The aim is to contribute to our understanding of teaching practices in order to relate them to students' learning and their performances. This analysis is oriented on the knowledge involved in these phenomena; it constitutes only a part of more complete studies. However, the proposed time scales are candidates to be used in studies choosing other perspectives.

First, to introduce to time scales, we present short analyses of two types of studies: the first type is focused on learning with a case study methodology and the second is focused on teaching with a large-scale statistical approach. Then we introduce our theoretical framework on which the time scales are based. We illustrate this perspective with data taken in two different physics classrooms at grade 10.

2. STUDIES OF LEARNING AND TEACHING: TWO TYPES OF INTERRELATED PHENOMENA

Both teaching and learning take place in the same space-time: the classroom. However, these two activities are those of different individuals (the teacher and the students) and have different aims. For teaching, aims consist of introducing new knowledge to the students, creating favourable situations for students' learning, evaluating students' acquisition; for learning aims are focused on understanding and acquiring the taught knowledge and being able to show that this knowledge is acquired (Tiberghien, 1997).

2.1 Evolution of Students' Conceptions with a Teaching Sequence

From the research studies on the evolution of students' knowledge as they go through a teaching sequence, we emphasize the difference between the knowledge that students acquire and the taught knowledge (Niedderer et al., 2005). The learning pathways, as shown by several researchers (Buty, 2000; Dysktra, 1992; Küzuközer, 2005; Petri & Niedderer, 1998), differ from the taught knowledge not only as regards the knowledge content but also in the rhythms of teaching and learning: the rhythm of introduction of new elements of knowledge throughout teaching is different from the rhythm of students' knowledge acquisition. The methodology used is mainly case studies with an analysis of students' productions during teaching sessions. For example, some studies analyse students' actions (verbal productions and gestures) in terms of ideas, therefore the time scale of the analysis is at the level of seconds. Then researchers look for the similarity in the students' ideas over time or the change in these ideas using constructs such as conceptions, learning pathways. They propose phenomena like stability or change of conceptions at the level of hours, days or weeks throughout the teaching sequence.

More or less explicitly, such studies involve at least two times scales (second and hours or more).

More recently some studies aim to establish relations between the specificity of the successive teaching situations in a sequence and the students' actions and understanding (Budde & Niedderer, 2005; Givry, 2003; Givry & Roth, 2006; Niedderer et al., 2007). Two approaches of relating teaching and learning components are taken. One consists of characterizing the elements of the teaching situations playing a role in the students' actions. In this case, the time scales for the teaching and learning components are the same (Givry, 2003). More generally the notion of learning opportunities leads the researcher to use the same time scale to take into account students' actions and the teaching elements used by the students. In the case of Givry's study, the process to generalize relations between teaching and learning phenomena involves categorization of learning environments fostering a type of evolution. For example it appears that manipulation of experiments particularly supports the establishment of links between several ideas, whereas new ideas about the meaning of words are supported essentially by the teacher's discourse and the activity work sheets used in the classroom.

The other, proposed by Budde & Niedderer (2005), is based on the notion of resonance between certain elements of taught content as part of the learning environment and the evolution of students' conceptions. In this case the relations are made between two analyses: students conceptions and the taught knowledge; the time scales are not defined a priori but depends on the observed phenomena of resonance (like congruence or discongruence resonances): the congruence between an element of the taught knowledge and its students' learning can happen after weeks of teaching or on the spot. Consequently to understand students' learning it is necessary to involve several time scales. Our perspective is to make explicit the time scales involved in the studies in order to understand how students' learning can emerge from classroom life.

2.2 Large-scale Studies of the Relations between Teaching Practices and Students' Performances

Nowadays, our educative systems are evaluated nationally and internationally. Assessment raises problems because it should be adapted to the goals of the situation. However, many works have been devoted in particular to the question of assessing conceptual understanding; at least to a certain extent they have shown that it is possible to evaluate conceptual acquisition. The obtained results are put in relation with other investigations on school, teacher, etc. and costs. Two large international evaluations are regularly done: TIMMS evaluates students at 8th grade whereas PISA evaluates 15-year-old students. Associated with these evaluations, the project "TIMSS video" is focused on teaching practices. The aims are (Hiebert et al., 2003)

- to develop objective, observational measures of classroom instruction to serve as appropriate quantitative indicators of teaching practices in each country in using complementary qualitative and quantitative approaches;

- to compare teaching practices among countries and to identify lesson features that are similar or different across countries;
- to describe patterns of teaching practices within each country.

This project is clearly on the teaching side; it develops measures on classroom instruction. With a pragmatic perspective, the authors intend to characterize teaching practices in different countries and to be able to describe teaching practices allowing students to get better performances. One main result consists of characterizing the particularity of each country in a mathematics lesson, what they call the lesson signature. Lessons were characterized with three dimensions: purpose of what is going on in the classroom (reviewing, introduction of new content, practicing content), classroom interaction, content activity (covered topics, nature of reasoning, mathematical relationships among the problems, etc.). The lesson signatures were constructed by studying what was happening along the three dimensions during each minute of every 8th mathematics lesson. It appears that, according to the countries, a lesson evolves differently with time.

For example, in Hong Kong SAR, "a number of lessons are initially focused on content activities such as non-problem-based mathematical work: presenting definitions, pointing out relationships among ideas, or providing an overview or summary of the major points in a lesson. [...] As students and teachers moved through the second half of the lesson, Hong Kong SAR mathematics lessons also began to focus on the practice of new content through a mix of independent problems and sets of problems (concurrent problems) assigned to students as whole-class or seatwork" (Hiebert et al., 2003, p. 132). Moreover, the classroom interaction dimension shows that the majority of the lesson is "public interaction", that is the whole class work together. On the reverse, in Netherlands, "the midpoint of the lesson is the time when a majority of Dutch lessons moved into private interaction, wherein students worked individually or in small groups with the help of the teacher and focused on sets of problems completed as seatwork" (Hiebert et al., 2003, p. 138). Moreover, the majority of lesson time was spent on new content (introduction of new content or its practice) and nearly one-quarter of lessons were devoted entirely to review.

We interpret this result by considering that the time scale of several minutes is relevant to differentiate teaching practices during a session at least for the dimensions (purpose, interactions, content activity) taken in this study. There is a change of time scales by a "cumulative treatment" of the first analyses of the videotapes at the level of seconds and the characterization of a lesson according to the countries at the level of minutes and hours.

However, the results do not demonstrate the characteristics of lessons which are important for learning and students' acquisition in mathematics or science. The authors stress the complexity of teaching practices: "More than anything, the findings of this study expand the discussion of teaching by underscoring its complexity. One thing is clear however: the countries that show high levels of achievement on TIMSS do not all use teaching methods that combine and emphasize features in the same way. Different methods of mathematics teaching can be associated with high scores on international achievement tests" (Hiebert et al., 2003, p. 149).

3. CONCLUSION

These two different types of research: students' evolution with a teaching sequence and large-scale studies show that nowadays the relation between teaching and learning is still an open question. Studies on teaching and learning involve several time scales and a given time scale of teaching is not particularly related to the same time scale in learning. For example, in TIMSS video studies, teaching practices are investigated from teaching sessions of about one hour, a different teacher for each session, the duration of events taken into account in the analyses is in seconds. However the lesson signatures show phenomena of the order of about 10 minutes; for example the comparison concerns the first or the second half of the lesson, etc. There is a change of scale between these phenomena and the "micro" events on which they are constructed: there is a collective behaviour, in the sense of emergence of something which cannot be produced by a single type of event but by interactions between events.

In order to take into account this complexity, we propose to cross over a step by making explicit the time scales of investigations, whatever they are, and emphasizing the importance of an intermediate scale (about 10 minutes) between micro (second) and macro scales (hours, week, month, years). In the following, the theoretical framework introduces concepts particularly relevant at this intermediate scale in addition to concepts at micro scale.

4. THEORETICAL CHOICES

We state that a determining teaching aspect for learning is the taught knowledge: its content and the way it is staged in the classroom. This statement is based on a hypothesis which is relevant only partially: "what is not taught is not learnt", radically different from what is taught is learnt. This hypothesis is too extreme in the sense that students learn a lot on the material world from their daily experience. However a large part of the scientific content taught at high school level does not belong to the common knowledge and the students have little chance of confronting the scientific knowledge elsewhere than in the classroom. We do not neglect scientific museum or other science-oriented cultural activities but we suppose that they are insufficient to allow conceptual learning for a large part of the students. Our theoretical choices involve a specific position on knowledge introduced below.

4.1 Knowledge

Our theoretical position on knowledge is based on Chevallard's work (1991). He dealt with knowledge by using the metaphor of life and ecology. Knowledge 'lives' within groups of people (a class in our case). The term "knowledge" in this perspective has a broad meaning: it is not limited to content, but includes the media involved on the one hand and the embedded epistemology on the other.

Within this framework, we introduce the concepts of "knowledge to be taught" and of "taught knowledge". These are constructs and not obvious productions of educators or students. The knowledge to be taught is not the same as scientific knowledge. It is elaborated by the group(s) in charge of the official curriculum at least in France. This knowledge to be taught is elaborated with reference to scientific knowledge. It differs from it, and also differs according to the teaching level. For example classical mechanics at middle school differs from mechanics at the end of the secondary school, and this in turn differs from mechanics at university level even though all of them refer to the same laws of physics. The concept of knowledge to be taught allows us to analyse the distance between these "knowledges" and the constraints which applied during their elaboration and which contributed to "shaping" them.

In this chapter, we mainly use the concept of taught knowledge which is necessarily associated to a particular class. It is a construct of the researcher based on the discursive productions of the class (including gestures), it corresponds to the knowledge staged in the classroom by the teacher and the students during the teaching sessions (Sensevy et al., 2007). This concept is integrated in a theoretical framework. The class is considered as a system constituted by the triad of the teacher, the students and the taught knowledge (Mercier et al., 2002). This system evolves with time. A characteristics of its evolution relative to the taught knowledge is called *chronogenesis*. This notion recovers the phenomena of evolution in time of the production of knowledge of the class.

4.2 Several Time Scales

Following Mercier et al. (2005) we consider that each system has own times in the sense that its behaviour is associated to specific rhythms. In other words it has its own time scale(s). Let us note that, in the following, time scales are referred to as the measure of time in terms of year, minute, seconds.

As regards science education, on the teaching side, we first consider the educational system of a country. Usually, at least in France, this system, which depends on the government, decides the succession of the academic years with an official curriculum for each year (or each stream): we call this the *scholastic time*. An example of this time is the way of naming the successive school levels; let us note that nowadays more and more researchers take the reference, used in several countries, of starting with 1 for the first elementary grade and going up to 12 (or 13 depending on the country) for the last grade of secondary schooling.

The official curriculum determines the knowledge to be taught in a given scholastic year, and specifies the topic sometimes with an approximate duration. However, the temporal progression of the teaching content is specific to a class and under the teacher's responsibility; this progression depends on the types of phases like reviewing, exercising, doing experiments, etc. and on the class organization, still under the teacher's responsibility. The rhythm of introduction of new elements of knowledge relevant to the official curriculum is up to the teacher who takes the decision according to the class. We call *didactical time* the rhythm of introduction of new elements of knowledge in a given class (Mercier et al., 2005). This time is specifically associated with the class as a system and not only with the teacher. The same teacher with two different classes of the same grade may manage different didactical times. These times are constrained by the official curriculum and by the class. The evolution of didactical time is associated with the "memory" of the class as a group. Like we mentioned above, the class, as a system, can also be characterized by several dimensions. For example, a first analysis of a video session consists of constructing what we call synopsis. This synopsis includes the following dimensions: (1) class organization (whole class, small groups, etc.); (2) themes; (3) resources (book, devices, slides, etc.); (4) didactical phases (introduction, development of lecture, oral assessment, etc.); (5) teacher and students' actions (reading, writing, asking/answering questions, etc.); (6) description of content. This synopsis consists of a table with a column for each dimension and with in addition a first column giving a physics time scale. Such synopsis clearly show that each dimension has a specific rhythm but that their time scale is the same: around several minutes (or about 10). This time scale is relevant of the system class as a group. It is interesting to note that the educative system introduces a fictional learning time since an evaluation done by the teacher supposes that the students have learnt what has been taught before. Like the teacher, the students also have a responsibility vis-à-vis the educational system to learn in order to give the correct answers during the successive evaluations made by their teacher. This fictional time is the same as the time of introduction of new elements of knowledge, and is also depending on the rhythm of teachers' assessments.

If we consider learning as a phenomena, according to the methodologies of the studies the time scales are very different; small elements of knowledge can be acquired at the scale of second or less whereas the acquisition of meaning of concepts like force or Newton laws can take months or years. As discussed above, research in science education shows that *learning time*, in the sense of students' rhythms of knowledge acquisition, may be very different from the didactical time and may differ from one student to another.

When the observation in a classroom is focused on individual (teacher or student) and not on the behaviour of the group, then events like a verbal interaction or a gesture are taken into account. These events are actions on a time scale at a micro level, seconds or less.

Table 1 shows different time scales going from years to seconds. It is not surprising that the action time scale associated to an individual is shorter than the didactical time scale associated to a group to the extent that the reaction time of a group needs a coordination/interaction of each individual's action.

The didactical time under the teacher's responsibility is situated at an intermediate time scale, the meso level. This meso scale is relevant to analyse the class group. We choose this scale to characterize the way in which taught knowledge is staged in the class.

TABLE 1. Time scales according to the systems; the time scale is referred to the physics time

Under the control of	Time of the system name	Time scales
Institutional system (country region)	**Scholastic time**	
	Academic year: Official curriculum according to the grades	Year, month
Teacher and class	**Didactical time**	
	Theme, sub-theme, rhythm of introduction of new elements of knowledge	Hour, minute
Individual: teacher or students	**Action time**	
	Talking, gestures, etc.	Minute, second, millisecond

5. CHARACTERIZATION OF A CLASS: DIDACTICAL TIME AND TAUGHT KNOWLEDGE

We present elements of characterization of a class at a meso time scale on the basis of analyses of two classes at 10th grade. These classes are observed during a part of mechanics teaching dealing with the introduction of force up to the introduction of the inertia principle. They are in different schools settled in areas corresponding roughly to a middle socio-economic level. Both teachers are recognized as very experienced. A main difference is that the teacher of class 1 (a woman) belonged to the group of research-development which designed the teaching sequence that she used, whereas the teacher of class 2 (a man) has no link with this group and constructed his own sequence.

The data consist of videotapes of the lessons taken with two cameras. One camera is focused on the teacher with a part of the class and the other on two students with also a large part of the whole class. Other data were collected (teacher's interview, students' written productions) but they are not used for this analysis. The analysis presented below is carried out from videos and their partial transcriptions.

5.1 Structuring the Taught Knowledge: Themes

The analysis in terms of themes aims to specify the taught knowledge staged in a class. The discursive productions of a class (Filletaz, 2001) can be divided into units on the time scale of about 10 minutes. These units have a structure, with frontiers and a thematic coherence. Most of the time, they include an introduction and a conclusion, and most of the utterances are related to a same theme. We call this unit "theme".

In some cases there are sub-themes or an inclusion to the extent that the class' productions are not completely related to the theme. For example, in class 2 the teacher introduces an experimental exercise with a ping-pong ball in water; most of the students consider that when the ball is in the water the Earth does not act on it. The teacher takes time to discuss this question (inclusion 6a) and goes back to the initial theme (theme 6).

From Table 2 differences between the two classes can be established; in particular on the content, the chronology and the duration of the themes. Except for themes 2a and 2b of class 2 corresponding to previous sessions of class 1, we recall that the themes presented in Table 2 are encountered nowhere else in the teaching sequences in the two classes.

With regard to the content and the chronology, the lines between the themes of classes 1 and 2 show that some themes are specific to a class (effects of force for class 1 and analysis of motion for class 2). They also show that a theme,

TABLE 2. Comparison of the succession of themes in two classes (grade 10) during the parts introducing forces until the introduction of the inertia principle. The bold line of the cells means a new session. When there is a general theme with sub-themes the duration is given by sub-theme. When there is an inclusion, the total duration is given and the duration of the inclusion is given between parentheses

Time (min)	Themes in class 1		Themes in class 2	Time (min)
	....			
1:25	*Introduction of the general theme of the notion of force*		1. Effects of force on the motion of a object	18
18:44	1. Determination of phases of motion of an object, direction of action on this object, variation of velocity		2. *Interactions*	
10:41	2. Analysis of interactions for different phases of motion of an object (case of a medicine-ball)		2a. Interactions = A acts on B then B acts on A	14:33
4:41	3. Introduction of the force and its vectorial representation and of the principle of reciprocal actions		2b. Interactions at distance and contact interactions	4:39
9:23	4. Using (exercising) force with and its vectorial representation from interactions (use of the full model of interactions)		3. Recalling of interactions	1:31
5:14	5. Interactions: relations between a symbolic representation and one or several material situations		4. *Modelling actions by the forces*	
10:10	6. Representation of force (with direction) modelling an interaction (not the length of the vectors)		4a. Representation of force	9:15
30:31	7. Representation of force modelling an moving object		4b. Measure of force	1:19
5:26	8. Introduction of the inertia principle		5. Forces and masses	10:35
22:21	9. Compensation of forces exerted on a motionless system		6. List of forces and compensation (or not) of forces	45:21
7:55	10. Non-compensation of forces exerted on a system of which the velocity varies		6a. *Inclusion*: terrestrial attraction on an object in water	(2:27)
32:27	11. Inertia principle applied according to horizontal and vertical directions of the motion		Total time: 1h45	
9:05	12. Influence of the mass on the motion			
Total time: 2h48				

like representation of force (class 2), can be involved in several themes of the other class. The comparison of the themes indicates difference in the content. For example, in class 1 the relation between interaction (concept largely introduced before the sessions discussed here) and motion is involved in several themes (1, 2, 5) and that between force and motion is introduced in the theme 7. In class 2 the teacher introduces the concept of force by its effects and directly relates force and motion. The concept of motion appears in five themes in class 1 and in one theme in class 2.

Table 2 shows also a global difference in the rhythm of introducing new knowledge. In class 1, the two shortest themes, 3 and 8 (total duration about 10 minutes), introduce the most important and rich concepts of the part of mechanics: force and its vectorial representation, principle of reciprocal actions and the inertia principle. In class 2, the introduction of new elements of knowledge is spread over time more regularly. This difference is due to a characteristic of the teaching sequence used in class 1 consisting of introducing a theoretical model explicitly in a written form given to students (one page) (Gaidioz et al., 2003, 2004). A part or the totality of the model is introduced by the teacher who reads and gives short comments, then the students have to use it in order to understand it during activities. The text of the model plays a double role: (1) on the spot when the teacher introduces it and also when the students use it during an activity or an exercise and (2) along the sessions since the same text (same sheet of paper) is used. This text of the model relates a long-term process to a short-term effect – this is a case of heterochrony (Lemke, 2000). Textbooks could have the same function, but in fact in both classes it is rarely used, mainly for problem statements.

The duration of similar themes is also different. For example in class 1 'representation of forces' intervenes in themes 3, 4, 5 and 6 with a total of 54 minutes whereas in class 2 the duration is 9 minutes. However, if we add the durations for these themes and those, involving the list of forces, that also make a "re-use" of representation of forces very recently introduced, we obtain 85 minutes with class 1 and 55 minutes with class 2.

Such an analysis allows the researcher to compare classes during several sessions or during a whole teaching sequence and contributes to characterize the effective taught knowledge of a given class and to construct hypotheses on students' knowledge acquisition.

5.2 Characteristics of a Theme

The analysis of each theme is developed to specify the elements of new knowledge and their epistemological status in the class.

Following the theory of didactical situations (Balacheff et al., 1997), we use the concept of "institutionalized knowledge" or "institutionalization" by the teacher. The task of institutionalization is under the responsibility of the teacher, who publicly and officially recognizes the emergence of a new element of knowledge compatible with the curriculum. Thus each new element of knowledge takes its place in a structured set, made up of all of the elements of knowledge previously produced in the class. This set forms the didactic memory of the class (Mercier et al., 2005). This deeper analysis of each theme includes the following characteristics: (1) new elements of institutionalized knowledge; (2) new elements of knowledge; (3) "re-using" of recent elements of knowledge; (4) knowledge processing from an epistemological perspective; it also includes how the knowledge is staged in the class: during an exercise, a laboratory work, a teacher's presentation, etc. what we call "scenery". To illustrate these characteristics, we choose the example of the theme "representation of force" that appears in the two classes; but the discussion is grounded on all the tables for every theme of the two classes. These two themes have the same title but are inserted in a different "chronogenesis". As we mentioned above, theme 4a of class 2 (left column) is related to three themes in class 1 (Table 2): theme 3 introduces force and its vectorial representation during almost 5 minutes; then the following theme 4 involves exercising for almost 10 minutes; and the third one, theme 6, again is a correction of an activity on these conceptual elements. In class 2 exercising these concepts is done during theme 6 where the concept of compensation of forces is also introduced.

First we specify "the *scenery*" in which a theme takes place in the class. This term means the environment in which the theme is involved in the life of the class. The example illustrates possible differences of sceneries between the two classes (or even between two themes of a class).

In class 1, this is a correction of an activity concerning a conceptual knowledge that has already been introduced. The class organization is typical of this class during correction.

To go further in the analysis we give how we account for the knowledge processing for this theme. The knowledge is at a formal level, more precisely it involves processes to go from a model of interactions with symbolic schemas (ellipses and arrows representing interactions) to a model of forces with vectorial representations. Before handing over to the student at the blackboard, the teacher recalls the rules of the model of interactions: "we studied the model of interactions which tells you how to proceed, we listen to you". Moreover, when the student produces an error on his schema, the teacher interprets the students' schema (direction of force) by applying it to a concrete case and asks the class to comment on it. From a solution proposed by a student, the teacher asks for an interpretation leading to an element of knowledge that has not been already taught (equality of vectors length representing forces in relation to compensation of forces and equilibrium). Moreover, the teacher does not institutionalize knowledge, but checks that students have the correct solution. The representation of forces done at the blackboard takes the place of institutionalized knowledge. Then in this theme, the taught knowledge consists mainly of symbolic representations that are shown.

The way of correcting the students' errors shows different modes of justification of the right knowledge: call to the rules of the model, interpretation of the symbolic representation in terms of a material situation or teacher's authority.

In class 2 the teacher develops new knowledge "step by step" with some students' contributions through short answers (about one word); he also shows how to make a vectorial schema of force representing a concrete situation: he makes a "*monstration* of procedure" (this word comes from the French word *montrer* (to show)) (Johsua & Dupin, 1993). At the end of the theme, he institutionalizes elements of knowledge and the rules of vectorial representation of forces. Generally, in class 2, the new elements are presented more frequently when the teacher talks and comments on an experiment to the whole class while asking short questions to the students.

As regards to the elements of new knowledge (Table 3, column 3), we take criteria of newness for the class, that is when an element of knowledge has never been introduced. Several cases happen depending on the ways of presentation: if the teacher dictates or gives a clear indication that an element is new and/or that students have to learn it, these elements are institutionalized (Table 3, column 2); if, during an activity or a correction, the teacher introduces new elements without specifying

TABLE 3. Analysis by theme of the taught knowledge; case of the theme "representation of force"

Theme duration	New institutionalized knowledge	Elements of new knowledge	"Re-using" of recent elements of knowledge	Knowledge processing
Class 1: Scenery: whole class, correction of an activity; a student at the blackboard; after questions teacher confirms or gives solutions				
Representation of force 10:10	Vectorial representation and other solutions written on the blackboard (Solution of an activity)	Equilibrium of forces Relation motionless – equilibrium Procedure: apply the model	Force represents and "models" actions of A on B. An interaction is modelled by two forces which have equal intensity and opposite directions. Diagram of interactions. Vector forces (direction, intensity)	Interpretation of a situation using concepts. Status of a model as a reliable element of knowledge
Class 2: Scenery: whole class, teacher at the blackboard, teacher asks some questions to the class				
Representation of force 9:15	Action exerted by object A on object B can be modelled by force F of A on B ≪ here is the way of representing forces≫ (vectorial representation on the blackboard)	System receiving action. On what system force is exerted. Force is a vector with direction, application point, numerical value or intensity (written on the blackboard)	If object A acts on object B, object B acts on object A simultaneously. When an object is in contact with others then it exerts a force on these objects	Definition (of a force). "Monstration" of characteristics of force

anything or telling that they will be studied later on, we consider that they are not institutionalized. For example in class 1, theme 6, the teacher, on compensation of forces, says: "it is not simple, we will talk about it later". In this case, this teacher introduces new elements of knowledge while making explicit that these elements do not belong to the structured set of the taught knowledge (Table 3, column 3). Moreover we use two types of institutionalized knowledge: (1) when the taught knowledge is standard in the discipline, like a principle or a general procedure (in bold, see Table 3), or (2) if the teacher introduces new elements of knowledge that are specific of an activity (exercise, experimental work, etc.) (Table 3, column 2).

The "re-use" of recent elements of knowledge corresponds to teacher's or students' productions (oral, gesture or written) that involve elements of knowledge, previously introduced in the same session or in prior sessions. Table 3 shows the absence of this type of elements of knowledge in class 2 and a teacher's explicit use of previous knowledge in class 1. There are different ways of re-using elements of knowledge. The reference to modelling process leads us to analyse knowledge staged in the class in terms of three main categories: (1) theory and/or model; (2) data, experimental facts, objects, events; (3) the relations between (1) and (2). It appears that, for both classes, the most frequent way of re-use is to change the experimental field; for example using the representation of forces for a new material situation. Another way, used in class 1, consists of interpreting an old element of knowledge with a new theoretical element ("A acts on B" (old one) is interpreted there is a force exerted by A on B).

Each theme can be analysed from an epistemological perspective, that is looking at how knowledge is processed in the class. Our main reference is based on our epistemological position on modelling: in physics we consider that the relationship between the theory and/or model on the one hand and the description of an experiment in terms of objects and events on the other is the basis of meaning in physics (Tiberghien, 1994; Tiberghien & Vince, 2005). Relationships between elements of these categories give meaning to concepts and allow the experimental validation of theories. We will not develop this analysis but just give some examples. In class 1, the teacher makes explicit the status of some statements through comments like "it is a principle" and explains what a physics principle is; the teacher also makes a clear distinction between facts and theories during interpretation of experiments. In class 2, as it is presented above, the teacher frequently uses a "monstration". This arises when the teacher uses an example or an experiment to show a phenomenon, and interprets the experiment using new conceptual elements (for the class) of physics without previously having given the theoretical model, and then presents this model as the conclusion. However, for some themes, the teacher of class 2 presents the physics theory first (for example the principle of reciprocal actions, and uses it in examples).

5.3 Discussion on the Analysis of the Taught Knowledge by Theme

Such an analysis leads us to compare the rhythm of introduction of new standard institutionalized knowledge in the two classes. It appears that the *new institutionalized elements of standard knowledge* are the same in the two classes except that "the effects of force" were introduced in class 2 only (in class 1 the concept of interaction had been introduced before the analysed sessions). *The rhythm of introduction of new standard elements* is higher in class 2 than in class 1 since the teaching duration of the themes introducing similar concepts is higher in class 1 (2h48 class 1 and 1h27 class 2). However, as we have already discussed, it appears that the elements of new knowledge are more regularly shared out in class 2 during the sequence, in particular when the students carry out an activity, new elements are frequently introduced whereas in class 1 this is the case for only some activities (this result is obtained from tables like Table 3 established for each theme).

The characteristic as regards the "re-using" of recent elements of knowledge may be very important for students' conceptual understanding. The results of studies on learning processes or pathways show that conceptual learning usually takes a longer time than teaching, then this "re-using" may play an important role as a continuous help in understanding concepts involved in the previously taught knowledge. However, this re-use can be done differently; in particular a conceptual element of knowledge can be re-used for different material situation or can be re-used to develop the conceptual network or new interpretation of material situations. We are in the process of analysing these differences. Globally, this "re-using" is a criterion of continuity of the taught knowledge.

This continuity of knowledge is situated between elements of new knowledge (before their institutionalization or not) and the "re-used" elements. It may be a factor in students' development of understanding as we have just discussed. Associated with this continuity, the epistemological analysis gives also an indication of the *potential ways in which students made meaning out* of the taught knowledge. This then determines the limits of possible construction of students' understanding.

In terms of time scales, our analysis in themes is at a meso scale; moreover the scale relative to the granularity of elements of knowledge can be questioned. Our position is that the formulation of theme (Table 2) comparing to the elements of knowledge involved in the analysis of each theme (Table 3) is at a higher scale. We would say that it is at meso scale for the formulation of theme and at micro scale for the elements of knowledge in each theme. The meso scale is relative to time and granularity of knowledge.

In summary, the analysis at the meso level of the taught knowledge includes the following characteristics: the order of the themes, their duration, and for each theme: the scenery in which the taught knowledge is staged, the new elements of knowledge (institutionalized or not), the re-used elements of knowledge, the epistemology involved, etc. In our example we showed that two classes can have similar institutionalized standard knowledge but be very different in the rhythm of introduction of this knowledge, in the teacher's processing of knowledge, and in the epistemology involved. Consequently we claim that it is necessary to differentiate several facets in the concept of "taught knowledge", in order to compare teaching practices. We consider this differentiation is important because the taught knowledge has a determining role in students' learning and knowledge acquisition. Moreover, from a methodological point of view, this analysis is applicable to a variety of teaching practices ranging from the teacher at the blackboard speaking most of the time, to small group activities.

6. CONCLUSION

The characteristics of classroom practices as regards the taught knowledge have to be related to learning hypotheses in order to establish relationship between teaching practice and learning. The analysis presented above has two characteristics: it focuses on taught knowledge, and it is conducted at a meso and partly micro levels for time and granularity of knowledge. The meso level analysis has been carried out because of the availability of the theory of didactical situations, which gives *concepts* like didactical time, chronogenesis *relevant for this meso scale*. The analysis at meso scale is not a re-composition of micro scale analyses. Obviously micro scale analysis can allow the researcher to interpret the higher scale but the meso analysis can be done directly from the data. Most of the time in science education studies, the taught knowledge is either decomposed at micro level via transcriptions or presented at macro level via standard labels (like laws of mechanics, etc.). Our meso scale analysis is not done straightforward; each theme has to be identified and summarized – its construction necessitates a researcher's interpretation.

This meso scale (about 10 minutes) fits with the rhythm of reactions of a group of individuals (taken as a group) whereas the micro scale is relevant to the reaction time of a person. Then this meso scale is particularly adapted to study teaching which is a phenomenon of the class group whereas it is not necessarily adapted to study learning which is associated to a person. Then it is not surprising that usually the micro scale is taken in studies on the teacher who is an individual. In this sense, studies on teacher and studies on teaching practice cannot be identified. For example, interactions between teacher and students or between students (Mortimer & Scott, 2003) use micro scale. In these approaches, re-composition of units of analysis can allow the researchers to reach the meso-level. Similarly, as we introduced at the beginning, studies on learning processes take a micro scale to analyse students' productions.

Further studies are necessary to coordinate these analyses. A series of hypotheses have to be constructed to relate the different characteristics of the taught knowledge and that of the conceptual learning: the extent to which the students' learning is more or less dependant on the rhythm of introduction of new standard knowledge, the environment in which it is staged (scenery), the way of processing knowledge and its epistemology or the relation between previous and new elements of knowledge. More generally, making explicit the time scales – macro, meso and micro – of investigations of teaching and learning phenomena can help us to go further in allowing questions of the relationships between teaching and learning.

REFERENCES

Balacheff, N., Cooper, M., Sutherland, R. & Warfield, V. (Eds.) (1997). *Theory of didactical situations in mathematics. Didactique des mathématiques*, 1970–1990 by Guy Brousseau: Dordrecht: Kluwer Academic Publishers.

Budde, M. & Niedderer, H. (2005). Influences of taught content on student learning in quantum atomic physics. In R. Pintò & D. Couso (Eds.), *Proceedings of the fifth international ESERA conference on contributions of research to enhancing students' interest in learning science* (pp. 461–463). Barcelona, Spain.

Buty, C. (2000). *Etude d'un apprentissage dans une séquence d'enseignement en optique géométrique à l'aide d'une modélisation informatique*. Unpublished Thèse, Université Lumière Lyon 2, Lyon.

Chevallard, Y. (1991). *La transposition didactique* (2ème ed.). Grenoble: La Pensée Sauvage.

Dykstra, D. I. (1992). Studying conceptual change: Constructing new understandings. In R. Duit, F. Goldberg & H. Niedderer (Eds.), *Research in physics learning: Theoretical issues and empirical studies. Proceedings of an international workshop* (pp. 40–58). Kiel: Institut für Die Pädagogik der Naturwissenschaften (IPN).

Filliettaz, L. (2001). Les types de discours. *Círculo de lingüística aplicada a la comunicación* (clac), 8(1), http://www.ucm.es/info/circulo/no8/filliettaz.htm; (http://www.ucm.es/info/circulo/index.htm).

Gaidioz, P. & Tiberghien, A. (2003). Un outil d'enseignement privilégiant la modélisation. *Bulletin de l'union des physiciens*, 850, 71–83.

Gaidioz, P., Vince, J. & Tiberghien, A. (2004). Aider l'élève à comprendre le fonctionnement de la physique et son articulation avec la vie quotidienne. *Bulletin de l'union des physiciens*, 866, 1029–1042.

Givry, D. (2003). *Étude de l'évolution des idées des élèves de seconde durant une séquence d'enseignement sur les gaz*. Unpublished Thèse, Université Lumière Lyon 2, Lyon.

Givry, D. & Roth, W.-M. (2006). Toward a New Idea of Conceptions and Conceptual Change: The Interplay of Talk, Gestures, and Semiotic Resources in the Setting. *Journal of Research in Science Teaching*, 43(10), 1086–1109.

Hiebert, J., Gallimore, J. H. R., Garnier, H., Bogard, K., Hollingsworth, G. H., Jacobs, J., et al. (2003). *Teaching Mathematics in seven countries. Results from the TIMSS 1999 video study*: National Center for Education Statistics (NCES) U.S. Department of Education.

Johsua, S. & Dupin, J.-J. (1993). *Introduction à la didactique des sciences et des mathématiques*. Paris: Presses universitaires de France.

Küçüközer, A. (2005). *L'étude de l'évolution de la compréhension conceptuelle des élèves avec un enseignement. Cas de la mécanique en 1ère S*. Unpublished doctorat, Université Lumière Lyon 2, Lyon.

Lemke, J. L. (2000). Across ths scales of time: artifacts, activities and meanings in ecosocial systems. *Mind, Culture, and Activities*, 7(4), 273–290.

Mercier, A., Schubauer-Leoni, M. L., Donck, E. & Amigues, R. (2005). The Intention to Teach and School Learning: The Role of Time. In A.-N. Perret-Clermont (Ed.), *Thinking time a multidisciplinary perspective on time*. USA, Canada, Switzerland: Hogrefe & Huber.

Mercier, A., Schauber-Leoni, M. L. & Sensevy, G. (2002). Vers une didactique comparée. *Revue Française de Pédagogie*, 141, 5–16.

Mortimer, E. F. & Scott, P. H. (2003). *Meaning making in secondary science classroom*. Milton Keynes: Open University Press.

Niedderer, H., Budde, M., Givry, D., Psillos, D., Tiberghien, A. & Mälardalens, H. (2005). Learning process studies. In R. Pintò & D. Couso (Eds.), *Proceedings of the fifth international ESERA conference on contributions of research to enhancing students' interest in learning science* (pp. 451–463). Barcelona, Spain.

Niedderer, H., Budde, M., Givry, D., Psillos, D., Tiberghien, A. & Mälardalens H. (2007). *Selected book of the fifth international ESERA conference on contributions of research to enhancing students' interest in learning science* Barcelona, Spain.

Petri, J. & Niedderer, H. (1998). A Learning pathway in high-school level quantum atomic physics. *International Journal of Science Education*, 20(8), 1075–1088.

Sensevy, G. & Mercier A. (2007). *Agir ensemble: Eléments de théorisation de l'action conjointe du professeur et des élèves*. Rennes: Presses Universitaires de Rennes (PUR).

Tiberghien, A. (1994). Modeling as a basis for analyzing teaching–learning situations. *Learning and Instruction*, 4(1), 71–87.

Tiberghien, A. (1997). Learning and teaching: differentation and relation. *Research in Science Education*, 27(3), 359–382.

Tiberghien, A. & Vince, J. (2005). Étude de l'activité des élèves de lycée en situation d'enseignement de la physique. *Cahiers du Français Contemporain*, ENS Editions, 10(numéro coordonné par V. Pugibet et N. Gettliffe-Grant.), 153–176.

MAJA BRÜCKMANN, REINDERS DUIT, MAIKE TESCH, HANS FISCHER, ALEXANDER KAUERTZ, THOMAS REYER, BERNHARD GERBER, BIRTE KNIERIM AND PETER LABUDDE

6. THE POTENTIAL OF VIDEO STUDIES IN RESEARCH ON TEACHING AND LEARNING SCIENCE

Abstract: Video technology has played a substantial role in research on teaching and learning in actual practice and in teacher professional development since the 1970s. However, the more recent advances of digital video technology and video analysis facilitate much deeper insight into the interplay of teaching and learning processes than the analog video technology available so far. It turned out that the new technology not only allows much more fine-grained coding than has been possible until now, it also enables investigating patterns of instructional scripts in rather large samples of classes and numbers of lessons

Three projects on investigating key patterns of lower secondary physics instruction in two countries (Germany and the German speaking part of Switzerland) closely cooperate. Major emphases are video-based analyses of dominating instructional patterns by employing common coding schemes and drawing consequences for improving actual instructional practices e.g. by teacher professional development. The following issues are the focus of the studies presented: (1) Standard basic coding of surface structures of instruction; (2) a coding scheme for investigating the practice of using experiments in science classes; (3) methods to investigate linkages of teaching and learning processes in instruction

Keywords: Coding of instructional videos; Instructional scripts; Intercultural differences; Linkages of teaching and learning processes; Studies on the practice of science instruction; Teaching and learning physics; Video analysis

1. VIDEO STUDIES ON THE PRACTICE OF PHYSICS INSTRUCTION IN GERMANY AND SWITZERLAND

In the following major features of the three research groups collaborating are briefly summarized.

1.1 The Video Study on Physics Instruction of the nwu Essen Group[1]

1.1.1 Aims and theoretical framework The study seeks to investigate everyday physics instruction in Germany, describing it on different structural levels to reduce the complexity of instruction and to explore relations between different aspects of instruction and students' performance and attitudes.

Oser and Baeriswyl (2001) developed a theory which may be viewed as a "manual for teaching more efficiently". In order to simplify the idea of specific intended learning processes, they identify a small number of specific learning

Reinders Duit, IPN – Leibniz Institute for Science Education, Olshausenstr. 62, 24098 Kiel, Germany

R. Pintó and D. Couso (eds.), Contributions from Science Education Research, 77–89.

processes and related teaching aims relevant for classroom learning. Modifying their theory for physics lessons, we propose ten different types of educational aims. These are: Learning by experience; conceptual change; problem-solving; theoretical knowledge; contemplation; training routines; transformation of affective excitement; social learning, constructing values and personality; analyzing outcomes.

The theoretical framework is employed to investigate teachers' conceptions of instruction. In order to analyze the lessons we distinguish the following four levels: (1) The level of conceptual structure represents any kind of teachers' general knowledge relevant for instruction; (2) the level of operation sequences addresses the transfer of pedagogical knowledge into interaction; (3) the level of apparent structure summarizes all activities of students and the teacher in a specific lesson; and (4) the level of students' learning processes covers the cognitive processes and their results concerning performance and skills.

1.1.2 Design, methods and samples We observe and analyze six teachers' conceptions of instruction and their actual instruction. Each of them teaches one class. There are two Grammar schools involved. One of the schools is regarded as the intervention group. Four teachers from this school are trained once a week to plan and to perform their instruction using the above approach of Oser and Baeriswyl (2001). Two teachers from the other school form the control group. They do not receive the treatment of the intervention group. Video documents are taken from instruction on the following three physics topics: Electricity, optics and Newtonian mechanics (force concept). The first topic was taught in fall 2000 (grade 8), the last topic in early 2002 (grade 9). The data base for investigating the teachers' concepts of teaching and learning, their ideas of physics instruction, and their estimation of their own role as teachers allow qualitative analyses of interviews and questionnaires.

1.1.3 Major findings Physics teachers at all school levels lack diagnostic and methods skills, though to differing extents. Teaching concepts and instructional goals tend to be dominated by the topic of the lesson. It is usually overlooked that variations in lesson content only go a short way to explaining differences in student performance. The analysis allows a distinction of two groups of teachers that show different types of instruction (Fischer et al., 2002). On the level of apparent structure these types can be described as (a) student oriented elaboration with experimental tasks and (b) teacher centred instruction with demonstration experiments. On the deep structure level of teachers' actions there are rather similar types, namely one group shows a strong tendency to reproductive learning and training routines, the other does not. The analysis of students' actions reveals that activities like comparing, differentiating and generalizing are rather seldom observed, activities like describing, observing and applying much more often. In addition the teachers are hardly aware of how they influence the students' learning processes by their activities and their design of the lesson.

1.2 The IPN Video Study Physics[2]

1.2.1 Phases 13 teachers participated in a pilot study (2000 to 2002). Another 50 teachers from randomly selected schools took part in the subsequent main study (2002 to 2004). This study is linked with a parallel study in the German speaking part of Switzerland (see below). The third phase of the project includes intervention studies that investigate possibilities to change teachers' ways of thinking about good instruction and their actual instructional behavior. One of the studies of this phase is closely linked with a major project to improve physics teaching and learning in German schools (www.physik-im-kontext.de)[3].

1.2.2 Aims and theoretical frameworks The video studies attempt to investigate the dominating instructional patterns (scripts) of German physics instruction and to identify relationships between these patterns and the development of student achievement and affective variables (like interests). The aim is to identify possibilities for improving practice. The script concept as introduced into video analysis of instructional practice by Stigler et al. (1999) provides a major theoretical frame. Constructivist views of teaching and learning play a significant role as a reference position of analysis.

1.2.3 Design of the main study Two subsequent lessons for each of the 50 teachers (on force or lenses/optical instruments; grade 9 students) were video-taped. Additional research instruments comprise: For the teachers a questionnaire on their views about teaching and learning physics before video-taping and a teacher interview with 20 teachers including discussion of video sequences after video-taping. For the students a questionnaire comprising items on various affective variables and test items on the topics force and lenses/optical instruments was provided at the beginning and the end of the school year of video-taping (September 2002 to July 2003). Additionally a questionnaire on students' views about the instruction video-taped, on certain issues of the IQ and on views about the nature of science was given the students right after video-taping. Various methods to code and analyze the videos have been developed (Seidel et al., 2005) such as: Basic coding (organization of classroom activities; Tina Seidel); Quality of class work and student oriented phases (Mareike Kobarg and Tina Seidel); Role and function of experiments (Maike Tesch), Constructivist oriented science classrooms (Ari Widodo); Reconstruction of content structure (Reinders Duit, Christoph T. Müller, Maja Brückmann).

1.2.4 Major findings There is a strong teacher dominance in German physics instruction. Students work in groups or individually for only 15% of the lesson time. A somewhat narrow kind of classroom discourse, called "developing-questioning" approach prevails. Video-data and the above additional student and teacher data reveal two major orientations of instruction: (1) *Instructional* – Oriented at physics; focus is on physics concepts; learning is viewed as knowledge transmission; (2) *Constructivist* – Focus is on student learning, in particular which conditions are necessary to support learning; learning is viewed as student construction. The instructional orientation predominates teaching behavior and teachers' views of good teaching. Most teachers are not informed about research findings concerning effective teaching and learning. They are not aware, for instance, that students' pre-instructional conceptions provide the interpretation schemata for everything presented in the classroom. If instruction and if teachers' views are more constructivist, there is a tendency for student achievement to be better.

1.3 The Video Study Physics in Switzerland[4]

The design and research methods of the above IPN Video Study Physics has been adopted in order to allow comparisons of major patterns of physics lower secondary instruction in Germany and the German speaking part of Switzerland. As various studies (e.g. the international monitoring studies TIMSS and PISA) have shown, the traditions of physics instruction are – partly substantially – different in the two countries. 32 teachers from randomly selected schools participate in the main study. There is an additional sample of 8 teachers who are engaged in "extended learning forms" aiming at more students' oriented teaching and learning. The data collection in Switzerland was carried out in the school year 2003/2004. Preliminary data characterizing Swiss physics instruction are available and presented in subsequent Sections of the present paper. It turns out that there are significant differences between physics instruction in Germany and the German speaking part of Switzerland, e.g. regarding the use of experiments and instructional methods. Whereas in Germany instruction is usually rather teacher oriented, in several Swiss classes students appear to have more opportunities for self-responsible learning than in German classes.

2. TOWARDS A "STANDARD BASIC CODING" OF VIDEO-DOCUMENTED INSTRUCTION

2.1 Aims

Videotaping classroom instruction has become part of the standard repertoire in educational research, at least by technical means. But one is still far from standards of a uniform coding. To compare the results of different studies, to recognize differences and similarities or to reach interdisciplinary and international conclusions about instruction, an agreement on common standards would be necessary. The three research groups briefly presented above have agreed on a common basic coding structure. This can be considered as a start into a wider international cooperation.

2.2 On the Basic Coding in Kiel, Berne and Essen

Due to close cooperation between the IPN Kiel and the University of Applied Science in Berne in both studies, the same observational scheme concerning the "surface structures" has been used for the analysis of videotaped physics instruction. Seidel (2003a) developed this observational scheme, based on the basic coding of videotaped mathematics instruction in the TIMSS video study (Stigler et al., 1999). Since the IPN Kiel researchers used additional and partially redundant observational schemes, Seidel (2003b) reduced the basic coding of "surface structures" to fundamental elements on the basis of the results of the first research phase. The videos of the research group located at the University of Essen are coded in a similar way, based on an observational scheme adapted by Reyer (2004) who partially modified and expanded the categories by Seidel. Therefore, it is possible to compare results of those three projects concerning the "surface structures" of instruction.

The observational scheme "surface structures" by Seidel (2003b) identifies patterns of instruction in the following three areas:

- Teaching time, i.e. the time on task.
- Organization of classroom interaction and student work activities, e.g. lecture, group work, individual seatwork, lab-experiments, etc.
- Instructional phases, e.g. repetition, introduction, instruction, practice, etc.

The coding is carried out on the basis of time sampling (analysis unit: 10 seconds in Berne and Kiel, 15 seconds in Essen), using the software "Videograph" (Rimmele, 2002). Video analysis of classroom instruction is a special method of qualitative content analysis. The core of such an analysis is the development of a category system and the respective coding rules. Applying it to the lesson-recordings is the "tool" that helps the coders assigning the observations from the video to categories of observations; doing this, the qualitative observations are "transformed" into quantitative data. There is an experienced practice of developing a coding system in a "cyclic process" of alternating deducing and inducing between theory and observations. But still there is the need to clarify standards of this process and of training the coders.

The main quality measure for qualitative analyses is reliability; it is usually calculated by determining the agreement of coded data from two (or more) independent coders. Sufficiently high degrees of agreement could be achieved for every coding system developed.

One should be aware that this coding can only serve for basic analyses of surface characteristics of instruction. More specific methods are needed to get to the heart of instruction and to assess its quality. Further observational schemes dealing with "deep structures" of instruction are currently being developed and tested in Kiel, Berne and Essen.

2.3 Preliminary Results of the Basic Coding

In Switzerland only about 30% of the lessons take place in whole class settings, whereas in Germany this part is significantly higher. In Switzerland lectures by the teacher predominate the plenary parts, whereas in Germany class discussions of the following somewhat narrow type prevail: Questions by the teacher are followed by students' answers. A similar kind of class discussion was also reported for German mathematics instruction (Stigler et al., 2000). In some Swiss classes, but in almost no German classes, students work for more than half of the time in group-work settings called learning cycles or workshops dealing with different tasks given by the teacher. In both countries group work (3 or more students in one group), partner work (2 students), and individual work (1 student) play an important role in the video-taped lessons. However in Switzerland it is significantly more frequent than in Germany. For example in the optics lessons 7 out of 24 Swiss, but only 1 out of 24 German classes spent more than 50% of the video-taped lessons on group, partner or individual work. In 50% of the German classes, but only in 20% of the Swiss classes whole class instruction, i.e. lecture by the teacher or classroom discussion, took more than 90% of the instructional time.

For the instructional phases we found for the Swiss sample that most of the time (85%) is devoted to the acquisition of new content. This is not surprising as the teachers had been asked to show an introductory lesson. Another phase that appeared in every videotaped lesson was some kind of introduction at the beginning of each lesson, varying from some seconds up to a maximum of 8 minutes.

3. ON THE ROLE OF EXPERIMENTS IN PHYSICS INSTRUCTION – DIFFERENT EMPHASIS IN DIFFERENT INSTRUCTIONAL TRADITIONS

3.1 Aims of the Study

Experiments are as essential for science education as they are for science. However, reviews criticize the limited use of experiments prevailing in science instruction. Harlen (1999) points out that "in practice" experiments are less effective than "in theory". They are not automatically "good practice" (Hofstein and Lunetta, 2004) and do not speak for themselves (Mortimer and Scott, 2003). A general focus for research in this area is to identify crucial features for the quality of practical instruction and to find out how these aspects might be best developed in practice. The video-studies on the practice of physics instruction in Kiel and Berne provide data that contribute to the body of research findings on the role of experiments so far available with a particular focus on comparing the practice in two partly different instructional traditions. There are the following major research questions:

- How do German and Swiss teachers organize practical activities in physics lessons (grade 9)? Are there common scripts and to what degree is the orchestration of lessons individual?
- Are there interrelations of certain patterns and students' development of achievement and affective variables (like interests)?
- Are there systematic differences between teachers from Switzerland and Germany?
- Is it possible to transfer a coding scheme developed for a specific instructional culture to another? What kind of intercultural reliability is possible and necessary when coding lessons from other countries?

3.2 Research Methods

The videos and the students' pre-post-tests are the data basis for this study. The coding scheme used for analyzing the videos of the second phase of the Video Study was developed and tested during the first phase of the IPN Video Study

Physics (Tesch, 2005). The major facets of the coding scheme comprise: phases of experiments, social organization, and openness of procedures, material used and functions of experiments in relation to theory.

Our focus is not just on the actual practical activity but also on other parts of the lesson linked to that activity. We decided to code the whole lesson based on a system of three phases of experimental activities:

- Introduction of the experimental activity,
- Actual experimental activity,
- Discussion of the experimental activity.

The first coding of the whole lesson involves identifying these three phases which are then coded separately according to further categories. The *level of openness* is coded with respect to the task or idea, the planning and the discussion of the observation. In addition, some classifications for the actual experimental activity are carried out. We further coded the function of the practical activity. This involves addressing the following questions: Is the practical activity the first approach to the phenomenon or is it an application/illustration of a known concept? Is this practical activity a test of a hypothesis? All German lessons were coded by one rater with the Software "Videograph" based on a time sampling unit of 10 seconds. The inter-rater reliability (German sample) was tested by coding 7 lessons. In a second step a rater in Switzerland was trained to code both German and Swiss lessons. The coding instructions were adapted and additional information about typical Swiss experiments was added. Then the inter-rater reliability was tested between two Swiss raters using 7 lessons. The raters both knew the context of Swiss instructional culture. This test shows that the coding scheme is generally capable of mapping experimental activities in Swiss instruction. In a third step a German rater coded 4 Swiss lessons to test the cross-cultural reliability. This test shows that the codings of an "insider" and an "outsider" of an instructional culture meet satisfactorily.

3.3 Preliminary Findings

It turned out that practical activities are a key part of the videotaped lessons. In both countries on average two thirds of the lesson time is connected with an aspect of practical work – however this amount varies substantially for the participating teachers. The actual experimentation phase is slightly higher in Switzerland (21% of the total instructional time as compared to 19% in Germany). The introduction phase is 12% in both countries, in German schools more time is spent for the discussion phase (35% as compared to 29% in Switzerland). A sufficiently long phase of discussion has proven essential. If practical activities are done they are often in a cookbook style. According to our findings, the students are allowed to plan some of their activities on their own, but the discussion of observations is mostly authoritative, with the teacher taking the lead. For the teachers it appears to be most important that the use of experiments is "correct" from the physics point of view. Although there is an emphasis on scientific processes in literature, "testing hypotheses" is hardly ever observed in German physics lessons. Teachers seldom explain specific scientific processes and do not make them explicit.

As already mentioned above, the Swiss teachers seem to teach more student oriented. There are more practical activities with a higher level of openness. There is, however, a group of teachers who teach nearly in the same way as most of the German teachers. In Switzerland eight teachers with a reputation for learner-oriented teaching ("extended learning forms") participated in the study – as an additional sample. Indeed, some of these teachers use experiments in a rather different way by posing open problems.

4. CUMULATIVE VERSUS ADDITIVE LEARNING – METHODS TO INVESTIGATE THE LINKAGES BETWEEN KNOWLEDGE ELEMENTS PROVIDED IN INSTRUCTION

4.1 Introduction

Concerning the quality of instruction it is essential that the "knowledge elements" presented are intimately linked. Often this appears to be neglected in actual instructional practice. Frequently, "additive learning" seems to predominate, denoting that pieces of knowledge that are only marginally linked are added up. "Cumulative learning" indicating that such links are provided is usually seen as an essential prerequisite for effective learning in recent quality development programs. Knowledge elements to be learned have to be linked with already learned issues and future learning. The quality of the linkage plays an important role in the nwu project in Essen and the IPN study in Kiel. Two methods of investigating the degree of linking knowledge elements are presented.

4.2 Theoretical Background

Competency level models provide a basis for the targeted development of curricula ensuring that the knowledge acquired in the classroom can be slotted into an existing conceptual network. The nwu essen project (Fischer et al., 2005) has the goal to develop a consistent model of conceptual linkage within a certain subject taking into account the results of research on learning processes, advance organizers and big ideas. This linkage is viewed as a central aspect of the teaching-learning process in classroom teaching. The teacher organizes the transformation of curriculum structure into students' knowledge structure. To describe and analyse teaching and learning in science lessons, concepts and relations from the biology, chemistry and physics curriculum are arranged in so-called logical chains to compare the transformation of the curriculum structure into the classroom by the teacher with the students' adaptation of the teacher's instruction on a cognitive level. The chains are a progression of concepts each linked to the following, chosen from the curriculum guided by a certain idea. The nwu essen project has its focus on both the cognitive activities the teacher wants to initiate and the students' actual (cognitive) response with the help of these logical chains (Reinmann-Rothmeier and Mandl, 2001).

The Kiel project developed a different way to display and analyze logical chains and the progression of those chains. The emphases here are on the content of the chain elements and their linkage and on coherence issues (complexity, sequence and integration of content).

4.3 Methods and Sample

The nwu essen project's goal is to discern several levels of logical chains according to their degree of "fragmentation". These are: isolated and disparate concepts, unconnected relations between concepts, completely linked concepts or concepts linked by a big idea. In order to observe the cognitive processes the teacher's instruction initiates the students' cognitive activities are categorized whether they remember, structure or explore the offered knowledge elements (so-called linking modus) as their reaction to this instruction. Each of the logical chains is described for each linking modus and for the purpose of distinguishing between the teacher's offers and the students' activities as response. As a result of the theoretical model 36 variables depict linkages in a biology, chemistry or physics lesson. To improve the perceptibility, several indicators for each variable have been developed.

Analyzing 50 videos each of physics, chemistry and biology lessons (N = 150) a couple of trained university students identified the indicators in these lessons. The analysis of the videos revealed two kinds of "extreme classes": Five "high linkage classes" per subject which have a high level of linkage; and five "low linkage classes" which have a low level of linkage. At the end of the school year (in June 2005) the students' knowledge structure is measured in these classes by concept maps and additional tests (30 classes).

The Kiel project has developed a way to display and analyze the linkages of "knowledge elements" presented by the teacher in class (see Figure 1).

Initially these content structure diagrams had been developed for the purpose of instructional planning (Duit et al., 1981). These maps, similar to the "Strand Maps" used in the AAAS Project 2061 "Atlas of Science" (Kesidou and Rosemann, 2002),

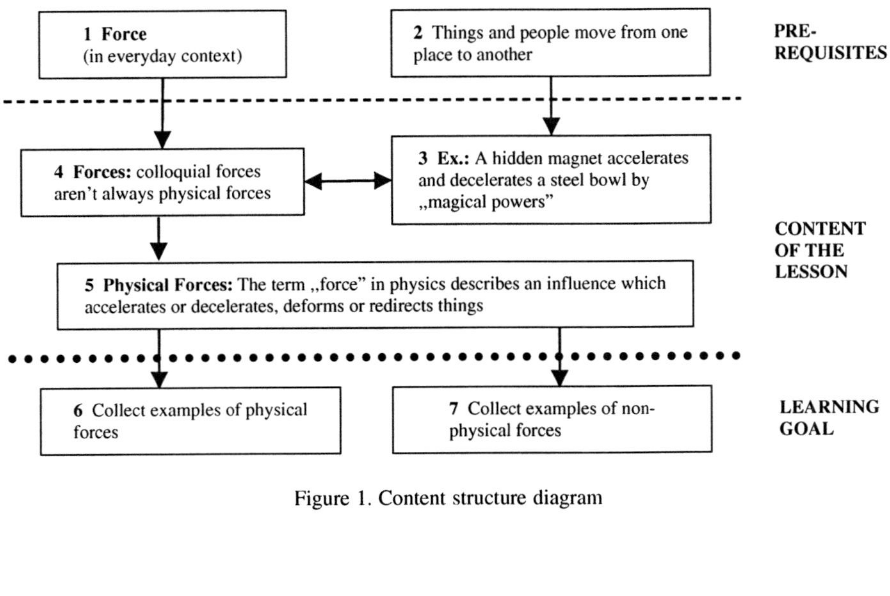

Figure 1. Content structure diagram

allow addressing or analyzing key issues of content coherence. The arrows in the maps denote which content elements are necessary to learn the content elements targeted by the arrows. They display content coherence and consistency – if used as planning instrument or as method to analyze the content structure of instruction. The reconstruction process is based on the videos of the lessons (including the transcript) and the materials (e.g. worksheets and textbooks, blackboard) used in class. Initially the reconstruction process proceeded as follows (Müller and Duit, 2004): (1) The Sections of instructions are identified; (2) the "content elements" are listed in chronological order and attached to boxes; (3) the relations between the boxes are identified and indicated by arrows. In order to increase the inter-rater-reliability more recently we have developed a different strategy for reconstruction. We start with the construction of a "reference map" for the science topic under inspection, e.g. introduction into the force concept. This map includes the key characteristics of the force concept from the physics and from the curriculum point of view. The following steps lead to the reconstruction:

- Time-based coding using a set of characteristics as coding categories. This step results in a "content score" displaying which characteristics are addressed during the lesson.
- The "score" allows identifying the sections in a more adequate way than merely drawing on viewing the video (see the initial strategy above).
- Content elements are identified, attached to boxes and linked with arrows in a similar manner as in the initial strategy.

4.4 Preliminary Findings

The nwu essen project shows that the logical chains are well perceptible. The quality measures for the research instruments used, e.g. inter-rater reliability, are fully sufficient. A student questionnaire asking for the perception of the linkages in the lessons revealed that students are able to assess the linking activities well but having problems assessing the level of linkages. Students' prior knowledge measured by TIMSS items shows no effect on the linkage presented in the classroom by the students or the teacher. Data about the students' knowledge structure are not available by now. However, a test developed on the presented model of linkage could reveal a relation between the level of fragmentation of logical chains and the difficulty of physics test items (Kauertz and Fischer, in print).

The content structure diagrams of the Kiel group and the "content score" of the lessons provide an insight into the consistency and coherence of content taught. Furthermore, these data allow analyzing the different pathways towards the content area (e.g. introduction into the force content) on the grounds of the aims of instruction and possible learning outcomes. For the data of the first phase of the IPN video study we also analyzed "formal" issues of the structure. It turned out, for instance, that a high degree of linkages between the boxes is related to better student achievement (significant correlation; Müller and Duit, 2004). This result is in accordance with the hypothesis underlying the research projects presented here that the linkage between knowledge elements is essential for efficient learning.

5. CONCLUDING REMARKS

The focus of the present paper is on examining methods of analyzing video documents of instruction by various means of coding. Recent advances of video technology and video analyses are employed. It turned out that video analysis methods provide valuable information on a large spectrum of essential features of instruction. It became also evident that the video-analysis methods presented fully meet the quality standards of empirical research on teaching and learning. This is not only true for analysis of apparent structures of instruction (like prevailing instructional settings) but also for methods on deep structure features like linkages of instructional elements and on reconstructing the content structure of instruction.

The results of the three research groups in Germany and Switzerland also show that video-analysis methods are well suited to uncover different instructional traditions in different countries and not only variations within one country (cf., Stigler et al., 1999; Clausen et al., 2003; Roth et al., 2006). However, developing and using coding schemes for different instructional traditions needs a particular attention. It is not only necessary that the research methods (e.g. a coding system) meet measures of reliability in the countries participating but also issues of cross-cultural reliabilities have to be regarded.

In a nutshell, the methods presented provide valuable information on key features of the practice of physics instruction in the participating countries. This information will provide a powerful basis for projects to improve instruction.

However, it has to be taken into account, that the studies presented are still preliminary in nature. First, analysis of data received by the methods employed is still in progress. Second, further development of the research methods on the basis of experiences gained in the studies presented appears to be essential. Finally, to investigate the power of video-analysis methods in research on instructional traditions in different countries needs a larger set of countries employing the methods. We therefore invite colleagues to use our methods and to cooperate for developing science education internationally.

NOTES

[1] The nwu essen Video Study Physics Team: Hans E. Fischer, Alexander Kauertz, Thomas Reyer, Georg Trendel, Knut Neumann. The Project is supported by the German Science Foundation (Deutsche Forschungsgemeinschaft) within the research units program nwu-essen (http://www.nwu-essen.de/).

[2] The IPN Video Study Physics Team: Manfred Prenzel, Tina Seidel, Reinders Duit, Manfred Lehrke, Rolf Rimmele, Maja Brückmann, Inger Marie Dalehefte, Constanze Herweg, Lena Meyer, Mareike Kobarg, Katharina Schwindt, Maike Tesch, Ari Widodo. The Project is supported by the German Science Foundation (Deutsche Forschungsgemeinschaft) within the priority program BiQua (http://www.ipn.uni-kiel.de/projekte/biqua/biqua_eng.htm).

[3] See the chapter by Duit, Mikelskis-Seifert and Wodzinski in the present volume. (pp. 119–130).

[4] Team members are: Peter Labudde, Bernhard Gerber and Birte Knierim. The project is supported by the Swiss National Science Foundation.

REFERENCES

Clausen, M., Reusser, K. & Klieme, E. (2003). Unterrichtsqualität auf der Basis hoch inferenter Unterrichtsbeurteilungen. Ein Vergleich zwischen Deutschland und der deutschsprachigen Schweiz [Instructional quality on the grounds of high-inferent assessment of instruction. A comparison between Germany and the German speaking part of Switzerland]. *Unterrichtswissenschaft*, 31(2), 122–141.

Duit, R., Häußler, P. & Kircher, E. (1981). *Unterricht Physik* [Physics instruction]. (Köln, Germany: Aulis).

Fischer, H. E., Klemm, K., Leutner, D., Sumfleth, E., Tiemann, R. & Wirth, J. (2005). Empirical research on science teaching and learning. *Journal of Science Teacher Education*, 16, 309–349.

Fischer, H. E., Reyer, T., Wirz, T., Bos, W. & Höllrich, N. (2002). Unterrichtsgestaltung und Lernerfolg im Physikunterricht [Instructional design and achievement in physics instruction]. *Zeitschrift für Pädagogik*, 45. Beiheft, 124–138.

Harlen, W. (1999). *Effective teaching of science*. (Edinburgh, UK: The Scottish Council for Research in Education (SCRE)).

Hofstein, A. & Lunetta, V. (2004). The laboratory in science education: Foundations for the twenty-first century. *Science Education*, 88, 28–54.

Kauertz, A. & Fischer, H. E. (2006). Assessing students' level of knowledge and analysing the reasons for learning difficulties in physics by Rasch Analysis. (In Xiufeng Liu (Ed.), *Applications of Rasch measurement in science education*, pp. 212–246. Minnesota: Maple Grove.

Kesidou, S. & Rosemann, J. E. (2002). How well do Middle School science programs measure up? Findings from Project 2061's curriculum review. *Journal of Research in Science Teaching, 39*(6), 522–549.

Mortimer, E. F. & Scott, P. H. (2003). *Meaning making in secondary science classrooms*. (Maidenhead, UK: Open University Press).

Müller, Ch. T. & Duit, R. (2004). Die unterrichtliche Sachstruktur als Indikator für Lernerfolg – Analyse von Sachstrukturdiagrammen und ihr Bezug zu Leistungsergebnissen im Physikunterricht [Content structure of instruction as an indicator for achievement – Analysis of content structure diagrams and their relation to achievement in physics instruction]. *Zeitschrift für Didaktik der Naturwissenschaften*, 10, 147–161.

Oser, F. K. & Baeriswyl, F. J. (2001). Choreographies of teaching: Bridging instruction to learning. (In V. Richardson (Eds.), *AERA's handbook of research on teaching – 4th edition* (pp. 1031–1065) Washington: American Educational Research Association.)

Reinmann-Rothmeier, G. & Mandl, H. (2001). *Unterrichten und Lernumgebungen gestalten* [Teaching and designing learning environments]. (In A. Krapp & B. Weidenmann, *Pädagogische Psychologie* (pp. 601–646). (Weinheim, Germany: Beltz.)

Reyer, T. (2004). *Oberflächenmerkmale und Tiefenstrukturen im Unterricht. Exemplarische Analysen im Physikunterricht der gymnasialen Sekundarstufe* [Surface structures and deep structures in instruction: Exemplary analysis of lower secondary physics instruction]. (Berlin: Logos).

Rimmele, R. (2002). *Videograph* (Kiel, Germany: IPN – Leibniz-Institute for Science Education).

Roth, K., Druker, S., Garnier, H., Chen, C., Kawanaka, T., Rasmussen, D., Trubacova, S., Warvi, D., Okamoto, Y., Gonzales, P., Stigler, J. & Gallimore, R. (2006). *Teaching science in five countries: Results from the TIMSS 1999 Videostudy. Statistical Analysis Report.* (Washington, DC: NCES – National Centre for Educational Statistics).

Seidel, T. (2003a). *Lehr-Lernskripts im Unterricht* [Teaching and learning scripts in instruction] (Münster, Germany: Waxmann).

Seidel, T. (2003b). *Sichtstrukturen – Organisation unterrichtlicher Aktivitäten.* [Surface structures – organisation of instructional activities]. (In T. Seidel et al. (Eds.), *Technischer Bericht zur Videostudie, Lehr-Lern-Prozesse im Physikunterricht"* (pp. 113–127). Kiel, Germany: IPN – Leibniz-Institute for Science Education.)

Seidel, T., Prenzel, M. & Kobarg, M. (Eds.) (2005). *How to run a video study – Technical report of the IPN video study.* (Münster, Germany: Waxmann).

Stigler, J. W., Gonzales, P., Kawanaka, T., Knoll, S. & Serrano, A. (1999). *The TIMSS Videotape Classroom Study. Methods and findings from an exploratory research project on eighth-grade mathematics instruction in Germany, Japan and the United States.* (Washington, DC: U.S. Department of Education).

Stigler, J. W., Gallimore, R. & Hiebert, J. (2000). Using video surveys to compare classrooms and teaching across cultures: Examples and lessons from the TIMSS video studies. *Educational Psychologist*,35, 87–100.

Tesch, M. (2005). *Das Experiment im Physikunterricht – Didaktische Konzepte und Ergebnisse einer Videostudie* [The experiment in physics instruction – conceptions and results of a video study]. (Berlin, Germany: Logos).

PART 3

SCIENCE TEACHERS' KNOWLEDGE, PRACTICE AND EDUCATION

JOHN LOUGHRAN, AMANDA BERRY AND PAMELA MULHALL

7. PEDAGOGICAL CONTENT KNOWLEDGE: WHAT DOES IT MEAN TO SCIENCE TEACHERS?

Abstract: This paper examines understandings of science teachers' pedagogical content knowledge (PCK) as conceptualized through CoRes (Content Representations) and PaP-eRs (Pedagogical and Professional-experience Repertoire). The consolidation of CoRes and PaP-eRs into Resource Folios that offer in depth, concrete examples of PCK in specific content areas (e.g., Particle Model, Chemical Reactions, etc.) was conceptualized as one way of articulating and portraying PCK that might be meaningful and accessible to science teachers. The study underpinning this paper is two-fold. First, it briefly considers the development of CoRes and PaP-eRs and the implications of these as both a methodology and a research product. Secondly, science teachers' understandings of the usefulness of Resources Folios through a pre and post test study in which the influence of CoRes and PaP-eRs on science teachers' thinking about science teaching and learning was explored. Results tentatively suggest that through a CoRe and PaP-eRs approach PCK becomes a more meaningful construct in terms of these participants' articulation of understandings of professional practice

Keywords: Pedagogical Content Knowledge, Science Education, Science learning, Science teaching, Teachers' professional knowledge

1. BACKGROUND

In recent times there has been growing interest in the notion of a scholarship of practice (Shulman, 2002). Scholarship is displayed through a teacher's grasp of, and response to, the relationships between knowledge of content, teaching and learning in ways that attest to practice as being complex and interwoven. A consequence of this work is the recognition that teachers' professional knowledge is difficult to define and categorize and therefore exceptionally difficult to articulate and document.

Attempts to articulate the critical links between practice and theoretical knowledge have proved to be exceptionally difficult because, for many teachers, their practice and the knowledge/ideas/theories that tend to influence that practice are often tacit. Additionally, definitions of knowledge, and distinctions between these definitions (Cochran-Smith & Lytle, 1999; Connelly & Clandinin, 2000; Fenstermacher & Richardson, 1993; Korthagen & Lagerwerf, 1996), have impacted what researchers have looked for, and valued, in attempts to describe a knowledge

John Loughran, Faculty of Education, Monash University, Wellington Rd., Clayton, 3800, Australia

R. Pintó and D. Couso (eds.), Contributions from Science Education Research, 93–105.

base that might be described as influencing teachers' approaches to, and practices of, teaching.

One aspect of teachers' professional knowledge that has received considerable attention over time has been that associated with the purposeful linking of subject matter knowledge and teaching knowledge that, when combined, highlights the skills and expertise of subject specialist teachers. This amalgam of subject matter knowledge and teaching knowledge has been described as pedagogical content knowledge and has been a major field of study for over two decades.

1.1 Pedagogical Content Knowledge

PCK is an academic construct that is based on the view that teaching requires much more than the simple delivery of subject content knowledge to students and, that quality student learning is not the simple recall of facts and figures. PCK, then, is knowledge that teachers develop over time, and through experience, about how to teach particular content in particular ways in order to lead to enhanced student understanding. Shulman (1986, 1987) described PCK as including, "the most powerful analogies, illustrations, examples, explanations, and demonstrations – in a word, the ways of representing and formulating the subject that makes it comprehensible for others" (1986, p. 9). It was, "the category [of teacher knowledge] most likely to distinguish the understanding of the content specialist from that of the pedagogue" (1987, p. 8). Shulman claimed that teachers needed strong PCK to be the best possible teachers.

PCK demands of teachers a rich conceptual understanding of the particular subject content that they teach: "This rich conceptual understanding, combined with expertise in developing, using and adapting teaching procedures, strategies and approaches, for use in particular classes, is purposefully linked to create the amalgam of knowledge of content and pedagogy that Shulman (1986, 1987) described as PCK" (Loughran et al., 2006, p. 9).

Shulman's notion of PCK has created many and varied responses and has been interpreted in many different ways (Geddis et al., 1993; Gess-Newsome & Lederman, 1999; Grimmett & MacKinnon, 1992; Grossman, 1990). Despite the way that PCK is sometimes described in the literature, it is not a single entity that is the same for all teachers of a given subject area. In various instances it may be the same or similar for some, yet incredibly different for others. Context and experience combine in powerful ways to shape a teacher's PCK, but importantly, PCK is a corner stone of a teacher's professional expertise; although it may well be that few teachers are able to articulate their PCK publicly for others. Regardless of its interpretation, there are few concrete examples of PCK in subject areas. Van Driel, Verloop, and De Vos (1998) highlighted this as a concern when they noted that although the research community embraced the notion of PCK, few science specific examples existed in the literature to illuminate this important aspect of teachers' professional knowledge. As a consequence, it is difficult for science teachers to access PCK in ways that might be meaningful for their practice. This means that for teachers who are an important group of end users of the construct, PCK remains somewhat elusive and is certainly not a term commonly used in their practice.

We suggest that one aspect of coming to better understand PCK as a useful construct is enmeshed in developing a shared language around the ideas that might allow greater access into and support for the ideas foundational to the concept. For PCK to be more meaningful in science teachers' work, we would argue that concrete examples need to be documented. However, uncovering, describing and portraying PCK is a difficult task. Our attempts (over a considerable period of time) to develop concrete examples of PCK have led to the conceptualization of CoRe and PaP-eRs (see Loughran et al., 2006). In this study we explore their value and utility for science teachers, with the purpose being to consider whether the research outcome is, in fact, meaningful and applicable in science teachers' work.

2. METHOD

The study reported in this paper has two distinct methodological aspects. The first is that which comprises the method of researching PCK through a CoRe and PaP-eRs framework (how that method was developed and refined). The second is linked to how Resource Folios (the final product that comprises CoRe and PaP-eRs on particular topics) are understood and interpreted by science teachers in relation to their thinking about their own practice.

2.1 CoRe & PaP-eRs

The method developed to uncover, document and portray science teachers' PCK comprises two tools: a CoRe (Content Representation); and, PaP-eRs (Pedagogical and Professional-experience Repertoires). The CoRe sets out and discusses science teachers' understandings of particular aspects of science specific PCK (such as, for example, an overview of the main ideas; knowledge of alternative conceptions; insightful ways of testing for understanding; known points of confusion; effective sequencing; and, important approaches to the framing of ideas. These are encapsulated in the prompts in column 1 of the abbreviated CoRe that is offered in Table 1).

Attached to the CoRe are PaP-eRs, which link to particular aspects of a content area/topic. The PaP-eRs (constructed as stories/vignettes/explanations of teaching practice/lesson plans/students' views of science, etc.) bring elements of the CoRe to life by illustrating how such knowledge might inform effective classroom practice. A PaP-eR offers a window into a teaching/learning situation where it is the content that shapes the pedagogical approach (a brief example is offered in Annex 1, PaP-eR: Playdough balls).

The method of CoRes and PaP-eRs was developed and refined over 3 years working exclusively with High School science teachers. Data were derived from individual interviews (n = 24), classroom observations (n = 12) and small groups (3 – 4 teachers per group; n = 10) of science teachers working in particular content areas through which, together, final *Resource Folios* were developed, validated and

TABLE 1. This highly abbreviated CoRe is The Particle Theory (the full version comprises 7 Big ideas and carries extensive responses to the prompts in the left hand column)

	Important science ideas/concepts		
This CoRe is designed junior secondary science, i.e., for years 7 – 9.	**A: Matter is made up of small bits that are called particles**	**B: There is empty space between particles**	**C: Particles are in constant motion**
What you intend the students to learn about this idea.	If we break up substances, the smallest bit of a substance we can get is a particle.	The relative distances between particles differ in solids, liquids and gases.	Particles of matter are always moving. The speed of particles can be changed (by heating/cooling, pressure change).
Why it is important for students to know this.	Because it helps to explain the behavior of everyday things e.g., diffusion.	Because it explains the ability to compress things and helps to explain events such as expansion and dissolving.	Because it explains what happens in phase changes, e.g., the need to contain gases is evidence the particles are moving.
What else you know about this idea (that you do not intend students to know yet).	At this stage 'particles' is used in a general sense without discriminating between atoms and molecules; Subatomic structure; Chemical reactions; ions; or More complex properties of materials. More complicated models of matter. Links to diffusion and thermal properties of matter.		
Difficulties/limitations connected with teaching this idea.	Particles are too small to see. The use of particular science models is not necessary to comprehend science in everyday life. It is difficult to decide when to introduce the labels (i.e., atoms, molecules) for different kinds of particles.	There is a big difference between macro (seen) and micro (unseen) levels, e.g., wood seems solid so it is hard to picture empty space between the 'wood' particles.	That macro properties are a result of micro arrangements is hard to understand. The commonly used term 'states of matter' implies that all things can be discretely classified as solid, liquid or gas.
Knowledge about students' thinking which influences your teaching of this idea.	Many students will use a continuous model (despite former teaching).	The notion of 'space' is very difficult to think about – most students propose there is other 'stuff' between the particles.	Students have commonly encountered 'states of matter' but do not understand the ideas in terms of particle movement.
Other factors that influence your teaching of this idea.	Maturity – stage of psychological development, readiness to grapple with abstract ideas. Dealing with many different student conceptions at once. Knowledge of context (students' and a teacher's). Using the term 'phase' suggests the idea of a continuum and may help to address the difficulties associated with the term 'state'		

Teaching procedures (and particular reasons for using these to engage with this idea).	**Probes of student understanding:** e.g., students draw a flask containing air, then re-draw the same flask with some of the air removed. Probes promote student thinking and uncover individual's views of situations.	**POE (Predict-Observe-Explain):** e.g., squashing syringe of air (ask students to predict the outcome based on different models of matter – e.g., continuous vs. particle).	**Translation activities:** e.g., role-play, modelling, drawing. For example, my life as a Carbon Atom; or, write about what you would see if you were inside a particle of water.

refined. (From this approach, a number of Resource Folios were then researched, developed and documented, see Loughran et al., 2006).

The CoRe should not be viewed as static or as the only possible representation of that content. It is a necessary but incomplete generalization resulting from work with a particular group of teachers that leads to a generalized view of the big picture ideas of that content/topic. The purpose of the CoRe is to help codify teachers' knowledge in a common way across the content area being examined and through this, to identify important features of the content that science teachers recognize and respond to in their teaching of that content.

PaP-eRs are linked to the CoRe to help to connect practice with the understanding of that particular content being taught, and are drawn from a particular quadrant(s) in the CoRe. PaP-eRs illuminate the decisions underpinning teachers' actions and offer insights into how teachers intend to help students better understand that content.

Importantly, one PaP-eR alone is not enough to illustrate the complexity of the knowledge around particular content thus a collection of PaP-eRs is attached to different areas of the CoRe. This is crucial in highlighting some of the different blends of elements that jointly are indicative of PCK in that content area (see Loughran et al., 2004).

The present study then sought to build on the development of a CoRe and PaP-eRs conceptualization of PCK by exploring with two different groups of science teachers their perceptions of the usefulness of the Resource Folios for their teaching.

2.2 Exploring the Usefulness of Resource Folios

This aspect of the research method was focused on answering the question: "What influence do Resource Folios have on science teachers' thinking about their own science teaching?"

In order to address this question a completed Resource Folio (Particle Theory) was examined by two groups of experienced science teachers. We established a pre and post inquiry approach in order to differentiate between these science teachers' thinking about their professional knowledge of practice prior to exploring the Resource Folio and then again afterwards, in order to determine any possible impact.

The pre and post inquiry was conducted on two separate occasions with two different groups of science teachers (n = 18; n = 32). A co-operative strategy, "think-pair-share" (Lymna, 1981) was employed to introduce participants to the ideas of the research. At the pre inquiry stage, participants were asked to construct an individual response to the question: "How do you think about/talk about your science teaching (i.e., what makes it thinking and talking about it easy, what makes it hard)?" then to share those responses in pairs, finally leading to a full articulation across the whole group.

After examining and discussing the Resource Folio, the think-pair-share process was again employed for the post inquiry stage. This time, participants responded to the questions: "How does the Resource Folio influence your thinking about your science teaching? What do you see as the advantages and disadvantages of a Resource Folio? Does the idea of PCK matter to science teachers?"

Through this process of teachers working with the Folio, it was anticipated that an initial (but still tentative) understanding of the value and usefulness of Resource Folios might emerge and that a major aim of the project (the development of concrete examples of PCK that could be useful to teachers) would be tested.

3. DATA ANALYSIS AND RESULTS

The methodology of CoRes and PaP-eRs was developed because of the inability of previous research to adequately capture, portray and codify science teachers' PCK. CoRes and PaP-eRs therefore have become both a methodology and a product as the process allows ways of gathering and portraying data in a useable form.

The program of research that comprises the development of CoRes and PaP-eRs has led to the development of *Resource Folios* that present and "re-present" PCK in science content areas. The Resource Folios offer new ways of sharing knowledge in meaningful ways within the profession through enhanced understandings of PCK and its impact on practice. In fact, in some of the content fields we have researched (e.g., chemical reactions) a number of CoRes are readily identifiable – and distinctly different – as different science teachers conceptualize the content in different, but equally valid, ways.

3.1 Pre CoRe & PaP-eRs Inquiry Phase

Data derived from these two groups of science teachers' discussions about their science teaching are encapsulated in two major themes: *doing science teaching* and *linking teaching and learning*. Each major theme is comprised of sub-themes that illustrate the range of views associated with that theme. The first theme, as the title suggests, illustrates a focus on the act of science teaching and is interesting in as much as it draws on the need for teachers to have successful teaching activities (perhaps mediated a little through knowledge of subject matter content), as opposed to a sense of professional knowledge of practice directing or influencing thinking about science teaching and learning. This is evident through the sub-themes of *doing science teaching* which were those of:

1. Activities that work. Indicative views of this response being: "Where there is plenty of scope for hands on experiences for kids."

2. Content familiarity. Indicative views of this response being: "I think my passion for the content definitely affects my teaching"; "If I am fascinated by it (content), it inspires me and helps me deliver better lessons. Some chem/biol topics which I am not so familiar with/inspired by, often up end quite disappointing for me in the way I teach/prepare lessons."
3. Fun/Engagement. Indicative views of this response being: "Teach it like it's a game or a challenge."

In each of these three sub-themes, teachers' responses carried a sense of a need for both themselves and their students to be involved in activities of doing science, or to be interested in the subject matter content in ways that might positively influence the development of such activities. Across this major theme, there was no obvious link between participants' thinking about science subject matter content and pedagogy in any way that might suggest that they used PCK as a framework to guide their thinking about practice. This theme was largely about being active in the classroom as opposed to simply delivering science as propositional knowledge (White, 1988).

In the second theme, *linking teaching and learning,* there were suggestions that better aligning subject matter content and pedagogy was important in more effectively shaping both teachers' teaching and students' learning; although not necessarily articulated in that way. For example, the sub-themes of *linking teaching and learning* were those of:

1. Social 'real world' relevance. Indicative views of this response being: "Look to media to get societal relevance and perception."
2. Sense of progression. Indicative views of this response being: "Look to past curricula to see what I'm building up and look to future curricula to see where I may be aiming."
3. Organisation. Indicative views of this response being: "Unfortunately if there is too much content to cover in too little time, my teaching becomes more conservative – chalk, talk, bookwork."
4. Teaching repertoire. Indicative views of this response being: "Must have time to talk about good ideas and to get access to them"; "use of role plays, good demonstrations, prac. etc."

Although the implication across this theme is that participants see the importance of better linking teaching and learning (particularly so in terms of relevance and progression) any suggestion that subject matter content and pedagogy might overtly be combined through the notion of PCK is still not clear. Rather, the suggestion is that a teacher's understanding of content might somehow shape views about how to make that content more engaging or interesting but not necessarily in the form of the amalgam implied by the academic understanding of the construct of PCK. However, the lack of a shared language to discuss the relationship between subject matter knowledge and pedagogical knowledge, much less the amalgam that is PCK, seems readily apparent. Hence, it could well be argued that it is not so much that a deeper knowledge of practice does not exist for these teachers, rather that it remains tacit and therefore highly elusive. This suggestion is supported by the unusual response in the pre inquiry phase by one participant who stated that: "Knowing the content is extremely important but knowing how to teach the content to particular students is also extremely important. It is necessary for me to have a large repertoire of various ideas (or different ways of teaching the same idea) so that students learn/understand the content." This well articulated position by one participant out of the total of 50 (i.e., 2% of the cohort) illustrates that notions of PCK do exist, but this result may be indicative of the very small number of teachers who might think about their practice in ways synonymous with understandings of PCK. This response also raises questions about teachers' expectations of the nature of "talk about science teaching and learning" and what a shared language might imply. Consideration of such implications begins to emerge through the post CoRe & PaP-eRs inquiry.

3.2 Post CoRe & PaP-eRs Inquiry Phase

Two major themes emerged from the data derived from these two groups of science teachers' discussions of their science teaching, after being familiarized with CoRe and PaP-eRs (using the Resource Folio: Particle theory).

The first theme, *Planning for teaching*, illustrates a focus on the way in which a CoRe and PaP-eRs conceptualization of PCK can lead science teachers to think differently about how they structure their understanding of subject matter content in terms of its relationship to learning. What this theme suggests is that through a consideration of the prompts and big ideas that comprise the CoRe (see Table 1), that these teachers started to see into subject matter content differently. The sub-themes imply that this deeper thinking about content could influence their practice and perhaps subsequently, students' learning. For example, the sub-themes of *Planning for teaching* were those of:

1. Planning/structure matters. Indicative views of this response being: "This is a very useful way to plan. Helps to organise the ideas and strategies. Good way to share and generate ideas."
2. Content organization. Indicative views of this response being: "CoRe is really great at revealing all major areas of content within the topic. The questions are really good at drawing out things I personally never have thought of before. It was interesting to see what other teachers considered when teaching this topic."
3. Content interactions. Indicative views of this response being: "Sometimes I had not fully formalized all the necessary concepts clearly for myself. The CoRe helps to ensure that all concepts could be covered – I should go back to doing some lesson plans"; "Makes topic/content less daunting."

This theme of *Planning for teaching* suggests that these participants found the CoRe useful and that, in some ways, it generated new ways of thinking about subject matter content in terms of conceptual understanding rather than propositional knowledge alone. The fact that planning emerged as an aspect of thinking about practice is interesting as it implies that a more reflective approach to constructing understandings of subject matter content may well be one way of moving forward or challenging more traditional views or expectations of teaching as doing "activities that work".

The second theme in the post CoRe and PaP-eRs inquiry is that of *Re-conceptualizing practice through professional learning*. The sub-themes that comprise this theme illustrate how, for these participants, PaP-eRs seem to have created new ways of looking into science teaching and learning such that what they considered as ways of building their own professional knowledge of practice were enhanced. Importantly, the language of the prompts and big ideas from the CoRe appear to have helped to shape ways of conceptualizing teaching and learning that might encourage engagement with the intentions of PCK (though the term itself was not specifically used; it was seen as academic jargon, not a term to be commonly used by teachers per se).

The sub-themes of *Re-conceptualizing practice through professional learning* also hint at the value of PCK as a way of better understanding science teaching, although do not explicitly state it that way:

1. What & Why of teaching. Indicative views of this response being: "A springboard for teachers to talk about teaching and learning"; "provides a link between thinking about engaging activities ... and how to tackle that through content by breaking down into big ideas & manageable sections."
2. Students' perceptions. Indicative views of this response being: "The PaP-eRs have many good ideas of ideas, stumbling blocks and students' perceptions that would assist my own or my department's approach to teaching." "Genuine/honest PaP-eRs of 3 pages in length are optimum."
3. What would happen in the classroom. Indicative views of this response being: "A great guide that also gives you scenarios of what could happen in the classroom with certain activities"; "Could build up a bank of student questions and teacher responses that would be available to all teachers in the faculty": "Use PaP-eRs to help teachers & students think about new dimensions to concept/big idea."

As these sub-themes suggest, the value of a shared language for discussing science teaching and learning is a clear entry point into developing science teachers' thinking about practice. Through becoming familiar with the conceptualization of CoRes and PaP-eRs, it seems as though these participants have started to find new ways of developing and sharing their knowledge of practice. What PCK appears to offer is a way of considering quality in science teaching and learning in ways that might be communicable and applicable in practice, but that the construct itself is perhaps too broad to convey specific advice to teachers; as is in accord with the lack of concrete examples in the literature. Therefore, it could be that conceptualizing PCK through a CoRe and PaP-eRs approach helps to make PCK more accessible for teachers; even though in this small study, they still appeared to avoid the term itself.

4. CONCLUSION AND IMPLICATIONS

This initial foray into the CoRe and PaP-eRs methodology suggests that it is a helpful way of capturing and portraying science teachers' pedagogical content knowledge for the profession. The importance of this methodology is also bound up in the enhanced valuing of PCK within the profession and how that might develop through a shared language of professional practice that goes beyond the notion of PCK itself.

An aspect of this type of study that is not so commonly noted in the literature, but is surely a feature of the researching the development of complex constructs such as that of PCK, is that in attempting to apply abstract academic constructs in the world of practice (such as this approach to CoRe and PaP-eRs), subsequent research is increasingly difficult. In this study, attempting to understand more about how such an intricate conceptualization of PCK might be "tested" with teachers, requires considerable resources, time and co-operation (particularly at the classroom level).

The nature of this study is a step towards exploring whether or not the outcomes of a major research program (i.e., attempting to articulate and portray concrete examples of PCK) are applicable to the perceived end users (science teachers). Results must be viewed tentatively; although they still carry positive overtones. What this means though is that research into science teaching and learning is perhaps approaching a new phase whereby "testing for applicability" in classroom practice might encourage new conceptualizations of the perceived value and purpose of educational research more generally. Inevitably, creating stronger links between theory and practice matters if constructs such as PCK are to genuinely impact practice in meaningful ways.

ACKNOWLEDGEMENTS

The research that comprises this paper has been possible through the support of the Australian Research Council through an ARC Discovery grant (Loughran, 2005–2007). The ongoing support of the ARC is very much appreciated by the research team.

ANNEX

Example of a PaP-eR. Playdough Balls: Concrete Models of Abstract Ideas

Many students in the early years of secondary school have difficulty grasping the abstract ideas of the Particle Model of matter. Providing students with concrete models which they can manipulate can be effective not only in promoting their understanding but also in enabling the teacher to diagnose possible sources of student misunderstanding. The parts that comprise this PaP-eR portrays how and why a teacher uses physical models with her class.

Part 1: The particle movement during phase changes needs to be made explicit to students

While students may know how the particles are arranged in solids, liquids and gases, they do not necessarily make the link that there must be particle movement to achieve a phase change. Recognising her students' difficulties in understanding particle movement as matter changes phase, the teacher responds by asking the students to make and manipulate model particles.

As she went through the Year 8's homework about solids, liquids and gases, it became clear that many of the students didn't really understand what it was that we were representing. They could draw dots as they were arranged in each phase, but I didn't feel they had a sense of the change, the transition between the phases.

So we all rolled little balls of playdough (that I had ready for the next part of the lesson). On a piece of butcher's paper divided into three Sections, each pair needed to be able to demonstrate how the balls, 'particles', would be arranged in each phase. As I came around I asked them to show me what happened as the substance changed from a liquid to a gas, or from a liquid to a solid. Thinking in terms of the particle model is such an abstract exercise – I think it benefited several students to actually manipulate the 'particles'. They had to show how they move; how they move more as they are heated; and how this results in the particles breaking away[1] from one another and moving further apart. It's like the roleplay[2] that students often take part in, but here they worked through it themselves to gain a sense of the transition between the states they see represented in their textbooks.

Part 2: Distinguishing between the particles in elements, compounds and mixtures can be confusing for students

This Part illustrates how one teacher's careful selection of an activity for her students that requires them to show the arrangement of particles in an everyday solution, using their playdough models, helps to serve as the basis for distinguishing the important difference between fixed ratio arrangements of atoms in a molecule and compound and the loose ratio of compounds in a mixture.

I gave the class a definition of an element and some examples (hydrogen, oxygen, chlorine and sodium) with their chemical symbols. The main idea was that all of the 'bits' in an element are the same as each other. Each pair of students used different coloured playdough balls to represent these elements.

Then I put the names and symbols for two compounds (water and sodium chloride (salt)) and a mixture (salty water) on the board and asked the students to use the playdough to work out models for these.

Different groups showed their representations and we used these to build up definitions for compound and mixture. Importantly, pairs had different proportions of salt and water in their salty water. So we could bring out that mixtures are not chemical arrangements and don't require fixed amounts of each particle.

The concepts of element and compound, represented using the playdough balls, required us to differentiate between atoms and molecules. The students then moved on to a worksheet that clarified these four concepts using nonsense examples and non-examples of each.

Three things I'd like to mention:

- I was not intending to make much of a link with the work on states of matter. It was interesting however that a number of students thought about whether the various substances were solids, liquids or gases, and tried to show this, as well as whether they were elements or compounds. On reflection I can see how using the same materials to represent the particles is a good way to reinforce that this model applies to all of the ways we ask them to think about matter.

- A few students had separated the sodium and chloride in their salty water. If they had also separated the hydrogen and oxygen then I knew that they didn't have the compound part right. If it was only the NaCl then I said something like: "You've shown what actually does happen to the salt in the water, but for now we want to keep the compounds together and show how the two compounds are mixed up." This is an example of where we teach misconceptions in order to get the concepts straight[3].
- Underlying all this is my belief that to grasp complex abstract ideas, students need to revisit them from different directions and manipulate them in different ways. If they just spout back the words that you give them, for example, just tell you back what an element is, then you can't be sure that they've learnt anything at all.

NOTES

[1] See PaP-eR: What is the smallest bit? Part 2

[2] In the roleplay, a group of students is asked to show how they would be arranged if each of them was a particle in a solid. They are then asked to show what they would be doing if they 'formed' (1) a liquid (2) a gas.

[3] See PaP-eR: Careful chemical reactions

REFERENCES

Cochran-Smith, M. & Lytle, S. L. (1999). Relationships of knowledge and practice: teacher learning in communities. In Iran-Nejad, A. & Pearson, P. D. (Eds.),*Review of Research in Education, 24* (pp. 249–305). Washington, DC: American Educational Research Association.

Connelly, F. M. & Clandinin, J. (2000). Narrative understandings of teacher knowledge. *Journal of Curriculum and Supervision, 15*(4), 315–331.

Fenstermacher, G. D. & Richardson, V. (1993). Elicitation and Reconstruction of Practical Arguments in Teaching. *Journal of Curriculum Studies, 25*(2), 101–114.

Geddis, A. N., Onslow, B., Beynon, C. & Oesch, J. (1993). Transforming Content Knowledge: Learning to Teach about Isotopes. *Science Education, 77*(6), 575–91.

Gess-Newsome, J. & Lederman, N. G. (Eds.) (1999). *Examining Pedagogical Content Knowledge.* Dordrecht, Netherlands: Kluwer Academic Publishers.

Grimmett, P. P. & MacKinnon, A. M. (1992). Craft knowledge and the education of teachers. In Grant, G. (Ed.), *Review of Research in Education, 18*, (385–456). Washington, DC: American Educational Research Association.

Grossman, P. L. (1990). *The Making of a Teacher: Teacher Knowledge and Teacher Education.* New York: Teachers College Press.

Korthagen, F. & Lagerwerf, B. (1996). Reframing the relationship between thinking and teacher behavior: Levels in learning about teaching. *Teachers and Teaching: Theory and Practice*, 2(2), 161–190.

Loughran, J. J., Berry, A. & Mulhall, P. (2006). *Understanding and Developing Science Teachers' Pedagogical Content Knowledge.* Rotterdam: Sense Publishers.

Loughran, J. J., Mulhall, P. & Berry, A. (2004). In search of pedagogical content knowledge in science: developing ways of articulating and documenting professional practice. *Journal of Research in Science Teaching, 41*(4), 370–391.

Lymna, F. (1981). The responsive classroom discussion. In Anderson, A. S. (Ed.), *Mainstreaming Digest*, College Park, MD: University of Maryland College of Education.

Shulman, L. (1987). Knowledge and teaching: foundations of the new reform. *Harvard Educational Review, 57*(1), 1–22.
Shulman, L. S. (1986). Those who understand: knowledge growth in teaching. *Educational Researcher, 15*(2), 4–14.
Shulman, L. S. (2002). *What is the Scholarship of Teaching? Lee S. Shulman explains his vision of the "scholarship of teaching".* http://www.carnegiefoundation.org/president/publications.htm
van Driel, J. H., Verloop, N. & de Vos, W. (1998). Developing science teachers' pedagogical content knowledge. *Journal of Research in Science Teaching, 35*(6), 673–95.
White, R. T. (1988). *Learning Science*. London: Blackwell Press.

ONNO DE JONG AND A. E. VAN DER VALK

8. SCIENCE TEACHERS' PCK AND TEACHING PRACTICE: LEARNING TO SCAFFOLD STUDENTS' OPEN-INQUIRY LEARNING

Abstract: The present study deals with a school-based in-service course for science teachers, aiming at guiding teachers how to scaffold students' open-inquiry learning for the topic of water quality. Seven experienced science teachers were involved in the study. Before teaching, the teachers developed several intentions for scaffolding, such as bringing students in a short initial stage of uncertainty regarding the issue how to design a realistic research question and plan, followed by a stage of classroom discussion and useful hints. They also wanted to make go/no-go decisions regularly, and to discuss their decisions with them for reflection purposes. In the classroom, in general, the teachers taught as intended, but some specific deviations were noted. After teaching, the teachers reported to have learnt a lot from their teaching practice, especially regarding scaffolding students by giving them a well-balanced combination of 'space' for own contributions at one moment and 'direction' at another moment. The implications for science teacher education are identified on the basis of the findings

Keywords: Guiding by scaffolding, In-service teacher course, Open-inquiry learning, Teacher knowledge, Teaching practice

1. INTRODUCTION

A new wave of science curriculum reform is going on in many countries. For many teachers, the implementation of innovations usually requires the development of sufficient knowledge of new curriculum issues and appropriate competence to teach in new ways. An important component of their knowledge base of teaching is often called 'pedagogical content knowledge' (PCK), a term coined by Shulman (1986). For developing PCK of new topics, teachers should get the opportunity to gain experience with teaching these topics in practice. This underlines the need for implementing in-service courses for science teachers, which pay attention to PCK development by including a strong relationship between course activities and teaching activities at school in order to bridge the gap between pedagogical (content) theory and teaching practice (De Jong et al., 1998).

The present study deals with a school-based in-service course for science teachers, focusing on guiding students' open-inquiry learning for the topic of water quality.

Onno de Jong, Centre for Science and Mathematics Education, Princetonplein 5, 3584 CC Utrecht, The Netherlands

R. Pintó and D. Couso (eds.), Contributions from Science Education Research, 107–118.

This course fits in the current interest in facilitating autonomous ('self-regulated') student learning. In the context of this course, we have tried to identify the development of science teachers' PCK and the use of their PCK in their lessons. The purpose of the study was to contribute to a better understanding of teachers' PCK of new science curriculum issues. Clearly, this understanding can contribute to the design of research-based in-service courses for science teachers.

2. BACKGROUND OF THE STUDY

2.1 Teachers' Pedagogical Content Knowledge

Elaborating on Shulman's work, various scholars have proposed different conceptualizations of PCK, in terms of the features included or integrated. Some describe PCK as a "mixture" of several types of knowledge needed for teaching, while others explain PCK as the "synthesis" of all knowledge elements needed in order to be an effective teacher (cf. Cochran et al., 1993). Magnusson et al. (1999) have presented a strong case for the existence of PCK as a separate and unique domain of knowledge. In any case, PCK, referring as it does to specific topics, is distinct from general knowledge of pedagogy, educational purposes, and learner characteristics. Moreover, because PCK is concerned with the teaching of specific topics, it may differ considerably from the subject matter knowledge of these topics. However, several authors have pointed out that it is not always possible to make a sharp distinction between PCK and subject matter knowledge (Tobin et al., 1994). Loughran and co-workers have defined PCK as "the knowledge that a teacher uses to provide teaching situations that help learners make sense of particular science content" (Loughran et al., 2001, p. 289). These authors argued that investigations of PCK should avoid reducing PCK to a mechanistic, technical description of teaching, learning, and content.

Magnusson et al. (1999) conceptualized PCK as consisting of five components: (a) orientations towards science teaching, (b) knowledge of the curriculum, (c) knowledge of science assessment, (d) knowledge of science learners, and (e) knowledge of instructional strategies. In the present study, we want to focus on the two latter components. In our view, these components are intertwined and should be used in a flexible manner: the better teachers understand their students' learning within a certain domain, and the more teaching activities they have at their disposal for the same domain, the more effectively they can teach in this domain.

Regarding the development of teachers' PCK, Magnusson et al. (1999) argued that the development of this knowledge is a complex process, which is determined, among other things, by the nature of the topic, the context in which the topic is taught, and the way a teacher reflects on teaching experiences. These authors concluded that a teacher education program could never completely address all the components of PCK a teacher needs. Up till now, the existing literature on PCK does not provide us with a complete and coherent research-based theoretical framework.

Teacher thinking and classroom behavior are related to each other in a reciprocal way. Teachers' PCK influences teaching practices, while in a reverse way teaching activities influence teachers' practical knowledge. This reciprocal relationship is complex, and, up till now, not understood very well.

2.2 Guiding Students' Inquiry Learning

In secondary science education, scientific inquiry is often not taught appropriately (cf. Gallagher & Tobin, 1987). However, bringing students in a more open-inquiry environment, that is, science laboratory teaching that leaves problems, answers, and methods of investigation open to students, stimulates students to learn much more autonomously how to do inquiry (Roth, 1995).

The open-inquiry approach, compared to traditional classroom laboratory settings, demands new roles and responsibilities from students and teachers. In the traditional 'cookbook' setting, students' main role consists of carrying out the prescribed activities; they have little control over problems and solutions. The teachers' role consists of *guiding by prescribing* the student activities, or, less restrictive, *guiding by modelling*, that is, by showing students how to handle experiments, how to interpret data, and how to come to conclusions. In the full autonomy setting, however, students gain ownership of their investigations, for instance, by framing research questions by themselves and looking for appropriate methods to find answers on their own. Then, the teachers' role is *guiding by laisser-faire*, which is, offering students full space for organizing their own activities. However, students have to learn to fulfil the autonomy role. When teaching them to take this role, the teachers' role consists of *guiding by scaffolding*, that is, the teachers take over those parts of the overall task that students cannot yet manage on their own, but give them responsibility for their own learning as soon as they can handle a task (Linn, 1998). Guiding by scaffolding means that the teacher knows when it has sense to use guiding by prescribing or by modelling or by laisser-faire. When scaffolding, teachers will stimulate their students to explain, justify, and revise their ideas as they examine their experiences with phenomena. Many teachers will find guiding by scaffolding difficult to carry out as they are not prepared for their new roles and do not see these roles exemplified by their colleagues at school. This underlines the need for supporting teachers who want to implement open-inquiry settings and to guide students by scaffolding.

3. RESEARCH QUESTIONS

Recent science curriculum reform in The Netherlands focuses on the implementation of new curriculum issues, more opportunities for autonomous student learning, and the need for appropriate professional development of teachers. A specific element of this reform for upper secondary level is the demand to students to execute a final open-inquiry (FOI) task as part of the new final school examinations. The FOI should cover more than one school science discipline. In this context, we developed a school-based in-service course for science teachers. This course aimed at guiding teachers to learn how to scaffold students in their pre-examination year to execute a mini-FOI task. This task has the same format as a full FOI, but it is smaller scaled in time and extent. The mini-FOI dealt with the science topic of water quality.

The course leaders provided the teachers new teaching strategies and materials as drafts ('half-finished products'), suitable for further elaborations by the teachers themselves. An important elaboration regarded the general notion of scaffolding students by giving them a well-balanced combination of providing 'space' for students' own contributions at one moment and giving 'direction' at another moment.

A study was launched that was guided by the following research questions:

(i) What intentions for scaffolding students' open-inquiry learning when executing the mini-FOI do the teachers report?
(ii) What scaffolding do the teachers realize in the classroom?
(iii) What reflections on their scaffolding do the teachers report?

4. METHOD

4.1 The Participants

The subjects in the study were seven experienced teachers (all M.Sc.) from two senior high schools. They taught the mini-FOI at their schools in teams to pre-examination classes of about 20 students (aged 16/17). In the first school, one team (TT1) consisted of a chemistry teacher and a biology teacher, while another team (TT2) consisted of one teacher in chemistry and two in biology. In the second school, there was only a team (TT3) of one chemistry teacher and one biology teacher. At all teams, one of the teachers, not always the same per team, was involved in plenary teaching, and the other(s) in scaffolding the student groups.

The course was guided by four experienced science teacher educators; two of them were also the researchers of the project (and authors of the present paper). For every meeting, a written script was developed in advance.

4.2 The Course

The course consisted of five institutional meetings and three lesson blocks.

4.2.1 The first and second meeting The teacher educators explicated the aims of the course in general terms. Subsequently, they introduced the mini-FOI package 'Water Quality', but they also emphasized that this package should be considered as a draft, and open for further discussions. The package covered 10 hours of study, split up in three lesson blocks of two hours each, and four hours of homework for students. The teacher part of the package consisted of a written general script for the lessons, including a description of a general structure of the project, accompanying instructional strategies, and student learning goals. An overview of this script is given in Table 1. The student part consisted of a booklet with a set of optional experiments for determining various features of water.

TABLE 1. Proposed script ('half-finished product') of the lessons

Structure of the project	Instructional strategy	Student learning goal
Lesson #1 (2 hours)	* Introducing aims and design	* Knowledge of the importance of an orientation
	* Presenting the research scope	* Knowledge of the importance of a relevant research scope
	* Scaffolding students' preparation of a research question and plan	* Knowledge of how to develop a research question and plan
Orientation & Preparation	* Making decisions about go/no-go	* Knowledge of criteria for realistic research questions and plans
	* Stimulating students' reflections on the lesson	* To be aware of how they learnt autonomously
	* Prescribing homework regarding the execution of a plan	* Knowledge of difficulties in preparing execution of a plan
Lesson #2 (2 hours)	* Scaffolding students' lab work	* Knowledge of lab work
	* Making decisions about go/no-go	* Knowledge of criteria for results
Execution of the inquiry	* Stimulating students' reflections on the lesson	* Awareness of how they learn autonomously
	* Prescribing homework regarding the preparation of a poster	* Knowledge of difficulties in preparing a poster
Lesson #3 (2 hours)	* Scaffolding students' preparation of a poster	* Knowledge of how to prepare a poster
	* Commenting the presentation	* Knowledge of criteria for a proper presentation
Presentation & Evaluation	* Stimulating students' reflections on the lesson	* Awareness of how they learn autonomously

The teacher educators invited the teachers to discuss how they would use the package in their classrooms, and gave them also the opportunity to revise the package. The teachers discussed the package extensively, expressed their enthusiasm to use the package, and indicated several places for revisions.

4.2.2 The third meeting The teachers discussed the elaboration of the script of the lessons. Their major attention was focused on teaching the first lesson block. Regarding this block, the teacher team TT1 had prepared (at school) a proposal for a sequential number of instructional strategies. The teacher discussed this proposal, and revised it at several places. They also started to adapt the set of student experiments for use in their classrooms. Finally, they discussed a realistic embedding of the project in the existing lesson schedules, and the opportunities for team teaching.

4.2.3 The fourth meeting The teacher educators introduced the issue of scaffolding: giving a combination of 'space' and 'direction' to students when inquiring water quality. The teacher educators and teachers discussed this issue by relating it to several parts of the elaborated script.

4.2.4 The lessons The teachers taught the package in their classrooms together with one or two colleagues during three lesson blocks of two hours each.

4.2.5 The fifth meeting The teachers reported and exchanged their classroom experiences. Subsequently, they reflected on these experiences, also in the context of an evaluation of the course.

4.3 Data Collection and Analysis

Data were collected at several moments of the course. Regarding the course meetings, the data collected consisted of audio tape recordings of the course meeting discussions, and written materials produced by the teachers. All lessons were observed by two teacher educators. This collection of data consisted of audio tape recordings of the lessons, observation notes, made by the observers, and written materials (forms) used in the classrooms.

For analysis purposes, the audio tape recordings were transformed to transcriptions on paper. All data were analysed by two of the science teacher educators independently by using the categories: (i) intended scaffolding, (ii) realized scaffolding, and (iii) reflections on scaffolding. They discussed their interpretations and presented the results to a third member of their team for a final check (investigator triangulation; Janesick, 2000). Issues raised were discussed till consensus was reached.

5. FINDINGS

5.1 Intended Scaffolding

The teachers started to talk about the issue of inquiry by exchanging experiences with students' contributions to an inquiry. They all have experienced that students are able to propose a topic for inquiry by themselves, but encounter difficulties as soon as they have to formulate an accompanying research question and designing a plan. As a teacher from TT2 expressed:

"A group has proposed a topic: alcohol (...). All right, boys, what alcohol, what are you intending to do? Yes, alcohol, yes, and then? Well, perhaps measuring alcohol concentration. Why? And then, silence again. (...) And this is what they have to learn: an initial competency to design a working plan by talking with each other in a cooperative way".

The teachers did not know how to guide their students properly with respect to this aspect of an inquiry, but they want to learn it. For that reason, they used the course meetings mainly for extensive discussions about intentions regarding the appropriate instructional strategies at lesson block 1: the orientation and preparation of an inquiry. As a consequence, there was a lack of time for discussing intentions regarding instructional strategies at both other lesson blocks. The teachers did not regret this very much, because, according to them, the students are much more familiar with carrying out experiments and presenting inquiry reports, because of prior experiences with lab work guided by prescriptions of experiments. In conclusion, the teachers' intentions for teaching the second and third lesson block were not elaborated beyond the proposed script of the lessons.

Regarding the first lesson block, the teachers wanted to use the instructional strategies that were initially proposed by the teacher team TT1 and revised during the third course meeting. These strategies fitted well with the script of the lessons, proposed by the teacher educators. An overview of these strategies, related to phases of teaching, is given in Table 2. An explication is given below.

After a short introduction of the aims and design of the three lesson blocks, the teachers wanted to present the research scope by using vague terms only, saying something like 'investigate water quality', without giving any explication of this task. Subsequently, the teachers wanted to stimulate their students to explore this scope by organizing a brainstorm session in small groups, without any guiding activities. In this way, they wanted to offer their students a lot of 'space' for

TABLE 2. Overview of instructional strategies for lesson block # 1

Phase of teaching	Intended instructional strategy	Realized instructional strategy per teacher team		
		TT1	TT2	TT3
1. Orientation on the project	* Introducing the project: aims and design	+	+	+
	* Presenting the research scope: water quality	+	+	+
2. Exploration of research scope	* Offering brainstorm session to groups	+	+	+
	* Lacking any guiding	+/−	−	+
	* Guiding classroom discussion about results	−	−	+
3. Specification of research scope	* Offering 'water jars' task	+/−	+/−	+
	* Guiding the groups	+	+	+
	* Guiding classroom discussion about results	+	−	+
4. Elaboration of research question and plan	* Offering students' booklet of experiments	+	+	+
	* Offering worksheets for noting question/plan	+	+	+
	* Guiding the groups	+	+	+
5. Assessment of question and plan	* Clarifying criteria for go/no-go	−	−	+
	* Making decisions about go/no-go	+	+	+
6. Termination of the lesson	* Stimulating students' reflections	−	−	+
	* Prescribing homework	+	+	+

TT = teacher team; + = teaching is according to intention; +/− = teaching is partly according to intention; − = teaching is not according to intention.

exploration. For that reason, the teachers called this part of the lesson, the 'no guidance phase'. The only limitation at the brainstorm task was given by the terms 'inquiry' and 'water quality'. By creating a 'no guidance phase', the teachers aimed at evoking a sudden awareness among students of their lack of knowledge how to start an inquiry (some of them used the term: shock effect). They also aimed at evoking feelings of uncertainty among the students, as being quite normal at the beginning of an inquiry. The teachers wanted to terminate the exploration phase by a guided classroom discussion, aiming at giving the students the opportunity to explicate their feelings of uncertainty. However, the teachers expected that the students would hardly be able to explore the research scope very much at that moment. This situation would give the teachers the opportunity to elucidate the underlying difficulty: lack of knowledge how to start an inquiry.

Subsequently, the teachers wanted to reduce the feelings of uncertainty by presenting a 'water jars' task, followed by a guided classroom discussion about the results. The task consisted of observing of four jars filled with different kinds of water (for instance, transparent or not, including small organism or not), and placing the jars in order of increasing water quality. In this way, they wanted to offer their students hints for searching to the meaning of the term 'water quality'. For that reason, the teachers called this part of the lesson, the 'direction phase'. By creating this phase, they aimed at giving their students the opportunity to develop criteria for water quality by themselves.

Then, the teachers wanted to ask their students to work in small groups on designing a research question and an accompanying plan. On order to help them, they wanted to give some hints by offering the students' booklet of experiments, a structured worksheet, and oral hints if necessary.

Subsequently, the teachers wanted to collect and review the filled-in worksheets for making go/no-go decisions. They wanted to explicate these decisions, aiming at promoting students' understanding of criteria for go/no-go decisions with respect to research questions and plans.

Finally, the teachers wanted to invite their students to reflect on the inquiry activities of this lesson, and to discuss their reflections. They also wanted to prescribe homework, especially the preparation of the execution of the plan, and, if necessary, looking for extra information about the scope.

5.2 Realized Scaffolding

In general, the teacher teams have taught according to the expressed intentions. Nevertheless, several differences between the intended and realized instructional strategies have been found. They are incorporated in the following report of the realized instructional strategies. A summary of correspondences and differences is given in Table 2.

The teacher teams started the *first lesson block* by introducing concisely the aims and design of the three lesson blocks, and presenting the research scope in vague terms, as intended. For instance, a teacher from TT2 announced: *"Inquiry in water quality: how do you have to do that?"*

During the subsequent brainstorm session, TT1 asked the students to make notes of their thoughts, and TT2 guided the groups extensively. TT3 did not guide the student groups, although they intended. After this session, TT3 started a classroom discussion, as intended, guided by the initial question: *"What was difficult for you at this assignment?"* TT1 and TT2 skipped the classroom discussion.

The teams offered the 'water jars' task, as intended, and asked the students to observe the jars, as intended. For instance, a member of TT2 only suggested: *"Perhaps this can give you a new idea"*. The teams TT1 and TT2 did not ask the students to place the jars in order of increasing water quality, while TT3 did it, as intended. At the end of the subsequent classroom discussions, TT1 and TT3 expressed conclusions. For instance, a teacher from TT3 indicated: *"You have to select the criteria that you will use later on"*. TT2 also guided the groups, but skipped the classroom discussions.

The teams offered the booklet of experiments to the students as well as a structured worksheet for noting their research questions and plans, as intended. In the beginning, they guided the groups by giving indirect answers to student questions, later on, they offered more specific hints. TT3 assessed the filled-in worksheets by go/no-go decisions, and clarified these decisions to the groups. The teams TT1 and TT2 skipped the clarification of the decision. TT1 only stated to their class: "*Most of you should get a no go*". The team did not explicate this decision. The team TT2 only stated: *"I propose to give them all a go*".

At the end of the lesson, TT3 interrupted students' activities for paying attention to reflection, as intended, by asking the students to note what they have learnt from the lesson about water quality and about the FOI concept, but he did not discuss their reflections. TT1 and TT2 let the groups till the very end of the lesson, and did not give opportunity for reflection. Finally, the teams indicated very quickly the possibility of preparing the next lesson at home.

At the beginning of the *second lesson block*, it appeared that most of the students have hardly prepared the execution of the inquiry. For instance, some groups did not have brought with them the needed samples of surface water that they wanted to investigate. Other groups did not have talked with the science technician in advance, in order to arrange practical materials in time. The teacher teams reacted on this event by telling the students that they consider this situation as a good learning experience for the students.

During the lesson, the teams guided the groups when they encountered problems at their lab work. In most cases, the teams did not offer direct answers to student questions, but asked for clarification first. The students had to fill in a structured worksheet with their experimental data. TT3 assessed the worksheets by making go/no-go decisions, and clarifying his decisions, but TT1 and TT2 skipped the assessment totally.

At the end of the lesson, the teams did not give opportunity for reflection on the lesson. They only emphasized the importance of new homework, especially the elaborating the empirical data and the preparation of the poster preparation.

At the beginning of the *third lesson block*, it appeared that nearly all student groups have done their homework. During the lesson, the teacher teams guided the groups when they encountered difficulties in preparing the poster. They gave comments on the group poster presentations. For instance, they checked if all members of a group agreed with the conclusions on the poster. At several groups, the members did not agree, because they did not have discussed the conclusions critically, and did not have contacted the student who wrote the poster after their discussions. The teacher teams discussed this issue, asking them to be aware of this potential problem later on.

At the end of the lesson, the teacher teams gave hardly opportunity for reflection on the lesson and own learning.

5.3 Reflections on Scaffolding

The teachers were aware that, in general, they have applied the instructional strategies as intended, but they were also aware that there have been specific deviations from their teaching intentions, especially skipped activities. The deviations that they reported correspond with the observed deviations (as given in the preceding sub Section). The teachers explained the skipping of instructional activities mainly by indicating the issue of time pressure. As one of the teachers from TT2 expressed: *"Even for a block of lessons, it was hustling and bustling at the time for finishing in time"*.

The teachers acknowledged that they have to learn to pay much more attention to a proper classroom time management. As one of the teachers from TT2 indicated:

"At the first lesson block, there was time pressure regarding the student evaluation: what do you have learnt? (...) I had to use a part of the break, and, obviously, that evoked resistance among the students. I did not encounter problems at the second lesson block. Then, they were ready in time. But then, I had announced: now, I will finish in time".

The teachers also acknowledged to have learnt a lot from their teaching practice, especially the fruitfulness of providing a combination of 'space' and 'direction' to students. They also valued the students' own contributions to the inquiry project, and the cooperation between students.

6. CONCLUSIONS AND DISCUSSION

The teachers have accepted the script of the lessons, proposed by the teacher educators, as a rough guideline for their teaching. They have discussed teaching that is aiming at stimulating students to learn by themselves, by offering them a sequence of instructional strategies, from giving them space for learning to guiding their learning in a more direct way. In line with this, the teachers adopted 'inquiry' as the leading issue of the lessons, related to the subject-matter topic of water quality, and they have adapted it for use at their own schools. Especially, they have elaborated some specific aspects of the proposed framework by developing a number of strategies of stimulating students to explore and specify the research scope of water quality in terms of a realistic research question and plan. In this context, the teachers stated that a crucial scaffolding strategy consists of bringing students in the beginning of the inquiry process to a state of 'uncertainty' for a while. They reasoned that this state is quite normal in science, and will also occur at the beginning of open-inquiry activities at projects later on. They wanted to reduce the uncertainty by showing the students how they could direct their development of a proper research question and plan.

In general, the teachers, working in small teams, taught as intended. Regarding the issue of inquiry, some deviations from their intentions have been found. The first one regards the guiding of a classroom discussion about the results of the brainstorm session. TT3 paid attention to it (as intended), but TT1 and TT2 skipped this strategy. The second deviation regards the go/no-go decisions. One team (TT3) paid attention to clarify the criteria for go/no-go decisions (as intended), but the teams TT1 and TT2 did not clarify the criteria. The third deviation regards stimulating the students to reflect on their inquiry activities. Only TT3 invited the students to do so, but he did not discuss their reflections. TT1 and TT2 did not pay attention to reflective activities.

Regarding the issue of a subject-related topic, the main deviation from the intentions regards a lack of classroom discussions to confront the groups with each other about the question what the meaning is of the term 'water quality'. This absence of instructional strategy was mainly found at one of the teacher teams (TT2).

Regarding the issue of classroom management, also several deviations from intentions have been found. The first deviation regards the go/no-go decision. One team (TT3) paid attention to making go/no-go decisions (as intended). The teams TT1 and TT2 made these decisions at the end of the first lesson block only, moreover, very quickly and in a very general way. The second deviation regards the homework. At the end of the first lesson block, the teams did not use much time for prescribing the homework (as intended). At the end of the second lesson block, they changed this strategy and used much more time to inform the students about the tasks they have to do at home.

In conclusion, through teaching open-inquiry in their classrooms, the teachers were able to develop their PCK at several places. They became more aware of student learning as well as several student difficulties. They also acquired knowledge of strategies for scaffolding students by giving them a well-balanced combination of 'space' for own contributions at one moment and 'direction' at another moment. However, this PCK development appears to be not so easy for every teacher.

From the present study, several implications for science teacher education can be formulated. They are given below in terms of hints.

- Acknowledge science teachers as professionals with a specific expertise. As a consequence, new teaching strategies and materials should be offered as 'half-finished products'. By doing so, science teachers get the opportunity to use their own PCK, for instance, when adapting the initial proposals for their own teaching practice.

- Create a discourse community in which science teachers and science teacher educators take part. This can be realized by giving science teachers the opportunity to become co-owners of important parts of new projects, for instance, by giving them space for contributing to the final version of 'half-finished products'.
- Create opportunities for science teachers to learn in cooperative settings. This can occur within the school (team teaching) or between schools (meetings of teachers from different schools).
- Apply the consequence principle ('teach as you preach'). This implies that teachers who are learning to guide their students by scaffolding should be supported by teacher educators who guide them by scaffolding too.

REFERENCES

Cochran, K.F., DeRuiter, J.A. & King, R.A. (1993). Pedagogical content knowledge: An integrative model for teacher preparation. *Journal of Teacher Education, 44*, 263–272.

De Jong, O., Korthagen, F. & Wubbels, T. (1998). Research on science teachers education in Europe: teacher thinking and conceptual change. In B. Frazer & K.G. Tobin (Eds.), *International Handbook of Science Education* (pp. 745–758). Dordrecht: Kluwer Academic Publishers.

Gallagher, J.J. & Tobin, K. (1987). Teacher management and student engagement in high school science. *Science Education, 71*, 535–555.

Janesick, V.J. (2000). The choreography of qualitative research design. In N.K. Denzin &Y.S. Lincoln (Eds.), *Handbook of Qualitative Research Design* (2nd edition; pp. 379–399). Thousands Oaks, CA: Sage.

Linn, M.C. (1998). The impact of technology on science instruction: historical trends and current opportunities. In: B.J. Frazer & K.G. Tobin (Eds.), *International Handbook of Science Education* (pp. 265–294). Dordrecht/Boston: Kluwer Academic Publishers.

Loughran, J.J., Milroy, P., Berry, A., Gunstone, R. & Mulhall, P. (2001). Documenting science teachers' pedagogical content knowledge through PaP-eRs. *Research in Science Education, 31*, 289–307.

Magnusson, S., Krajcik, J. & Borko, H. (1999). Nature, sources, and development of pedagogical content knowledge. In J. Gess-Newsome & N.G. Lederman (Eds.), *Examining Pedagogical Content Knowledge* (pp. 95–132). Dordrecht/Boston: Kluwer Academic Publishers.

Roth, W.-M. (1995). *Authentic School Science: Knowing and Learning in Open-inquiry Science Laboratories*. Dordrecht: Kluwer Academic Publishers.

Shulman, L.S. (1986). Those who understand: Knowledge growth in teaching. *Educational Researcher, 15*, 4–14.

Tobin, K., Tippins, D.J. & Gallard, A.J. (1994). Research on instructional strategies for teaching science. In D.L. Gabel (Ed.), *Handbook of Research on Science Teaching and Learning* (pp. 45–93). New York: Macmillan.

REINDERS DUIT, SILKE MIKELSKIS-SEIFERT
AND CHRISTOPH T. WODZINSKI

9. PHYSICS IN CONTEXT – A PROGRAM FOR IMPROVING PHYSICS INSTRUCTION IN GERMANY

Abstract: The theoretical framework and preliminary results of various evaluation measures of a German program to improve the quality of physics instruction are presented. The major emphasis of the program is to develop teachers' thinking about good instruction as an indispensable prerequisite for improved teaching behaviour. It turns out that students' development of affective variables (such as their self-assessed competence) appears to be more pleasing for the "Physics in Context" group than for the control group. Instruction within the program seems to include a significantly higher amount of inquiry activities than for the control group. The teachers rate their participation in the project rather positively

Keywords: Context based physics instruction, Constructivist oriented teaching and learning; Quality development, Teacher professional development, Teacher thinking, Teaching and learning physics

1. INTRODUCTION

Improving science instruction on a national or regional level has played a significant role during the past few years – especially (but not exclusively) in countries that did not do well in the international monitoring studies TIMSS and PISA (Beeth et al., 2003). All over the world various quality development programs have been initiated to improve science instruction – often by ministries of education. The key to the success of these programs has proven to be teacher professional development. Developing teachers' ways of thinking about "good" instruction as well as their views of the teaching and learning process are generally seen as essential for improving teaching behaviour and implementation of more efficient teaching and learning settings (Anderson and Helms, 2001, Peterson et al., 1989). The present paper provides the framework and preliminary findings of the German physics quality development project "Physics in Context" (piko; www.physik-im-kontext.de) that explicitly draws on this position.

The project is funded by the German Ministry for Education and Research. The team comprises members of the Leibniz Institute for Science Education (IPN) in Kiel and of four other universities in Germany.[1] The three-year program started in September 2003. In the first phase, the conception of the program was piloted in about 80 schools. In the second phase, the conception will be further developed and afterwards more widely implemented. The program is part of a larger set

Reinders Duit, IPN – Leibniz Institute for Science Education, Olshausenstr. 62, 24098 Kiel, Germany

R. Pintó and D. Couso (eds.), Contributions from Science Education Research, 119–130.

of attempts to improve science instruction in Germany. In particular, it draws on the experiences of a national quality development program of science and mathematics instruction (Prenzel et al., 2002) and on findings of a nationwide video-study on investigating dominating patterns of introductory physics instruction in Germany (Duit et al., 2004).[2] It further closely cooperates with the German program "Chemistry in Context"[3] and a project funded by the German Science Foundation on investigating the effects of video-based interventions on the development of teachers' thinking about good physics instruction.[4]

2. THEORETICAL FRAMEWORKS AND AIMS

The key goal of the project is to improve the range and quality of teachers' thinking about teaching and learning physics as well as their teaching behaviour by developing and introducing new teaching and learning methods, new media and new content structures of certain topics. The process of developing teachers' thinking is deliberately supported. "Context" in the title of the project denotes primarily that the teaching and learning context provided addresses issues of students' learning and interests. However, it also includes taking into account that teaching and learning physics in a certain class is embedded in the broader context of the schools and the society (e.g. media, peers, parents). Furthermore, it comprises key features of "context-based science instruction" (Nentwig and Waddington, 2005) that puts major emphasis on embedding learning science concepts into contexts that make sense to the students.

The theoretical frameworks applied comprise a wide range of approaches that draw on constructivist epistemological perspectives of the teaching and learning processes of both teachers and students. More specifically, approaches for teacher professional development (Anderson and Helms, 2001; Borko, 2004; Peterson et al., 1989) are taken into account.

With regard to constructivist views of teaching and learning we draw on the literature on the role of students' pre-instructional conceptions and conceptual change (Duit and Treagust, 2003). We do not restrict our attention to exploring and taking students' prior concepts into account. We also take into consideration that to enable students to build transferable knowledge, learning has to be situated in contexts that are as close as possible to students' everyday life (Lave, 1991). Authentic contexts are seen, moreover, as a useful way to improve students' interest and motivation. Furthermore, learning should include "multiple contexts" and "multiple perspectives". Last, but not least, learning should be fostered systematically. Students should be cognitively active, and instruction should provide insight into the scientific view step by step as laid out by the cognitive apprenticeship approach (Collins et al., 1989). For teachers, this perspective means a change from knowledge transmission to counselling learning processes.

To develop teachers' ways of thinking about "good" instruction we draw on the literature of teacher competencies, especially Shulman's (1987) approach with a particular focus on his idea of developing teachers' content specific pedagogical thinking. Like West and Staub (2003), we see two features of improving teachers' ways of thinking as essential; (1) to develop teachers' views of teaching and learning towards actual constructivist views as outlined above, and (2) to make them familiar with Shulman's issue of content specific pedagogical thinking. For the latter, we draw on the Model of Educational Reconstruction (Duit et al., 2005). The basic idea of this model is to give equal attention to science content issues and students' perspectives (including cognitive and affective variables).

The basic aims of the project read as follows:

(1) Developing a new (constructivist) culture of teaching and learning.

Instructional practice and teachers' views about teaching and learning in Germany appear to be oriented at constructivist ideas only to a limited extent (Duit et al., 2004; Widodo, 2004). A teacher-dominated mode of "questioning-developing" teaching style prevails. A transmission view of teaching and learning predominates. Teachers' thinking about instruction is rather topic-oriented, neglecting issues of student learning. Their spectrum of instructional methods is limited. The aim of the project is to address these limitations.

(2) Improving students' competencies of thinking and working like scientists and to use physical knowledge in everyday life contexts.

Current school practice in Germany tends to narrow down physics content to just the concept level, and to give little attention to the process level including views of the nature of science. Practice often also fails to show students how science concepts and processes may be used in real life contexts (Duit et al., 2004). Students should, however, not only learn physics concepts and principles as facts, but also how science theories and models are constructed, which role hypotheses and experiments play in physics inquiry and how physics knowledge can, for instance, be used to understand every day life problems such as appropriate use of safety belts.

(3) Integrating topics of modern physics and technologies.

Most curricula in Germany do not venture beyond the state of development that physics had reached at the end of the 19th century. Aspects of so-called modern physics (i.e. the physics of the 20th century; e.g. quantum and relativistic physics, non-linear and complex systems, cosmology) are rarely to be found in courses addressing scientific literacy for all students. In other words, most students leaving school do not have established ideas of recent views of the world. Aspects of modern technology are also usually ignored. The project "Physics in Context" attempts to develop contexts that are connected to students' everyday life and that provide the chance to understand the fundamental principles of modern physics and modern technologies (such as the function of sensors in various everyday life appliances).

3. IMPLEMENTATION STRATEGY

Teams of some ten teachers and a science educator are working together. At this time ten teams in nine of the 16 German federal states have been established. The cooperation between teachers and physics educators is viewed as "symbiotic", i.e. teachers are seen as experts for the practice of teaching physics and science educators as team members who are familiar with recent research literature and also with the literature on new instructional methods. The role of the physics educators is to serve as coaches guiding the work of the "community of practice" composed of teachers and physics educators. Brief summaries of research findings and theoretical perspectives (piko-letters) as well as workshops (partly video-based) play the major role in supporting the development of teachers' thinking about good physics instruction and their actual practice.

The "piko-letters" are short texts (about four pages) on important research topics like situated learning, cognitive apprenticeship or improving students' interest in physics. There are two basic piko-letters, namely (1) providing an overview on the role of students' conceptions in the learning process including an introduction into the constructivist view and (2) on the above "Model of Educational Reconstruction". The letters are illustrated with examples and oriented to instructional practice. They provide theoretical perspectives for reflecting actual classroom practice as well as a base for developing new concepts and materials.[5]

The workshops also address major topics of the "piko-letters" and make the teachers familiar with alternative forms of classroom work like learning circles or other group methods that foster students' cognitive activity. The aim of the set work is to develop new concepts and materials for physics instruction that take into account new methodological elements as well as different contexts and students' everyday life experiences.

4. EVALUATION

Programs like "Physics in Context" have to be evaluated carefully. In our project, the aims of the evaluation are

- to document the cooperation of teachers and researchers in school sets and
- to evaluate the development of teachers' content-specific pedagogical knowledge, the changes in their instructional practice and their students' cognitive and affective development.

The project has been evaluated in various ways (see Figure 1).

Formative evaluation serves both to support and to document the processes within the sets. Teachers document their work in portfolios. The physics educators, who are the coaches, produce field-notes.

Summative evaluation includes student and teacher measures. At the beginning and the end of the intervention students are asked how they perceive their instruction and which activities take place in their classroom. The questionnaire also addresses the development of affective and cognitive variables (like interests and views of the nature of science). Teachers are provided with a questionnaire that allows us to investigate the development of their subjective theories (their thinking) about physics instruction at the beginning and at the end of the intervention. There are also items providing information on teachers' acceptance of the project, its aims and outcomes. The questionnaire includes Likert type items. To allow a deeper insight into teachers' familiarity with the aims of the program, interviews are carried out

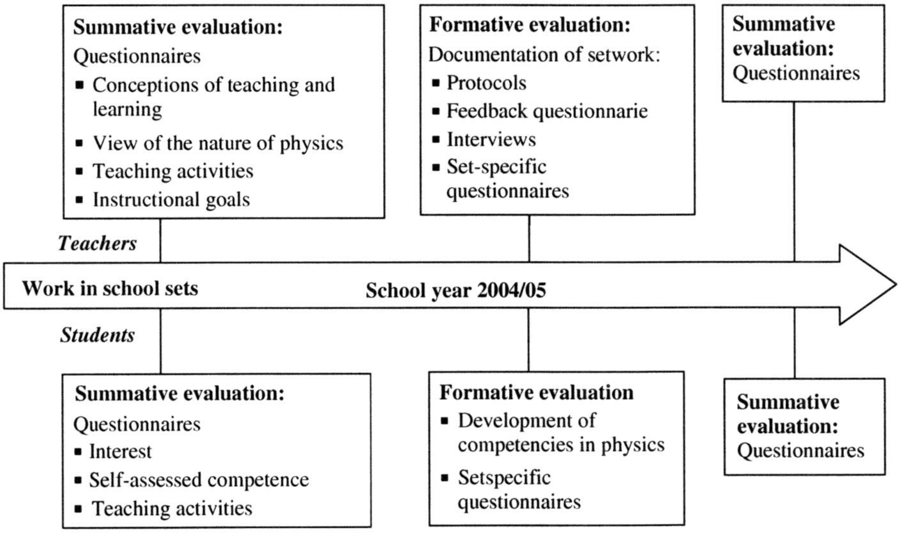

Figure 1. Evaluation design

with about 30% of the whole sample after the first and later after the second year of attending the program.

Additional research studies investigate key issues of the above aims. In total, six PhD studies are being written, dealing, for instance, with teaching and learning nano-science (Kiel), the role of evidence in student argumentation (Kiel), the development of teachers' instructional patterns towards using a larger spectrum of tasks (Paderborn), learning physics with sensors in various situated learning contexts (Ludwigsburg), the development of teachers' views of other teachers (Kassel), and student learning basic concepts of the force concept in settings that deliberately draw on students' bodily experiences (Berlin).

5. HYPOTHESES AND RESULTS OF TEACHERS' QUESTIONNAIRES

As mentioned above, "Physics in Context" aims at improving teachers' content-specific pedagogical knowledge. It is expected that this improvement will lead to changes in teachers' instructional practice that result in a more pleasing development of students' interests and learning. The main aim of the development of teachers' content-specific pedagogical knowledge is to focus teachers' thinking on methodological implications of constructivist theories, that means to foster teachers to use a broader variety of instructional methods and materials to provide better opportunities for students to think and work independently. This kind of instruction is meant to foster not only students' interest in physics and their self-assessed competence but their knowledge achievement as well. On the side of the teachers, these processes should lead to a more pleasing development of their professional confidence.

The teacher questionnaire used in the project consists of two parts. The first part assesses teachers' ideas of teaching and learning, the second part deals with their perception of their own instruction. The questionnaire draws – for example – on the teachers' constructivist views, their perspectives on the nature of physics and their way of negotiating students' pre-instructional concepts. The questions concerning the actual teaching practice assess the amount and frequency of students' activities and opportunities for students to think and work independently. The reliabilities of the scales of the teacher questionnaire range from $\alpha = 0.65$ to $\alpha = 0.82$.

The results presented here are taken from the teachers' pretest. Figure 2 shows the teachers' conceptions of teaching and learning. The Likert-scale reaches from 0 ("not important") to 3 ("very important"). The teachers participating in piko express a relatively appropriate view of the nature of science (scale 1: teachers' views of the nature of physics), that means they claim building theory driven hypotheses and testing hypotheses with experiments are necessary. They also see constructivist and student-oriented perspectives as important (scale 2: constructivist views about teaching and learning), as well as taking students' preconceptions and interests into account (scale 3: views of teaching oriented towards students' interest; scale 7: consideration of students' preconcepts). From their point of view, instruction should be well-structured (scale 5: structuring of lessons), that means that they see the necessity of integrating students' activities and teachers' guidance in instruction, and goal-oriented (scale 4: goal orientation in teaching), that means that they try to inform students about the goals of instruction. Furthermore, the teachers think that instruction should make use of students' experiments to foster students to think

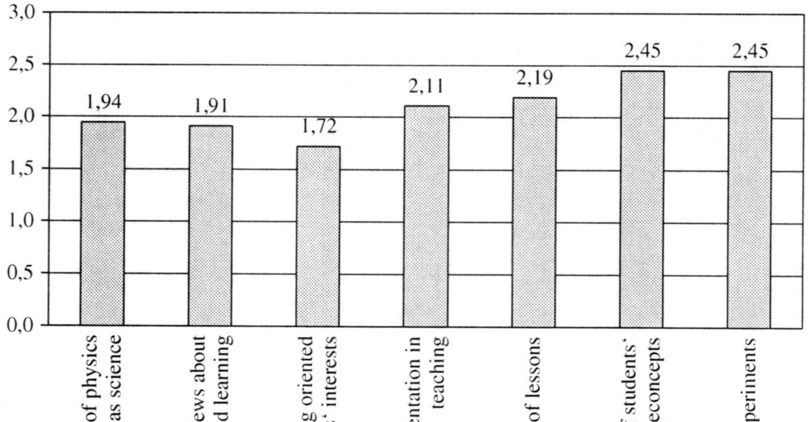

Figure 2. Teachers' views of teaching and learning

and work independently, to test theory driven hypotheses and to use their achieved knowledge in new contexts (scale 8: aims of experiments).

These "subjective theories" of instruction are strikingly different from the teachers' self-reported classroom practice that is shown in Figure 3. Their instruction hardly provides opportunities for students to work and think like scientists (scale 2: working like scientists), that means that students do not get the opportunity to develop their own hypotheses, experiments or conclusions. The teachers sometimes try to take students' interests into account (scale 1: teaching oriented at students' interests), but they hardly use students' oriented teaching methods (scale 3: methods oriented towards student activities, e.g. experiments, learning cycles, group work, etc.). Teachers' methods to engage students in a kind of scientific thinking (how to build hypotheses, how to draw experiments and conclusions) are somehow limited (scale 4: teachers' methods for promoting scientific thinking), they hardly use teaching methods like protocols or students' discussions.

Briefly summarized, the teachers' questionnaires at the beginning of their work in the project (N = 63) show that our teachers hold conceptions of teaching and learning that are – at least at a certain superficial level – mainly in accordance with constructivist perspectives. Their self-described instructional practice, however, meets characteristics of constructivist classrooms only to a rather limited extent. In accordance with our hypotheses, we find a gap between teachers' conceptions of teaching and learning and their classroom practice.

At the moment, the results from the teachers' questionnaire at the end of the school year are not available as the assessment is still in progress.

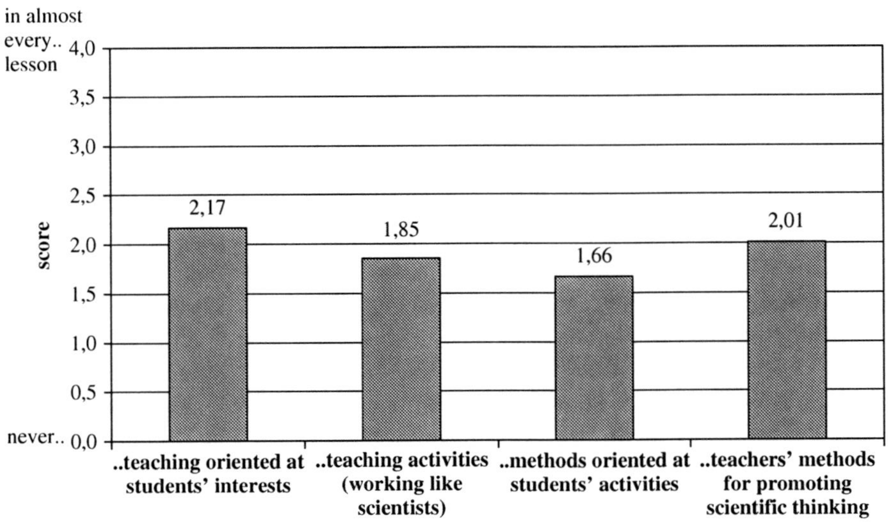

Figure 3. Teachers' self-assessed lesson activities

In addition to the teachers' questionnaire, we provided a short questionnaire on the work in the school sets. The results of this questionnaire (N = 40) show that the teachers participating in "Physics in Context" value the cooperation with researchers very highly. The help of the science educators in providing both information on physics and technologies as well as on educational issues (like the role of students' conceptions in learning) is very much appreciated. The theoretical input (that means workshops and piko-letters) is seen as highly useful. Teachers value the guidelines of the project too, but not to the same extent as the cooperation with researchers. The homepage of the project is seen as useful to a more limited extent.

The teachers claim that their personal use of participating in the project is relatively high. They experience significant changes in their thinking about instruction as well as in their instructional practice. From their perspective, their students' learning and interests change also.

Concerning the quality of the whole program, teachers think that piko is clearly better than other programs, and most of them would recommend it to colleagues without reservation.

Taken together, teachers participating in "Physics in Context" describe their work in piko as personally useful. They value the interaction with educational researchers very highly, and they perceive an improvement not only for piko-lessons, but for their whole instruction.

From our point of view, these results indicate that programs like "Physics in Context" are useful to foster the development of teachers' content-specific pedagogical knowledge as well as the development of a new culture of teaching and learning that provides better opportunities for students' learning and that fosters students' interest in physics.

6. HYPOTHESES AND RESULTS OF STUDENTS' QUESTIONNAIRES

In accordance with the guidelines of the project, teachers and researchers develop – in the above "symbiotic" cooperation – instructional environments which should overcome the deficits of traditional somewhat narrow questioning-developing approaches.

This instruction is more closely oriented towards students' learning opportunities and interests, and should be suitable to avoid the often observed loss of interest during physics instruction. Fostering students opportunities to think and work independently is meant to foster students' self-assessed competence in physics. Moreover, the new instructional concepts should mediate a better view on the nature of physics, it should improve the achievement of scientific competencies, and it should mediate a more positive view of learning.

At the moment, the results of student assessment in the German state of Brandenburg (N = 200) are available to test these hypotheses. In the Brandenburg school set, a unit on scientific ways of thinking and working was developed. After introducing these competencies over three months the students obtained instruction on scientific ways of thinking and working in the field of mechanics and optics.

The questionnaire was provided at the beginning and the end of the whole unit. Additionally, an intermediate test was conducted after the first three months of introducing scientific ways of thinking and working. The results of the pre- and post-test can be compared to results of a control group (N = 100). The teachers of the control group did not participate in the project, and the students of the control group did not receive piko instruction.

The students' questionnaire consists of four scales on their interest in physics, their leisure interest in physics, their self-assessed pleasure[6] and their self-assessed competence.[7] Additionally, the questionnaire includes two scales on the students' view of their actual instruction. On these scales, students' are asked to estimate the amount and frequency of "traditional" activities (like questioning-developing talks, teacher presentations, demonstration experiments, etc.) on the one hand and "inquiry activities" (like independently planning and conducting experiments, students' discussions, etc.) on the other. The reliability of the scales ranges from $\alpha = 0.75$ to $\alpha = 0.81$ in the pre- as well as in the post-test.

The rating of the instructional activities shows that the instruction of all teachers is predominated by "traditional" activities (see Figure 4). Comparing the results of pre- and post-test reveals that for teachers participating in "Physics in Context" the amount of "traditional" activities decreases significantly, while the amount of "inquiry activities" increases slightly, but not significantly. For the control group, the instruction at the end of the school year is predominated by "traditional" activities to the same amount as at the beginning of the school year, while the amount of "inquiry activities" decreased significantly.

These results allow us to conclude that the piko program leads to a change of instructional practice towards providing better opportunities for students to think and work independently. The findings presented in Figure 5 show that these changes

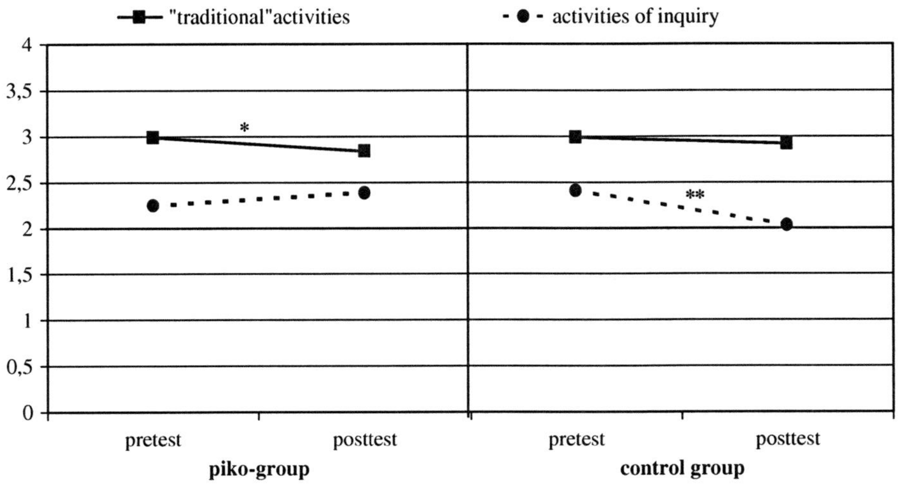

Figure 4. Students' estimations about instructional activities

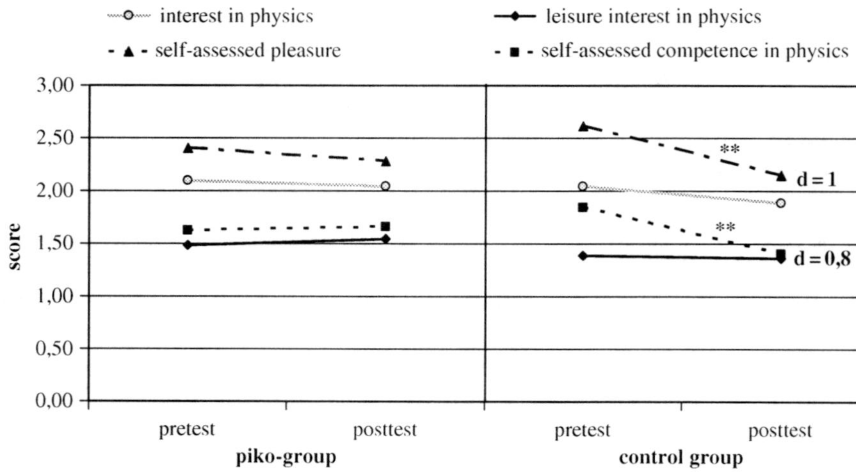

Figure 5. Development of students' affective variables

appear to be linked with a more pleasing development of affective variables for the students of the piko-group.

While the affective variables in the piko group are only marginally different for the pre- and post-test, the results in the control group show significant decreases of students' pleasure in doing physics (standard deviation d = 1) as well as in their self-assessed competence (d = 1).

There are interesting gender differences. For girls in the piko group as well as for girls in the control group, the pleasure in doing physics decreases significantly, but not to the same degree (girls in piko: d = 0.3, girls in the control group: d = 0.8). Remarkably, the self-assessed competence of girls in the piko-group does not change from pre- to post-test, while we find a significant decrease for girls in the control group (d = 1).

For boys in the piko group we find no differences from pre- to post-test. For boys in the control group, the results show dramatic decreases of the variables pleasure in doing physics (d = 1.6) as well as self-assessed competence (d = 0.8).

Taken together, the results show that in piko instruction only the girls' pleasure in doing physics decreases slightly, whereas in "traditional" instruction for girls as well as for boys the pleasure in doing physics and the self-assessed competence decrease dramatically.

7. CONCLUSIONS

On the teacher side, there are convincing findings that the teachers rate their participation in the project rather positively. They experience the work as personally very useful. Interestingly, they particularly appreciate the chance to closely cooperate with colleagues and physics educators as well as the theoretical input by workshops and "piko letters". They also welcome the guidelines of the project but to a slightly minor extent.

Unfortunately, only the results of the teacher questionnaire provided at the beginning of the project are available so far. They reveal that teachers' views of teaching and learning are fairly in accordance with constructivist ideas but that their self-assessed instructional activities are still rather traditional. However, students' estimations of instruction in one of the sets shows that there is a development of instruction towards inquiry-oriented activities. It will be interesting to see whether these findings are supported by the teacher questionnaires at the end of the first year of participation in the project.

In general, the preliminary results of the evaluation of "Physics in Context" provide the impression that piko instruction is – compared to "traditional" instruction – characterized by a higher amount of "inquiry activities". The results also indicate that this kind of instruction leads to a pleasing development of students' affective variables, whereas the students of the control group show significant decreases in their pleasure in doing physics as well as in their self-assessed competence. A very impressive finding is that piko-instruction seems to foster the self-assessed competence of both boys and girls to the same extent.

The preliminary results presented here appear to support the rational of the program that development of teachers' content-specific pedagogical knowledge towards more constructivist views of teaching and learning is linked to the development of instructional activities in this direction also and is additionally linked with a pleasing development of student affective variables like interests and self-assessed competence in physics. Whether instruction is also linked to better student achievement needs to be investigated in further studies.

NOTES

[1] The "Physics in Context" Team: Reinders Duit, Manfred Euler, Silke Mikelskis-Seifert, Christoph T. Wodzinski, Thorsten Bell, Gunnar Friege, Michael Komorek (IPN – Kiel); Raimund Girwidz (University of Education, Ludwigsburg); Peter Reinhold (University of Paderborn); Lutz-Helmut Schön (Humboldt University Berlin); Rita Wodzinski (University of Kassel)

[2] http://www.ipn.uni-kiel.de/projekte/video/Videostudie_eng.htm

[3] http://www.ipn.uni-kiel.de/abt_chemie/ChiK/sites/english.htm

[4] http://www.ipn.uni-kiel.de/projekte/vint/index.html

[5] They may be downloaded from: www.physik-im-kontext.de

[6] The scale "self-assessed pleasure" includes items like: I like to do experiments; dealing with physics is fun.

[7] The scale "self-assessed competence in physics" evaluates similar issues as "self-concept". It includes items like: I am gifted to do physics; I am able to achieve good results in physics.

REFERENCES

Anderson, R.D., & Helms, J.V. (2001). The ideal of standards and the reality of schools: needed research. *Journal of Research in Science Teaching, 38*, 3–16.

Beeth, M., Duit, R., Prenzel, C., Ostermeier, R., Tytler, R., & Wickman, P.O. (2003). Quality development projects in science education. (In D. Psillos, P. Kariotoglou, V. Tselfes, E. Hatzikraniotis,

G. Fassoulopoulos, & M. Kallery (Eds.), *Science education research in the knowledge based society* (pp. 447–457). Dordrecht, The Netherlands: Kluwer).

Borko, H. (2004). Professional development and teacher learning: Mapping the terrain. *Educational Researcher, 33*, 3–15.

Collins, A., Brown, J.S., & Newman, S.E. (1989). Cognitive apprenticeship: Teaching the crafts of reading, writing, and mathematics. (In L.B. Resnick (Ed.), *Knowing, learning, and instruction* (pp. 543–494). Hillsdale, NJ: Erlbaum).

Duit, R., Gropengießer, H., & Kattmann, U. (2005). Towards science education research that is relevant for improving practice: the model of educational reconstruction. (In H.E. Fischer (Ed.), *Developing standards in research on science education* (pp. 1–9). London: Taylor & Francis).

Duit, R., Müller, C.T., Tesch, M., & Widodo, A. (2004). A video study on the practice of German physics instruction. (In D. Zandvliet (Ed.), *NARST Conference 2004, Conference Proceedings*. Vancouver: NARST – National Association for Research in Science Teaching) (CD Rom).

Duit, R., & Treagust, D. (2003). Conceptual change – A powerful framework for improving science teaching and learning. *International Journal of Science Education, 25*, 671–688.

Lave, J. (1991). Situating learning in communities of practice. (In L.B. Resnick, J.M. Levine, & S.D. Teasley (Eds.), *Perspectives on socially shared cognition* (pp. 63–82). Washington, DC: American Psychological Association).

Nentwig, P., & Waddington, D. (Eds.) (2005). *Making it relevant: Context based learning of science*. (Münster, Germany: Waxmann).

Peterson, P.L., Carpenter, T., & Fennema, E. (1989). Teachers' knowledge of students' knowledge in mathematics problem solving: correlational and case analyses. *Journal of Educational Psychology, 81*, 558–569.

Prenzel, M., Ostermeier, C., Duit, R., & Geiser, H. (2002, April). *Improving science and mathematics education in Germany – The concept of a national quality development program and research on its implementation*. (Paper presented at the annual meeting of the National Association of Research in Science Teaching (NARST), New Orleans).

Shulman, L.S. (1987). Knowledge and teaching: Foundations of the new reform. *Harvard Educational Review, 57*(1), 1–21.

West, L., & Staub, F.C. (2003). *Content-focused coaching: Transforming mathematics lesson*. (Portsmouth, NH: Heinemann/Pittsburgh, PA: University of Pittsburgh).

Widodo, A. (2004). *Constructivist oriented lessons*. (Frankfurt A.M., Germany: Peter Lang).

ANNEMARIE M. ANDERSEN, SØREN DRAGSTED, ROBERT H. EVANS AND HELENE SØRENSEN

10. THE RELATIONSHIP OF CAPABILITY BELIEFS AND TEACHING ENVIRONMENTS OF NEW DANISH ELEMENTARY TEACHERS OF SCIENCE TO TEACHING SUCCESS

Abstract: This study's goal was to examine the relationship between the personal capability beliefs of new elementary-level science teachers in Denmark and their teaching environments for their association with teaching success and science teacher retention. Results suggest that pre-service training can increase the self-efficacy of elementary teachers of science as they enter the profession. In addition, the higher the teacher's assessments of their teaching environments were, the greater the positive changes in their self-efficacy were from pre-service until the end of their first year of teaching. Importantly, there is a positive relationship between these changes in self-efficacy during the first year of teaching and whether teachers continue to teach science. This knowledge may help identify new teachers who are more likely to drop teaching science from their teaching load and therefore enable schools to provide the support needed to ensure the viability and retention of new science teachers. The study encourages those who have found that higher teaching self-efficacies are associated with greater success

Keywords: Capability beliefs, Elementary teachers of science, Self-efficacy, Teacher retention, Teaching environments

1. OBJECTIVES OF THE STUDY

This study's overall goal was to examine the relationship between the personal capability beliefs of new elementary-level teachers of science in Denmark[1] and their teaching environments. Our objective has been to use both quantitative and qualitative means to discover for two cohorts of new teachers whether a relationship exists between measurements of two factors: their self-efficacies taken before, during and at the end of their first year of teaching and their various science-teaching environments. A further goal was to assess the effects on capability beliefs of a new Danish program specifically aimed at preparing elementary teachers to teach science and technology and experienced only by our second cohort of teachers. A follow-up study of these same new teachers after several years of teaching was used to assess their teaching environments and whether they have continued to teach.

Annemarie Møller Andersen, The Danish University of Education at Århus University, Tuborgvej 164, 2200 Copenhagen, NV, Denmark

R. Pintó and D. Couso (eds.), Contributions from Science Education Research, 131–142.

2. SIGNIFICANCE

This work hypothesizes a mechanism by which school environments can interact with and modify new teacher's beliefs about science teaching self-efficacy and consequently, the quality of science teaching and learning. Since self-efficacy may be related to successful science teaching (e.g. see Ashton & Webb,1986; Gibson & Dembo, 1984; Enochs & Riggs, 1990; Guskey, 1988; Woolfolk & Hoy, 1990), relating changes in self-efficacy to school environmental factors may be useful in eventually helping new elementary teachers of science become more successful. A focus on new teachers has not only practical importance, since early successes shape their teaching habits for years, but also because the self-efficacy of experienced teachers is less amenable to change (Roberts et al., 2001). The ease of use, high reliability and validity of the instruments available for measuring self-efficacy and science teaching contexts, make them possible candidates for monitoring new teacher adjustments to their schools (see Enochs, 1999; Lumpe et al., 2000). Since new teachers undergo rapid development in their first year in the classroom, this study chose to sample the changes in their self-efficacies before they began to teach, in the middle of their first year and at the end. This longer view of changes in self-efficacy has been a key call by other self-efficacy researchers (e.g. Cantrell et al., 2003).

This study also chose to follow two groups of new teachers, one before and one after a new science teacher[1] training program was instituted in Denmark. The second group had significantly more content and science teaching methods preparation before entering the classroom. In light of work by Morrell and Carroll (2003), which found a positive effect on self-efficacy from methods coursework and no effect from content coursework for pre-service elementary teachers, we are interested in finding the extent to which the self-efficacies of our second cohort of teachers was affected by their additional training.

By taking a fourth look at some of these teachers through a follow-up questionnaire two or four years after they began teaching, we aim to understand their longevity as teachers and their teaching environments.

3. THEORETICAL UNDERPINNINGS

In his original work, Bandura (1977) showed that, based on life experiences, people have specific expectancies about their action-outcome contingencies. They act not only because they believe their actions will result in specific outcomes but also because they believe in their own ability to perform those actions.

In the years following Bandura's original writing, the natural relationship between self-efficacy and teaching was widely explored by a number of researchers (e.g. Ashton & Webb, 1986; Czerniak, 1990; Gibson & Dembo, 1984; Enochs & Riggs, 1990; Guskey, 1988; Woolfolk & Hoy, 1990). Some of these studies showed that teaching behaviours such as persistence at tasks, risk taking and the use of innovations were all related to levels of self-efficacy (Ashton & Webb, 1986). For example, in science teaching, teachers with high self-efficacies were found to be more likely to use inquiry and student centered teaching methods while those with low efficacies were more likely to be teacher directed (Czerniak, 1990). Research has also shown that teacher self-efficacy beliefs strongly influence the nature of a teacher's role, planning and consequently curriculum and student learning (Tobin et al., 1994).

More recently, Bandura (1997) expanded upon the mechanisms by which teacher self-efficacy may be influenced, to include modelling of other teachers, performance feedback and supportive environmental factors. He looked at the relationship of individual self-efficacy to that of collective groups and suggested that confidence is not only a personal but a social construct, so that systems such as classrooms, teams of teachers, schools and school districts develop a sense of collective efficacy which influences the nature of teaching in those systems. He, and others, have suggested that schools with a strong sense of collective efficacy result in effective schools (Bandura, 1997; Smylie, 1990). Their collective efficacies mediate the effects of students' socio-economic status, prior academic achievement and teachers' longevity on the academic achievement of students.

Another important contribution to our understanding of motivation and control beliefs was made by Ford (1992), who proposed a comprehensive 'person-in-context' model to describe the motivational processes associated with individual achievement and competence. He postulated that "...a motivated, skillful, and biologically capable person interacting with a responsive environment..." is essential for personal achievement and that if any of these parts are diminished or absent, competence will be compromised (Ford, 1992, p. 70). Intimately associated with this individual motivation are personal agency beliefs, which are composed of both capability and context beliefs. These personal beliefs about an individual's assessment of his/her ability to perform a given function (capability) and the helpfulness of the environment in that performance (context) are essential precursors to action.

Both Ford (1992) and Bandura (1997) suggest that an understanding of the role that motivation plays in the performance of teachers requires knowledge of their goals, emotions and personal agency beliefs (capability and context beliefs). Over the years, instruments to assess personal agency beliefs among teachers of science have been developed (Enochs & Riggs, 1990; Lumpe et al., 2000). Since Bandura's (1997) self-efficacy is a form of personal capability, the subscale of the Science Teaching Efficacy Beliefs instrument (STEBI) designed to assess self-efficacy (Enochs & Riggs, 1990) is appropriate for assessing the capability part of personal beliefs in the Ford (1992) model. The Context Beliefs about Science Teaching (CBATS) instrument developed by Lumpe et al. (2000) has been specifically designed to measure the context beliefs part of Ford's personal agency beliefs (Ford, 1992).

Riggs & Enochs 1990) originally developed their instrument to specifically measure the self-efficacy and outcome expectations of in-service elementary teachers. Subsequently they adapted the original to assess pre-service teachers (Enochs & Riggs, 1990). The instrument developed by Lumpe et al. (2000) is specifically aimed at assessing teachers' context beliefs about their science teaching environments. This inventory, entitled CBATS (Context Beliefs about Teaching Science), complements measures of self-efficacy in that it correlates at a low level with self-efficacy (Lumpe et al., 2000), yet provides a measurement of the science teaching environment in which self-efficacy is expressed. CBATS provides a better look at the contextual beliefs component of Ford (1992) and Bandura's (1997) view that personal agency beliefs determine teacher behaviour.

Based on the theoretical usefulness of self-efficacy in affecting teacher behaviour and the potential importance of both personal and social environments in affecting teaching beliefs, our research looked at the self-efficacies and teaching contexts of new Danish elementary teachers using special versions of STEBI and CBATS, which were developed for use in a different country from where they were developed (Andersen et al., 2003). Uniquely, this study examines beginning teachers whose self-efficacies are not only crucial to their success, but may be more malleable due to their relative lack of experience (Roberts et al., 2001). Like other recent work it relates the self-efficacy of these new teachers to teaching environments. In line with the work of Haney et al. (2002), this study takes a yearlong view at two groups of new teachers by sampling their beliefs before they begin to teach and two times during their first year of teaching and attempts to associate self-efficacy and context beliefs through statistical correlations as well as follow-up case studies a few years later. It also compares the self-efficacies of a group of teachers after a significant increase in both content and methods preparation to those without such a program.

4. METHODS AND SAMPLE

4.1 Differences between the Cohorts

Two cohorts of new Danish elementary teachers were studied. These two groups had significantly different preparation for teaching science (Natur/Teknik or N/T), due to changes in Danish standards instituted between our two years of data collection. The first group which began teaching in August of 2000, had only a general science course as one of the common core subjects included in their study. The course, which counted for up to about five percent of their total teacher education coursework, was aimed at a general understanding of science for all pre-service teachers and was not designed specifically to prepare future science teachers. As a consequence the first group had a general science subject matter preparation, but they had not studied pedagogical issues or methods related to teaching science at elementary level.

However, educated under the new standards, our second cohort which began teaching in August of 2002 did not have science as a common core subject, but instead had chose a program of study in science, technology and related pedagogical issues (Natur/Teknik or N/T) as one of four main subjects in which to specialize, accounting for about fourteen percent of their total teacher education coursework. The aims of studying science and technology for the elementary school, which were not a part of the preparation of Danish elementary teachers up to and including those in our first cohort, included the following:

- Acquisition of a broad knowledge about scientific and technical phenomena and interrelationships relevant to classroom science content. This included substantial studies of physical and biological science as well as applications to health and environmental issues which are relevant to children.
- Experience with practical and experimental work in science. Coverage of standard science processes was included here, as well as work with the design and construction of simple instruments.
- Knowledge about children's development of language and concepts based on experiences and investigations in their surroundings.
- Knowledge about how science may contribute to children's development of interest in inquiry, responsibility towards the environment, practical skills, creativity and collaboration.

Recent work by Morrell and Carroll (2003) has shown that science content courses don't have an effect on pre-service elementary teachers' self-efficacy whereas science methods courses have significant positive changes. Since the new Danish preparatory standards combine both of these elements, it is possible that our second cohort of teachers began their teaching with higher self-efficacy scores than did our first group.

5. INSTRUMENTS

We used quantitative data gathered with questionnaires about self-efficacy and school environmental together with qualitative data from an open-ended Internet survey to assess overall teaching situations and continuance. The goal was to use the environmental information to help explain any changes in self-efficacies over each group's first year of teaching and to follow the teachers into their first several years after teacher preparation.

For both groups of new teachers, the Danish version of the Science Teaching Efficacy Beliefs instrument (STEBI) (Andersen et al., 2003) was used to measure teacher self-efficacies three times, once before they began their first year of teaching, but after their professional preparation (summers 2000/2002), again in the middle of their first year (winters 2000–01/2002–03) and finally at the end of their first year in late springs 2001/2003. This Danish STEBI instrument has a reliability of $\alpha = .93$ for the self-efficacy scale, compared to $\alpha = .90$ for the instrument from which it was modified, STEBI-B (Andersen et al., 2003; Enochs & Riggs, 1990).

During the final administration of the Danish version of STEBI, we also asked each participant to complete a Danish version of Context Beliefs about Teaching Science (CBATS), to find out which 'enabling' factors in a school environment would be desirable for good science teaching and then the 'likelihood' that these same factors were available to them during their first year of teaching. The Danish CBATS, while mainly a translation of the English version, was also adjusted to accommodate Danish schooling differences. Table 1 provides a sample of the twenty-six items on CBATS which were assessed by each beginning teacher during both years of this study.

TABLE 1. Sample factors from the Danish version of CBATS, mainly derived from the original CBATS (Lumpe et al., 2000; Lumpe & Chambers, 2000)

- Support from other teachers (advice, mentoring, informal discussions, etc.)
- Hands-on science kits (activities and equipment)
- Field trip possibilities (e.g. waterworks, business, forest)
- Planning time
- Permanent science equipment (magnifiers, glassware, thermometers, aquariums, etc.)
- Classroom physical environment (room size, proper furniture, sinks, etc.)
- Expendable science supplies (paper, plastic cops, seed, chemicals)
- Support from school administrators
- Science curriculum materials (textbooks, lab manuals, activity books, etc.)
- Technology (computers, software, Internet)
- Parental involvement
- Involvement of university professors

With quantitative measures of self-efficacy taken at three different points in their first year of teaching (STEBI) and their own assessments of their teaching environments (CBATS), we obtained an idea of how these two cohorts of new teachers faired during their first year in the profession.

Because we were able to measure self-efficacy three times over a year's period, we were able to compare the changes in self-efficacy to the teacher's perception of their science teaching environment. Specifically, we tested correlations between the degrees of self-efficacy changes and the teaching environments. Because we made identical measurements for both cohorts, we were able to compare self-efficacies both with and without the special N/T teacher training which the second group had.

During the spring of 2005 we invited each teacher from both cohorts to complete an Internet administered survey about their teaching status and conditions. Our interest in this follow-up was to discover whether these teachers of elementary science were still teaching science and under what advantages and/or difficulties they were operating. We were also interested in any relationship between their first year self-efficacy changes and teaching environments to their subsequent experiences in teaching.

6. FINDINGS

6.1 Changes in Self-efficacy

For the first rounds, out of original responding samples of 66 during the first year and 56 during the second, we had 49 and 36 continue to the second assessment and 39 and 30 complete all three rounds. Typically self-efficacies declined between the first and second rounds and rebounded somewhat by the third depending on the teaching environment.Suggestions such as those of Cantrell et al. (2003) that self-efficacies in general do not increase significantly during early teaching experiences perhaps due to the weight of school environmental factors, gain support from these findings. The correlation between new teachers' self-efficacies and their judgments about the actual contents of their environments before they began to teach were, of

TABLE 2. Correlations of changes in self-efficacy from pre-service, winter and spring measures with the likely presence of teaching environmental factors for both years

		Changes from pre-service SE to winter SE	Changes from winter SE to spring SE	Changes from pre-service SE to spring SE
CBATS likelihood	Pearson Correlation Sig. (2-tailed) N	.125 .308 69	.32** .006 69	.35** .002 69

** Correlation is significant at the 0.01 level (2-tailed)

course, not significant since these teachers had not yet even visited their schools ($r = .057$; $p = .640$; $n = 69$). The winter assessments of self-efficacy were still not correlated with teaching context ($r = .126$; $p = .302$; $n = 69$), however, by the end of their first year both cohorts of new teachers had significant correlations between self-efficacy and the chance of supportive features in their teaching contexts ($r = .281^2$; $p = .019$; $n = 69$) (see Table 1 for examples of these features). Some of this correlation may be due to the overlap in the STEBI self-efficacy scale and CBATS that Lumpe et al. (2000) noted when reporting the development of the CBATS instrument. They found CBATS to be slightly correlated with STEBI ($r = .19$; $p = .002$).

More importantly, we found that the higher the context assessments, the greater the positive changes in self-efficacy from the pre-service measures until the end of the first year (see Table 2) and that these correlations increased over the course of the year.

Since Ford (1992) and Bandura (1986) have both strongly linked the context within which personal capability beliefs are formed, it is reasonable to suspect that for these teachers, such a link was operant. Since teaching environmental factors can only explain part of the variance in self-efficacy (approximately 11% for changes from winter to spring measures and about 13% from pre-service to spring), factors other than the teaching environment are likely to be affecting teacher's self-perceptions about their ability to teach science. Possible causal variants include students, content preparation and school issues not assessed by CBATS.

6.2 *Differences between the Cohorts*

Of particular interest, in light of the onset of training for teaching science at the elementary level between the two cohorts of this study, all of the self-efficacies of the second group, which received the N/T content and pedagogy training, were significantly higher than those of the earlier group ($t(67) = 2.947$, $p = .004$; $t(67) = 2.973$, $p = .004$; $t(67) = 2.660$, $p = .010$) (see Table 3). These results coincide with those of Morrell and Carroll (2003) who found that pre-service N/T methods coursework increases self-efficacy. Noteworthy in our data is the stability of the

TABLE 3. Self-efficacy mean scores for first year and second year cohorts for each of the three data collection points. Independent samples t-tests of differences between each set of means were highly significant

	Group statistics				
	Sample and	N	Mean	Std. deviation	Std. error mean
Pre-service Self-efficacy	Second Cohort	30	53, 4667	4, 51613	, 82453
	First Cohort	39	47, 8718	9, 60312	1, 53773
Winter Self-efficacy	Second Cohort	30	53, 2000	5, 55474	1, 01415
	First Cohort	39	47, 8974	8, 45980	1, 35465
Spring Self-efficacy	Second Cohort	30	53, 5667	3, 98863	, 72822
	First Cohort	39	49, 6923	7, 16039	1, 14658

elevation for the second year's cohort from pre-service all the way through to the end of their first year of teaching.

6.3 Follow-up Questionnaire

The follow-up questionnaire administered via the Internet in the spring of 2005 was used to find out which of these teachers were still teaching science and how they were faring in their schools after from two to four years of teaching. Nineteen (28%) of the 69 students from our original two cohorts of participants answered our follow-up survey. As can be seen from Table 4, they seem to be reasonably representative of their overall cohorts since as a combination of the two cohorts (eight from the first cohort and 11 from the second) their mean scores fall between those of the original first and second groups.

A summary of the original self-efficacy change, teaching environment and current teaching situation is given in Table 5 for these nineteen respondents. Most interesting is that of this sample of the 2000/01 cohort only two are currently teaching science

TABLE 4. A summary comparison of the original changes in self-efficacy and CBATS likelihood results between each of the cohorts and the group which responded to the follow-up questionnaire

Group	Year	Number	Mean SE change during first year	Mean CBATS likelihood during first year
First Cohort	2000/01	39	−1.820	77.0256
Follow-up Group	2005	19	−1.3684	80.1053
Second Cohort	2002/03	30	−1.00	84.9000

TABLE 5. The nineteen students from the 2000/01 and 2002/03 cohorts who responded to the 2005 follow-up survey. The respondents are ordered from top to bottom from most negative changes in self-efficacy during their first year of teaching to the most positive. First year teaching environment scores are also listed as well as whether they were still teaching and if so, whether they were teaching N/T

ID	Began teaching	Change in self-efficacy	Teaching environment	Still teaching?	Teaching N/T now?
1	2000	−9	76	yes	no
2	2000	−9	95	yes	no
3	2000	−6	76	yes	yes
4	2002	−5	81	yes	yes
5	2002	−4	58	yes	no
6	2000	−3	53	yes	no
7	2000	−3	62	no	no
8	2000	−3	78	yes	yes
9	2002	−2	73	yes	no
10	2002	−2	77	yes	yes
11	2000	−1	94	yes	no
12	2002	−1	108	yes	no
13	2000	0	93	no	no
14	2002	1	84	yes	yes
15	2002	2	91	yes	yes
16	2002	3	66	yes	yes
17	2002	3	93	yes	yes
18	2002	4	91	yes	yes
19	2002	9	73	yes	yes

while from the 2002 group, eight are. The added instruction in teaching N/T was introduced between the time when these two cohorts prepared for teaching. Also of interest are the changes in self-efficacy during their first year of teaching. These changes are the most negative for those who are no longer teaching N/T and for those who did not have the extra N/T teaching preparation.

Six of the follow-up group of 19 (see Table 6) were chosen on the basis of their original self-efficacy change scores for a more detailed examination. The mean self-efficacy scores for the follow-up group, out of a possible 65, were as follows: before teaching first cohort = 48, second cohort = 54; winter first cohort = 48, second cohort = 53; spring first cohort = 50, second cohort = 54. The total possible CBATS score is 130 and the mean was 77 for the first cohort and 84 for the second. Two with the most positive self-efficacy changes during their first year of teaching (+9 and +4); two with no change or a very small change in self-efficacy (0 and −1) and two with the most negative changes during the three measures taken during their first year (−9 and −9) were selected. Details about their original responses to teaching via their self-efficacy and likelihood CBATS scores as well as information about their teaching status and teaching conditions after two or four years are also presented.

TABLE 6. Six selected case studies from both cohorts (SE = Self-efficacy, CBATS = Likelihood)

	Before teaching	Winter	Spring	Self-efficacy change
Case #1 SE (first cohort)	40	32	31	−9
Case #1 CBATS	–	–	76	
Case #1 Current Status: teaching, but not N/T				
Reasons: "Does not feel qualified to teach N/T."				
Case #2 SE (first cohort)	50	41	41	−9
Case #2 CBATS	–	–	95	
Case #2 Current Status: teaching, but not N/T				
Reasons: "It is far too much work when you are not educated in N/T."				
Case #12 SE (second cohort)	53	49	52	−1
Case #12 CBATS	–	–	108	
Case #12 Current Status: teaching, but not N/T				
Reasons: "It didn't fit into my schedule."				
Case #13 SE (first cohort)	51	46	51	0
Case #13 CBATS	–	–	93	
Case #13 Current Status: teaching, but not N/T				
Reasons: "I liked N/T, but there are no materials and no chance for investigations."				
Case #18 SE (second cohort)	51	54	55	+4
Case #18 CBATS	–	–	91	
Case #18 Current Status: teaching N/T				
Reasons: "Children are very fascinated with and interested in n/t and it is an important part of their general education."				
Case #19 SE (second cohort)	51	25	60	+9
Case #19 CBATS	–	–	73	
Case #19 Current Status: teaching N/T				
Reasons: "Pupils are very interested. Close to a beautiful natural area."				

7. IMPLICATIONS

This study's suggestion that pre-service training can increase the self-efficacy of N/T teachers as they enter the profession encourages those who have found that higher self-efficacies are associated with greater success. The follow-up study showing that changes in self-efficacy during the first year of teaching may be related to whether teachers will continue to teach science may help identify new teachers who are more likely to drop teaching science from their teaching load. If with further study, the relationships between teaching environment and new teacher self-efficacy are also substantiated, then not only may new elementary teachers be educated for higher self-efficacies, but also followed to assess their progress during their first years, providing chances for support. More importantly, factors from the school environments can be ameliorated to enhance efficacies, environments and perhaps teacher effectiveness and retention. Opportunities for teacher educators to follow new teachers from their preparatory institutions into their first years of teaching, to both assist them and benefit from feedback about their success and problems, may also be provided by these measures. Convincing research about these potential benefits will be essential to shift policy in this direction.

NOTES

[1] In Denmark teachers who teach elementary school science also teach geography and technology. This domain is called 'Natur/Teknik' or nature/technology. We use the terms 'science teaching' and 'Natur/Teknik' (N/T) interchangeably in this article.

[2] Correlation is significant at the 0.05 level (2-tailed)

REFERENCES

Andersen, A.M., Dragsted, S., Evans, R.H. & Sørensen, H. (2003). Transforming the Standard Instrument for Assessing Science Teacher's Self-Efficacy Beliefs (STEBI) For Use in Denmark. (In Psillos, D. et al. (Eds.), *Science education research in the knowledge-based society* (pp. 395–403). Dordrecht/Boston/London: Kluwer Academic Publishers.)

Ashton, P.T. & Webb, R.B. (1986). *Making a difference: Teachers' sense of efficacy and student achievement.* (New York: Longman)

Bandura, A. (1977). Self-efficacy: Toward a unifying theory of behavioral change. *Psychological Review*, 84, 191–215

Bandura, A. (1986). *Social foundations of thought and action: A social cognitive theory.* (pp. 390–454), (Englewood Cliffs, NJ: Prentice-Hall)

Bandura, A. (1997). *Self-efficacy: The exercise of control.* (New York: W.H. Freeman and Company)

Cantrell, P., Young, S. & Moore, A. (2003). Factors affecting science teaching efficacy of preservice elementary teachers. *Journal of Science Education*, 14(3), 177–192

Czerniak, C.M. (1990, April). *A study of self-efficacy, anxiety, and science knowledge in pre-service elementary teachers.* (Paper presented at the annual meeting of the National Association of Research in Science Teaching, Atlanta, GA)

Enochs, L.G., Posnanski, T. & Hagedorn, E. (1999, April). *Science teaching self-efficacy beliefs: measurement, recent research, and directions for future research.* (Paper presented at the National Association of Research in Science Teaching conference in Boston, MA)

Enochs, L.G. & Riggs, I.M. (1990). Further development of an elementary science teaching efficacy belief instrument: A preservice elementary scale. *School Science and Mathematics*, 90, 695–706

Ford, M.E. (1992). *Motivating humans: Goals, emotions, and personal agency beliefs.* (Newbury Park, CA: Sage Publications, Inc.)

Gibson, S. & Dembo, M.H. (1984). Teacher efficacy: A construct validation. *Journal of Educational Psychology*, 76(4), 569–582

Guskey, T.R. (1988). Teacher efficacy, self-concept, and attitudes toward the implementation of instructional innovation. *Teaching and Teacher Education*, 4, 63–69

Haney, J.J., Lumpe, A.T., Czerniak, C.M. & Egan, V. (2002). From beliefs to actions: The beliefs and actions of teachers implementing change. *Journal of Science Teacher Education*, 13(3), 171–187

Lumpe, A.T., Haney, J.J. & Czerniak, C.M. (2000). Assessing teacher's beliefs about their science teaching context. *Journal of Research in Science Teaching*, 37, 275–292

Lumpe, A.T. & Chambers, E. (2000, April). *Teacher's beliefs about technology utilization in an engaged learning environment.* (Paper presented at the National Association of Research in Science Teaching conference in New Orleans, Louisiana)

Morrell, P.D. & Carroll, J.B. (2003). An extended examination of preservice elementary teachers' science teaching self-efficacy. *School Science and Mathematics*, 103, 246–251

Riggs, I.M., Enochs, L.G. & Posnanski, T.J. (1990, April). *The teaching behaviors of high versus low efficacy elementary science teachers.* (Paper presented at the annual meeting of the National Association of Research in Science Teaching, San Diego, CA)

Roberts, J.K., Henson, R.K., Tharp, B.Z. & Moreno, N.P. (2001). An examination of change in teacher self-efficacy beliefs in science education based on the duration of inservice activities. *Journal of Science Teacher Education*, 12(3), 199–213

Smylie, M.A. (1990). Teacher efficacy at work. (In P. Reyes (Ed.), *Teachers and their workplace: Commitment, performance, and productivity* (pp. 48–66). Newbury Park, CA: Sage Publications.)

Tobin, K., Tippins, D.J. & Gallard, A.J. (1994). Research on instructional strategies for teaching science. (In D. Gabel (Ed.), *Handbook of research on science teaching and learning* (pp. 45–93). New York, NY: Macmillan Publishing Company.)

Woolfolk, A.E. & Hoy, W.K. (1990). Prospective teachers' sense of efficacy beliefs about control. *Journal of Educational Psychology*, 82, 81–91

11. A BEGINNERS' MODULE OF INTEGRATED NATURAL SCIENCE FOR SECONDARY TEACHER STUDENTS: THE RESULT OF AN EDUCATIONAL RECONSTRUCTION PROCESS OVER THREE ITERATIONS

Abstract: In Central Switzerland, pre-service teacher education is intended to reflect the integrated concept of science education in school. The project presented in this chapter assists the development of a research-based unit on integrated science for first-year students. The theoretical background underpinning the research process is provided by the model of educational reconstruction adapted to the settings of universities. The result of this process reflects three important aspects of educational reconstruction: content, concepts and methods. On the level of content, it provides ten suitable topics, which successfully integrate human biology and physics. On the level of concept, it reveals a characteristic self-concept of the involved students and also stimulated conceptual change in questions of integrated science education. On the level of methods, the concept of 'Unterrichtsminiaturen' evolves into the new concept of educational miniatures, a method that proves to be especially effective in the pre-service education training of science teachers

Keywords: Integrated natural science, teacher education, educational reconstruction, educational miniatures, secondary school level, natural science topics

1. INTRODUCTION: HISTORICAL AND CONCEPTUAL BACKGROUND OF THE PROJECT

In Switzerland, an interdisciplinary approach to science education at the secondary school level is common (Labudde 2003) – especially in the central part of Switzerland, the so-called integrated science education has had a long tradition of more than ten years.

We speak about integrated science education, when a topic such as "water as a basis for life" is the focus of the teaching unit, while different disciplines of natural sciences are involved in its understanding (Defila & Giulio 2002). In Central Switzerland, the content of science education at the secondary one level is predominantly organised around issues regarding science-techniques-society (STS). These issues have an important impact on the pupils' ability to cross borders into the scientific subculture (Aikenhead 1994). The classical disciplinary approach

Dr. Albert Zeyer, Institute for Teacher Education, University of Zurich, Beckenhofstr. 31, CH-8006 Zurich

R. Pintó and D. Couso (eds.), Contributions from Science Education Research, 143–156.

of biology, chemistry and physics on the other hand is not of concern in this curriculum, which favours the discussion of biological and environmental topics (Kyburz-Graber 2003).

In 2003 – motivated by the foundation of new Swiss universities of education – the formation of secondary science teachers in Central Switzerland has been completely reorganised. Pre-service teacher education is now intended to reflect the integrated concept of science education in school, i.e. the training itself should be developed from an integrated basis. Therefore, teacher students are not educated in biology or physics or chemistry specifically but rather in integrated science. Moreover, their studies should provide practical experience with science teaching in school right from the start.

2. THE AIMS OF THE PRESENTED RESEARCH

This concept is unique in Switzerland. Therefore, a research-based development of the new curriculum for science teacher education has been established. The project presented in this chapter is part of this investigation. It is aimed at assisting the development of a research-based unit on integrated science for first-year students, which is

I. Adequate, interesting, and authentic for the participating students.
II. A reliable means of self-assessment. The idea behind this concept is that there is no special entrance test for these beginner students, but the module as a whole provides the opportunity to evaluate their faculties. It is expected to test all the different kinds of theoretical and practical skills of a prospective science teacher during the whole term and should also provide an opportunity for self-testing. This formative assessment process should avoid punctual testing and result in a fair and comprehensive entrance examination for the beginner students. If they fail in this formative assessment, they are not allowed to enter the formation.

3. THEORETICAL BACKGROUND OF THE RESEARCH PROCESS

Research in education should be undertaken with a view to improving educational provision (Black & Wiliam 2003). Meanwhile, there exists extensive literature about the practical application of research in education (Gorard & Taylor 2003) and of research in science education (Millar 2000). One result of great practical importance is that assessment can not only measure learning but also support it (Black & Wiliam 2003). In this context, the term 'formative evaluation' was coined by Scriven (1967) and defined in its generally accepted current meaning by Bloom et al. (1971). It contrasts summative evaluation tests – given at the end of episodes of teaching for the purpose of grading or certifying students – with another type of evaluation where assessment supports and improves learning (Winter 2003). There is strong evidence that improving the quality of formative assessment would raise standards of achievement in each country in which it has been studied

(Natriello 1987; Crooks 1988; Black 1993). According to Dann (2002), self-assessment can be considered to be at the heart of the idea of assessment *as* learning. Especially, in teacher training, self-assessment seems to be the most adequate way of formative evaluation, since teacher students often have negative feelings towards summative entrance tests (Keogh et al. 2001) and profit from reflective assessment (Lang 2001).

The theoretical background underpinning the research process itself is provided by the model of educational reconstruction developed by Kattmann et al. (1997) and elaborated by Komorek et al. (2001). The model has been adapted to the settings of universities, in particular universities of education (Theyssen 2001; Neumann 2004).

It is crucial to specify that this model does not intend to reduce the complexity of a given topic during the teaching process. Rather, educational reconstruction 're-constructs' the science content in the context of students' lives and thus establishes a different kind of complexity. With this concept in mind, we organised our research according to three key questions (adapted from Kattmann et al. (1997)):

I. The question of content. What is the scientific content of the topic, i.e. what scientific disciplines are involved, what sources of information exist and how can they be made available for the teaching process?
II. The question of concepts. What perspectives do students hold about the topic? How do they proceed to learn the scientific concepts? What conceptual consequences should be drawn and how do they affect the introduction and the use of scientific and technical terms?
III. The question of methods. What educational methods (Lijnse 2000) could be useful to teach the given topic? Which actions of the students and interactions between lecturer and students could be helpful? Do they reflect the goals of teaching?

Using these three questions as pillars, we can now integrate both, the scientific content and the individual learning processes of the students (Welzel et al. 1999), to implement an authentic learning environment (Roth 1995). In the case of teacher education, the process of educational reconstruction is especially demanding because one has to simultaneously achieve two completely different goals. One goal is to improve teacher students' content knowledge in natural sciences. The other goal is the acquisition of professional skills required for science teaching, as soon as possible. The educational reconstruction of the presented beginners' module had to respect each of these two important goals simultaneously.

An important aspect of the model of educational reconstruction is the iterative character of the research process. Therefore, our results from the investigation were continuously reinvested back into the study to further the evolution of the reconstructed unit. The 'classical' model of educational reconstruction mainly uses content analysis and interviews for investigation. However, we considerably enlarged the spectrum of used methods (see below).

4. RESEARCH QUESTIONS

We wanted to investigate if the theoretical framework for the model of educational reconstruction, as outlined in the previous Section, was a suitable instrument for developing a beginners' module for the pre-service education of secondary science teachers. Having the three key questions of educational reconstruction in mind, we asked:

What will be the outcome of the process of educational reconstruction in terms of

I. The contents of the module.
II. The concepts of students about integrated science and their own situation as teacher students.
III. The methodical structure of the module.

5. METHODS AND SAMPLE OF THE RESEARCH PROCESS

The first unit of the evaluated module was taught to teacher students in Lucerne (Switzerland) during the winter term 2003–2004. This was a pilot study with 12 students. The second unit took place in summer 2004. The third unit was taught during the summer of 2005. The number of participants was 36 and 32 respectively. Each of these units was subject to research. The results were reinvested in the shaping of the module. According to the model of educational reconstruction, we pursued a process of continuous iteration in our development and research. For the reconstruction process, we used several empirical methods, which were combined through triangulation. We started with a science content analysis. Using tests on students' content knowledge we explored the students' scientific background. Pre-post questionnaires on students' expectations and experiences, and question-guided feedback interviews with students completed the data basis. To get specific information about the teaching-learning-process participating observation was also implemented. The activities of the students during 'educational miniatures' (see below) were videotaped and analysed through CBAV (Niedderer et al. 1998).

6. FINDINGS

Every cycle of research iteration produced its own results. For the purpose of presentation, we gathered these results according to our three key research questions, which reflect the model of educational reconstruction.

6.1 The Reconstruction of Content

It is known that the integration of physics and human biology within a science curriculum is attractive to students (Müller 1999). Female students in particular prefer this combination because in the context of human biology the physical questions become more meaningful to them (Häußler & Hoffmann 1998). Therefore, subjects of health and disease were chosen for the beginners' module. The selected subjects were tested during the iteration process. Some of them were cancelled, for instance the topic 'physical and biochemical energy', because it proved to be too abstract and too distant from students' own interests. New ones were added during research, for instance 'renal function and the pH of blood'. This new subject proved to be far more appropriate to students' mental dispositions. At the end of the reconstruction process, a list of ten topics resulted that integrated both aspects of 'health and disease' and physics in an attractive way:

1) Growth and velocity. Students learn to see paediatric percentiles as time-way-diagrams and to understand them as an instrument for diagnosis.
2) Bones and forces. The trabeculae of human bones are from a physical point of view a materialisation of a force field. Their dynamical change gives an insight into biological reactions on a cellular level.
3) Static of the human body and lever law. Physics explains why using a false technique of weight lifting can damage spinal discs.
4) Lungs and flow velocity. Asthma is a disease of small vessels. The law of Hagen Poiseuille explains why a tiny reduction of vessel diameter has a dramatic impact on flow resistance.
5) Inner ear and gravity. Why do we feel weightless for a short time in an elevator? Is it the same in a basin of water?
6) Lungs and surface tension. Why do preterm babies have hard lungs and how can this respiratory disease syndrome be treated with so-called surfactants?
7) Walking and weight loss. How physical work is linked to biochemical energy.
8) Blood and pressure: blood pressure. This topic gives a deep insight into hydrostatic and dynamic pressure and their link to health. The process of measurement is attractive and simple to demonstrate.
9) Blood sedimentation and viscosity. This subject is an opportunity to solve differential equations and to talk about rheumatic diseases.
10) Eye and optics. This is a classical topic raising opportunity for the integration of physics and biology.

A more detailed list of these topics and of their realisation has recently been published (Zeyer & Welzel 2006b). Furthermore, we practised topic 6 and 8 in great detail to demonstrate how deep an insight into nature can be, when an integrated perspective is used (Zeyer & Welzel 2005a, b).

The final list of topics proved to be very attractive for the students. Eighty-nine percent of them said that their expectations about content had been fulfilled. The integrated concept of natural science was broadly accepted. During the interviews, students said that physics and chemical issues become more meaningful in the context of questions of health and disease. Two students commented (during a focus group interview):

"It was the best and most interesting access module of all." "For the first time in my life, I realised how interesting and meaningful physics can be."

However, there were also difficulties to manage:

It was difficult to find appropriate background literature. A lot of scientific and textbook sources exist about both the physical and the biological aspects of the topics. However articles or textbooks covering the integrated aspects are rare. Still useful materials are available on some websites (for example, see Wiesner 2005).

Pertaining to textbooks, there is a lack of basic texts of anadequate level for teacher students. Either they are too simple (secondary level) or too complex (scientific level). During our research process, the used textbooks were changed twice – and the course was completed without resulting in a suitable text to work from.

The problem of lacking textbook sources for integrated topics is linked to the scientific limitations of the students (cf. Section 6.2). Physics subjects in particular immediately showed the limits of their knowledge and the extent of their understanding. The process of integration itself also proved to be very demanding for the students. This may be illustrated by two commentaries of the focus group interviews:

"This access module was the hardest and the most difficult one of all access modules." "I did not intend to study physics. So why should I understand physical questions in detail."

One consequence of these problems was that many content mistakes were made during presentations (cf. Section 6.3). For example students confused centripetal force and centrifugal force, or they drew false time-way-diagrams. They mixed up anatomical and histological details and did not correctly link physiological functions to physics explanations. These problems became most salient on the occasion of colleagues' questions during the miniatures (cf. Section 6.3). Students tended to bypass questions they could not answer or invent 'fantasy explanations'. Two examples derived from the video analysis are the following:

Question: "What is the difference between extrinsic and intrinsic asthma?"
Answer : "This is not known so far. More research has to be done."
Question: "What is viscosity actually?'
Answer : "It is sort of density of the substance. But actually I do not know it either."

6.2 Reconstruction of Concepts

Figure 1 shows the distribution of second course students concerning their specialisation during secondary school time.

The distribution for the students of the third course is similar. As can be seen, these students are not mainly interested in natural science. Only 13 of 47 students chose natural science as a specialisation during their secondary school education (27%). Only 7 students (15%) were specialised in mathematics/physics during their secondary school time. The responses showed a broad spectrum of chosen disciplines, including music, economics and modern languages. The same picture results when these students were questioned about their hobbies. Twenty-three students indicated that sports were their favourite hobby. Other favourites were music and dancing (16 times), travelling (10 times) and nature and animals (10 times). Only one student reported to have a technical hobby. These results may be astonishing

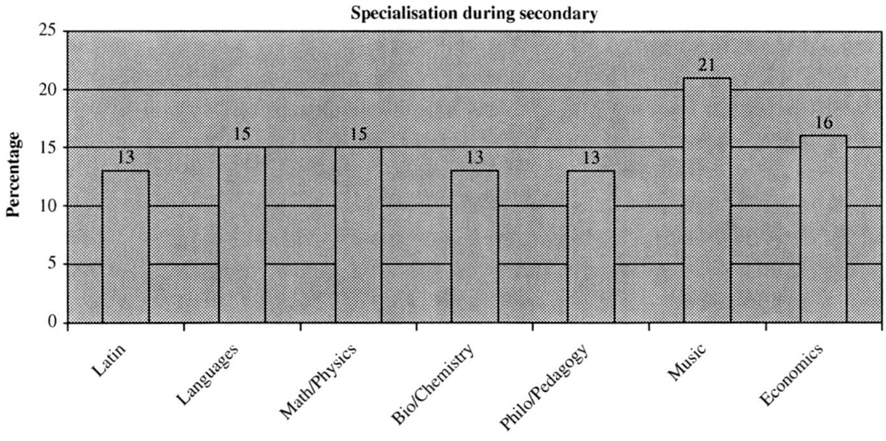

Figure 1. Specialisation during secondary school time. Second course, 47 students

at first glance. But in fact they are not untypical of the population of a university of education which attracts students who show a rather mixed profile of interests.

For our educational reconstruction process it was crucial to keep these figures in mind since they represented the self-concept of the students. In this context, Figure 2 is also interesting, because it shows the self-estimation of the students concerning their aptitude in physics, chemistry and biology.

Figure 2 shows a salient difference between students' self-estimation in physics and biology. Whereas 80% of them indicate that their knowledge of biology is good or even very good, only 23% of them believe in the quality of their physics skills. The results of chemistry lie in between. In summary, these figures indicate

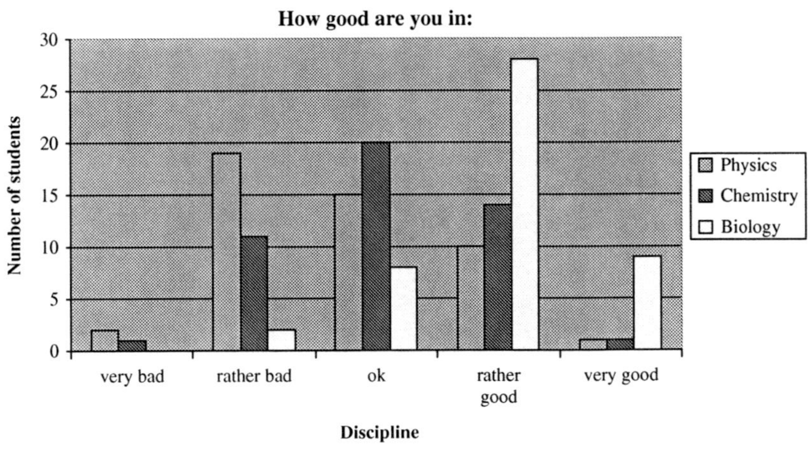

Figure 2. Self-estimation of abilities in disciplines of natural science

a rather poor self-esteem of our students concerning their skills in natural science, especially in physics.

Feedback interviews showed, in fact, that students were clearly aware of this. Then, one might ask, why they chose natural science at all as their discipline to teach. But we would be mistaken to judge these students as being uninterested in science. On the contrary, in the focus group interviews, they proved to be keen on improving their knowledge and skills. However they expected the instruction to be focused on issues relevant to their later profession and were not interested in scientific and technical details, but much more interested in obtaining an overview and a general structure of scientific knowledge. Their self-concept was not the one of a scientific professional, but the one of a teacher who is going to teach science – and not only science, but other subjects as well, such as languages or music.

These findings partially explain the problems with content and scientific knowledge and skills described in Section 6.1. It was important for the process of educational reconstruction that we were aware of this characteristic composition of our students' group. Physics had to be embedded in problems of general interest and must not predominate too much. The level of abstraction and the depth of knowledge had to be carefully 'titrated' in order not to provoke stress and complaints. On the other hand, it was important to point out that science education does not consist only of biology.

Figure 3 shows another result that proved to be important for the reconstruction process. Here are the answers of the students at the beginning of the second course to the question: What is integrated science education?

Most students (60%) interpreted integrated science education simply as an additive combination of the three disciplines of physics, chemistry and biology.

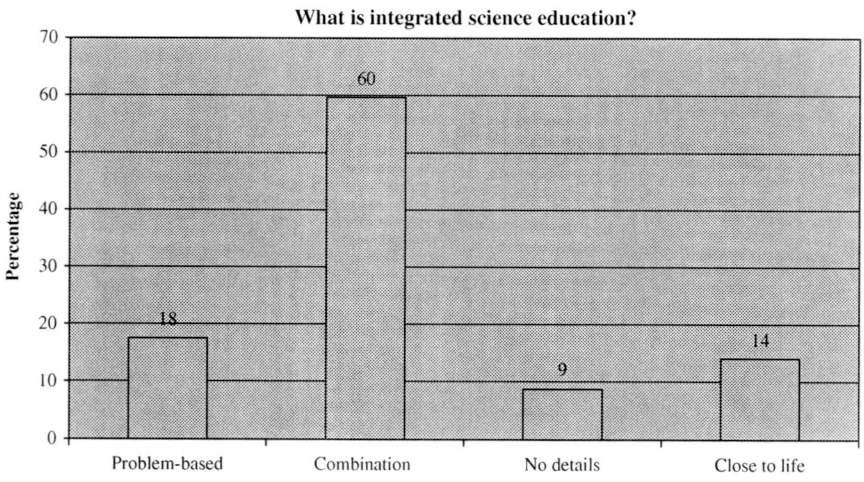

Figure 3. What do you think is integrated science education?

Some other students gave a vague and unspecified definition of integration as abstaining from details (9%) or being close to life (14%). Only 18% of the students came close to the definition of Defila and Giulio (2002), who interpret integration as a problem-based application of different disciplines. The question was asked again at the end of the module. The results were more or less the same. We concluded from these results that students' concept of integration did not change during the module.

This was why we included explicit explanations about this important aspect into the curriculum of the third course. The outcome of this procedure was clearly much better, as shown in Figure 4.

After having attended the module, most students reproduced the intended concept of integration. More than that they described it in their own words and quite precisely. Here are two remarkable answers quoted from the second questionnaire:

To each issue, there exist different answers out of different scientific disciplines. Together they lead to a comprehensive understanding of the facts. The result is more than the sum of the different disciplines.

6.3 Reconstruction of Method

For the pilot study, we started with a combination of lectures and practical courses. During the lectures, integrated topics were demonstrated and the lecturer provided background information. The laboratory courses were devoted to the preparation of animal organs (heart and lungs) and to the analysis of human blood in a hospital laboratory. In general, the concept of the module was highly esteemed by the students. They liked, in particular, the combination of theoretical and practical parts and the mixture of learning and doing. However, other concerns came up such as:

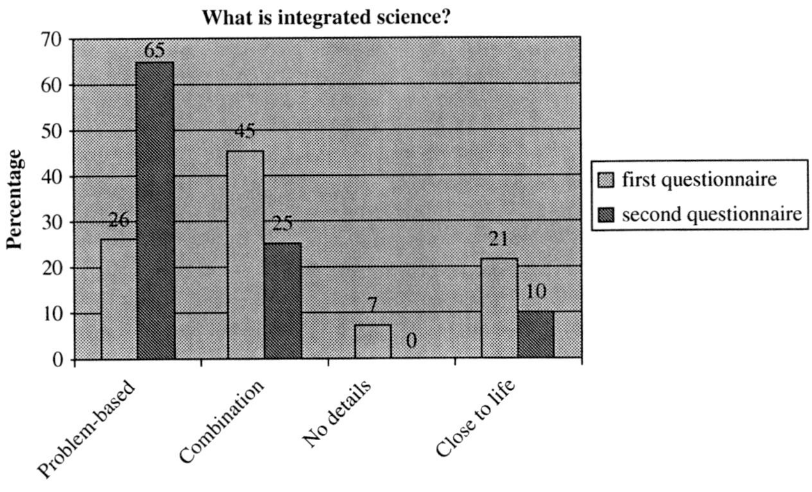

Figure 4. What do you think is integrated science education?

- The lecture was too difficult.
- Students preferred practical work. In a feedback interview they mentioned that practical work was much more suitable to their later profession than a theoretical form of lecture.
- The module was not reliable as an assessment procedure: There was no distinction between different levels of theoretical fitness, because most of the students were overtaxed. There was no distinction between theoretical and practical skills, because practical work was underrepresented. Two students failed to pass.

In response to these results, we looked for an alternative educational method and found it in the concept of the so-called miniatures (Labudde 2002). Students' talks as a method of pre-service teacher training are well known. Labudde pointed out that they are especially useful in pre-service development of science teachers. He called them 'Unterrichtsminiaturen', what may be translated into 'mini-lessons'. Students prepare their own little educational units and perform them in front of their colleagues. Miniatures seemed to be an adequate educational method to improve the structure of our module. For the second course, we decided to reduce lecture time and introduced 'miniatures' instead.

The new concept of miniatures earned a very positive feedback from the students. More than 90% of the students said that the content was interesting or very interesting and that the structure of this part of the module was clear and adequate. In the feedback interview, students judged the miniatures to be attractive, informative and especially fitting to their later profession.

Participant research showed, however, that students' performance suffered from important methodical deficits. They tended to imitate the teaching style they had experienced during their own school time. Being under pressure of performance in front of their colleagues, they hid themselves behind strict frontal lectures and fully prepared teacher experiments. They showed a lack of time management and made plenty of beginners' mistakes. Here are two examples of our video analysis:

One student had prepared an elaborate PowerPoint presentation. However, the slides were overloaded with information and the student was so absorbed in handling them that he never made eye contact with his audience.

Another student had provided calf eyes to demonstrate optical phenomena in biology. She put a plastic bag containing five eyes on the table in front of the students. On the video screen, the astonishment and shock of the students nearby can easily be observed.

To improve this situation, we decided to integrate educational questions right from the beginning of the preparation of the miniatures until their conclusion and evaluation. We called this new and improved approach 'educational miniatures'. Educational miniatures – as we define them – are mini-lessons conducted by students and characterised by a two-level concept: scientific content and educational reconstruction that are considered equally important. Thereby, the second level is the innovative one because it explicitly enforces the position of educational and professional issues in student activities, which have previously been neglected.

Educational miniatures were introduced into our beginners' module during the third course. Due to this new concept needing much more time for preparation, performance and evaluation, the frontal lectures were completely cancelled. In our research, educational miniatures proved to be a very useful tool for appropriate pre-service teacher education. In fact, they are a salient illustration of the fact that the process of educational reconstruction may lead to new methodical approaches, which perfectly reflect the investigated teaching situation. Elsewhere we described in detail the management of educational miniatures and their educational potential (Zeyer & Welzel 2006a).

The introduction of educational miniatures considerably improved the assessment process. During the second course, 18 percent of the students failed. However, merely one student failed only because he had not passed the written exam. The other unsuccessful students realised during the course of the module that they did not match the expected skills.

Finally, we present in Table 1 a schema that summarises the general process of collection of data, analysis and uses in the process of reconstruction.

TABLE 1. Schema of the general process of collection of data, analysis and use in the process of reconstruction

	First iteration	Second iteration	Third iteration
Type	Start Winter term 03/04: A pilot project consisting of lectures and practical laboratory	Summer term 04: Introduction of the element of Miniatures	Summer term 05: Downsizing of lectures. Introduction of the concept of **educational** miniatures.
Collection of data	Interviews	– Pre-post-questionnaires – Pre-post-tests – Videographs	– Pre-post-questionnaires – Pre-post-tests – Videographs
Results	– too much lecture – lecture was too difficult – lack of teaching training – practical work underrepre-sented	1) Questionnaires: – topics (with some excep-tions) and integrated concept broadly accepted – students not primarily interested in science – bad self-esteem in science – no change of students concepts about integration 2) tests: – limits of knowledge and understanding 3) Video graphs: – many scientific mistakes – lack of teaching skills	1) Questionnaires: – concept change concerning integration 2) tests: – valuable assessment procedure 3) Video graphs: Continuous improvement of teaching skills during the module **Final Result:** A beginners module consisting of Educational Miniatures combined with self-studies using adequate textbooks and practical laboratory experience

7. CONCLUSIONS AND IMPLICATIONS

Our study shows that the application of the model of educational reconstruction leads to a suitable beginners' module for pre-service secondary science teacher student education. The result of the research process reflected three important aspects of educational reconstruction: content, concepts and methods. On the level of content, it provided ten suitable topics which successfully integrated human biology and physics. On the level of concept, it revealed a characteristic self-concept of the involved students and also stimulated conceptual change in questions of integrated science education. On the level of methods, the concept of 'Unterrichtsminiaturen' evolved into the new concept of educational miniatures, a method that proved to be especially effective in the pre-service education training of science teachers.

The resulting module is based on educational miniatures, self-studies and practical laboratory work. Each of these elements of the module must be shaped and tested in an appropriate way. This provides a reliable assessment process that is not reduced to a traditional knowledge test. The integrated approach of science teaching was highly estimated by the students but it is demanding for beginners with a sparse background of scientific knowledge.

REFERENCES

Aikenhead, G. (1994). What is STS in science teaching? STS education: international perspectives on reform. *J. Solomon and G. Aikenhead.* New York, Teachers College Press : 47–59.

Black, P. & Wiliam, D. (2003). In praise of educational research: formative assessment. *British Educational Research Journal* 29(5): 624–637.

Black, P. J. (1993). Formative and summative assessment by teachers. *Studies in Science Education* 21, 49–97.

Bloom, B. S., Hastings, J. T. et al. (Eds.) (1971). *Handbook on the Formative and Summative Evaluation of Student Learning.* (New York: Mc Graw Hill)

Crooks, T. J. (1988). The impact of classroom evaluation practices on students. *Review of Educational Research* 58, 438–481.

Dann, R. (2002). *Promoting Assessment as Learning.* (London: RoutledgeFalmer)

Defila, R. & Giulio, A. D. (2002). *Interdisziplinarität in der wissenschaftlichen Diskussion und Konsequenzen für die Lehrerbildung. Interdisziplinäres Lehren und Lernen in der Lehrerbildung. Perspektiven und innovative Ausbildungskonzepte.* A. Wellensiek and H. -B. Petermann. (Weinheim: Beltz Verlag)

Gorard, S. & Taylor, C. (2003). Editorial-in praise of educational research. *British Educational Research Journal* 29(5), 619–621.

Häußler, P. & Hoffmann, L. (1998). Chancengleichheit für Mädchen im Physikunterricht – Ergebnisse eines erweiterten BLK-Modellversuchs. *Zeitschrift für Didaktik der Naturwissenschaften* 4(1), 51–67.

Kattmann, U., Duit, R., Gropengiesser, H. & Komorek, M. (1997). Das Modell der Didaktischen Rekonstruktion – Ein Rahmen für naturwissenschaftsdidaktische Forschung und Entwicklung. *Zeitschrift für Didaktik der Naturwissenschaften* 3(3), 3–18.

Keogh, B., Naylor, S. et al. (2001). Formative assessment using concept cartoons: Initial teacher training in the UK. *Research in Science Education – Past, Present, and Future.* Behrendt, H., Dahnke, H., Duitet, R., et al. (Dordrecht: Kluwer Academic Publishers)

Komorek, M., Stavrou, D. & Duit, R. (2001). Nonlinear physics in upper physics classes: Educational reconstruction as a frame for development and research in a study of teaching and learning basic ideas of nonlinearity. *Proceedings of the Third International Conference on Science Education*

Research in the Knowledge Based Society, Vol. 2. Psillos, D., Kariotoglou, P., Tselfes, V., Bisdikian, G., Fassoulopoulos, G., Hatzikraniotis, E., Kallery, M., Thessaloniki, Greece, Aristotle University of Thessaloniki: 483–485.

Kyburz-Graber, R. (2003). Integrating interdisciplinarity in curriculum development. *Interdisciplinary Challenges in Teacher Education in the Light of Environmental Education*. R. Kyburz-Graber, P. Posch and U. Peter. (Wien: Studienverlag)

Labudde, P. (2002). Situiertes Lernen in fachdidaktischen Lern-Lehr-Veranstaltungen. *CD zur Frühjahrstagung des Fachverbandes Didaktik der Physik in der Deutschen Physikalischen Gesellschaft Bremen 2001*. (Berlin, Lehmanns Fachbuchhandlung)

Labudde, P. (2003). Fächer übergreifender Unterricht in und mit Physik: Eine zu wenig genutzte Chance. *Physik und Didaktik in Schule und Hochschule* 1(2), 48–66.

Lang, M. (2001). Teacher professionalism and change: developing a professional self through reflective assessment. *Research in Science Education – Past, Present, and Future*. H. Behrendt, H. Dahnke, R. Duitet et al. (Dordrecht, Kluwer Academic Publishers)

Lijnse, P. (2000). Didactics of science: the forgotten dimension in science education research? *Improving Science Education*. J. L. Robin Millar, Jonathan Osborne. Philadelphia, Open University Press: 308–326.

Millar, R., Leach, J. & Osborne, J. (2000). *Improving Science Education*. (Buckingham: Open University Press)

Müller, W. (1999). Physik und Medizin – ein themenorientiertes Unterrichtskonzept für fachübergreifende physikalische Bildung. *Beiträge zur Didaktik und Methodik des Physikunterrichts*. (Berlin: Volk und Wissen Verlag)

Natriello, G. (1987). The impact of evaluation processes on students. *Educational Psychologist* 22, 155–175.

Neumann, K. (2004). *Didaktische Rekonstruktion eiens physikalischen Praktikums für Physiker*. (Berlin: Logos Verlag)

Niedderer, H., Tiberghien, A. et al. (1998). Category based analysis of videotapes from labwork (CBAV) – the method and results from four case studies. *http://didaktik.physik.uni-bremen.de/niedderer/projects/labwork/papers.html* Working Paper 9 from the European project Labwork in Science Education (Targeted Socio-Economic Research Programme, Project PL 95-2005).

Roth, W. -M. (1995). *Authentic School Science*. (Dordrecht: Kluwer Academic Publishers)

Scriven, M. (1967). The Methodology of Evaluation. *American Educational Research Association*. (Washington, DC)

Theyssen, H., v. Aufschnaiter, S. & Schumacher, D. (2001). A laboratory course in physics for medical students as a result of a theory-based development with accompanying evaluation. *Proceedings of the Third International Conference on Science Education Research in the Knowledge Based Society, Vol. 2*. D. Psillos, Kariotoglou, P., Tselfes, V., Bisdikian, G., Fassoulopoulos, G., Hatzikraniotis, E., Kallery, M. Thessaloniki, Greece, Aristotle University of Thessaloniki: 652–654.

Welzel, M., v. Aufschnaiter, C. et al. (1999). How to interact with students? *The Role of Teachers in a Learning Situation. Practical Work in Science Education: Recent Research Studies*. J. Leach and A. C. Paulsen. Roskilde, University Press: 313–327.

Wiesner, H. (2005). *Didaktik der Physik*. http://www.physik.uni-muenchen.de/didaktik/.

Winter, J. (2003). The changing prepositions of assessment practice: assessment of, for and as learning. *British Educational Research Journal* 29(5), 767–772.

Zeyer, A. & Welzel, M. (2005a). Was Seife mit dem ersten Schrei des Neugeborenen zu tun hat. *Praxis der Naturwissenschaften. Physik in der Schule* 7(54), 40–44.

Zeyer, A. & Welzel, M. (2005b). Was Viskosität und Senkung miteinander zu tun haben - Die Blutsenkung. *Praxis der Naturwissenschaften. Physik in der Schule* (Praxis-Magazin).

Zeyer, A. & Welzel, M. (2006a). Didaktische Miniaturen (Educational Miniatures) – Eine methodische Alternative für den naturwissenschaftlichen Unterricht. *Lernen im Physikunterricht.* R. Girwidz, M. Gläser-Zikuda, M. Laukenmann and T. Rubitzko. (pp. 147–169) (Hamburg: Verlag Dr. Kovac)

Zeyer, A. & Welzel, M. (2006b). Lernen, um das Gelernte zu kommunizieren. Didaktische Miniaturen als methodische Alternative im naturwissenschaftlichen Unterricht. *Physik und Didaktik in Schule und Hochschule* 1(5), 54–61.

PART 4

LEARNING AND UNDERSTANDING SCIENCE

HANS NIEDDERER, MARION BUDDE, DAMIEN GIVRY,
DIMITRIS PSILLOS AND ANDRÉE TIBERGHIEN

12. LEARNING PROCESS STUDIES

Abstract: Learning processes will be analysed as "evolution of student's ideas" or as "conceptual change" on a timescale of several hours. The idea is to follow a single student's constructions "during" the whole process of learning in more detail, including analyses of learning effects. Part two discusses theoretical and methodological issues of such learning process studies. Here, the focus is on the unit of analysis (expressed ideas or conceptions) and on the relation between teaching and learning. In part three, a more recent study about the evolution of students' ideas about gases is presented, describing learning by using three categories: A student expresses a new idea, a student increases the domain of validity of an idea, or a student establishes a link between several ideas and develops a network. In part four, another more recent study about analysing learning effects of the learning environment on single students learning is presented. Here, different types of resonances are used as categories, e.g. congruent or disgruent resonance, spontaneous or retarded resonance. Both studies come to grounded hypotheses how to improve teaching. In part five, more general issues about learning process studies are discussed

Keywords: Atomic physics, Conception, Gas, Idea, Learning, Learning effects, Learning pathway, Methodology

1. INTRODUCTION

In 1991, an international workshop held in Bremen was working on "Research in Physics Learning - Theoretical Issues and Empirical Studies" (Duit et al., 1992). The intention of this workshop was to develop a new research goal, which was to study learning processes with data from *during* the learning process.

The main point, which often makes learning process studies very important and interesting, is the following: students' actual constructions are often different from taught knowledge. In other words: there is a gap between what we teach and what is learnt (McDermott 1991), the knowledge to be taught is different from students' steps of learning (Tiberghien 1997).

In this paper, we are presenting some theoretical and methodological issues together with some results from new studies (Givry 2003; Budde 2004).

Hans Niedderer, Mälardalens Högskola, Institutionen för Matematik och Fysik (IMa), Box 325, S 63105 Eskilstuna, Sweden

R. Pintó and D. Couso (eds.), Contributions from Science Education Research, 159–171.

2. THEORETICAL FRAMEWORK

2.1 Conceptions and Expressed Ideas

Many authors use the term "conception" to denote their basic concept of thinking and learning (see Duit 2007). A conception is seen as a hypothetical set of statements, skills, procedures, that the researcher attributes to one or more students in order to account for students' behaviour in a set of given situations (Tiberghien 1997, 365). Here, we distinguish between two cases:

(1) One is to consider that a conception intends to be part of modelling students' mind. It is a construction of a researcher to describe typical use of elements of knowledge and ways of thinking of students. A conception has to be stable and must appear in more than one context and point in time.

(2) The other is to consider that the researcher infers only ideas, which are expressed in the students' productions without making hypotheses on students' mind (Givry & Roth in press). These inferences will be called "expressed ideas" to distinguish them from the previous approach.

These two approaches show that, even if the points of view on the students' mind modelling are different, the analysis and the results on learning are compatible and mutually reinforced. Both approaches are inferred from students' productions (utterances, gestures, writing) by the researcher and can show some stability over time (Petri & Niedderer 1998) and through several situations (Givry 2003). The set of these situations represent their domain of validity, this domain can be reduced to a single situation or be stable in several situations. In both cases there is a construction of a researcher, which describes the core of several ideas of students in the researchers own words using the most distinctive features of those ideas for this description. This procedure is also aiming at a considerable reduction of data, describing the core of a set of ideas in a set of situations with one conception.

2.2 Learning Processes

Often, learning processes[1] can be represented as a sequence of conceptions developed by students during instruction. These conceptions do not exclude each other; a student can have more than one conception in parallel at the same time (Petri & Niedderer 1998, 2003; Taber 2000; Hartmann & Niedderer 2005). Here conception is taken in the larger sense given previously and is specified in each study. For example, some authors (Petri & Niedderer 1998; Psillos & Kariotoglou 1999) use the term "conception" or "concept of a student" to describe a learning process as a series of conceptual changes (Dykstra 1992). Taber (2001) in his case study of conceptual development as competition between alternative conceptions about bonding uses "evolving explanatory principles" as one form of conception to describe a learning process of one student in two years of an A level course in chemistry. Cleme͡nt & Steinberg 2002) in their study on mental models of electric circuits use a series of successive modified models to describe their student's learning process all along a teaching unit.

Givry (2003; part 3 of this paper) uses a slightly different approach to define learning, using the model of "expressed idea", with fewer generalisations than "conceptions". This model considers a learning process as "evolution of students' ideas" and identifies three kinds of evolutions to describe some aspects of learning:

- A student expresses a new idea.
- A student increases the domain of validity of an idea.
- A student establishes a link between several ideas and develops a network.

This last type of evolution is very similar to learning as forming new conceptions, whereas the former two aspects can be seen as first important steps towards learning.

Budde (2004; part 4 of this paper) understands learning to be the development or change of conceptions as well as the modification of the status and strength (Petri & Niedderer 1998) of co-existing conceptions. Learning is described as the interaction between the cognitive system of the student (the student's ideas and conceptions) and the contents of the learning environment. She uses both "idea" and "conception" to describe learning. An idea is a description with which a single statement of the student is described in the researcher's own words. A conception is able to explain two or more of one student's statements.

2.3 Intermediate Conceptions

The idea of "intermediate notions" or "intermediate conceptions" as building of bridges between children science or formal science goes back to Driver's work in 1989 (Taber 2001, 735). It has been used in similar ways by many authors. We stated earlier that we should expect actual constructions being often different from intended conceptions by the teacher, even if they are influenced by the taught knowledge. Often, these resulting actual constructions are described as intermediate conceptions or intermediate explanatory models, and as unexpected or not at all intended by the teacher. Taber (2001) writes about his student's "Coulombic forces explanatory principle": "Although such ideas were formally incorrect, they had greater potential for being developed into explanations in terms of physical forces: i.e. they were able to act as intermediate conceptions on an appropriate conceptual trajectory." Clement & Steinberg (2002) in their study use also a similar term "intermediate explanatory model". So, intermediate conceptions are different from prior conceptions, and also from the intended conceptions and likely to be developed in students' learning processes (Petri & Niedderer 1998; Psillos & Kariotoglou 1999).

2.4 Some Methodological Issues of Learning Process Studies

Using data from *during* the teaching and learning processes is perhaps the critical instance of learning process studies. All studies reported followed single students, either one, two or three. The new studies reported here, Givry (2003) and Budde (2004), follow two students each. Of course we need to know about the learning processes of all students in a class. But the amount of qualitative data seems not to allow this. Clement & Steinberg (2002) have been very explicit on this problem: "Attempts to track learning processes at this level of detail in groups of students have been frustrating for us because we do not hear enough from each student to follow the process without large gaps." Then they argue that nevertheless these studies could be of some value: "... such studies can be an important source for generating grounded hypotheses about learning processes that have a substantial initial level of plausibility and that are worth investigating in larger samples." Perhaps, this problem can be solved best with a combination of a qualitative learning process study *and* a quantitative evaluation study.

Givry (2003; part 3 of this paper) collected, during one month, different types of data about students' points of view on gas behaviours. These data include: (1) videotaped lessons, (2) videotaped interviews and (3) written questionnaires. Students' learning processes are studied through the videos and students' ideas about gas are reconstructed by the researcher based on: (a) the context of the situations and (b) students' speech, gestures, manipulation, and written works.

In the study of Budde (2004; part 4 of this paper) the data collected consisted mainly of audiotape records of the regular school lessons. Before and directly after instruction the students provided written responses to questions probing their views and understanding of atomic models. To investigate the longer-term stability of the students' conceptions, control interviews were also held two years after the end of the instruction. Alongside the evaluation of the learning of the full cohort of students (26 students in total), exploratory case-studies, focussing on two 18-years old students, were carried out to obtain a better understanding of the influence of the teaching on the development of the individual student's conceptions.

2.5 What is the Relation between Teaching and Learning?

This is finally the most important question. Givry (2003; part 3 of this paper) uses the French concept of "*milieu*" to study the factors, which support the evolution of students' ideas. It is composed of several elements with which students could interact to learn (Brousseau 1998). Based on a socio-constructivist approach and literature from didactics of physics, the milieu is defined with three kinds of interaction (Figure 1).

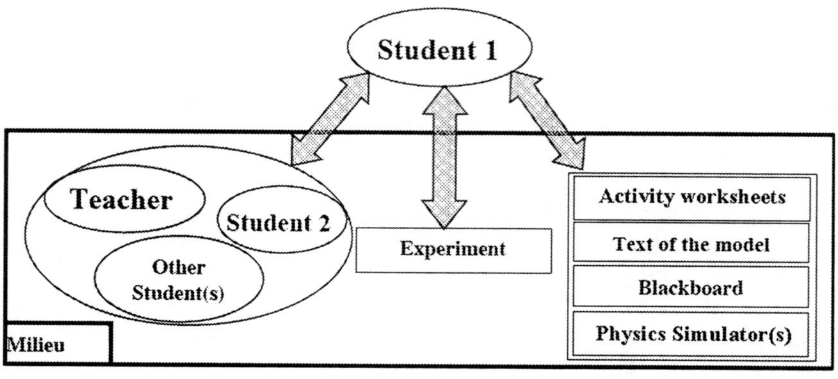

Figure 1. *Milieu* composed of different elements (social, experimental, teaching material), with which students could interact to learn

Then the influence of teaching is studied through elements (social, experimental, teaching material), which are explicitly used by students to support their ideas.

Budde (2004; part 4 of this paper) uses a notion from von Glasersfeld (1992) to describe the relation between teaching and learning as a resonance between learning environment and students' mind. Hypotheses about resonances between certain elements of taught content as part of the learning environment and the evolution of students' conceptions as part of their cognitive system are formulated as teaching hypotheses (see part 4 for two examples). By carefully following the talk of individual students and relating it to the content of teaching, it is possible to test these teaching hypotheses. Observations are made to ascertain whether, and after how many repetitions and discussions, the students construct the intended conceptions. In this way, the opportunities and difficulties presented by different teaching approaches can be evaluated against each other, and principled decisions about the planning and design of teaching can be made. The analysis was based on definitions of different types of resonances (see Budde 2004). Here for illustration, we present a few of the defined categories:

- Congruent resonance: the student constructs in his own independent thinking conceptions, which are essentially equal to the content presented in the learning environment before.
- Disgruent resonance: the student constructs in his own independent thinking conceptions, which are essentially different to the content presented in the learning environment before.
- Spontaneous (congruent) resonance: Already after the first presentation of content a congruent resonance is happening.

The results are summarised diagrammatically (see Figure 3) in such a way that differences between the contents that are taught (on the left side) and the conceptions that the students construct (on the right side) are made explicit.

3. EVOLUTION OF STUDENTS' IDEAS ABOUT GASES

3.1 Context of the Research

The purpose of our study is to give some elements to follow students' conceptual change and the factors that support this change during a teaching unit about gas behaviour. Providing empirical results from a longitudinal study in a French physic classroom, we answer the questions: How do students' ideas about gas evolve during a teaching unit? What are the elements of the *milieu* that support this evolution? This last question deals with a specific aspect of the relation between teaching and learning.

3.2 Theoretical Background

As presented above, a model "expressed idea" has been developed based on a previous theoretical model about "facets" (Minstrell 1992). Furthermore, we use the French concept of "*milieu*" (Brousseau 1998) to study the elements of the teaching situations, which support the evolution of students' ideas (Figure 1).

3.3 School Context and Participants

A group of physics teachers and researchers (including two authors) designed the curriculum to be taught over a one-month period (6 lessons) for students at the upper secondary school level (equivalent to tenth grade [15-year-old students]). The teaching unit[2] consisted of six lessons: two sessions (1 hour, whole class) and four laboratory sessions (1 hour 30, half-class). It adopted a socio-constructivist approach with respect to three main dimensions: (a) modelling activity (Tiberghien 2000), (b) semiotic registers (Duval 1995), and (c) students' conception of a gas (i.e. Benson et al. 1993). The purpose of the unit was to allow students to use (a) macroscopic variables (pressure, volume, temperature and quantity of matter), and (b) their interpretation at the microscopic level (molecules' collisions and velocity) for describing and explaining gas behaviour.

3.4 Methodology

Different types of data sources were collected including videotaped interviews and lessons as well as questionnaires, sampling students' responses to questions about gas behaviour. All in all, the database comprises 420 questionnaires, one-hour taped interviews with 14 students prior to and after the unit in videotaped sessions, 48 hours of classroom video featuring the same 14 students, with two students particularly observed, and approximately 160 pages work sheets. Following the precepts of the Learning Process Studies (Niedderer et al. 1992), fourteen students were videotaped *continuously* during one month of the lessons in classroom. Furthermore, these 14 students were selected according to their achievement levels in physics – based on information given by the teacher – to be representative for the whole class. Two low level students were particularly observed. Video recordings were analysed on the basis of video extracts systematically selected on the basis of two observable criteria: (1) use of words about gas by the teacher or students (i.e. air, gas, molecules) or (2) hands-on experiments by students. We transcribed each video extract and our transcription included: (a) the context and (b) a students' speech, gestures, manipulation, and written works. Then, we constructed a student's ideas about gases expressed through: (a) written language and (b) oral language: speech, gestures and salient elements of the setting. For each idea, we tried to identify, what elements of the *milieu* (social, experimental, teaching material) are explicitly used by the student to support his/her idea.

3.5 Results

With our methodology described above, we characterized according to our theoretical framework the different kinds of evolution of two students' ideas about gases (called Anne and Ellen) during the entire teaching unit. We identify several explicit elements of the milieu, which support this evolution.

3.5.1 Expressing a new idea A student expresses an idea through language, oral or written statements; and an idea is new, when it appears in our data for the first time. In the case of oral language, an idea could be expressed through speech, gestures and salient element of the setting; e.g. Anne expresses that "there is no air in an open bottle" by using simultaneously these three modalities (Figure 2).

Figure 2. Anne's idea expressed simultaneously by talk ("here you've got no air"), gestures (her hand is pointing to the open bottle), and semiotic resources in the setting (glossed as "the open bottle")

New ideas about gases are expressed 12 times by Anne and 9 times by Ellen during the whole teaching unit. This kind of evolution could be supported by some specific elements of the milieu. Indeed, new ideas about the meaning of words (like macroscopic pressure) are supported essentially by the activity worksheets and the work with hands-on experiments. Furthermore, the text about the model of a gas supports new ideas about the properties of a gas (i.e.: homogenous distribution, collision of molecules).

3.5.2 Increasing the domain of validity of an idea This kind of evolution appears when a student expresses the same idea in another situation. A situation is new, when (1) the question changes (e.g. a question about the pressure of gas and a question about the gas distribution) or (2) the material (objects or events) changes (e.g. use of an open bottle, then use of a balloon). This kind of evolution appears 19 times with Anne and 20 times with Ellen during the teaching unit. This kind of evolution is more frequent for the two students, compared to "express a new idea" or "make a link between ideas". However, our results show that each element (social, experimental, teaching material) of the *milieu* supports the increase of the domain of validity; no single element of the milieu alone seems to support this kind of evolution.

3.5.3 Decreasing the domain of validity of an idea This kind of evolution is given, if a student realises that an idea doesn't work in a specific situation and stops to use it. This kind of evolution is particularly important; it appears however only once with Anne and three times with Ellen. Decreasing the domain of validity appears only a few times during the teaching unit and necessitates the simultaneous use of several elements of the milieu. These elements are essentially social and experimental; e.g. if students work with a hands-on experiment and at the same time discuss it.

3.5.4 Establishing a link between several ideas and developing a network This link can be expressed by a student through the language, either oral or written; e.g. Anne wrote: There is space between molecules; consequently we can add air into the bottle. Here, she makes a link between two ideas: (1) "there is space between molecules" and (2) "we can add air into the bottle". This kind of evolution was found 9 times with Anne and 7 times with Ellen during the teaching unit. This kind of evolution is supported essentially by the manipulation of experiment.

3.6 Conclusion

According to our case study methodology, we consider our results as grounded hypotheses to be investigated and tested in other studies. Our results lead us to propose hypotheses about (a) evolution of student's ideas and (b) elements of the milieu, which could support this evolution.

With regard to our model of expressed idea, our results on the evolution support the following hypotheses according the frequency of the kinds of evolution chosen in our theoretical framework: (1) the most frequent kind of evolution is "increase the domain of validity of an idea", (2) the next frequent kind of evolution is to "express a new idea" and to "establish a link between several ideas", (3) students rarely use the evolution "decreasing the domain of validity".

Our results show that our characterisations of a teaching situation with the *milieu* and of student's evolution lead us to propose some specific hypotheses on the relations between teaching situations and learning: (1) the activity worksheets and the hands-on experiments support new ideas about the meaning of words (like macroscopic pressure); (2) the text of the gas model supports new ideas about the properties of a gas (i.e.: homogenous distribution, collision of molecules); (3) hands-on experiments supports the establishment of links between several networks of ideas; (4) the simultaneous use of social elements (especially students' discussions) and experimental elements (particularly hands-on experiments) support the decrease of the domain of validity. Moreover, our result concerning the most frequent kind of evolution - to increase the domain of validity - supported by any element (social, experimental, teaching material) of the *milieu* should also be further investigated to formulate more specific grounded hypothese.

4. INFLUENCES OF TAUGHT CONTENT ON STUDENT LEARNING IN QUANTUM ATOMIC PHYSICS

This study was a part of a project to evaluate the Bremen teaching approach in quantum atomic physics. Some very important qualitative results of Budde (2004) and Budde et al. (2002) could in fact be validated quantitatively by Niedderer and Deylitz (1999). The teaching approach used a more abstract probability density interpretation alongside with a visual charge cloud/electronium interpretation of the ψ-function for bound states of an atom (Herrmann 2000). Implicit in the approach to teaching the charge cloud/electronium model were a number of teaching hypotheses, which were developed from an analysis of previous research in various domains including quantum atomic physics.

When teaching the probability model two main learning problems emerge (e.g. Bethge 1992; Fischler & Lichtfeldt 1992; Mueller & Wiesner 1999):

- Students tend to retain their preconceptions (mainly planetary orbit or shell conceptions) or revert to their preconceptions after teaching, thus there is no long-term learning effect.
- Students construct alternative conceptions, which differ significantly from the intended models. In particular, they retain the belief that the electrons are moving on trajectories in the atom.

The introduction of the charge cloud/electronium model bases on the following starting hypothesis: The visual appearance of the charge cloud/electronium model may support the acceptance of a quantum atomic model, based on the Schrödinger equation.

Two students, Thomas and Klaus, were selected for a detailed analysis of their learning process, because additional data were available for them. In the following, some results concerning the starting hypothesis stated above are presented.

For Thomas the charge cloud/electronium model shows a congruent resonance from the early stages of teaching. After the pre-questionnaire, in which the models were introduced briefly, Thomas took the initiative in using the charge cloud/electronium model in an intuitive way.

Although Klaus the teacher preferred a probability model throughout the whole course, he also introduced the charge cloud/electronium model. After the probability and charge cloud/electronium models were discussed in detail, both Thomas and Klaus, and all their classmates, agreed that they preferred the charge cloud/electronium model. One repeatedly expressed reason, from both students, for this preference of the charge cloud/electronium model focussed on its substantial, visual appearance. Thus Thomas commented:

"Me too. I also rather prefer the model of Friedrich Herrmann. It is more descriptive. It is easier to imagine. In this model, the electron does not disappear and appear again without one knowing, how it managed this [like in the probability model]."

For all nine students in the class, the charge cloud/electronium model achieved a high acceptance. Furthermore both Klaus and Thomas were still able to outline the charge cloud/electronium model in the control-interviews two years after the end of the instruction, which proves the stability of the charge cloud/electronium conception.

There were two further aspects of the charge cloud/electronium model, which were frequently and spontaneously referred to by both students and thus constituted a strong congruent resonance for the students. The two aspects are:

- Concerning the charge distribution in the ground state (1s-state): The charge density is the highest at the nucleus and decreases to a higher radius.[3]

- Concerning the change of the charge distribution in the case of a transition between two stationary states: The charge will move away from the nucleus (the charge will be distributed more distant from the nucleus) if energy is added.

The majority of all students (17 of 26) mentioned spontaneously the decreasing charge density to a higher radius in the post-questionnaire. It is assumed that the high acceptance outcome results from the fact that the new ideas build upon students' preconceptions. Students gave many explanations for this preference of the charge distribution model.

The findings for Klaus and Thomas are summarised in Figure 3. They explained the charge distribution in terms of attractive or repulsive electrostatic forces. The students also used analogies between atoms and their ideas about the atmosphere or water, where the density decreases with height or increases with depth. The effect of the electrostatic force is seen as being equivalent to the effect of pressure, which is interpreted as compression: the higher the force or pressure, the more the substance is compressed, the higher is its density.

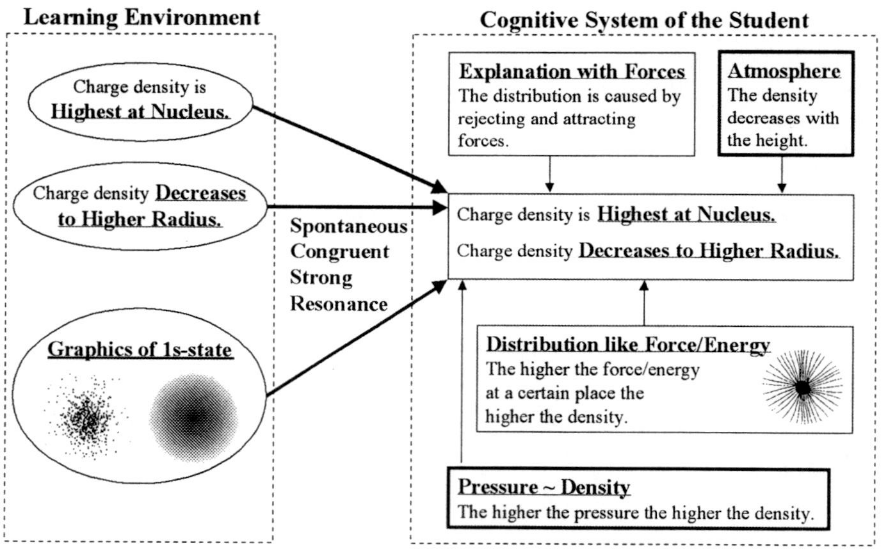

Figure 3. Examples of spontaneous congruent strong resonances (see part 2)

The main argument for the characteristic charge distribution given by the two students is, that it is simply logical that the charge is distributed like the electrostatic force or field.

These results lead to the following final grounded hypotheses:

- The visual appearance of the charge cloud/electronium model may support the acceptance of a quantum atomic model.

- The specific distribution of the charge density in the ground state and the increased radius of higher states may be plausible to students, which increases the acceptance of the new atomic model.

It appears that if the teaching focuses on the aspects that are plausible to the students (distribution in the 1s-state; the uptake of energy causes a more distant distribution) and easily to visualize (charge cloud/electronium), the students accept an atomic model that bases on the solutions of the Schrödinger equation and do not give up their new quantum atomic conception after instruction.

5. CONCLUSIONS AND DISCUSSION

Despite differences in theoretical and methodological aspects, it appears that researchers of different learning process studies share implicitly or explicitly certain assumptions that are reflected in the present and the previous studies. One assumption, which has already been stated, is that teaching does not necessarily involve learning by students; a second assumption is that advances in our understanding of teaching learning interactions can be considerably facilitated by developing models of the teaching and learning process; a third assumption is that such models can draw on individual students' conceptual evolution during a teaching learning sequence. The advantage of such a modelling process is two fold. At first, it provides for dynamic aspects of students' conceptual evolutions leading to detailed descriptions of different states and therefore to clarifications of what is involved in learning science in specific domains. Second, it relates such evolutions to certain aspects of the respective learning environment notably the local demands on students during the course of a teaching sequence. In other words, it allows for developing grounded hypotheses and humble theories about domain specific learning in relation to teaching. This allows for the development of local hypotheses concerning particular aspects of teaching leading not only to modelling of learning but to modelling of teaching, which at times is rather neglected in published studies.

Having said the above it appears that there are certain limitations, at least at present, which need to be dealt with in future research. One is the development of a common terminology, which is shared by researchers and applied to describe students' developments; a second is that the units of analysis of individual student development is subject to contextual constraints and a third is that the analysis and description of students' intermediate steps involves craft expertise by researchers, and therefore is difficult to be validated in different contexts. Such limitations have appeared in several published investigations in science education research. They are more pertinent though for learning processes studies aiming at gradually developing consensus models about teaching and learning.

Despite such limitations the presented studies provide considerable insights not only on students' learning but also on modifying the corresponding teaching in the domain of teaching gases and quantum atomic physics, as in previous topics. By modifying we mean either eliminating or constructing activities and steps, which are dynamically adapted to students' constructions. One consequence could be that certain teaching aims are not realistic and could be adapted to what seems learnable as stepping stones (Brown & Clement 1992; Tiberghien 1997; Psillos & Kariotoglou 1999). Another consequence would be to find hypotheses for learning process studies for larger groups (Clement & Steinberg 2002). Such information would be valuable for research-based design of teaching in a number of domains and for guiding future research.

A quite new research idea as introduced in this paper is to determine learning effects of special elements of the learning environment during a learning process and thus helping to improve learning environment by curriculum development.

NOTES

[1] The use of the term process here can be justified by stating that it describes a series of steps, in which students approach the scientific taught concepts; these steps are a student's way – learning process – of going from initial to final conceptions ("learning pathway").

[2] The teaching unit can be download at URL: http://pegase.inrp.fr/

[3] Although this description is correct for all states (if the region with nodal areas is disregarded), it is assumed that the students especially imagine the 1s-state. One indicator for this is that the students always draw the 1s-state when asked for their image of an atom.

REFERENCES

Benson, D. L., Wittrock, M. C. & Baur, M. E. (1993). Students' preconceptions of the nature of gases. *Journal of Research in Science Teaching*, 30, 587–597

Bethge, T. (1992). Schülervorstellungen zu grundlegenden Begriffen der Atomphysik. (In H. Fischler (Ed.), *Quantenphysik in der Schule* (pp. 215–233). Kiel: IPN.)

Brousseau G. (1998). *Théorie des situations didactiques*. (Grenoble: La Pensée Sauvage editions)

Brown, D. E. & Clement, J. (1992). Classroom teaching experiments in mechanics. (In R. Duit, F. Goldberg, H. Niedderer (Eds.), *Research in physics learning: Theoretical issues and empirical studies* (pp. 380–397). Kiel: IPN.)

Budde, M. (2004). *Lernwirkungen in der Quanten-Atom-Physik. Fallstudien über Resonanzen zwischen Lernangeboten und SchülerInnen-Vorstellungen. (Learning effects in quantum atomic physics – case studies on resonances between content-specific elements of the learning environment and the evolution of students' conceptions).* Doctoral dissertation University of Bremen. (Berlin: Logos)

Budde, M., Niedderer, H., Scott, P. & Leach, J. (2002). The quantum atomic model 'Electronium': a successful teaching tool. *Physics Education*, 37, 204–210

Clement, J. J. & Steinberg, M. S. (2002). Step-wise evolution of mental models of electric circuits: a "Learning-Aloud" case study. *The Journal of the Learning Sciences*, 11(4), 389–452

Duit, R. (2007). *Bibliography – STCSE (Students' and Teachers' Conceptions and Science Education).* (Kiel: IPN – Leibniz Institute for Science Education) Retrieved 2006 from IPN Web site: http://www.ipn.uni-kiel.de/aktuell/stcse/stcse.html

Duit, R., Goldberg, F. & Niedderer, H. (Eds.) (1992). *Research in Physics Learning – Theoretical Issues and Empirical Studies*. Proceedings of an International Workshop in Bremen. (Kiel: IPN)

Duval, R. (1995).*Sémiosis et pensée humaine, registres sémiotiques et apprentissage intellectuels [Semiosis and human thought, semiotic registers and intellectual learning].* (Berne: Peter Lang)

Dykstra, D. (1992). Studying conceptual change: Constructing new understandings. (In R. Duit, F. Goldberg, H. Niedderer (Eds.), *Research in physics learning: Theoretical issues and empirical studies* (pp. 40–58). Kiel: IPN.)

Fischler, H. & Lichtfeldt, M. (1992). Modern physics and students' conceptions. *International Journal of Science Education*, 14(2), 181–190

Givry, D. (2003). Étude de l'évolution des idées des élèves de seconde durant une séquence d'enseignement sur les gaz. (Study of the evolution of students' knowledge during a teaching sequence on gases at the upper secondary school level). Doctoral dissertation, Université Lyon II

Givry, D. & Roth, W. -M. (2006). Toward a new conception of conceptions: The interplay of talk, gestures, and semiotic resources in the setting. *Journal of Research in Science Teaching* 43(10), 1086–1109

Hartmann, S. & Niedderer, H. (2005). Parallel conceptions and learning in the domain of force and motion. (In K. Boersma, H. Eijkelhof, O. de Jong, M. Goedhart (Eds.), *Research and the quality of education* (pp. 471–482). Dordrecht: Springer.)

Herrmann, F. (2000). The Karlsruhe physics course. *European Journal of Physics*, 21, 49–58.

McDermott, L. C. (1991). What we teach and what is learned: Closing the gap. *American Journal of Physics*, 59(4), 301–315

Minstrell, J. (1992). Facets of students' knowledge and relevant instruction. (In R. Duit, F. Goldberg, H. Niedderer (Eds.), *Research in physics learning: Theoretical issues and empirical studies* (pp. 110–128). Kiel: IPN.)

Mueller, R. & Wiesner, H. (1999). Students' Conceptions of Quantum Physics. (In D. Zollman (Ed.), *Research on teaching and learning quantum mechanics*. Papers presented at the annual meeting of the National Association for Research in Science Teaching, Boston, March 1999 (pp. 20–22). Retrieved 2006 from http://www.phys.ksu.edu/perg/papers/narst/)

Niedderer, H., Goldberg, F. & Duit, R. (1992). Towards learning process studies: A review of the workshop on research in physics learning. (In R. Duit, F. Goldberg, H. Niedderer (Eds.), *Research in physics learning: Theoretical issues and empirical studies* (pp. 10–28). Kiel: IPN.) (Retrieved 2006 from http://www.idn.uni-bremen.de/pubs/Niedderer/1992-HB-HNFGRD.pdf)

Niedderer, H. & Deylitz, S. (1999). Evaluation of a new approach in quantum atomic physics in high school. (In D. Zollman (Ed.), *Research on teaching and learning quantum mechanics*. Papers presented at the annual meeting of the National Association for Research in Science Teaching, Boston, March 1999 (pp. 23–27). Retrieved 2006 from http://www.phys.ksu.edu/perg/papers/narst/)

Petri, J. & Niedderer, H. (1998). A learning pathway in high-school level quantum atomic physics. *International Journal of Science Education*, 20(9), 1075–1088

Petri, J. & Niedderer, H. (2003). Atomic Physics in Upper Secondary School: Layers of Conceptions in Individual Cognitive Structure. (In D. Psillos, P. Kariotoglou, V. Tselfes, E. Hatzikraniotis, G. Fassoulopoulos, M. Kallery (Eds.), *Science education in the knowledge-based society* (pp. 137–144). Dordrecht: Kluwer Academic Publishers.)

Psillos, D. & Kariotoglou, P. (1999). Teaching fluids: Intended knowledge and students' actual conceptual evolution. *International Journal of Science Education*, 21(1), 17–38

Taber, K. S. (2000). Multiple frameworks? Evidence of manifold conceptions in individual cognitive structure. *International Journal of Science Education*, 22 (4), 399–417

Taber, K. S. (2001). Shifting sands: A case study of conceptual development as competition between alternative conceptions. *International Journal of Science Education*, 23(7), 731–754

Tiberghien, A. (1997). Learning and teaching: Differenciation and Relation. *Research in Science Education*, 27(3), 359–382

Tiberghien, A. (2000). Designing teaching situations in the secondary school. (In R. Millar, J. Leach, J. Osborne (Eds.), *Improving science education: The contribution of research* (pp. 27–47). Buckingham, UK: Open University Press.)

von Glasersfeld, E. (1992). A Constructivist View of Learning and Teaching. (In R. Duit, F. Goldberg, H. Niedderer (Eds.), *Research in physics learning: Theoretical issues and empirical studies* (pp. 29–39). Kiel: IPN.)

CARLOS REIGOSA AND MARÍA P. JIMÉNEZ-ALEIXANDRE

13. MEANING CONSTRUCTION AND CONTEXTUALIZATION WHILE SOLVING A DYNAMICS TASK IN THE LABORATORY

Abstract: This paper presents a case study with a group of 11th grade students solving a problem about the centripetal force that an object needs in order to follow a circular motion. The focus is on the process of development by the students of (a) the meanings in terms of semantic relationships among concepts and of these with the context; and (b) a purposeful heuristic during the process of solving the task. This process is examined in terms of contextualizing practices (Lemke, 1990; Jiménez-Aleixandre & Reigosa, 2006), creation of meanings through the connections established among actions and their context. The results are interpreted as transformations by the students of concepts as centripetal force or speed into knowledge tools for solving the problems

Keywords: Contextualizing practices, Dynamics, Heuristics, Laboratory, Meaning-making

1. INTRODUCTION: HOW ACTIONS BECOME MEANINGFUL IN A CONTEXT

The goal of this study is to document and analyze the meanings constructed by a group of 11th grade students (16–17 years) while solving in the laboratory a task part of a sequence of seven physics and chemistry open problems, and how these shared meanings are used in order to solve the task. As a frame for the analysis we draw on Lemke's (1990) notions of contextualizing practices and indexical context, and on Collins et al.'s (1989) perspective of cognitive apprenticeship.

The cognitive apprenticeship perspective assumes that learning is a process of construction highly tuned to the situation or context in which it takes place: learning is situated. As a consequence, Collins et al. suggest that the learning tasks should be related to the tasks performed by practitioners in the field – scientists – requiring from the students active integration and adequate application of skills and conceptual knowledge. This means that laboratory tasks oriented to mere illustration of a theory or to follow instructions are not adequate or, in the terms of Brown et al. (1989), are framed in the school – not in the scientific – culture. These authors draw a distinction among authentic activities, framed in the culture of a domain, and archetypal school activities, implicitly framed by one culture – school – while attributed to another – scientific. In all the seven assignments the students were

Carlos Reigosa, High School Fuentesnuevas, c/ Las Dehesas 31, 24411 Fuentesnuevas, León, Spain

R. Pintó and D. Couso (eds.), Contributions from Science Education Research, 173–185.

given an open problem, which required them to take epistemological and procedural decisions, integrating theoretical and methodological tools, practices belonging to the scientific culture. The distinction between scientific and school cultures can be viewed through the notion of procedural display (Bloome et al., 1989), belonging to school culture, which is the enactment by teacher and students of procedures that count as the accomplishment of a lesson, although not necessarily related to learning, but rather to the set of cultural meanings and values held by the local education community. So a lesson could take place without involving real learning, if it is constructed as an event with the cultural meaning of a lesson or, in our case, laboratory session.

The cognitive apprenticeship perspective emphasizes the relevance of the social context in the development of the practices of a culture, paying attention to the participants' interactions. Wells (1999) and Mortimer and Scott (2003) point to language as the most powerful social tool in meaning-making, used by teacher and students to structure the development of ideas; for Mortimer and Scott meaning-making is a dialogic activity. For Tobin et al. (1997) all students in a community of practice should have access to and should be able to appropriate a shared language to communicate with one another. In a study about the creation of an intellectual community in a classroom, Kelly et al. (2000) explore how the learning of science was constructed as a social accomplishment, as the intellectual community – created through the social practices that came to define science for these classroom members – afforded opportunities for students to access particular features of science. In the present study we explore some of the social practices that came to define science in a laboratory setting in this particular classroom.

A frame of discourse analysis coherent with a social perspective about learning is Lemke's (1990), which assigns a central role to language in use in a community. Belonging to a community involves the appropriation of its language, its resources system to create meanings, its semiotic system. The objective of this study is to analyze the process of meaning construction, and for it we have used Lemke's notion of *contextualizing practices*. For Lemke (1990) meanings are made or constructed by people, rather than being something 'built-in' in words or actions. He defines meaning-making (semiotic) practices as the ways in which members of a community perform meaningful actions. Actions are made meaningful, through contextualizing practices, by placing them in a context: "Making meaning is the process of connecting things to contexts" (Lemke, p. 187). The students' actions in the laboratory are connected to the indexical contexts or social contexts associated with an action in a particular community. For Lemke what matters in indexical contextualization is the pattern of actions, the systematic relations of actions to each other and to other sort of contexts, the particular ways that members of a social group share of talking about a subject, constructing in their speech thematic patterns, specific relationships among meanings or semantic relationships.

In Jiménez-Aleixandre & Reigosa (2006) we have used this frame to explore the construction of meanings for the concepts of concentration and neutralization, through connection to the social context of a school laboratory, the transformations from definitions to tools used to perform a titration. In the present study our foci are two dimensions of the indexical context, here the social context of the school laboratory, on the one hand the construction by the students of the semantic relationships among concepts needed to solve the task; on the other the development by the students of an heuristic. According to Jiménez-Aleixandre and Díaz (1997), the heuristics are strategies put into practice to solve a problem, the intentional planning and performing of sequences of actions directed to a goal. We take the discussions about the use of equipment and of other resources as books or notebooks, and their use itself, as part of this dimension.

The purpose of the study is to document these processes of development by the students, not to compare the task with other laboratory settings, but to progress toward the understanding of the difficulties experienced by students in open tasks. Although open tasks and problems had been advocated in the literature as being coherent with scientific culture, with grounds in learning theory or epistemology, more studies are needed about the difficulties entailed by their implementation in actual classrooms. In particular we analyze:

- How do students build semantic relationships among the concepts, through contextualizing practices, transforming them from definitions to usable tools, in the setting of an open task?
- How do students progress in the design and implementation of a heuristic, connecting it to the purpose of the task?

2. METHODS, PARTICIPANTS, TASK

The participants are a group of four 11th grade students and their teacher (first author). They are identified by pseudonyms respecting the gender. The task required them to study empirically the relationship among centripetal force and both the speed and the radius of an object following a circular motion. They were not given particular steps and they spent four lab sessions (see in annex 1 the handout given to the students). There was only one video camera placed on a bench, and the students who chose to sit there were asked to be recorded, which they accepted, so they were not selected by the teacher-researcher.

The task could be viewed as pertaining to the scientific culture, being oriented to increase their understanding about a certain phenomenon, for example circular motion. Some of its features are the need for students to take methodological and epistemological decisions, as scientists do; to design themselves the path for its completion, to carry it empirically, or its openness (in the sense of not requiring a fixed sequence of steps to complete it).

The equipment provided included a circular platform that could rotate and that was connected to a small electric engine through a rubber band. They were also given dynamometers, rules for measuring the radius, clamps, holders, etc. The students were expected to assemble the equipment as they did it (Figure 1), or in any other way allowing to measure the centripetal force in relation to v and r. The students had worked in the classroom, both in grades 9 and 10 and in the previous lessons, with

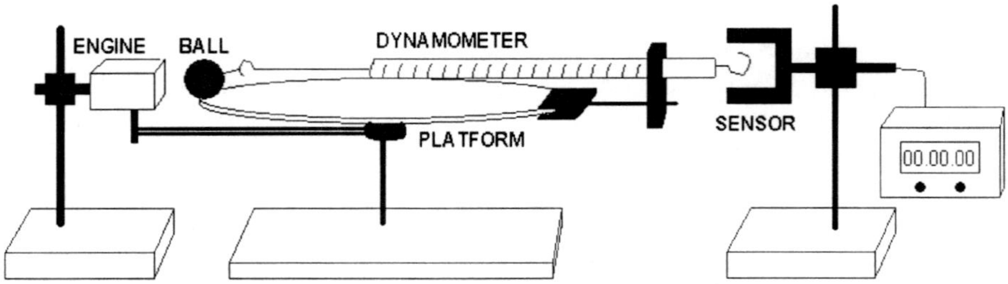

Figure 1. System built by the students

kinematics and dynamics, with the magnitudes used to describe circular motion (angular speed and its relation to linear speed, or angular, tangential and normal acceleration), as well as with the formula $F_c = mv^2/r$ and its connection to the first and second Newton's laws.

The teacher's purpose was that they would assume an increasing responsibility in the process of solving the task (Reigosa & Jiménez-Aleixandre, 2007), so he observed their progress, noting their difficulties and when they were not able to proceed, he helped them to master the resources needed to carry the task, either intellectual – as centripetal force, speed – or material – as dynamometer, branching the engine, etc., but avoiding giving the students precise instructions about the steps to follow.

Data collection involved the audio and videotaping of the students and a reflective journal kept by the teacher-researcher. The analysis focuses on the data from the audio and video transcriptions, and the interpretations of data were validated between both researchers. The categories for the analysis emerged from the interaction with the data (Lincoln & Guba, 1985), so the analysis was not oriented to confirm a priori hypothesis, but rather an iterative process of data immersion by both authors and negotiation of the interpretations.

The analysis focuses on the meaning of the actions and dialogues of the students, in connection to the context, and how it changed, because of the relevance of the language in the culture of a community. The purpose was to arrive at a detailed knowledge of how dialogue allowed the participants to make discourse and thinking progress. Conversations and events in classrooms and laboratories may seem confuse, but detailed analysis can uncover complex reasoning. With the objective of studying the discourse – not only from the microanalysis perspective, focusing on instances of particular dimensions, but also from the perspective of the bigger phases or steps in the process of solution – each session was segmented into episodes. Segmentation is a problematic issue in verbal data analysis, as units of meaning may not have definite boundaries. Drawing from Gee (1999), we have used two concurrent criteria in the segmentation of sessions into episodes: activities performed or activity building, that is what activity is going on, and/or theme or topic discussed, in other words what situated meaning is constructed. A set of turns is considered a different episode either when students are engaged in a different activity and/or when they discuss a different topic or theme.

The construction of the semantic relationships and the development of heuristic are discussed in the results section with instances of some dimensions in the four sessions. The physical actions are in *italics* and comments from the authors in `courier`; and = indicates overlap of speakers' utterances. Some turns have been omitted to reduce the length.

3. RESULTS

3.1 Building Semantic Relationships between the Concepts

In order to be able to perform the task, the students need to build appropriate relationships among specific concepts, for instance circular motion, centripetal force or speed. These concepts and relations are needed to give meaning to students' actions in assembling the equipment and carrying the experiences. Although the concepts had been introduced in the classroom, our hypothesis is that the students may know the literal definition, but they need to construct a more complete meaning for them, connected to the experimental situation. This is apparent at the beginning of session 1, when they are trying to grasp the meaning of centripetal force.

Session 1, episode 4 Discussing concepts: circular motion, centripetal force

270	Pedro	In a curve yes, but there is a force pointing outwards, to begin with it is what... =what= *(pointing to the circular platform)*
271	Paula	=Go ahead=
272		*(Pause)*
273	Pedro	At least yes, centripetal, centripetal, pulling inwards
274	Pili	We don't know ... [`inaudible`] centripetal force directed inwards [`doubtful intonation`]
275	Pedro	No, here (*points to the circular platform*) it doesn't make sense
276	Pili	But =it is=
277	Pedro	=Here the= force that goes it is outwards, like that
		(...)
315	Paula	Here [`notebook`] the normal force depends on square vee and on er

Paula (turn 315) seems to be appealing to previous knowledge, by means of a notebook, reading the formula for centripetal force. But, although the recovered knowledge is correct, it seems not a significant definition of these concepts, as they are not placed in a conceptual structure related to Newton's laws. This is shown in Pedro's confusion (turns 270, 275, 277) about the force, shifting in considering whether it was directed outwards, an incorrect idea, or inwards. Later, in episode 7 they have problems with the relationships among centripetal force and speed. They place a spherical weight in the platform without fastening it, and when they switch on the engine, the weight falls.

Session 1, episode 7, Performing first actions: F_c relation to v

561	Pedro	(*switches the engine on at high speed and the weight falls*)
562	Pedro	Do you see? Outwards. It threw them out. It doesn't hold them
563		(*the students laugh*)
564	Pedro	I thought it was, it was the centripetal force
592	Pedro	That thing ... [the rotating platform] place an object like that, do you see? Knowing the speed... it pulls stronger the outward force than the inward force, so... place it... it pulls a little, doesn't pull enough (*he places a weight on the platform*) (...)
637	Pedro	We'll ask him [the teacher] and see what does he say
638	Paula	[centripetal force] it is em by square vee equals... square vee we can have it. The speed we have to see what it is... then we have all of them. How was it? Could it be... the mass

Pedro (592) interprets the falling of weights in terms of forces and speed, but he does not establish adequate connections between speed and centripetal force, to which he still refers to as "outward" force. Later Paula recalls the definition of centripetal force, but they do not consider measuring it simultaneously with speed, what is required for the task as both change together. In subsequent episodes they discuss kinematics magnitudes as angular speed and its measure, without noticing this relationship. Our interpretation is that, at this stage, they view the relationship among F_c and v as a formal one, part of a formula.

The building of semantic relationships needed to measure speed continues in other episodes and in the second session when Paula is explaining what they are planning to do.

Session 2, episode 4, Designing and planning how to measure speed

383	Paula	Could we do it with the formula speed equal to space by time? And then in a circumference, we look how many centimetres are they and we divide it by... by the time spent in taking a round and that gives you the speed
384	Pedro	[to the teacher] We need in order to measure...
385	Teacher	Yes
386	Teacher	(*goes to find a chronometer*)

Paula relates speed to space covered and time, part of the concepts making part of the thematic context of the task. These lines are also related to heuristics, what equipment to use, and Pedro asks for a chronometer. The process of building specific relations between concepts continues in the third session: in the second episode they talk with the teacher about the centripetal force and the problem of the falling weights.

Session 3, episode 2, Which one is the centripetal force in this case?

117	Teacher	Look: Why this one [`flat weight`] doesn't fall? [`when it is on the platform and this is rotating`]
118	Pedro	Because there is a centripetal force pointing inwards
119	Teacher	And which one is the centripetal force?
120	Teacher	(*places a spherical weight over the circular platform, switches the engine on, the ball falls and he catches it in the air*) And why does that one [`the ball`] fall?
122	Pedro	Because it has no friction... it rolls
123	Teacher	OK. Because there is a centripetal force, the friction and with this one [`the ball`] there is not, or if there is, it is not enough. OK? So for something not to fall, there must be a centripetal force directed inwards, because otherwise it tends to follow their motion in a straight line. And it would fall.

The teacher, realizing that they have problems with the meaning of centripetal force in this situation, asks why a flat weight does not fall and the ball does, to which Pedro (118, 122) answers with an appropriate interpretation in terms of centripetal force and friction. This shows how the construction of the semantic relationships is a process involving different stages, and one step may be being able to use the concepts as tools to interpret physical phenomena. They are beginning to use, with the teacher's help, some concepts as centripetal and friction forces to interpret empirical situations, and overcoming confusions, as Pedro's about centripetal force. In the next episode, they plan the measurements:

Session 3, episode 3, Deciding how to measure the centripetal force

178	Pili	It was better with a round weight [`than with a flat weight`], hey
179	Paula	It is the same
180	Pili	But this one [`the flat weight`]... being still it is not going to pull [`of the dynamometer`]
181	Paca	Over it I believe it is going
182	Pili	No, leave it separate. And this like that: What? What are you going to measure?
		(...)
226	Pedro	Sure, it [`the flat weight`] is not going to move from here because... the force of friction is bigger than that one. Then it is not going to move

The students are beginning to realize that the flat weights are not appropriate for measuring the centripetal force. Pili says it in turn 180, even if she does not mention the force of friction acting as centripetal force, but she realizes that the centripetal force is not to be exerted by the dynamometer, so the measurement cannot be taken. Several turns later Pedro (226) explains it in terms of the force of friction. That had been discussed before with the teacher, although at that moment they did not realize its relevance for the use of equipment, as they do now.

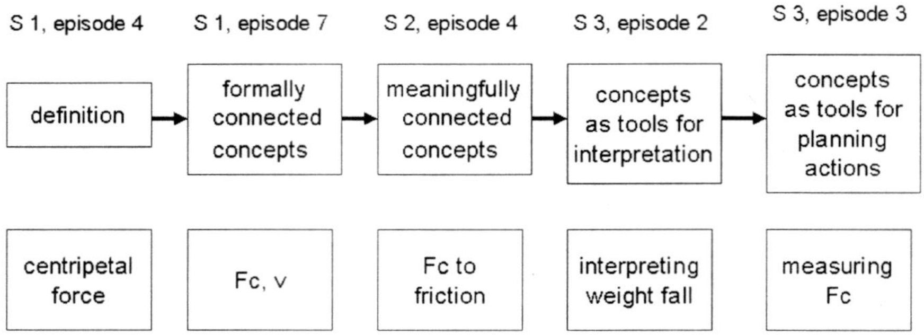

Figure 2. Changes in the meaning of the concepts

In order to be able to perform the task, the students need to develop a conceptual frame that they can use as a tool to take decisions. This conceptual frame has to include appropriate meanings for centripetal force, friction force, circular motion, speed, space and time and their relationships. These meanings change during the process, they know some relevant literal definitions, circular motion, centripetal force, friction or Newton's first law, but at the beginning they experience difficulties with their meaning and establishing appropriate relations among them, for instance the requirement of a centripetal force (directed inwards) for a circular motion to take place. These semantic relationships are also connected to physical actions performed during the task. One instance of the difficulties in connecting concepts between them and with the empirical actions could be seen when they use flat weights, seeming unaware that they would cause a force of friction interfering with the measurement of centripetal force. Sometimes they seem to understand the concepts, but not their relationships in a scheme or network. Later, they use these two concepts, friction and centripetal force, to *interpret* the different behaviour of flat and spherical weights, but they do not use them as a context into which to insert *decisions and actions* directed to complete the task until some time after. This process is summarized in Figure 2.

3.2 Progress in the Design and Implementation of a Heuristic

To be able to develop a heuristic, to connect it with the purposes of the task, the students need to appropriate the thematic pattern, using the concepts as tools for planning actions. The process is holistic, so discussing different dimensions separately does not suggest that they are constructed independently; it is just a matter of systematizing. But there are also other aspects in their performances that are relevant for the heuristic development, for instance the views that the students have of their own role in the school laboratory. For instance, at the beginning the students were just manipulating the equipment without any apparent purpose and episode 3 in session 1 is the first attempt to relate the equipment to the purpose of the task.

Session 1, episode 3, Handling equipment, attempt to restate problem

222	Pedro	(*manipulates an utility holder*)
223	Paula	Put it there and then hang it
225	Pedro	(*hangs tongs from the utility holder*)
228	Pili	Yes, I mean, wait: What do you want this hanging there for?

They manipulate the different pieces of equipment in these turns (and in previous ones not reproduced), but Pili's (228) question points to the lack of relation among this random assembling and the purpose of the task. This random assembling can be interpreted as procedural display (Bloome et al., 1989). Pili is beginning to acknowledge that they need to insert the actions in some context, that the actions should have a meaning, serve for the purpose of the task. This step means a change, both in their own role, assuming responsibility, and in the conception of the heuristics itself, from a sequence of instructions whose meaning does not need to be questioned, to actions decided by the students themselves, with a clear purpose. It can be considered as a move from a stereotyped school culture to a scientific one (Brown et al., 1989). Further steps in the construction of appropriate heuristics can be seen in the second session, as in episode 3 when they discuss about measuring speed.

Session 2, episode 3, Discussing which magnitudes to measure

198	Paula	This [speed] is equal to space divided by time
		(...)
225	Paula	(*reading a notebook looking for the formula of speed*)
230	Pili	Yes, space: And that what do you put?
231	Paula	The whole circumference
		(...)
254	Paula	Then... but look in the notebook
255	Pedro	Two pi [multiplied] by the radius I believe it was, or was it not?
257	Pedro	Two pi by the radius. The radius even if you don't have it, measure it, see if it you do it well... Two pi by the radius... ten point three

The students are looking for information in the books and notebooks, which constitutes not a frequent pattern in standard laboratory work, when students mainly follow instructions, but is not so uncommon in open laboratory tasks, when students need to relate the actions to theory (Jiménez & Díaz, 1997). For the calculus of speed, that cannot be measured directly, they decide to measure the two magnitudes that can be, space and time. Then they solve the problem of how to measure the space covered. The heuristics dimension is intertwined with the conceptual dimension, because they need to use appropriately concepts such as speed or space.

There are instances that highlight the interactions between the heuristic and the students' epistemological commitments, which according to Hewson (1985) are:

"the standards which a person holds which he or she uses to judge knowledge. Before such a person is prepared to trust knowledge, react to it, or depend on it, he or she has to see that it meets certain criteria".

The epistemological commitments are their implicit theories about how scientific knowledge is constructed, what is scientific work, or the purpose and methods of experimental work (Sandoval & Morrison, 2003). For contextualizing practices, for connections between their actions and the decisions in an experimental procedure, this dimension is important as it influences, for instance, the quality of the data obtained by the students. Two instances are reproduced, when they are planning the measure of speed.

Session 2, episode 4, Designing and planning how to measure speed

296	Pedro	Another question, to do several turns [`of the platform`], in order to diminish the error, because we are going to have the angle higher
297	Pili	But it... that... And can you know the turns that it takes in this direction?
298	Pedro	Yes, you could know it, man
299	Pili	And how is it... Set it going a little, to see how...
300	Pedro	Plug in [`the engine`] there, Paca
301	Pili	It has to go very slowly

They are trying to measure the speed, and they are even capable to propose actions to improve the quality of the data obtained (turn 296), diminishing the error by measuring bigger magnitudes. This is related to the epistemological dimension, showing a commitment to obtain data of better quality. Another epistemological commitment shown by the students is the recognition of the necessity of rigour in the control of variables.

Session 2, episode 4, Designing and planning how to measure speed

342	Pedro	Although you don't touch the equipment, sometimes it [`the speed`] is higher
343	Paula	How?
349	Pili	It [`the engine`] has just accelerated
350	Paula	But we look...
351	Pedro	Yes it even changes alone, without touching it
354	Paula	But if we look only one turn [`of the platform`] I don't think that the speed changes too much

As they detect some irregularity in the movement of the engine, Paula (turn 354) proposes to measure the time of only one round (opposite to the proposal of Pedro in 296) to make sure of having a motion closer to a circular uniform one. They worry about changes in the variables and take decisions to control them. To have meaning, the actions of the students need to be inserted in a context in which the epistemological aspects are taken into account. In episode 5 they encounter a problem related to the engine, which sometimes gets locked and at low speed does not go smoothly, so they decide that the speed has to be higher, subsequently finding it difficult to chronometer.

Session 2, episode 5, Measuring speed empirically

519	Paula, Pili	(*Paula switches on the engine and Pili is ready to chronometer, but the motion is too fast; the students laugh*)
521	Pedro	Don't you see it?

522	Paula	How could one see?
523	Pedro	It is not so easy, wait let's see even if here... but we should put it
524	Pili	And there is not something to count the turns? ... like the sensors there... the time it took a turn, to count

They knew the sensors and used them in a previous task, and we interpret that the epistemological commitment to obtain reproducible data leads to asking for appropriate material resources. Using better equipment in connection with the requirements of the process is an action belonging to the scientific culture. These instances show the relevance of the epistemological dimensions for the heuristic, both dimensions being intertwined.

4. CONCLUSIONS AND EDUCATIONAL IMPLICATIONS

This chapter explores the shared meanings constructed by students solving a problem in the laboratory. This process is interpreted as contextualizing practices (Lemke, 1990), construction of meanings for the concepts of centripetal force, speed or force of friction, establishing relationships among them, and connections to their context of use, an empirical situation in the laboratory. The progression of the students is studied as they attempt to connect the empirical data with the thematic pattern that they are building: for instance time and radius measurements to the concept of speed. This contextualization is a slow process, and it experiences advances and steps back either in aspects as semantics relationships among concepts, for instance centripetal force and speed, or heuristics, for instance decisions about appropriate use of equipment. In the frame of Lemke's (1990) notion of contextualizing practices, we interpret the performances of the students as a process of building meanings; in other words, we do not see the concepts of centripetal force and its relationship to speed and radius as something meaningful for the students, that they only have to put into practice, but as *meanings being constructed* and reconstructed along the process.

About the semantic relationships among concepts, we have identified a progression in its construction: a concept as a mere definition, formal connections among concepts, meaningful connections, and finally concepts used as tools for interpreting empirical situations and for taking decisions and actions (summarized in Figure 2). What we mean by saying that the meanings of the concepts have changed is that the students can view them as knowledge tools for solving empirical (or theoretical) problems, as explored in another paper (Jiménez-Aleixandre & Reigosa, 2006) for titration. These are only a few examples and we need more studies to explore this dynamic construction.

The concepts' relationships are related to the development by the students of a heuristic, or intentional strategies put into practice to achieve a goal. We found that the heuristic for this task differs from the usual ones in standard labwork, when students follow instructions. Open tasks require the students to adopt an active role, to understand the motivation of the actions performed. In this case the differences with standard performances are shown for instance in the use of books and notebooks by students, an evidence of connections among theoretical and empirical knowledge. In the development of the heuristic a progression can be seen from initial random behaviour as handling equipment without purpose, to acknowledging that the empirical actions need to be inserted in an appropriate context.

The importance of other dimensions has to be noted: this task gives the students the opportunity for using epistemological commitments adequate for scientific work such as rigour in the control of variables and search of quality data.

In this study, the application and further development of Lemke's ideas allowed us an interpretation of the students' activity that may be useful to understand the difficulties that they experience, a requirement to give them an assistance adjusted to their needs. This help is necessary, as it is shown for instance in session 3, episode 2. An educational implication would be that even laboratory experiences conceived as "illustration" of theory are not a one-step process: without precise instructions or teacher's support it takes a long time for students to give the concepts the appropriate meanings.

The need for open tasks in the laboratory has been advocated as being more coherent with scientific culture. The results of this study point that to engage in them the students need an adequate indexical context, as a frame to which they can meaningfully connect their actions. We do not see this frame as mere conceptual lenses through which they would interpret the results, but rather as combining several dimensions that need to be developed by the students as, on the one hand, the contextualization of concepts until they are transformed (for the students) in cognitive resources or tools appropriate for the task, which requires to construct semantic relationships among them; on the other hand to develop heuristics, intentional strategies, planning and performing actions, using material resources or equipment. Intertwined with these dimensions are the students' epistemological commitments and their views about the construction of scientific knowledge. To engage in open tasks some changes are needed, for instance related to the students' role in the laboratory, assuming responsibility for the actions. The complexity of these dimensions poses difficulties, so we consider that the teacher's assistance is needed, creating an environment that would motivate students to take an active role, to take decisions and to assume the responsibility in their own learning.

ACKNOWLEDGEMENTS

Work supported by the Spanish Ministerio de Educación y Ciencia (MEC), grant SEJ2006-15589-C02-01/EDUC, partly funded by the European Regional Development Fund (ERDF).

REFERENCES

Bloome, D., Puro, P. & Theodorou, E. (1989). Procedural display and classroom lessons. *Curriculum Inquiry*, 19, 265–291

Brown, J.S., Collins, A. & Duguid, P. (1989). Situated cognition and the culture of learning. *Educational Researcher*, 18, 32–42

Collins, A., Brown, J.S. & Newman, S.E. (1989). Cognitive apprenticeship: teaching the crafts of reading, writing and mathematics. (In L. Resnick (Ed.), *Knowing, learning and instruction. Essays in honour of Robert Glaser*, pp. 453–494. Hillsdale: Lawrence Erlbaum.)

Gee, J.P. (1999). *An introduction to discourse analysis: theory and methods* (London: Routledge)

Hewson, P.W. (1985). Epistemological commitments in the learning of science: examples from dynamics. *European Journal of Science Education*, 7, 163–172

Jiménez-Aleixandre, M.P. & Díaz, J. (1997, March). *Analysing classroom discourse: practical work in the biology laboratory*. (Paper presented at the AERA meeting, Chicago)

Jiménez-Aleixandre, M.P. & Reigosa, C. (2006). Contextualizing practices across epistemic levels in the chemistry laboratory. *Science Education*, 90, 707–733

Kelly, G.J., Brown, C. & Crawford, T. (2000). Experiments, contingencies, and curriculum: providing opportunities for learning through improvisation in science teaching. *Science Education*, 84, 624–657

Lemke, J. (1990). *Talking science: language, learning, and values*. (Norwood: Ablex)

Lincoln, Y.S. & Guba, E.G. (1985). *Naturalistic Inquiry*. (Newbury Park: Sage)

Mortimer, E.F. & Scott, P. (2003). *Meaning making in secondary science classrooms*. (Maidenhead: Open University Press)

Reigosa, C. & Jiménez-Aleixandre, M.P. (2007). Scaffolded problem-solving in the physics and chemistry laboratory: difficulties hindering students' assumption of responsibility. *International Journal of Science Education* (in press). DOI: 10.1080/09500690600702454

Sandoval, W.A. & Morrison, K. (2003). High school students' ideas about theories and theory change after a biological inquiry unit. *Journal of Research in Science Teaching*, 40, 369–392

Tobin, K., McRobbie, C. & Anderson, C. (1997). Dialectical constraints to the discursive practices of a high school physics community. *Journal of Research in Science Teaching*, 34, 491–507

Wells, G. (1999). *Dialogic inquiry. Toward a sociocultural practice and theory of education*. (Cambridge: Cambridge University Press)

ANNEX

The Task Circular Task Motion

The study of circular motion is interesting both from a theoretical perspective and from a practical one, since many appliances follow circular motion: the drum of washing machines, wheels, mixers, etc. In order for a circular motion to take place, it is required that a centripetal force is directed inwards. The analysis of that centripetal force is important for an adequate design of household appliances that involve circular motion. In this task you have to study empirically the relationship between the centripetal force and the speed and radius of the movement, for an object rotating in a horizontal surface.

14. DEVELOPMENT OF A MODEL OF FORMATIVE ASSESSMENT

Abstract: This chapter describes the formulation of a model of formative assessment as it has emerged through a series of collaborative research and development activities with teachers. The aspiration is to articulate our emerging understanding of a complex set of interacting factors in an integrated and coherent manner. Though necessarily tentative, the model will serve to guide our future research activities as well as offering a perspective on the research of others. A specific approach to conceptual change is at the heart of the model, which attempts to extrapolate from constructivism as a theory of learning to its applications to the practice of teaching. Our approach integrates domain-specific elements – hypothesised conceptual trajectories that re-occur in science learning – with domain-general elements. The latter describe recurring cycles of interaction between teachers and pupils that might occur in any curricular domain. Some of the cost-benefits to teachers in their approach to the implementation of formative assessment practices are discussed. It is suggested that an implication of the suggested approach is the need to conduct and share more empirical research into pupils' conceptual development, nationally and internationally

Keywords: Conceptual change, Formative assessment, Pedagogy, Science education

1. INTRODUCTION

This paper does not describe a single piece of research but reviews the tentative emergence, from a series of studies, of a model of formative assessment in science classrooms. The evidence base drawn upon is some years of work between the authors and teachers as collaborative researchers. We attempt to set out the major features of formative assessment in such a manner that the research enterprise can be pursued systematically on the basis of this analysis.

The activity of 'formative assessment', in an educational context, has the purpose of collecting feedback on the process of teaching and learning that is useful in shaping the further actions of teachers and pupils. 'Formative assessment' and 'assessment for learning' are used synonymously and interchangeably. Black et al. (2003) point out that there is a frequent misunderstanding that formative assessment refers to any assessment carried out by teachers, but emphasise the *purpose*:

"Assessment for learning is any assessment for which the first priority in its design and practice is to serve the purpose of promoting pupils' learning." (*op. cit.*, p. 3).

Terry Russell, CRIPSAT, University of Liverpool, 126 Mount Pleasant, Liverpool L69 3GR, U.K.

R. Pintó and D. Couso (eds.), Contributions from Science Education Research, 187–198.

Formative assessment is something done *with* pupils rather than *to* them. Involving them in the process is critical to negotiating personal learning targets, to reviewing where they are in relation to those targets and what they have to do in order to progress towards the desired outcome. Summative assessment, by contrast, generates outcomes that inform a statement of achievement when the teaching-learning episode is completed. One and the same assessment tool may have the potential to inform *either* formative *or* summative descriptions of learners' achievements. It is the *purpose* of the assessment that varies. For summative purposes, the question to be answered will be, 'Pass or fail?', 'What grade has been achieved?', or 'How well did this student perform?' For formative purposes, the assessment results must invite answers to the question, 'What teaching-learning experience comes next?' The desirability of promoting such an approach was confirmed by Black and Wiliam's (1998) meta-analysis which identified significant enhancement of pupil achievement associated with formative assessment practices.

2. MODE OF ENQUIRY

Bruer (1996) points to the importance of making explicit the links between instructional innovation and underlying theory.

> "Without theory or first principles, instructional innovations can be divorced from their purpose and run the risk of becoming meaningless, superficial rituals."

Drawing on research situates the model and the assessment practices it describes within an explicit set of theoretical relationships and also sheds light on the practical realisation of the model in classrooms. We are aware of other approaches to modelling formative assessment, (Wilson and Sloane, 2000; Black and Wiliam, 2006). Our formulation shares some assumptions with the BEAR assessment system, in particular, the developmental perspective and the principle that assessment and teaching should be "seamlessly integrated".

> "That is, assessment is not merely tacked on at the end of instructional units, but is embedded in normal classroom activity and may even be, from the student's point of view, indistinguishable from instruction." (Kennedy, 2005, p. 4)

We differ in that we attempt to use assessment in a more diagnostic manner, rather than pursue a more formal, psychometric model. Drawing on our own research (Russell et al., 1998) and the body of constructivist and conceptual change literature (Driver, 1985), several iterations have led to the current relatively robust realisation of the model. The model has gradually evolved with the accumulation of evidence assembled in the course of research with teachers into the practices associated with constructivist teaching and learning.

A particular feature of our model is its attempt to incorporate in a specific manner the notion of conceptual change and development. Our approach attempts to reconcile constructivism as a theory of learning with its application to the practice of teaching. One specific and tangible driver has been our attempts at the development of materials to support teachers' classroom use of formative assessment. The approach integrates

i) domain-specific elements, in the form of hypothesised conceptual trajectories or 'Teaching and Learning Sequences';
ii) a domain-general element, describing recurring cycles of interaction between teacher and pupils.

Because the pedagogical processes we wish to discuss are complex, we use the mechanical metaphor in Figure 1 to support our description of the structures and processes involved.

A pupil's conceptual progression is represented by the linear movement of the rack while the teeth of the pinion stand for the recurrent phases within teacher-driven formative assessment cycles that are critical to the process of teaching and learning.

The teeth on the linear rack can be thought of as representing key points of understanding in a science curriculum and in an individual's knowledge and understanding of key science concepts within any given domain.

As with any model, there are limitations in this representation. Of course, formative assessment cycles in school are not the sole drivers of conceptual progression, but school experiences offer a focused interaction directed at particular science curricular targets. Children's ideas develop as the result of exposure to the family, peers and wider social environment. With maturation, children gain capability in dealing with greater complexity. Nor are students' ideas assumed to develop in a simple, linear, uni-dimensional, predetermined and consensual manner. One implication of the rack in the mechanism is that the progression pathway is clearly laid out, whereas in practice, curricula are often ambiguous (or even silent) about progressive conceptual steps between the 'big ideas' in science education. Even optimal classroom processes can only approximate to the ideal. To use the metaphor more directly, some of the teeth in the rack may be missing. The catalogue of constructivist research offers the evidence to fill those

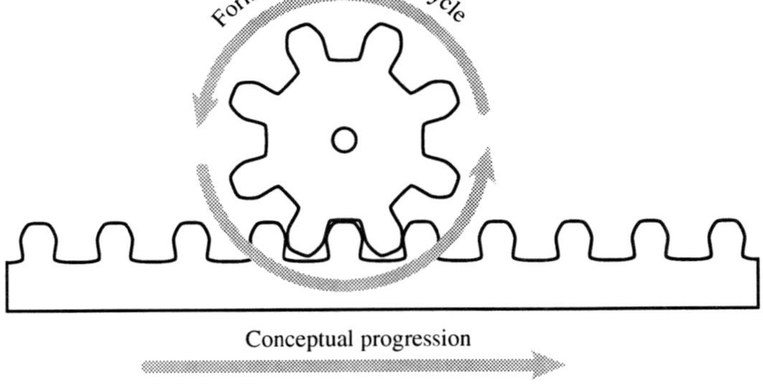

Figure 1. A rack and pinion metaphor for the formative assessment process

gaps; teachers can be provided with indications of pupils' likely conceptual trajectories that help them to negotiate the next steps in learning. Formative assessment has to do more than reveal pupils' uncertain understandings. We suggest that reference to pupils' likely conceptual trajectories offers significant support to teachers in deciding how to move ideas forward from current expressions of pupils' thinking.

Another limitation of the rack in our metaphor is its lack of overt multi-dimensionality. Students are capable of holding mutually discrepant multiple representations. While the aspiration might be to encourage them to reconcile such inconsistencies, toleration of a lack of closure is required for teachers and students alike. It would be naïve to assume that posing a one-dimensional question will elicit reliable evidence of a student's total grasp of a particular concept. In acknowledging this, Duschl and Gitomer (1997) advocate "assessment conversations" that promote a portfolio process. That is, students are encouraged to engage with the collection of diverse representations likely to be generated by a class engaging with a novel concept, using evidence gathering and argumentation processes in order to support their ideas. The portfolio idea evokes notions both of a collection of ideas expressed by a *group* of pupils (to which members of that group will be exposed by means of the teacher's management strategies) as well as the sub-set of viewpoints capable of being held simultaneously in different representational forms by each *individual*. Ideas will need to be checked against evidence, including empirical investigation, and arguments rebutted or supported accordingly. Tiberghien (2000) takes a similarly analytic stance towards unpacking knowledge construction into various representational forms for teaching – or the process of 'didactic transposition' as she calls it, using the language of semiotics:

> "... a hypothesis on learning is that an individual's understanding of a concept (or, more generally, an idea) develops when relations are established between different semiotic registers associated with the idea." (Tiberghien, *op.cit.* p. 4)

This multi-representational aspect of knowledge construction is important in describing micro-developmental change. The fine grain conceptual change process is an aspect we have elaborated elsewhere (Russell and McGuigan, 2002), drawing on the theoretical description offered by Karmiloff-Smith (1992). The theory assumes that conceptual development occurs as a result of re-describing representations – representational re-description or 'RR'. Through RR, the pupil becomes increasingly aware and in conscious control of their own understanding. According to this view, cognitive development is characterised by individuals accruing implicit information of which they lack conscious awareness. Coming to understand involves intuitive knowledge becoming increasingly explicit and available knowledge to the pupil. Within a multi-modal understanding of genetic epistemology, language serves an organising function, mediating access and enabling links between different representations that otherwise remain inchoate.

3. DIFFICULTIES WITH AND RESISTANCE TO FORMATIVE ASSESSMENT PRACTICES

Turning to the pinion in the metaphor, there may be difficulties associated with this aspect of the interaction between teachers' actions and pupils' developing understanding. For example, there may be uncertainty on the teacher's part as to what to do next in the assessment cycle. Duschl and Gitomer (1997) suggest that teachers' assessment of cognitive goals tend to be ignored or subsumed in favour of activity goals – management routines - in the classroom. Their case history report reveals the considerable tensions involved in teachers changing their practices, even when supported with structured activities and the wherewithal to elicit students' ideas. In a cited instance, there was clear evidence that the major impediment to changing pedagogy was the teacher's "reluctance to allow the students to take ownership of the problem conceptualization – to do it themselves." Instead, "the teacher was still in control of all conceptual orchestration." (Duschl and Gitomer, *op. cit.* p.64), displaying what the authors call "get through it" teaching behavior.

This kind of teacher behaviour became familiar to us in our collaborative research and warrants further comment and interpretation. Firstly, it is important to appreciate that, for many teachers, a 'top-down', teacher-led didactic style might be considered a core responsibility in their self-identity as teachers. In this style of teaching, listening to pupils' unsophisticated ideas *and* taking them seriously as a starting point for more managed activities involving rigorous thinking might be considered eccentric. Telling students the 'facts' might seem a more responsible and rational approach. For such teachers, 'wait times' after posing questions are likely to be of limited duration. Black et al. (2003) comment on teachers' difficulties in extending the time they waited for a pupil's response to a question. According to Rowe (1974), some teachers may leave less than one second for a pupil response after posing a question. Such short wait times, it is argued fail to encourage reasoning and result instead in dialogue comprising only questions which can be answered quickly. (See also, Alexander, 2004). A second observation is that teachers feel a responsibility to their pupils to achieve curriculum coverage; to do otherwise is to give students less than their educational entitlement. A pedagogical approach that emphasises students' active construction of meaning requires the pace of their reasoning activities to be taken into account in the management of teaching and learning. Evidence-based learning will inevitably take longer than 'get through it' teaching.

Given the desirability of encouraging formative assessment practices, how is practice to be influenced? By 2003, Black et al. were able to suggest at least a partial response to this question:

> "It seemed to us that teachers would need a variety of living examples of implementation, by teachers with whom they could identify and from whom they could both derive conviction and confidence that they could do better." (Black et al., 2003, p. 3.)

4. MODE OF ENQUIRY: FROM 'CONCEPTUAL TRAJECTORIES' TO 'TEACHING AND LEARNING SEQUENCES'

Our focus has been the dynamics of conceptual change, both micro-genetic (describing conceptual change in real time) and macro-genetic, as it occurs over several years. This enterprise of describing 'conceptual trajectories' is a long-term programme for pooled research rather than an undertaking that can be resolved rapidly by individual researchers. While there may be considerable agreement as to the major concepts to be addressed by science curricula, and even, in broad terms, which ideas should be addressed earlier and which later, much of the fine grain of conceptual progression remains to be described. The U.S. Project 2061 (http://www.project2061.org/), attempts to set out a long-term agenda to meet this need by developing, on the basis of research information, 'conceptual strand maps':

> "Learning goals should be considered not merely as a collection of ideas organised under various topic headings, but as a progression of understanding that includes what leads up to each specific goal and where it then leads (AAAS, 2001)."

The domain-specific, conceptual progression elements that we use to illustrate our teacher support materials are validated by published research relating to pupils' conceptual understanding. In many cases, it is possible to adapt the reported research elicitation procedures for teachers' use in finding out pupils' ideas for classroom formative assessment purposes. Our position with respect to our posited 'teaching and learning sequences, (the equivalent of the Project 2061 'strand maps') is tentative. We assume that current formulations of the likely sequences of ideas that pupils move through are a) broad descriptions of trends and b) provisional, capable of further refinement by teachers as well as capable of being made obsolete by cultural shifts. The sequences we describe are not assumed to be exhaustive or final. Nor do we assume that every pupil follows an identical pathway. The premise that broad conceptual migration routes can be mapped is useful for practitioners, we believe, both to guide teaching of specific science topics and in the approach to conceptual development that it encourages.

5. FORMATIVE ASSESSMENT AS RECURRING CYCLES OF TEACHER-PUPIL INTERACTIONS

How are these constructivist-inspired researched sequences to be applied to a teaching approach which incorporates formative assessment? Perhaps because our research into pupils' thinking has been mostly collaborative with teachers, links to classroom practice are clear. Support materials incorporating the principles described above were first realised as Optional Assessment Materials for Wales, (Russell et al., 2001). The first incarnation was exposed to trials with 24 collaborating ('hot') teachers organised into pairs within each of twelve schools representing the 5-14 age range. Researchers initially mapped out hypothetical learning trajectories for different areas of the curriculum. Teachers, together with researchers, used these trajectories to develop a series of 'concept probes' – formative assessment devices to elicit pupils' ideas about key concepts – largely influenced by published research instruments. Following teachers' classroom experimentation with these probes, or 'ideas sheets' as they came to be known, the instruments evolved. Their viability as formative assessment activities within a range of classroom environments was confirmed. Further iterations of the materials were used to refine the product by reference to qualitative feedback assembled through teacher group meetings and written commentaries. For example, teachers' concerns about repeated use of the same concept probes – albeit designed to be used two or more years apart – led to the abandonment of the idea of revisiting the same concepts using the same instrument. While valued by researchers as a device to track longitudinal change, from a teaching perspective the repeated assessment strategy was regarded as repetitious and potentially alienating to pupils. Some teachers also opposed ideas sheets deemed to be too open-ended in the pupil responses they generated. The provision of a range of illustrative responses with commentaries eased their anxieties in this regard. The deeper problem probably related to the distinction between researchers seeking to elicit qualitative information and teachers' concerns about what they perceived as marking burdens. At the current point in development of teachers' assessment experience, it is possible that many were thinking in terms of the more traditional practice of using a series of summative measures to monitor pupils' progress, rather than probing the quality of pupils' understanding in order to inform teaching.

A second group of 24 unsupported teachers trialled 'final draft' versions of the materials, referred to as 'cold' teachers because they provided an 'arms-length', evaluation. Their role was to check whether the materials were capable of classroom use without further support – emulating the circumstances of any teacher picking up the materials 'cold'. Apart from minor changes of text and artwork to clarify some messages, the feedback was that the materials were comprehensible and classroom-viable. We emphasise that our formative evaluation exercises were used to confirm the *classroom viability* of the materials within a range of classroom ecologies. Issues surrounding *sustained use* or *changes in core beliefs and practices* associated with the use of these materials remain to be addressed. The further development of a coherent model of practice – a framework for which this paper attempts to describe – is a necessary pre-condition to enable such 'outcomes' research to be systematically addressed.

6. DOMAIN-GENERAL PHASES WITHIN FORMATIVE ASSESSMENT

Building on the experiences gained and adding a domain-general expression of the classroom implementation processes, a more elaborated version of the materials was developed, (Russell et al., 2003). This elaboration paid more attention to the subtleties of teacher behaviour associated with phases within the formative assessment cycle.

Several theorists refer to a formative assessment as a series of teacher-pupil interactions involving several phases of activity. Some distinguish more than one kind of formative assessment: Bell and Cowie (2001), for example, refer to "planned" and "interactive" formative assessment.

Ruiz-Primo and Furtak (2004) refer to "formal" and "informal" practices. While there appears to be broad agreement about the general nature of the phases, the vocabulary used to label them is less than categorical. The lack of common vocabulary may be the result of the challenge of imposing order on a complex set of practices rather than significant differences in describing formative assessment processes. Table 1 assembles some major contributions to this issue of phases in the formative assessment process.

The distinction between 'using' and 'acting' upon ideas is elusive and it may be that the different verbs describe one and the same teacher behaviour, while 'formal' and 'informal' are more commonly used to refer to general styles of teaching. There is more evident commonality in the three sequential phases when expressed in the more direct language of (i) finding out what ideas a pupil or pupils hold; (ii) locating those ideas within a teaching-learning context and (iii) taking action, informed by those ideas. The distinction between 'receiving' and 'eliciting' pupils' ideas might be more closely defined by referring to contexts in which the teacher is *proactive* in finding out ideas and contexts when pupils' ideas emerge incidentally, the teacher being *receptive*. To be proactive, the teacher will need to have a clear teaching-learning agenda in mind, possibly centred on strategic 'key concepts' and perhaps using pre-prepared elicitation instruments with a whole class or group of students. In this sense, it might be appropriate to describe the teacher as using formative assessment in a (relatively speaking) 'formal' manner. In receptive mode, the teacher deals with the fine grain issues that arise more spontaneously from classroom interactions, often at an individual pupil level, when techniques such as questioning are likely to come into play. 'Reception' must be understood to encompass 'reaction', in the sense of the teacher responding adaptively to pupil feedback – and perhaps adjusting plans in a bespoke manner.

TABLE 1. Ideas about kinds of, and phases within, formative assessment

Published source	Kind of assessment	Named phases		
Duschl and Gitomer, (1997); Duschl, (2003)	Formative assessment	receive	recognise	use
Bell and Cowie (2001)	Planned formative assessment	elicit	interpret	act
	Interactive formative assessment	notice	recognise	respond
Ruiz-Primo and Furtak, (2004)	Formal formative assessment	gather	interpret	act
	Informal formative assessment	elicit	recognise	use

7. FINDINGS AND OUTCOMES

The model of formative assessment that summarises the current realisation of our thinking is shown in Figure 2. This domain-general aspect of our formulation shares much common ground with the positions summarised in Table 1 – though we emphasise *cycles* of activity rather than *sequences*. We also add details describing the preliminary steps of planning and orientation at the beginning of a cycle and reflection and evaluation at the other end. A more significant difference is the manner in which the model and associated support materials (Russell and McGuigan, 2005) treat the phase described in Table 1 as 'recognising' or 'interpreting', for it is at this point in the process that reference to the domain-specific aspects of the model – the teaching and learning sequences – come into play. Specifically, teachers are invited to map elicited pupil responses to as close a correspondence as judgment permits to the suggested teaching and learning sequences. By locating a pupil's current understanding on such a sequence, the next step to be negotiated becomes apparent.

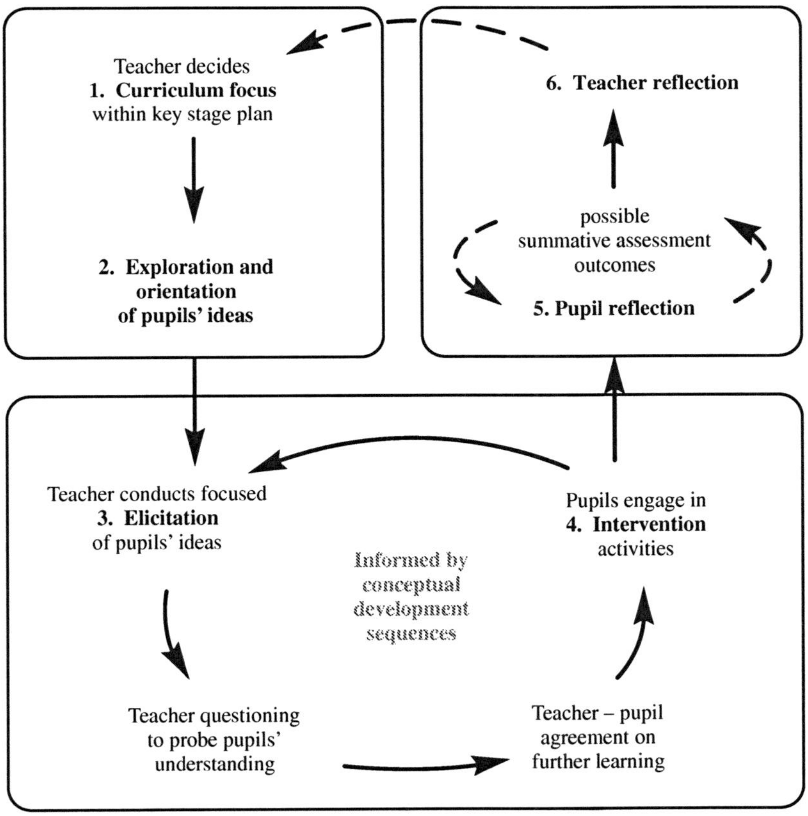

Figure 2. Domain-general model of formative assessment

Bearing in mind that the model requires the teaching and learning to be interactive and responsive to pupils' emerging ideas, teachers' plans must have sufficient flexibility to allow details to be modified on the basis of whatever evidence emerges. The details of the phases have the following characteristics.

1. *The curriculum focus* indicates the learning agenda set by the teacher (or any particular scheme to which the teacher refers). While assessment for learning has some characteristics of a 'child-centred' approach, the teacher is assumed to be clearly directing the learning agenda.
2. *Exploration and orientation* may include a review of previous teaching, perhaps some direct experiences that remind pupils of their prior knowledge and personal experiences. Learning objectives might be encouraged to be re-stated by pupils in their own words. The key criterion of this phase is that pupils are given the opportunity to orientate their thinking vis-à-vis the teacher's agenda. The teacher's objective must be to minimise ambiguity and enable pupils to connect with the new topic by drawing on direct or previous personal experience.
3. *Elicitation* or 'finding out ideas' is the process of enabling pupils to make explicit their ideas when they have had time to reflect in a considered manner on their own beliefs. Concept probes designed to elicit ideas provide the possibility of children representing their ideas in different ways. They are relatively open-ended so that a diverse range of understandings might be represented.
4. *Intervention* may be specific to individuals or strategic, for groups. It must be tailored to pupils' emerging ideas and guided by conceptual development sequences. Target-setting is negotiated between teacher and pupils. Multi-modal interactions drawing on a full range of representational formats enable pupils to triangulate, share and justify their ideas. Reflective teachers and learners will review their targets and achievements. Summative feedback may also be available. The two kinds of formative assessment suggested earlier could be accommodated within Figure 2 in the sense of the proactive version that moves through all of steps 1-6, and successive cycles at a more conversational level consisting of iterations between steps 3 and 4.
5. *Pupils' reflection* on their understanding, in a metacognitive sense, is expected: they are increasingly encouraged to think about what they know and how they know it and how they may move their understanding forward. (A hermetic distinction between steps 4 and 5 is not assumed, but a more focused and deliberate encouragement of reflection is posited within step 5.) Duschl (2003) emphasises the scope for exploring epistemic aspects of pupils' understanding. Our model focuses explicitly on conceptual understanding, with epistemic understanding being treated as an incidental – a lack of specific attention that warrants further consideration.
6. *Teacher reflection* acknowledges that classroom strategies are adopted and developed when they prove to be positive and need to be adjusted when they reveal problems.

8. CONCLUSIONS AND IMPLICATIONS

The implementation of the formative assessment approach described above has particular costs and benefits for teachers. The approach may require:

i) *the development of new skills*. For some, taking pupils' ideas seriously, using those ideas as starting points for their teaching, using more open-ended questions and negotiating learning targets with pupils, will require significant reorientation and the acquisition of new skills.

ii) *investment of time*. Assessment for learning involves finding out pupils' ideas before what many see as the real job of 'teaching' begins. Some teachers may have to convince themselves that additional time is merited because it is linked with later time savings in the form of more precisely targeted teaching and more effective, self-motivated learning.

iii) *flexibility of planning*. Formative assessment requires teachers to have a broad plan of the area to be covered and an awareness of the likely level of difficulty which will provide pupils with the necessary challenges. However, there needs to be sufficient flexibility within plans to allow teachers to decide the next steps for pupils' learning on the basis of information as it emerges. Teachers may need to plan to be surprised. Benefits may include:

a) more effective and efficient teaching, because formative assessment leads to targeted teaching which addresses learners' precise needs in order to progress. Time may be saved when well-understood curricular areas can be set aside because the use of formative assessment strategies tells the teacher that the ground is well-understood.

b) improved attainment is strongly associated with assessment for learning (Wiliam et al., 2004). Pupils' achievements associated with teachers' use of formative assessment in its various forms can be checked empirically, both in terms of eventual summative performances and changes in learning styles towards greater autonomy.

The model of formative assessment described in this paper has evolved through working directly with teachers. As every curriculum must describe the ideas to which pupils are to be exposed, the notion of research-informed concept development sequences appears to be capable of generalisation across national systems. The correspondence between such formulations across diverse educational environments is itself of research interest. Some domain-specific practices developed by teachers to support the recoding of representations across modes have been identified (Russell and McGuigan, 2002). Further studies across different domains could examine the impact of such approaches on micro-genetic change. Our research has also pointed to some ways in which children can be encouraged to self-assess and provide feedback to each other about their learning. The success of such approaches on pupil confidence has been identified by Flutter and Ruddock (2004). The impact of peer and self-assessment on the planning of next steps and learning outcomes suggest fruitful avenues for further enquiry.

REFERENCES

Alexander, R. (2004). *Towards dialogic teaching: Rethinking classroom talk.* (Second edition. Cambridge: Dialogos)

AAAS (2001). *Atlas of scientific literacy. Project 2061.* (American Association for the Advancement of Science. National Science Teachers' Association, Co-publisher)

Bell, B. & Cowie, B. (2001). *Formative assessment and science education.* (Dordrecht, Kluwer Academic Publishers)

Black, P. J. & Wiliam, D. (1998). Assessment and classroom learning. *Assessment in education: Principles, policy and practice.* (Vol. 5, No. 1, pp. 7–74)

Black, P. J. & Wiliam, D. (2006). Developing a theory of formative assessment. (In J. Gardner (Ed.), *Assessment and learning* (pp. 81–100). London: Sage)

Black, P. J., Harrison, C., Lee, C., Marshall, B. & Wiliam, D. (2003). *The nature and value of formative assessment for learning.* (American Education Research Association. Chicago, April 2003)

Bruer, J. T. (1996). Classroom problems, school culture, and cognitive research. (In K. McGilly (Ed.), *Classroom lessons: Integrating cognitive theory and classroom practice* (p. 290). Cambridge, MA: MIT press)

Driver, R., Guesne, F. & Tiberghien, A. (1985). *Children's ideas in science.* (Milton Keynes: Open University Press)

Duschl, R. (2003). Assessment of inquiry. (In J. M. Atkins & J. Coffey (Eds.), *Everyday assessment in the science classroom* (pp. 41–60). Arlington, VA: NSTA Press)

Duschl, R. A. & Gitomer, D. H. (1997). Strategies and challenges to changing the focus of assessment and instruction in science classrooms. *Educational Assessment*, 4(1), 37–73

Flutter, J. & Ruddock, J. (2004). *Consulting pupils: What's in it for schools?* (London: Routledge Falmer)

Karmiloff-Smith, K. (1992). *Beyond modularity: A developmental perspective on cognitive science.* (Cambridge, MA: MIT press)

Kennedy, C. (2005). *The BEAR assessment system: A brief summary for the classroom context.* Retrieved October 3, 2006, from http://bearcenter.berkeley.edu/publications/pubs.php

Rowe, M. B. (1974). Wait time and rewards as instructional variables, their influence on language, logic and fate control. *Journal of Research in Science Teaching*, 11, 81–94

Ruiz-Primo, M. & Furtak, E. M. (2004). *Informal formative assessment of students' understanding of scientific enquiry.* (Centre for the Study of Evaluation (CSE) Report 639)

Russell, T., McGuigan, L. & Hughes, A. (1998). *Forces. Science process and exploration research reports.* (Liverpool: Liverpool University Press)

Russell, T., McGuigan, L., Roberts, D. & Hughes, A. (2001). *Science optional assessment materials for KS1, 2 and 3.* (ACCAC)

Russell, T. & McGuigan, L. (2002). Promoting understanding through representational redescription: an illustration referring to young pupils' ideas about gravity. (In D. Psillos, P. Kariotoglou, V. Tslfes, E. Hatzikraniotis, G. Fassoulopoulos & M. Kallery (Eds.), *Science education research in the knowledge-based society* (pp. 61–68). (Dordrecht: Kluwer Academic Publishers)

Russell, T. & McGuigan, L. (2003). *Assessing progress in science KS1, 2 and 3.* (Qualifications and Curriculum Authority)

Russell, T. & McGuigan, L. (2005). *Assessing progress in science INSET Toolkit KS1, 2 and 3.* (Qualifications and Curriculum Authority)

Tiberghien, A. (2000). Designing teaching situations in secondary school. (In R. Millar, J. Leach & J. Osborne (Eds.), *Improving science education: the contribution of research.* (pp. 27–47). Buckingham, UK: Open University Press)

Wiliam, D., Lee, C., Harrison, C. & Black, P. J. (2004). Teachers developing assessment for learning: impact on school achievement. *Assessment in education: principles, policy and practice*, 11(1), 49–67

Wilson, M. & Sloane, K. (2000). From principles to practice: An embedded assessment system. *Applied Measurement in Education*, 13 (2), 181–208

15. MEMORISATION OF INFORMATION FROM SCIENTIFIC MOVIES

Abstract: This study deals with students working with a hypermedia that links a bank of chemical movies. Two aspects were examined: (i) the factors (such as surface features) that prevail to the choice of movies by students that use the hypermedia; (ii) the kinds of information that students are able to memorise. The research questions were formulated as (i) How chemical movies are chosen by students who have to answer questions about chemistry? and (ii) How difficult is it to memorise conceptual information from such movies? The films were specifically produced for this research, several of them with two possible narrations: one with a description as perceptible as possible of the pictures of the movie, the other with a maximum of chemical terms. In a first experiment, six pairs of 17-year-old students were given a set of questions to answer with the hypermedia. In a second one, four students were given a series of seven movies to be used, and were questioned after 7 days. The most occurring factors for choosing movies were surface features, but we observed students making their choice after reformulating a question. It also happened that a pair of students chose a movie more than once. Low memorisation seems to occur with movies that display several events, such as chemical reactions. After 7 days, students remembered more pictures of the movie than words of the narration, and from the picture, more icons than animations or photos, and almost no symbols

Keywords: Acid, Bases, Chemical education, Hypermedia, Memorisation, Narration of movies, Surface feature, Use of movies by students

1. INTRODUCTION AND THEORETICAL BACKGROUND

Chemical movies have been used in teaching situations since the late 1950s (Slabaugh and Hatch, 1958), and huge efforts were taken to produce more and more movies. The use of such movies has evolved and diversified during the second part of the 20th century. Movies to prepare laboratory work have been proposed to students (Rouda, 1973). Later, movies of experiments have been produced for demonstrations in several domains of chemistry (Jacobsen et al., 2002 and references herein). In most cases, students had to watch one movie prescribed by their teacher. In this study, we decided to provide students with a bank of movies, and to let them chose the movie(s) they needed according to a task they had to accomplish. The use of such a bank is the aim of the present research and the factors influencing

Jean-François Le Maréchal, Ecole Normale Supérieure de Lyon, 46 allée d'Italie, 69364, LYON CEDEX 07, FRANCE

R. Pintó and D. Couso (eds.), Contributions from Science Education Research, 199–210.

students' choices of chemical movies are analysed. Among these factors, surface feature of symbols, drawings, situations, etc. have frequently been put forward. Surface features have been described as the point from which novices start to develop knowledge from a situation; however, unlike experts, without relating them to deep knowledge structure (Kozma and Russell, 1997). Without an appropriate prior knowledge, novices are both enabled and constrained by surface features.

Learning is not only dealing with information, but also remembering it in the process of building new knowledge. Most theories about memory – see for example the cognitive load theory (Pass et al., 2004, and references herein) – consider that a long-term memory with an infinite storage capacity is coupled with a working memory. The latter has a limited capacity when dealing with novel information, and can easily be overloaded (Chandler and Sweller, 1991). Mayer has shown that students who listen to (or read) explanations that are presented solely as words are unable to remember most of the key ideas and they experience difficulties in using what was presented to solve new problems (Mayer, 1997, 1999). Having text and spoken word produces a large multimedia effect on deep learning, which has been tested on problem-solving transfer. This effect can be interpreted by the cognitive theory of multimedia learning, which states that multimedia causes learners to use both visual and verbal channels and not only one of these channels.

2. RESEARCH QUESTIONS

This contribution tries to answer two questions: (i) How chemical movies are chosen by students interacting with a task? For example, we will look for the uses of the surface features as one of the factors students use to choose a movie during a task. (ii) How difficult is it to memorise conceptual information from such movies? Memorising information from a movie comes down to dealing with a large amount of information that may prove difficult for novices. The answers to these questions are to be presented in three steps: (i) description of the research tool (a hypermovie), (ii) interaction of students dealing with a hypermovie and a series of questions, and (iii) memorisation of concepts from movies after seven days.

3. DESCRIPTION OF THE HYPERMOVIE

Hypermovie is a neologism that we adopted to name a bank of movies connected by links. The uses of this neologism can be traced in a search engine as Google™ and always stands for interactive movies (see for example Melzer et al., 2004).

We created two ways for navigating in our hypermovie (Figure 1). One is based on a tree diagram with three headers (acids, bases, acid–base pairs); each header may open several sub-headers and each of them may open several titles of movies (Pekdağ and Le Maréchal, 2003a). Activation of a title leads to a movie-page with a button on the left and a list of words on the right. The button can activate a movie and the list of words consists of chemical concepts extracted from the narration of the movie. The second possibility of navigation uses these chemical concepts that

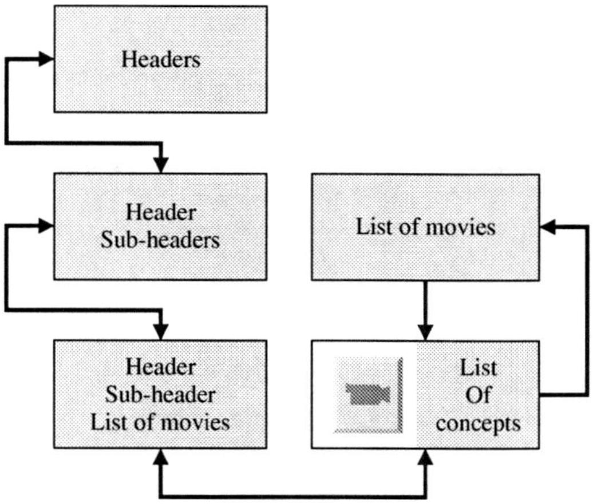

Figure 1. Representation of the two ways of navigating the hypermovie: left from the tree diagram, right from conceptual links

form a list of links. Each of these conceptual links leads to a new page with a list of movies the narrations of which contain this concept.

The 34 one- to three-minute long movies of the bank were home-made after a conceptual analysis of both students' curriculum and students' conceptions. They are short to be perceived as a one-piece unit best studied in a single session (Capsi et al., 2005). They can contain the video of a chemical experiment, an animation, commentaries about data, etc.

Movies were created with a careful relation between narrations and pictures, respecting as much as possible Mayer's four conditions (1989): to be helpful (1) the text must describe a cause-and-effect system; (2) the illustrations must reasonably depict the system or process under consideration; (3) appropriate outcome measures must be selected; and (4) the learners must be inexperienced with respect to the targeted content domain.

The movies were created using the main kinds of picture styles: (i) video, which is appropriate for describing events such as chemical experiments; (ii) animation, much adapted for molecular display; (iii) slide show when data or photos are to be commented. Pictures were designed as representational (e.g. pictures such as an experiment that seems to be described by the narration), organisational (e.g. pictures that provide a useful structural framework for example in the case of tables of data) and interpretational (e.g. pictures that help to clarify a difficult narration), in Levin's terminology first used to describe storybook illustration (Carney and Levin, 2002; Levin, 1981). The two other functions of picture: decorational (pictures simply decorate the screen, bearing little or no relationship to the narration) and transformational (pictures that provide mnemonic aids) do not appear in our movies.

4. WORKING WITH THE HYPERMOVIE

This part presents a first experiment in relation to both research questions.

4.1 Methodology

The method to collect data and the way to analyse them are detailed here. Although the hypermovie is a new research tool, proven methods were used in our work.

4.1.1 Collecting data A series of concept-oriented open questions relative to acids and bases and an access to the hypermovie were given to six pairs of 17-year-old students (11th grade) who were video recorded. Each pair of students produced a handout that was collected. The one-hour discussions of two pairs were transcribed and analysed. One pair is named E/F and the other one M/B below, and in Table 1. The experiment was conducted several months after the students had their regular courses on acids and bases. It took place after regular school time, without the teacher who had looked for volunteer students.

Asking questions to students was a way to make them discuss. The resulting discussion was recorded and served as data for our analysis. Asking questions was also a way to have them watching the movies with an active role (Gourgey, 1998). A few questions are given below as examples with the results. They have no "whys" nor "whats" relations to the movies. They would have provoked an interrogative elaboration of the meaning of the movie (Carney and Levin, 2002), which was not intended. Instead, questions were open for students to process information on the movies.

4.1.2 Analyzing data The first analysis of data consisted on finding relations between the choice of a movie and students' activities. Students' choices were categorised as: (i) choice of a movie that had already been watched at least once; (ii) choice of a movie prescribed by the task; (iii) choice of a movie with a surface feature; (iv) choice of a movie after a reformulation of the question of the task, or for a new question (see Table 1).

We considered that students chose a movie from a surface feature when there was at least one common word in the question and in the title of a movie. Using surface features to solve problems is a classic novices' behaviour, unlike experts who group them by their underlying principles based on laws of physics (Gourgey, 1998).

TABLE 1. Number of Movies Chosen by Students' Pairs E/F and M/B in Each Category as Defined in Section 4.1.2

Students	Re-watching of a movie	Prescribed movies	Surface feature	Reformulation or new question
E/F	2	4	7	2
M/B	5	4	4	1
Total	*7 (24%)*	*8 (28%)*	*11 (38%)*	*3 (10%)*

Novices often use the surface features (such as colour, motion, labels, etc.) of the displays to try to build an understanding of chemical phenomena (Kozma, 2003).

Telling apart the reformulation of the question, and the occurring of a new question, is important as questions generated by students trace an active thinking and a better learning (Hartman, 1994).

4.2 Results and Discussion

As expected, students could not answer the questions without watching movies. They watched movies that were prescribed as well as others. The numbers of movies in each category are given in Table 1.

An action can be done either to accommodate local conditions or with an objective (Thévenot, 1998). In our case, the local condition was the prescription of a movie and the objectives correspond to our other categories. Students did not watch movies for other reasons than those listed in Table 1. Whatever the reasons of the choice of a movie may be, students did not watch a movie by chance; the choice was individually/collectively led by a teleological need. Providing movies as a resource for students seems therefore pertinent.

The high percentage of movies chosen from surface features is not surprising, as it has for long been known that it is the way novices construct mental representations of physical situations (Chi et al., 1981; Larkin et al., 1980). However, the case of the choice of a movie has probably not been described before. The following example helps to understand how such a choice may happen during students' discussions. The question students had to answer is: "Chlorhydric acid is added to a sodium acetate solution. Explain why a vinegar smell appears?". The choice of the movie "Effect of adding acid on pH" is surface feature–driven indeed, as the question deals with addition of acid, such as indicated by the title of the movie. Moreover, students had spent 5 min 5 s exploring 17 other concept links before they clicked on "Effect of adding acid on pH", and during this time, no other concepts than those of the links had been put into the discussion.

F yea/in the acid–base pair (F mentioned the link "acid-base pair")
E why for/wait
F here for/click on it once here (F showed the link "release of H^+")
E here close the link release of H^+
F let's go here (F mentioned the link "properties of acids")
E it was the first menu
F yeah
E click on the menu
F here in (F showed the link "acid/base pair")
E acid/base pair (E read the link)
F let's go here/then we may have the pair (F showed the link)
E chlorhydric acid (E read the question and clicked on the link)
F look at these link/conjugated form (F read the link "Conjugated form")/go in half equation (F mentioned the link "half equation")
E hm no/where do you want to go here

F I don't know may be in conjugated form (F read the link "conjugated form") Such behaviour toward concepts that are read on links and never related to any personal knowledge clearly belongs to the surface feature category (Kozma and Russell, 1997). Reading or mentioning 17 links within 5 min 5 s (one every 18 s as an average) before deciding to watch one film is another criteria of inefficiency of such behaviour. Nevertheless, these students explored the hypermovie with an objective. They did not follow a prescription and were in the teleological Thévenot's level (1998).

Rewording a question is a better way to answer it. Therefore, after reading the question "How is it possible to represent what happens in solution when sodium acetate is introduced in water?", one student suggested writing an equation, which is one of the possible representations indeed. She could have proposed a drawing at the molecular level or at the macroscopic one, but she decided to propose the symbolic one (Johnstone, 1993). Such a reformulation of the question is not deep, but it led to the consultation of the movie "General notation of half equations". In this case, students used their knowledge to transform the question to find a movie, unlike they did in the cases of surface features. These students therefore actively interacted with the question and tried to clarify the purpose of reading (Gourgey, 1998). Such behaviour reveals the beginning of an understanding.

An interesting but broad question is how much knowledge can be reused from a movie. The number of times a given movie was watched is interesting in this respect. Table 1 shows that one of every four movies was watched at least twice. One of the movies "Effect of adding acid on pH", although it was not longer than others, was watched several times by the six pairs of students. Two factors can be proposed to understand the reason of its low memorisation. The first one is the number of non-perceptible events, such as reagents of the chemical reactions reacting together, and a gas transfer from a flask to a beaker by means of a rubber hose. Although all of these non-perceptible events had a visible counterpart (a liquid addition for the reaction $H^+ + HCO_3^-$ – Figure2b; a bubbling during CO_2 transfer – Figure 2d; a pH change for the reaction $CO_2 + OH^-$ – Figure 2e; and a colour change for the reaction of pH indicator – Figure 2f), the capacity of decoding and processing these perceptible events in the students' working memory must have been overloaded (Baddeley, 1992; Chandler and Sweller, 1991). Constructivist theory can help us to understand the effect of overloading memory on learning. It tells us that meaningful learning requires the learner to actively select relevant information, organise them into coherent representation and integrate them with other knowledge (Mayer, 1996). The flow of new information from the video did not allow these selections, organisation and integration processes that could have led to memorise the non-perceptible events.

A second interpretation of low memorisation may be given from the dual coding theory. This theory states that visual and verbal materials are processed in different processing systems (Clark and Paivio, 1991; Paivio, 1986). Eyes and Ears produce pictorial and verbal representations respectively, and a more meaningful learning occurs when both representations interact with prior knowledge (Mayer

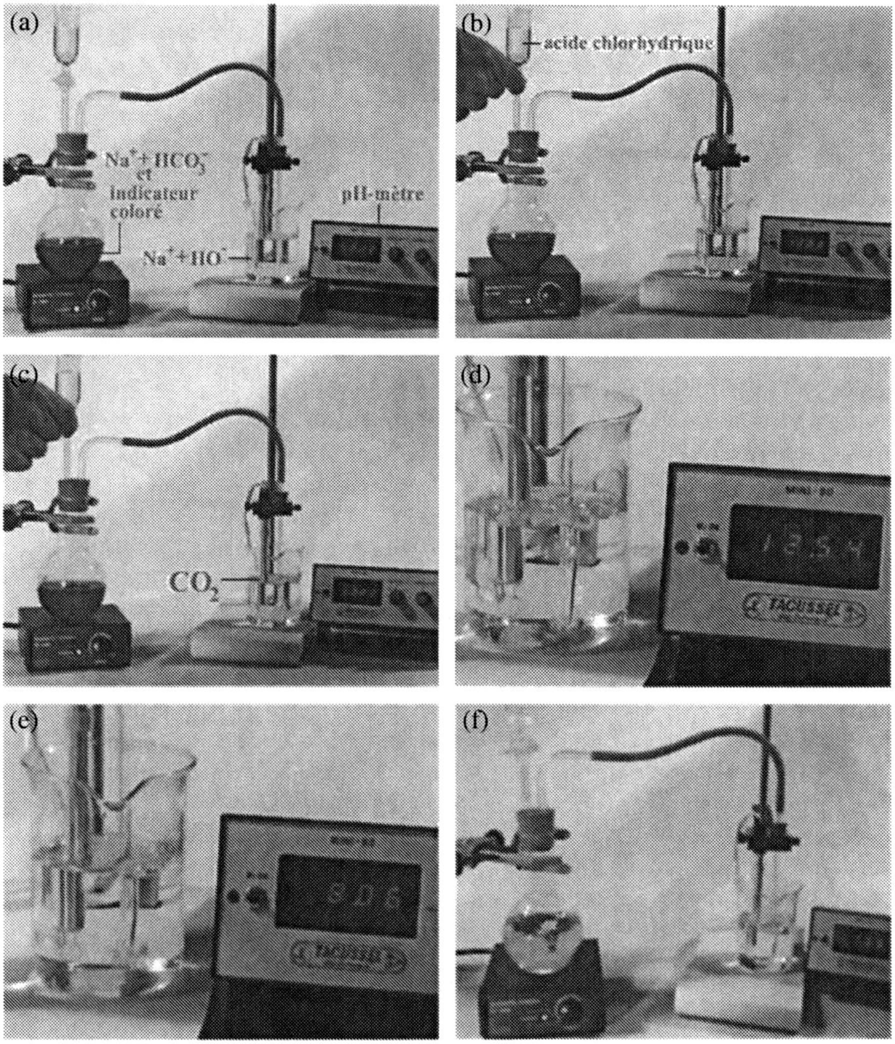

Figure 2. Pictures from the movie "Effect of adding acid on pH". From left to right (a) presentation of the set up; (b) chlorhydric acid addition; (c) CO_2 bubbles in the beaker; (d) close-up on the bubbles; (e) close-up on the decrease of the pH; (f) colour change of the pH indicator in the flask

and Moreno, 2002). In our case, the non-perceptible events displayed in the movie that was repeatedly watched by all students were commented by a narration. Nevertheless, only the verbal processing system was activated, as the chemical reactions and gas transfer were not directly perceptible. Therefore, students could not build an immediate representation from the visual processing system.

5. SEVEN-DAY MEMORISATION

Another memory-oriented experiment was performed to better understand the memorisation difficulties in the case of chemical movies.

5.1 Method

Seven movies from the hypermovie and a few questions were given to two other pairs of students (Gl-Gu and Ga-Je) of the same level as those of the previous experiment. They spent approximately 45 min with this other task.

Movies were of different kinds: animations ("Dissociation – microscopic animation"; "HCl, a Brønsted acid"; "pH of Vinegar"), video ("Vinegar and sodium hydroxide") or slide show ("Acid rains"; "Air pollution"; "Dissociation of HCl and CH_3COOH"). For each movie, two slightly different narrations were made. Each narration started and finished with the same sentences. The same ideas were then developed in both narrations, but one used a descriptive vocabulary of whatever is perceptible on the screen, whereas the other frequently used the chemical vocabulary. For example, the narrations of the movie "Dissociation of HCl and CH_3COOH" that went along a slide show, the first slide of which is given in Figure 3, were:

Perceptible narration – Acids may release a hydrogen ion. The connected white and green balls represent a hydrogen chloride molecule HCl. The separation of the white ball represents the release of a H^+ ion. It illustrates the acidic property of HCl represented by a half equation written under the balls. This other formula is a more complex acid: acetic acid. One of the white balls can also be separated from the rest of the assembly. It is again represented by a half equation written under the balls.

Non-perceptible narration – Acids may release a hydrogen ion. It is the case of hydrogen chloride. When a H^+ ion is released, there is simultaneously formation of a chloride anion. It is also the case of acetic acid. When a H^+ ion is released, the acetic acid releases also an acetate ion.

The use of perceptible and non-perceptible vocabulary in modelling has previously been described and can affect students' comprehension (Le Maréchal, 1999). Such a modelling approach also provides keys to analyse movies with educational perspectives (Pekdağ and Le Maréchal, 2003b). In our case, the perceptible features of narrations were used to test their influences on memorisation.

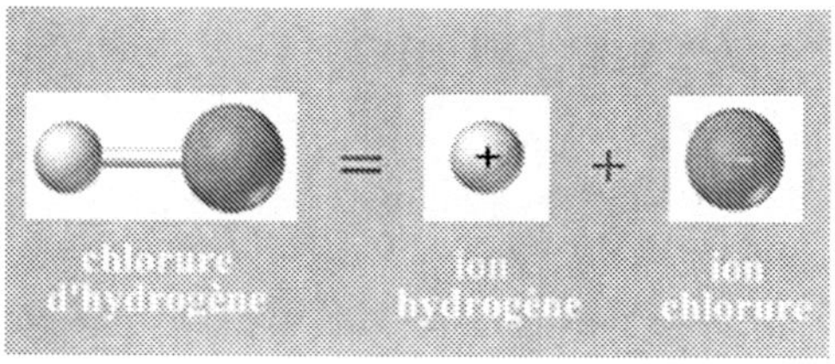

Figure 3. One of the slides of the slide show "Dissociation of HCl and CH_3COOH"

Seven days after Gl-Gu and Ga-Je answered the questions that went along with the seven movies, we individually asked these four students to put down what they recalled from the movies with the following questions:

Which movies do you remember and what were their topics (answer with as many details as possible)? Briefly depict with drawing and/or with sentences what you remember from the movies.

From students' answers, we could infer which movies were individually recalled. We could also notice whether students used words of the narrations, and/or sketched pictures from the screen, and/or used words to describe these pictures. Then recalls of pictures were categorised as iconic when a drawing was recalled, as video when a dynamic picture was recalled, as photo when a photo was recalled, and as symbol when symbols where recalled.

In this experiment the students watched the same movies in the same order, whereas in the previous one, as described above in Section 4.1, they watched the movie they needed. Interpretation of the memorisation is therefore simpler. For example, it is now possible to test the influence of the order in the casting of the movies.

5.2 Results and Discussion

Table 2 shows the results of the 7-day memorisation analysis. From Table 2, it appears that the casting order (recency effect), the kinds of pictures (animation, slide show or video), and the kinds of narrations (perceptible or not) have no real influence on the 7-day memorisation. We therefore turned our analysis to the way students recalled the movies. We found out that students used information from the picture more than from the text to answer our questions. Students produced description of pictures with their own words, using only some of the concepts that were in the narrations and few others. For example, student Je produced the following comments about his recall of the movie "*Acid rains*":

The second movie talked about **acid rains** *and their effects on nature. We noticed that* **acid rains** attack ***leaves of trees*** *and* erode ***statues****. We were also explained how rain becomes acidic (it comes from* emissions *of* factories *then* clouds *get the acidity of* gas*).*

Bold face words came out of the narration of the movie, but italic words were not. Instead, factories with their smokes were shown twice, and once with SO_2 and NO_2 symbols. This example is typical that students did remember part of the picture they commented with no precise recall of the corresponding narration.

For all the four students, 39 words such as chemical concepts, and nouns related to them (i.e. leaves, trees or statues in the above example), belonged both to students' descriptions and narrations, whereas 51 were used to describe pictures of the movie. Such results confirm that memorisation of pictures is better than memorisation of narrations (Dubois and Tajariol, 2001). We also found that 58% of the 51 recalls of pictures came out of iconic representations, 23% from videos, 15% from photos and 4% from symbols. This information may be used by educational movie producers for choosing the more efficient way to represent concepts to be

TABLE 2. Memorisation of Movies According to their Kinds of Pictures and Narrations. P Stands for a Perceptible Narration and R for a Non-perceptible One. Upper-case P and R with no Bracket Stands for Movies that were Recalled. Lower Case r and p Stands for Movies that were not Recalled. Movies had been Cast in the Order of the First Column

Casting order	Title of movies	Kind of movie	Gl	Gu	Ga	Je
1	Dissociation – microscopic animation	Animation	P	(p)	(r)	(r)
2	Acid rains	Slide show	P	P	R	R
3	Air pollution	Slide show	P	P	(r)	R
4	Dissociation of HCl and CH_3COOH	Slide show	(r)	R	P	P
5	HCl, a Brønsted acid	Animation	(r)	R	(p)	P
6	pH of Vinegar	Animation	(r)	R	P	P
7	Vinegar and sodium hydroxide	Video	R	R	(p)	P

recalled. For example, we suggest symbols not being overused although it is the most common way to represent chemistry.

The better memory of pictures over narrations does not mean that pictures alone are responsible for memorisation. Pictures alone would not even be correctly understood due to their polysemy (Joly, 1994). The narration specifies the meaning of the pictures. The remembering of pictures in the case of movies can be associated to the role of pictures in storybooks; it establishes the setting and reinforces the text, it motivates the reader and serves as a mental scaffold (Fang, 1996). Details of the pictures of the movie such as the factories could be remembered although they were not mentioned in the narration, just as pictures in storybooks may go beyond this role by adding details (e.g., Stewig, 1992).

6. CONCLUSION AND IMPLICATION IN TEACHING

Instructional movie designers and teachers who may involve movies in their teaching should know basic principles that rule the uses of movies in education. One of these principles is that surface features are often the starting point for grasping new knowledge and should be selected before being proposed to students. Another principle tells us that the use of movies in teaching should be coupled with tasks that force students to use the information presented by the movies. Such information should therefore not be related to too many non-perceptible events to avoid low memorisation. Instead, important information should better be both presented in the picture and in the narration to take advantage of the dual coding and multimedia effects. Although symbols are paramount in chemistry, their uses in movies are questionable as it leads to little recall by learners. Now that many chemical movies are available, they may be organised on hypermedia for students to use them with profit. The links of these hypermedia have, therefore, to be informative as the knowledge that can be embedded in the links is usable by learners even though surface features cannot be avoided.

7. REFERENCES

Baddeley, A. (1992). Working memory. *Science*, 255, 556–559.

Capsi, A., Gorsky, P. & Privman, M. (2005). Viewing comprehension: Students' learning preferences and strategies when studying from video. *Instructional Science*, 33, 31–47.

Carney, R. N. & Levin J. R. (2002). Pictorial illustrations still improve students' learning from text. *Educational Psychology Review*, 14(1), 5–26.

Chandler, P. & Sweller, J. (1991). Cognitive load theory and the format of instruction. *Cognition and Instruction*, 8(4), 293–332.

Chi, M., Feltovich, P. & Glaser, R. (1981). Categorization and representation of physics problems by experts and novices. *Cognitive Science*, 5, 121–152.

Clark, J. M. & Paivio, A. (1991). Dual coding theory and education. *Educational Psychology Review*, 3(3), 149–210.

Dubois, M. & Tajariol, F. (2001). Présentation multimodale de l'information et apprentissage. In E. De Vries, J. -Ph. Pernin & J. -P. Peyrin (Eds.), *Hypermédias et Apprentissages*, 5, 197–209. Actes du cinquième colloque (Grenoble, 9–11 avril 2001). Paris, INRP.

Fang, Z. (1996). Illustrations, text, and the child reader. What are pictures in children's storybooks for? *Read Horizons*, 37, 130–142.

Gourgey, A. F. (1998). Metacognition in basic skills instruction. *Instructional Science*, 26(1–2), 81–96.

Hartman, H. J. (1994). From reciprocal teaching to reciprocal education. *Journal of Developmental Education*, 18(1), 2–8.

Jacobsen, J. J., Moore, J. W., Zimmerman, J. F. & Browne, L. M. (2002). Chemistry comes alive! Volume 6: Abstract of special issue 30, a CD-ROM of laboratory techniques. *Journal of Chemical Education*, 79, 1381–1384.

Joly, M. (1994). *L'image et les signes. Approche sémiologique de l'image fixe*. Paris, Nathan.

Johnstone, A. H. (1993). The development of chemistry teaching: A changing response to changing demand. *Journal of Chemical Education*, 70(9), 701–705.

Kozma, R. (2003). The material features of multiple representations and their cognitive and social affordances for science understanding. *Learning and Instruction*, 13(2), 205–226.

Kozma, R. B. & Russell, J. (1997). Multimedia and understanding: Expert and novice responses to different representations of chemical phenomena. *Journal of Research in Science Teaching*, 34(9), 949–968.

Larkin, J., McDermott, J., Simon, D. & Simon, H. (1980). Expert and novice performance in solving physics problems. *Science*, 208, 1335–1342.

Le Maréchal, J. -F. (1999). Modelling student's cognitive activity during the resolution of problems based on experimental facts in chemical education. In J. Leach & A. C. Paulsen (Eds.), *Practical Work in Science Education*, 195–209. Dordrecht, Kluwer.

Levin, J. R. (1981). On functions of pictures in prose. In F. J. Pirozzolo & M. C. Wittrock (Eds.), *Neuropsychological and Cognitive Processes in Reading*, 203–228. New York, Academic Press.

Mayer, R. E. (1989). Systematic thinking fostered by illustrations in scientific text. *Journal of Educational Psychology*, 81, 240–246.

Mayer, R. E. (1996). Learning strategies for making sense out of expository text: The SOI model for guiding three cognitive processes in knowledge construction. *Educational Psychology Review*, 8, 357–371.

Mayer, R. E. (1997). Multimedia learning: Are we asking the right questions? *Educational Psychologist*, 32(1), 1–19.

Mayer, R. E. (1999). Multimedia aids to problem-solving transfer. *International Journal of Educational Research*, 31, 611–623.

Mayer, R. E. & Moreno, R. (2002). Aids to computer-based multimedia learning. *Learning and Instruction*, 12(1), 107–119.

Melzer, A., Hasse, S., Jeskulke, O., Schön, I. & Herczeg, M. (2004). The interactive and multi-protagonist film: A hypermovie on DVD. In M. Rauterberg (Ed.), *Entertainment Computing – ICEC 2004*, 193–203. Third International Conference Eindhoven, The Netherlands.

Paivio, A. (1986). *Mental representations: A dual coding approach*. Oxford, UK: Oxford University Press.

Pass, F., Renkl, A. & Sweller, J. (2004). Cognitive load theory: Instructional Implications of the Interaction between information Structures and Cognitive Architecture. *Instructional Science*, 32, 1–8.

Pekdağ, B. & Le Maréchal, J. -F. (2003a). Hyperfilm: Un outil de recherche en didactique de la chimie. In C. Desmoulins, P. Marquet & D. Bouhineau (Eds.), *Environnements Informatiques pour l'Apprentissage Humain*, 547–550. Actes de la conférence EIAH (Strasbourg, 15–17 avril 2003). Paris, ATIEF et INRP.

Pekdağ, B. & Le Maréchal, J. -F. (2003b). Influence of the relations between picture and text of chemical education films on conceptual change. In D. Krnel (Ed.), *Proceedings of the Sixth ESERA Summer-school* (Radovljica, Slovenia, 25–31 August 2002), 204–211. Ljubljana, University of Ljubljana Press.

Rouda, R. H. (1973). Student-produced videotapes in a physical chemistry laboratory course. *Journal of Chemical Education*, 50, 126–127.

Slabaugh, W. H. & Hatch, C. V. (1958). General chemistry via television. *Journal of Chemical Education*, 35(1), 95–96.

Stewig, J. W. (1992). Reading pictures, reading text: Some similarities. *New Advocate*, 5(1), 11–22.

Thévenot, L. (1998). Pragmatiques de la connaissance. In A. Borzeix, A. Bouvier & P. Pharo (Eds.), *Sociologie et connaissances, nouvelles approches cognitives*, 101–139. Paris, CNRS Editions.

PART 5

TEACHING AND LEARNING SCIENTIFIC CONCEPTS

16. MICRO-ORGANISMS: EVERYDAY KNOWLEDGE PREDATES AND CONTRASTS WITH SCHOOL KNOWLEDGE

Abstract: Some ways of looking at – and seeing – micro-organisms by lower secondary school pupils (ranging in age from 11 to 14) have been singled out by means of the administration of a questionnaire and by encouraging pupils to reason out their answers. The study confirms the need to induce pupils to express their prior knowledge in order to improve teaching; allows the collection of a number of very popular misconceptions (mould infecting food after it goes bad and why; micro-organisms travelling throughout the body and why; good and bad bacteria facing-off in combat in our body); and indicate mass media (mainly advertising spots and TV programmes) as pupils' elective source of information in the absence of any contribution at Italian school

Keywords: Lower secondary school, Micro-organisms location and action, Pupils' mental representations, Pupils' sources of information, Science education

1. INTRODUCTION

Comparing common knowledge with scientific knowledge, as Bronowsky did all the way back in 1953, is the underlying educational methodology of constructivism, based on the need to consider not only the previous disciplinary knowledge of those desiring to learn, but everything else – words, examples, personal experiences – that can also help teachers tease out possible mental representations as alternative to the scientific ones.

Although some limitations have been pointed out in the constructivist approach (Solomon, 1994), it continues to be the dominant paradigm for science teaching/learning (Neisser, 1976; Driver, 1989; Carey and Smith, 1993). Inspired by this approach studies carried out in the 1980s and 1990s in all fields of experimental sciences sought to identify the most popular and pernicious misconceptions (see Grimellini and Segrè, 1991; Pfundt & Duit, 1991), and the surprising data often emerging from these have proved very useful for teaching. Consequently, placing science into the context of students' everyday experience has been increasingly recommended in an effort to ensure effective learning (Lave, 1993; Campbell and Lubben, 2000).

Milena Bandiera, Biology Department, "Roma Tre" University, Viale Marconi 446, 00146 Roma, Italy

R. Pintó and D. Couso (eds.), Contributions from Science Education Research, 213–224.

At present the younger the students are the more essential such a strategy is considered (Harlen, 2002). The science classroom should refer to pupils' everyday life, starting with their mental representations and, where possible, noting their sources of information. This was precisely what was undertaken when two teacher-training courses on micro-organisms ("Micro-organisms and human health" and "Micro-organisms around us") were planned jointly in 2004 by the National Health Institute and the Biology Department of "Roma Tre" University. Lower secondary school teachers were informed of this requisite from the very beginning of the programme and accordingly asked to explore pupils' ideas about micro-organisms. The questionnaire designed was aimed at providing answers to the following research questions:

– *What deep-rooted conceptions exist regarding the presence and action of micro-organisms?*
– *What are students' non-school information sources?*

At the same time the questionnaire was designed to provide teachers with data useful in planning their own educational interventions concerning micro-organisms, and to give methodological and subject-matter suggestions to colleagues.

2. PRELIMINARY ASCERTAINMENT ON EDUCATIONAL CONTEXT

The Italian National Curriculum (DM 9 febbraio 1979) was carefully examined in an effort to identify the subjects and educational objectives that could potentially be correlated with the knowledge of micro-organisms. Neither of the presentations of the two out of five Italian national curriculum topics concerned with biological subjects – "Structure, function and evolution of living beings" and "Man and the environment" – even explicitly refers to micro-organisms; the first implies their presentation in the process of describing "matter cycles", and the second in the process of listing "diseases" (sub-topic: "Health education").

The ten texts (30 volumes in total) most often used in the schools of Rome and its provincial suburbs were analysed with reference to five keywords: micro-organism (microbes, germs), bacterium, virus, yeast, mould. Both subjects (the location and the space dedicated to them), and pictures (subject and consistency) were catalogued. All pay minimal attention to micro-organisms, merely mentioning them or else devoting *a one or two-line definition* of them, while, nevertheless, treating (a) the origin of life (primordial, prokaryotic cell), (b) living organism classification (bacteria as "Monera" and moulds as "Fungi"), (c) matter cycles and food chains (decomposition), (d) digestion (tooth-plaque, bacterial flora), and (e) health care and pollution (diseases, personal hygiene, drugs and vaccines, food maintenance). Expanded references are found in appendices (from one to three in each book) devoted to individual cases of disease (AIDS, cholera, flu, infectious diseases, smallpox), hygienic-sanitary measures (antibiotics, cleaning, disinfection, oral hygiene, toxinfections, vaccinations), food production (fermentation) and the history of science (almost exclusively Pasteur).

Needless to say, it is understood that these appendices are only considered, and students urged to read them, when the individual teacher voluntarily chooses to introduce that particular subject.

On the whole, each book offers from two to three pictures (usually bacterial cell, cell suspension or colonies; apical mycelial segment with conidiophore or moulded food; or virus particles), whose decidedly non-informative nature is witnessed by the fact that in none of the 30 books is the degree of enlargement specified.

3. METHODS AND SAMPLE

An eight-item questionnaire aimed at identifying and documenting the peculiarities of student beliefs and reasoning was designed and will be quoted, item by item, along with the presentation of data, with the exception of item 8, which was not pertinent with regard to the research objectives:

> 8. Your insights: indicate which of the following subjects concerning micro-organisms you would deal with: what particular features micro-organisms have; how they are formed and fed; how they reproduce and spread; their preferred locations and what they do there; how their presence can be detected and how they can be recognized.

Twenty-five teachers were first given instructions on how to administer the questionnaire (in particular, that they should avoid any interference with spontaneity and the free expression of personal conceptions), and subsequently on how to stimulate more in-depth reflection and explanation in order to clear up pupils' uncertain conceptions and to clarify which experiences and contexts have been responsible for the emergence of odd conceptions.

A total of 502 questionnaires were filled out by 135 students of the first lower secondary grade level (9 classes), 251 students of the second grade level (16 classes), 64 students of the third grade level (5 classes), and 52 students of unspecified grade levels (2 classes).

Each of the participating teachers — selected from 13 different lower secondary schools — involved at least one of the three classes they were teaching. Pupils ranged in age from 11 to 14.

4. DATA ANALYSIS AND RESULTS

4.1 The First Item of the Questionnaire

> 1. Describe the object on the desk (*a piece of fruit, bread or cheese on which mould is growing*) and pay special attention to its unusual features.

This item allows assessment of how students observe and describe, but essentially points out some leading indications about how students represent the effect of mould and micro-organism on food. Moulds look like cobwebs, *show* the presence of and *are made up* of micro-organisms, and develop *when* food goes bad or expires: "There is a very thick cobweb, which perhaps is made of micro-organisms", "There is a substance called mould that forms when bread goes bad", "Mould begins to form after a food's expiry date", "Mould are bacteria which develop on resting food", "Mould forms when a product expires and becomes inedible", "Mould is produced by decomposed bacteria", "The white fur is mould made up of micro-organisms", "Mould is made up of very small micro-organisms which cannot be seen with the naked eye", "Mould consists of microscopic fungi". Exceptionally: "Micro-organisms breaking down the food."

Assertions regarding mites as micro-organisms were frequently reported by teachers following supplementary discussion: a popular conviction openly ascribed to a highly recurrent TV advertisement for "Eminflex" mattresses, which claims to be "absolutely free of mites, moulds and bacteria" thanks to a permanent hygienizing treatment.

Recurring conceptions and noteworthy aspects, which seem to be especially significant with regard to the state and origin of knowledge of micro-organisms, were inferable from the answers to the subsequent items of the questionnaire.

4.2 Items 2 and 3 Concern the Presence of Micro-organisms in a Healthy Body and Sick Body

2. When you are in good health, are there any micro-organisms in your body?
   - If your answer is "no", indicate at least two ways by which you prevent micro-organisms from catching up with or getting into your body.
   - If your answer is "yes", where are the micro-organisms? What are they doing there?
3. When you are ill, are there any micro-organisms in your body?
   - If your answer is "no", indicate what illness you are referring to. Do you think there is any way of avoiding micro-organisms catching up with or getting into your body?
   - If your answer is "yes", indicate what illness you are referring to. Where are the micro-organisms? What are they doing there?

The data are summarized in Tables 1 and 2 respectively, referring to the entire sample, due to the evidence that, according to the optional nature of treating the subject in the curriculum, no meaningful differentiation exists between the three grade levels.

TABLE 1. The presence of micro-organisms in a healthy body (item 2). The item was covered in 477 filled-out questionnaires. The grey lines directly refer to the per cent values of answers; the white lines refer to the per cent values within the previous "grey" category (for instance: among the 62% of students who believed that there were micro-organisms in a healthy body, 24% located them in the digestive organs)

	Per cent values
"NO" micro-organisms in the body	**38**
"YES" there are micro-organisms	**62**
– in digestive organs	24
– in the blood	17
Micro-organisms are useful	**59**
– digestion and "proper functioning"	29
– protection and defence	62
Micro-organisms are harmful	**25**
Actions and operations aimed at getting rid of micro-organisms*	
– proper diet	23
– good hygiene practices	27
– avoiding "contact"	12
– using medicines and vaccines	12

*Answers were expressed by 258 students.

TABLE 2. The presence of micro-organisms in a sick body (item 3). The item was covered in 443 filled-out questionnaires. The grey lines directly refer to the per cent values of answers; the white lines refer to the per cent values within the previous "grey" category (for instance: among the 94% of the students who believe that there are micro-organisms in a sick body, 44% name the disease by one of its symptoms)

	Per cent values
"NO" micro-organisms in the body	**6**
"YES" there are micro-organisms	**94**
– mentioned disease (*absolute value*)	*626*
– symptom as disease	44
where micro-organisms are	
– all over the body	33
– in the blood	4
when we are sick micro-organisms	
– protect and defend us	9
– circulate and reproduce	13
– injure and attack us	25
– attack immune defences	9
– induce diseases or symptoms	40

TABLE 3. Foods or drinks that could not exist if micro-organisms did not exist (item 4). The item was covered in 393 filled-out questionnaires. The grey lines directly refer to the per cent values of answers; the white lines refer to the per cent values within the previous "grey" category (for instance: of the 51% of foods and drinks indicated as produced with the intervention of micro-organisms, bread and pizza represented 30%)

	Per cent values
Because micro-organisms intervene in their production[a]	**51**
– bread and pizza	30
– yoghurt	26
– cheese	21
– wine	10
Because they unavoidably contain micro-organisms[b]	**19**
– water	42
– milk	33
Because they can easily be contaminated	**14**
– fruit, legumes	57
– meat, fish, processed meats such as "salami"	28
Foods and drinks which are indigestible[c]	**12**
– coca cola	42
– cold soft drinks	29

Some foods and drinks were occasionally mentioned in addition to the ones listed: (a) alcoholic drinks, beer, vinegar; (b) "actimel", yeast; (c) sweeties, fried foods, snacks.

TABLE 4. The presence of micro-organisms at home (**A**) and in the garden (**B**) (items 5 and 6). The items were covered in 462 filled-out questionnaires. The grey lines directly refer to the per cent values of answers; the white lines refer to the per cent values within the previous "grey" category (for instance: of the 47% of answers of students placing micro-organisms in rooms, 76% indicated precisely bathroom and kitchen)

		Per cent values
A	**In rooms**	**47**
	– bathroom and kitchen	76
	In places	**15**
	– on the floor and behind/under furniture	86
	In furnishings	**11**
	– in the refrigerator	26
	In food	**5**
	In dust	**5**
	Everywhere, in the air	**5**
B	**In the ground**	**27**
	On plants and grass	**52**
	Everywhere, in the air	**7**
	Dead leaves, compost	**0.3**

TABLE 5. Actions and operations aimed at getting rid of micro-organisms (item 7). The items were covered in 361 filled-out questionnaires. The grey lines directly refer to per cent values of answers; the white lines refer to the per cent values within the previous "grey" category (for instance, 83% of the 36% of students who consider the possibility of ridding the body of micro-organisms indicate medicines as the proper means for doing so)

	Per cent values
From the body	**36**
– medicines	83
From the environment	**20**
– ridding of pests	36
– disinfection, sterilization	42
Prevention (body)	**27**
– diet	16
– personal care and hygiene	51
– vaccines	23
– "actimel", "enterogermina", ferments	10
Prevention (environment)	**12**

Two significant aspects should be emphasized: more than one student out of three thinks that no micro-organisms can be found in a healthy body, and micro-organisms present in a healthy body are mainly concerned with protection and defence.

This last idea emerges clearly: "There are good bacteria moving around in my body that protect the organism"; "Also when we are ill our micro-organisms stay in our body and defend us against microbes"; "A lot of benign micro-organisms defend us against bad ones"; "Micro-organisms protect the intestine from microbes"; "Micro-organisms move around in order to protect and to cure us"; "Our micro-organisms fight against bacteria and viruses that threaten health"; "They defend us: they are called antibodies"; "They are in our bodies in order to defend us against microbes"; "They stimulate immune defences."

Explored in the course of the post-administration discussions were the concepts recognizable in statements like these: defence and circulation. Many teachers agreed in their reports of the conception of a permanent war taking place everywhere in the body between "good" and "bad" micro-organisms, and of students' often explicit references to TV advertisements for yoghurt and ferments. "Actimel", a milk derivative rich in probiotic bacteria, takes a centre-stage position thanks to both the images (the product is compared with a toy-soldier) and the messages conveyed ("reinforces good bacteria in our intestines and sustains their struggle against aggressors"; "helps protect our bodies"; "reinforces the immune system"), leading to some tendentious interpretations of current knowledge about probiotics (Tannock, 1999).

Concerning the second conception, which will reappear later, micro-organisms are supposed to be able to move throughout the body ("Micro-organisms travel in the belly"; "They travel all through the body"): the good ones doing their rounds and the bad ones attempting to produce illnesses. This outlook of road traffic was proven to descend from the very well-known cartoon “Exploring the human body” (De Agostini, from “Il ètait une fois … la Vie”, Procidis, 1985). Over the past 20 years all Italian children have been exposed to this cartoon, with the laudable intention of introducing them to science, and have absorbed a great deal of misleading information, among which is that the inside of the human body looks like a roadway system on an uneven landscape.

Some competent — or nearly competent — opinions were also expressed: “Micro-organisms make the stomach work well”; “Micro-organisms help digestion”; “Bacterial flora are parasites”; “Micro-organisms are part of our bodies (blood,saliva).”

Only a minor percentage of students — while indirectly confirming the opinion that a healthy body is free of micro-organisms and stating that there are no micro-organisms in a sick body — seem to take non-infectious diseases into consideration.

In general, students know about many different diseases (mainly children’s and old people’s diseases), and name them mostly in relation to their most striking symptom (fever, sore-throat, vomiting, etc.).

If micro-organisms are thought to be present, they are in every part of the body (“Micro-organisms are present: they are malignant and spread throughout the body”; “They fill up the whole body”) and obviously have bad intentions (“Micro-organisms tend to make us ill”; “They try to make contact with us, to sneak in our bodies and to attack us in order to produce the substances they need”; “They attack our immune defences (and kill them)”; “They attack internal micro-organisms”; “They attack us in the weakest point”; “They live and multiply at the host’s expense”).

4.3 Item 4

> 4. Indicate at least two foods or drinks that could not exist if micro-organisms did not exist.

Pupils’ open answers have clearly shown that the statement was interpreted in four different ways (Table 3): (1) food and drinks could not exist because micro-organisms intervene in their production (the intended “right” version), (2) food and drinks could not exist because they unavoidably contain micro-organisms; (3) food and drinks could not exist because they can easily be contaminated; (4) food and drinks which are indigestible (due to the presence of micro-organisms?).

Both the systematic coexistence of the four interpretations and the slightly better performance of third-level students should be noted, as well as some clues concerning misconceptions: “Lactic ferments are one of the foods produced by micro-organisms.”

4.4 Presence and Action of Micro-organisms in Familiar Environments was Investigated by Means of Subsequent Items

5. You are at home: indicate at least two places where micro-organisms can be found. What are they doing there?
6. You are in the garden: indicate at least two places where micro-organisms can be found. What are they doing there?

At home (Table 4) micro-organisms such as bacteria and moulds (and mites) are present in the more frequently dirty rooms (bathrooms and kitchens), and where dust collects (spaces difficult to clean). "Micro-organisms are created with dirt." "Micro-organisms create dust."

In the garden micro-organisms are present mainly in the soil (which is *dirty*) and on plants (afflicted by *parasites*). Decomposition was seldom taken into consideration: "They help decompose waste substances."

4.5 The Last Item

7. Indicate at least two actions or operations aimed at getting rid of micro-organisms.

Students' prevailing concern seems to be ridding the body of micro-organisms and preventing infectious diseases (Table 5).

Lack of scientific information yields contrasting opinions: "Perhaps they can never be entirely eliminated!", "They do nothing in our bodies and in the environment."

5. CONCLUSIONS AND IMPLICATIONS

Though the relevance of pupils' alternative conceptions in microbiology was pointed out more than half a century ago (Naguy, 1953), the Italian national curriculum and textbooks corroborate the subordinate role played by this field in lower secondary school education (Hilge and Kattmann, 2003).

Due to the very poor attention dedicated in lower secondary teaching to the real and potential role of micro-organisms in the environment and in the human body, no significant increase in competence is noticeable from first to third lower secondary grade levels when students express their representations and opinions regarding the nature and behaviour of micro-organisms, and the advisability of making the most of them or of defending against them. Some random evidence of detailed knowledge is likely attributable to family background or to the initiative of a particularly willing teacher.

As far as disciplinary knowledge is concerned, students' ideas about micro-organisms were one of the first subjects of the research dedicated to common knowledge and mental representations when constructivist thought began to require this type of information (Brumby et al., 1985; Prout, 1985; Vasquez, 1985) and it still offers rich terrain for teacher consideration (Gillen and Williams, 1993; Bazile, 1994; Simonneaux, 2000; Hilge and Kattmann, 2003).

This study, which involved the administration of a questionnaire to lower secondary school pupils (from 11- to 14-year-old), after which they were encouraged to reason out their answers, confirms the need to induce pupils to express what they are thinking, and also confirms some previously described ideas – such as micro-organisms essential harmfulness to humans (Naguy, 1953) and their association with poor hygiene (Vasquez, 1985) – and the glaring neglect for the role of micro-organisms as decomposing agents (Palmer, 1999; Simonneaux, 2000). It also adds some new ideas that were not previously so evident: the confusion between the different types of micro-organisms (in particular mould and single-cell organisms), between micro-organisms and dust or dirt, between micro-organisms and animal parasites (of plants); the presence of micro-organisms inside and on objects (not free in the air); micro-organisms lying in ambush in foods and drinks; and, on the positive side, the importance played by diet and personal care, and not only of medical treatment such as vaccines, in disease prevention.

Moreover, the ideas that emerged indicate some incorrect ways of looking at (and seeing) micro-organisms of which school-teaching is unaware, and indicate mass media (mainly advertising spots and TV programmes) as pupils' elective source of information. In the absence of any contribution at school, some alternative representations begin to take shape: moulds are not micro-organisms, but *are made of* micro-organisms; food goes bad and *then* micro-organisms infect it; food expiry is a "timing" device that disregards micro-organisms. Moreover, some biased messages help build other descriptive notions: mites (not insects, which are equally "micro") are equated with bacteria and mould thanks to a mattress advertisement; good (beneficial) and bad (harmful) bacteria face-off in combat in our bodies according to an ad for a milk derivative: this misleading ad has done harm in other European countries also (Hőrsch and Kattmann, 2004); a so-called educational cartoon suggests that bacteria walk around the entire body, and leads to mistaking them for antibodies since differentiated cells are shown travelling in speedways from cell-factory to peripheral organs, and proteins and salts are shown walking together with cells, and all of these are approximately the same dimension.

It is easy to arrive at the recommendation that lower secondary schools cover basic microbiology, at least highlighting the ubiquity of micro-organisms, their role as human body microflora and their role in matter cycles. At this school level pupils' experiences and language could be integrated fruitfully into teaching programmes (as advertising spots do) and could thereby pave the way to an upper secondary school's approach to all those topical and enticing microbiology-related subjects such as ecology, genetics, biotechnology, bioethics, and so on (Simonneaux, 2000; Harms, 2002).

ACKNOWLEDGMENTS

Thanks are due to doctor Cecilia Bedetti and her staff of the Ufficio Relazioni Esterne of the National Health Institute (Istituto Superiore di Sanità) who provided for the organization of the courses and handled the distribution and collection of the questionnaires.

REFERENCES

Bazile, J. (1994). Conceptions des opérateurs de bas niveau de qualification du secteur agroalimentaire en matière de microorganismes. *Didaskalia*, 4, 23–37

Bronowsky, J. (1953). *The common sense of science*. (London, UK: Harvard University Press)

Brumby, M., Garrard, J. & Auman, J. (1985). Students' perceptions of the concept of health. *European Journal of Science Education*, 7, 307–323

Campbell, B. & Lubben F. (2000). Learning science through contexts: helping pupils make sense of everyday situations. *International Journal of Science Education*, 22, 239–252

Carey, S. & Smith, C. (1993). On understanding the nature of scientific knowledge. *Educational Psychologist*, 28, 235–251

Driver, R. (1989). Students' conceptions and the learning of science. *International Journal of Science Education*, 11, 481–490

Gillen, A.L. & Williams, R.P. (1993). Dinner with a microbe. *The American Biology Teacher*, 55, 268–274

Grimellini Tomasini, N. & Segrè, G. (Eds) (1991). *Conoscenze scientifiche: Le rappresentazioni mentali degli studenti*. Firenze. (Italy: La Nuova Italia)

Harlen, W. (2002). Links to everyday life: the roots of scientific literacy. *Primary Science Review*, 71, 8–10

Harms, U. (2002, December 15). Biotechnology Education in Schools.*Electronic Journal of Biotechnology*, ISSN 0717-3458. Retrived from http://www.ejbiotechnology.info/content/vol5/issue3/teaching/01

Hilge, C. & Kattmann, U. (2003, August). *The significance of microbes for biology teaching – A study of scientific and students' conception*. (Paper presented at the 4th ESERA Conference, Noordwijkerhout, the Netherlands)

Hőrsch, C. & Kattmann, U. (2004, September) *Micro-organisms and microbial processes in the human body – A contribution to Educational Reconstruction*. (Paper presented at the 5th *ERIDOB* Conference, Patras, Greece)

Lave, J. (1993). The practice of learning. In S. Chaiklin & J. Lave (Eds), *Understanding practice. Perspective on activity and context* (pp. 3–32). Cambridge: Cambridge University Press)

Naguy, M.H. (1953). The representation of germs by children. *Journal of General Psychology*, 83, 227–240

Neisser, U. (1976). *Cognition and reality: principles and implications of cognitive psychology*. (New York: W.H. Freeman)

Palmer, D.H. (1999). Exploring the link between students' scientific and non-scientific conceptions. *Science Education*,83, 639–653

Pfundt, H. & Duit, R. (1991). *Bibliography of students' alternative frameworks in science education*. (3rd edition, Kiel, Germany: IPN)

Prout, A. (1985). Science, health and everyday knowledge. *European Journal of Science Education*, 7, 399–406

Simonneaux, L. (2000). A study of pupils' conceptions and reasoning in connection with 'microbes', as a contribution to research in biotechnology by education. *International Journal of Science Education*, 6, 619–644

Solomon, J. (1994). The rise and fall of constructivism. *Studies in Science Education*,23, 1–19

Tannock, G.W. (1999). *Probiotics. A critical review.* (Norfolk: Horizon Scientific Press)

Vasquez, E. (1985). Les representations des enfants sur le microbes. *Feuilles d'epistemologie appliquée et de didactique des sciences*, 7, 31–36

JENARO GUISASOLA, JOSÉ L. ZUBIMENDI, JOSÉ M. ALMUDÍ AND MIKEL CEBERIO

17. USING THE PROCESSES OF ELECTRICAL CHARGE OF BODIES AS A TOOL IN THE ASSESSMENT OF UNIVERSITY STUDENTS' LEARNING IN ELECTRICITY

Abstract: The research study which is set out here constitutes one stage of a project whose goal is to develop situations for the teaching and learning of electricity at university level. The subject of electricity continues to be widely regarded by students as difficult and therefore unattractive. A particular problem is the relation between electrostatics and electrokinetics in calculus-based physics courses in first-year university courses. This research study has allowed us both to check the university students' learning of the basic concepts of electricity, in particular electrical potential and capacitance, and to define the important conceptual difficulties that students came across when studying electricity and more precisely processes of electrical charge of bodies. Difficulties in analyzing the processes of charge of bodies from a systemic and energetic point of view are discussed. These results will be used to elaborate contents and situations for the teaching and learning of electricity at university level

Keywords: Electrical capacitance, Electrostatics to electrokinetics, Potential difference, Students' difficulties, Ways of reasoning

1. INTRODUCTION

The subject of electricity continues to be widely regarded by students as difficult and therefore unattractive. A particular problem is the relationship between electrostatics and electrokinetics in introductory physics courses in the last year of secondary school and the first years of university, as it involves extremely complex and highly abstract concepts and is thus totally dependent on macro–micro relationships. Research consistently shows very poor student understanding following the teaching of electricity (Duit et al. 1985; Psillos 1998).

From a constructivist perspective, knowledge about the way in which students reason is an essential element in the didactical reconstruction work of teaching objectives and contents. In the field of electricity, many studies have pointed out some of the difficulties encountered by young pupils or teenagers in order to explain simple electrical circuits of continuous current (Mulhall et al., 2001).

Jenaro Guisasola, Departamento de Física Aplicada I, Escuela Universitaria Politécnica, Plaza de Europa 1, Donostia 20018, Spain

R. Pintó and D. Couso (eds.), Contributions from Science Education Research, 225–236.

In this respect, recent electrostatic studies show that students have difficulties in applying scientific explanations to 'elementary' electrostatic phenomena (Park et al. 2001; Furió et al. 2004). Few studies focus on analysing students' conceptions in the processes involving electrostatics and electrokinetics. These studies analyse specific aspects such us the role of the battery (Benseghir and Closset 1996) or relationships between capacitance and electrical current (Thacker et al. 1999). However, they do not deal with the students' ideas regarding the first stage of the transition between electrostatics and electrokinetics, more precisely the processes of a body's electrical charge. These processes have no easy scientific explanation and require significant knowledge of the basic concepts of electricity (charge, electrical potential, capacitance, etc.). Thus, in this research we should try to analyse students' difficulties in learning the scientific explanations of these basic processes at university level.

2. RESEARCH QUESTIONS AND PROBLEM

The analysis of students' mistakes and first preliminary interviews led us to study the method of reasoning linked to the use of electrical potential and capacitance in the processes of electrical charge of a body. We have chosen this type of phenomena because their interpretation by means of the theoretical concepts of electrical potential and electrical capacitance cannot be simply induced from the evidence. The scientific interpretation of the processes of charge of a body imply a significant learning of the basic concepts of electrostatics such as electrical charge, electrical potential and electrical capacitance which are necessary prerequisites for a comprehensive learning of electrical theory. Therefore, the correct interpretation by students of this type of process may serve as an indicator of the learning achieved in the first lessons of an electricity programme at university level. We present in this chapter the results obtained as well as the students' reasoning strategies linked to alternative explanations of these phenomena. So the questions research which arise from this problematic are the following:

- What do we mean by a good understanding of the phenomena of electrical charge of bodies?
- What conceptions and forms of reasoning do students use when dealing with the processes of electrical charge of bodies?

In an attempt to answer the first research question we made a bibliography review regarding the physics theoretical core and, in summary, we obtained the following scientific explanation. To charge a body is to do work in the system body-environment, which is stored in the form of potential electrical energy in the charged body. The body will charge until the potential difference between it and the battery is equal. Each body possesses an electrical capacitance that measures the greater or lesser ability to charge itself with respect to the work done to create it. The concepts of potential and electrical capacitance are interesting for physicists, in that it is only necessary to know the macroscopic variables of the process (the potential difference in the system battery-body and the geometrical of the body to be charged). It is not necessary to know the value of other variables such as the current which passes through the wire, the electrical field, etc.

In accordance with the above, the questions which are used to test the students' representations about the concepts of potential difference and electrical capacity involve processes of electrical charge of a body, which are made up of a source of energy, conducting wire and bodies. An electrical current takes place between the highest potential and the body with the lowest potential. Such a process can be regarded as an electrical charge of a body when the difference of potentials, which is the cause of the transformation, tends towards zero.

3. EXPERIMENTAL DESIGN AND SAMPLE

The population studied is made up of students attending the first and third year of the Industrial Engineering Degree. 116 students completed the questionnaire; they belonged to groups from the first year of university (53 students) and the third year (63 students). To devise the nine questions of the current questionnaire (see Annex), an earlier study was conducted with small samples of students from each level. This confirmed that, in general, students (even those in the lowest level) had no problem in understanding the meaning of the questions. We interviewed a small sample of 11 students from different university levels (7 from the first year and 4 from the third year), in order to go deeper into their explanations when filling in the questionnaire. The questionnaire and interviews were completed by students after the teaching of electricity in a calculus-based physics course in the last part of the second semester of the course.

4. RESULTS AND DATA ANALYSIS

The preliminary questionnaires led us to consider the working hypothesis that in situations where a body is charging the students have clearly understood that the process of charge consists of the passing of charge from one part of the system to another but they have not taken into account the basic concepts of electricity (charge, potential difference, capacitance, etc.) to explain the process correctly. We then produced nine questions (see Annex) to find out about the students' ideas. The questions which were proposed refer to processes involving electrical charge of bodies related to concepts such as potential difference (questions 1 to 5) and electrical capacitance (questions 6 to 9), which are taught in class during an introductory physics course. More than one question was used to check each concept. Checking the same topic in different contexts also makes it easier to elicit the students' knowledge characteristic (Viennot 1996).

Next, we describe the process followed in order to analyse the answers. We derived a set of categories for each question on the basis of the established goals and the exhaustive analysis of the answers of questionnaire, by a member of the research team. The students' answers were classified according to the categories defined. Once the categories were discussed with two other components of the research team, we held a training session in which we examined 10% of the sample. Next, the members of the research team went on to analyse all questionnaires independently. The level of agreement between the reviewers on classifying the answers was more than 85% for each question. In the cases of disagreement, the definitive categorization was made by means of discussion and consensus between the three reviewers.

We will present the results of the first five questions and the categories devised in Table 1 followed by an explanation of some of the results.

The goal of questions 1 to 5 is to highlight what ideas and reasoning the students use to account for the fact that a body will charge until the potential difference between it and the battery is equal. For the educational levels involved we have considered a correctly grounded justification to be the one that uses the concept of potential difference to explain the process of charge in a body in the different situations presented in these five questions. We called this group of answers category B "Based on potential difference". We grouped the other answers into two categories. If the students explain that electrical charge of a body depends exclusively on the size of the body and that charges move from where there is a 'greater quantity of charge' to where there is a 'lesser quantity of charge', the answer is classified as category A "Based on the charge quantity". Finally, the answers that are descriptions of a formula or the justification giving cause and effect reasoning based on the formula are classified as category C "Based on the formula". The number of non-codable and non-answers is very low, under 10% in most of the questions.

Next, we will describe some of the questions in detail. Question 3 asks for an explanation of the process of charge of a body within the system constituted by the body and a battery. Both body and battery interact and produce changes which are quantified by means of the potential difference which exists between the bodies. The potential of the battery is written explicitly and this information may stimulate the student to use the said concept. However 83% of the students from year 1 and 77.7% of those from year 3 include reasoning based on the magnitude of the electric charge included in Category A and that involves the following characteristics:

TABLE 1. Percentage of Students Answering Questions 1 to 5 for All Groups

Questions	Percentage of answers %					
	Category A "Based on the charge quantity"		Category B "Based on potential difference"		Category C "Based on the formula"	
	1st Year	3rd Year	1st Year	3rd Year	1st Year	3rd Year
Q1	73.5	50.7	22.6	28.6	–	–
Q2	92.4	87.3	3.7	4.7	–	–
Q3	83.0	77.7	13.2	14.3	–	–
Q4	37.7	34.5	24.5	19.0	26.5	31.7
Q5	41.5	35	20.7	14.2	20.7	28.5

- The explanations of the students describe electricity as a property that depends exclusively on the size of the bodies and that moves from where there is a 'greater quantity of charge' to where there is a 'lesser quantity of charge'.
- The explanations of the students consider only the body to be charged. Thus, the body is considered as a recipient that admits a certain quantity of charge depending on its size. For example: "It will charge until the charge in both bodies is the same or the body is full" (1st year)
- The large majority of explanations indicate that the conductor body will allow electricity to pass through it and it will charge, whilst the insulating body will not allow the charges to pass and it will not charge. For example:

I think that the metal sphere will acquire more charge, given that it is a conducting material and the plastic will not charge as it is not such a material. (year 3)

Within this category there are an appreciable number of explanations (28% in year 1 of Engineering and 30% in year 3 of Engineering) that show explicitly that an insulator cannot be charged in any circumstance whatsoever.

13.2% of the students from year 1 and 14.3% of those from year 3 indicate in their explanations that the concept of difference in potential existing between the battery and the body is what allows us to analyse and know about the process of charge (Category B). For example:

The sphere will charge until the battery and body have the same potential, at which time the charge will stop passing along the wire. In addition, there will be a limit to the charge of the ball depending on its radius and its dielectric constant (1st year).

The metal sphere will acquire more charge, given that the plastic sphere charges because of an irregular distribution that concentrates the charge, with less voltage, and therefore, with less electrical energy (3rd year).

Question 5 asks explicitly about the role of the electrical potential in the process of charge. The objective is to ascertain the meaning that students attribute to the electric potential magnitude in relation to the process of charge. 41.5% of the students from year 1 and 35% of those from year 3 attribute to the electrical potential the meaning of "indicator of quantity of charge" which the body can contain (Category A). The explanations of the students establish identification between charge and potential. For example:

As the body is charged, the potential increases lineally because they are directly proportional. The greater the potential of the battery the greater the quantity of charge that reaches the body, which in turn increases the potential of the body (3rd year).

20.7% of the students from year 1 and 14.2% of those from year 3 base their explanations on descriptions of the formula $C=Q/\Delta V$ (Category C). Although in the statement of the question it is pointed out that formulae should not be used, these students establish a cause-and-effect interpretation of the formula. For example:

the potential of a conductor whilst it is charging is inversely proportional to the capacity of the conductor: $\Delta V=Q \,.\, 1/C$ (1st year).

20.7% of the students from year 1 and 28.5% of those from year 3 mention in their explanations that the process of charge depends on the difference of potential battery-body (Category B). The process is detained either when the potential of the sphere coincides with that of the battery or the charge in the conductor, being very high, manages to polarize the medium. In addition, they establish that the sources of potential are the charges (not forgetting how they are distributed). For example:

The conductor will charge very easily at the beginning because there is no charge to repel it and the potential will increase as the charge accumulates in an ordinate way on the surface. Once the potential of the body reaches that of the battery, the charge will be completed (3rd year).

The results of questions 1, 2 and 4 (see Table 1) are convergent with the aforementioned comments and they seem to indicate that the students have no scientific explanation for the concept of difference of potential and fail to establish a significant relationship between potential, charge and electrical capacitance.

The second part of the questionnaire, questions 6 to 9, aims to investigate the significance that the students attribute to the capacitance concept. The questions present situations in which students must consider the capacitance of a body as its facility to acquire charges in relation to the work necessary to charge it. The answers that use this type of reasoning were considered as correct answers and grouped in category B. The answers that wrongly point out that capacitance only exists when there is charge or that confuse potential and capacitance are grouped in category A. Finally, the answers that are descriptions of the formula were grouped in category C. The percentage of non-codable or non-answers is very low and it gives us an idea of the interest with which students completed the questionnaire. The results obtained are presented in Table 2.

Question 8 asks specifically about the electrical capacitance of a body in relation to its charge. A correct interpretation of the question would use the concept of potential depending on the geometry of the body, the type of material and the energy stored in the process of charge. The correct answer to the situation considers the capacitance of a body as its facility to acquire charges in relation to the work necessary to charge it, independently of whether or not it has a net charge (Category B). The incorrect answers will tend to identify null net charge with null electrical capacitance.

TABLE 2. Percentages of Students Answering Questions 6 to 9 for All Groups

Questions	Percentage of answers (%)					
	Category A "Based on the charge quantity"		Category B "Based on meaningful distinction between V, Q and C"		Category C "Based on the formula"	
	1st year	3rd year	1st year	3rd year	1st year	3rd year
Q6	41.5	58.7	3.7	3.1	50.9	30.1
Q7	69.8	71.4	15.0	20.6	–	–
Q8	22.6	36.5	11.3	19.0	52.8	33.3
Q9	49.0	47.6	7.5	6.3	32.0	31.7

The 22.6% of the students from year 1 and 36.5% of those from year 3 will consider only the magnitude of electric charge in their explanations (Category A). In these explanations the electrical capacitance is identified with the charge, as if the body were a container of charges and therefore the electrical capacitance were independent from the potential of the body. For example:

The electrical capacitance is the quantity of charge that a body can admit (1st year).

Other answers of Category A understand the electrical capacitance as the facility to drive electrical charge (not to store it) and, therefore, an identity between electrical charge and electric current is established. In this way, they associate electrostatic with electrodynamic situations, mistaking the reason for which the bodies have capacitance (their constitution and their geometry and, consequently, the charge that they are capable of storing under a particular potential). For example:

It has no capacitance because as it is not charged there can be no transport of charge nor electrical current (1st year).

It has no capacitance because the capacitance is directly related to the presence of charges in movement (3rd year).

52.8% of the students from the first year and 33.3% of those from the third year justify their affirmations giving cause-and-effect reasonings based on the formula C= Q/V (Category C). For example: "If $Q = 0 \Rightarrow C = Q/\Delta V = 0/\Delta V = 0$ and if $Q \neq 0 \Rightarrow C = Q/\Delta V \neq 0$".

A minority of explanations, 11.3% of the students from year 1 and 19% of those from year 3, respond with a correct use of the electrical capacitance concept and defining in a meaningful way the relation between charge, potential and electrical capacitance (Category B). For example:

The capacitance of a body, is the facility with which it can store energy, depending on its size and form and also the potential difference between the battery and the body. Therefore, a body without charge also has electrical capacitance (1st year).

Adequate analysis of the situation presented in Question 9 implies considering that to charge the conductor equally in both cases, less work would be carried out for the case of the oil (the conductor's potential is less). In order to reach this conclusion it is necessary to take into account the polarization of the medium where the conductor is found, or in other words by significantly applying the concepts of electrical potential and the electrical nature of the material.

In category A answers (49% in the first year and 47.6% in the third year) the students indicate that as the conductor's charge does not vary then its electrical capacitance does not vary either. Some representative examples are shown below:

The capacitance would be the same seeing as it is the same conductor in situations where they are not surrounded by any type of charge, so the capacitance shouldn't vary (1st year).

It will depend on whether the oil is good or bad or the air as a dielectric as some electrons can flow through dielectrics (air, oil...) (3rd year).

In Category C "Based on the formula" (32% for the first year and 31.7% for the second year) the majority of the answers use the formula which is obtained from calculating the capacitance of a condenser with flat and parallel plates, with the approximations which are generally made at this level. Let us see some examples.

The capacitance varies depending on the permittivity ε. If the permittivity of the oil is greater than that of the air, then the capacitance will increase and if it is less then it will decrease, because the capacitance is directly proportional to ε (3rd yr ETSII).

$C = \varepsilon_0 \varepsilon_r$ A/d where $\varepsilon_r > 1$ for the oil, then we would have the situation where the capacitance of the conductor is greater in oil than in the air (1st yr EUITI).

The explanations based on polarization of the medium and on the potential have been grouped into Category B (7.5% for the first year and 6.3% for the third year). The students mention the change in the conductor's potential due to the influence of the medium surrounding it. For example:

The separation of charge which exists in the oil when charging the conductor is the cause of the reduction in the conductor's potential and so therefore the increase in capacitance in the second case (1st yr EUITI)

Although question 9 asked directly about the influence of the medium in the charging process, we must note that approximately a quarter of the students in both years explicitly mention that the medium is irrelevant in relation to the capacitance to accumulate conductor charge. For example:

It would be the same because the capacitance of a conductor does not depend on the material surrounding it, but it depends on its size and the material it is made out of (3rd year).

...in both situations the capacitance is the same. This is due to the fact that the electric field which exists in the conductor is not affected by the medium in which it is submerged (3rd year).

4.1 Discussion of the Results

The results obtained confirm that students fail to use basic concepts of electricity such as electrical potential and electrical capacitance to explain processes involving charge of a body. This would confirm poor learning of electricity in university physics courses. In this sense, it is necessary to highlight that the chi-square statistic was calculated for the first- and third-year group in category B ('correct answer') for all questions, obtaining no significant differences. These results indicate that, although students receive more instructions, their explanations do not become correct in terms of applying the concepts of electricity to obtain a scientific explanation.

In accordance with the results obtained different aspects can be highlighted concerning students' difficulties in interpreting the processes of charge of a body and using the concepts of electrical potential and electrical capacitance in a significant way:

- A majority of the students explain the processes of charge of a body as the passing of charge which goes from the generator to the body to be charged. This passage of charge is because of the difference in quantity of charge between both (about 80% of the answers to questions 1, 2 and 3). In these explanations it is not necessary to discuss the difference of potential between the bodies to explain the process of charge (about 40% in question 5). Likewise, the magnitude electrical capacitance is extended as the quantity of charge that can be stored in a body and, therefore, this magnitude has no significance for the bodies without charge (about 30% in question 8 and more than 40% in question 6).

- A significant percentage of the answers are based on a literal description of a formula or on the incorrect causal analysis of a formula. This type of explanation increases in the questions related to the meaning of concepts such as electrical potential (about 50% in question 4) and of electrical capacitance (about 30% in question 9). It seems that when failing to make any sense of the concepts the students take refuge in the operative definitions and reason on the basis of these. The results are convergent with other studies in different areas of science and at other educational levels (Viennot 1996, Furió & Calatayud 2001).
- Only a minority of explanations analyse the process of charge of a body as a system between the source and the body to be charged and relate meaningfully the concepts of electrical charge, difference of potential and electrical capacitance to these processes. The results indicate that only a minority of explanations mention the potential difference as the magnitude which allows us to analyse the process of charge of a body (less than 25% in question 3). Even when asked specifically for the role developed by the magnitude difference of potential in the process of charge, the answers remain in the minority (about 20% in question 5). As a consequence very few explanations relate meaningfully the concepts of electrical charge and potential difference by means of the concept of electrical capacitance (less than 20% in questions 6 to 9).

5. CONCLUSIONS AND IMPLICATIONS

What emerges from the analysis of students' responses to the questionnaire is the realization that students are not able to tie concepts from electrostatics into their description of phenomena occurring in the processes of charge of a body. As already noted, this leads to a number of difficulties. First, the concepts of voltage and capacitance remain vague; its formal definition is not utilized operationally. Second, most students do not use the concepts of potential difference and electrical capacitance, and therefore they do not create a consistent picture of the mechanism to explain the phenomena. We note in passing that this situation does not necessarily represent misconceptions, but rather the lack of any clear concepts. Consequently, we encountered contrived explanations based on the quantity of charge. Third, we believe that this absence of a electrostatics–electrokinetics link impedes students' ability to conceptualize the process of charge of a body as a system and to appreciate the functional relationships between its parts.

These results will be used to elaborate activities at senior high school (ages 16–18) or early college level in electricity. We think an accurate use of concepts (potential, field, capacitance, etc.) cannot be expected before these levels.

After students study of electrostatics, including a rigorous clarification of central concepts such as electric field and potential, the study of processes of charge of bodies and d.c. circuits should be tied explicitly to electrostatics. For example, in the study of charge of bodies, students could be asked to predict the behaviour of different bodies with different batteries, could observe it experimentally or in a simulation, and arrive at a simple microscopic model to explain the phenomena. Such activities can be repeated for a variety of situations during the introduction of the study of electric circuits.

It is essential to relate fields and potential – those studies in electrostatics – and mechanisms of movement of charges through the bodies or wires studied in electrokinetics. These activities will provide students with the opportunity to become aware of the contradictions they express when asserting that the charge quantity is the magnitude that controls the processes of charge of bodies. Thus, students will be led to regard the processes of charge as a system involving work and energy transformations.

REFERENCES

Benseghir, A. & Closset, J.L. (1996). The electrostatics–electrokinetics transition: historical and educational difficulties. *International Journal of Science Education*, 18(2), 179–191.

Duit, R., Jung, W. & Rhoneck, C. (1985). *Aspects of understanding electricity: The proceedings of an international workshop*. (Kiel: I.N.P.).

Furió, C. & Calatayud, M.L. (2001). Functional fixedness and functional reduction as common sense reasoning in chemical equilibrium and geometry and polarity of molecules. *Science Education*, 84, 545–565.

Furió, C., Guisasola, J. & Almudí, J.M. (2004). Elementary electrostatic phenomena: historical hindrances and students' difficulties, *Canadian Journal of Science, Mathematics and Technology Education*, 4(3), 291–313.

Mulhall, P., Mckittrick, B. & Gunstone, R. (2001). A perspective on the resolution of confusions in the teaching of electricity, *Research in Science Education*, 31, 575–587.

Park, J., Kim, I., Kim, M. & Lee, M. (2001). Analysis of students' processes of confirmation and falsification of their prior ideas about electrostatics. *International Journal of Science Education*, 23(12), pp. 1219–1236.

Psillos, D. (1998). Teaching introductory electricity. (In A. Tiberghien, E.L. Jossem & J. Barojas (Eds.), *Connecting research in teaching in physics education with teacher education*. International Commission on Physics Education. Published electronically at URL http://www.physics. ohio-state.edu/ jossem/ICPE/ BOOKS .html)

Thacker, B.A., Ganiel, U. & Boys, D. (1999). Macroscopic phenomena and microscopic processes: Student understanding of transients in direct current electric circuits. *Physics Education Research (A supplement to the American Journal of Physics)*, 67(7), S25–S31.

Viennot, L. (1996). *Raisonner en physique. La part du sens commun*, (De Boeck Université París). Translated to English in Kluwer Academic Publishers.

ANNEX

QUESTIONNAIRE

Q1. A metal marble is charged when it is put into contact with another charged body by means of a metal wire. So:

a) The marble will accept the charge indefinitely until there is no more charge in the other body.
b) A moment will come when the marble will not accept any more charge, even though the other body continues to charge.
c) Another possibility.

Q2. If the marble in the previous question was made of wood, would the charging process vary?

Q3. Two spheres of equal radius R, one of metal and the other of plastic, are connected separately (see Figures) to a 15 volt generator. Which of them will acquire more charge? Why?

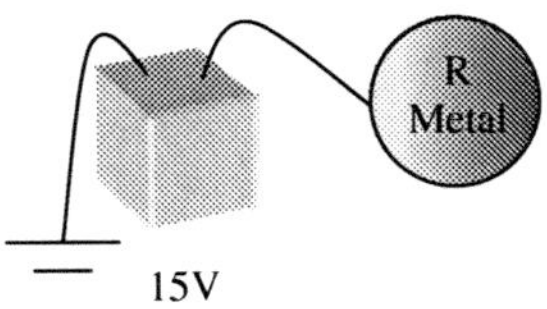

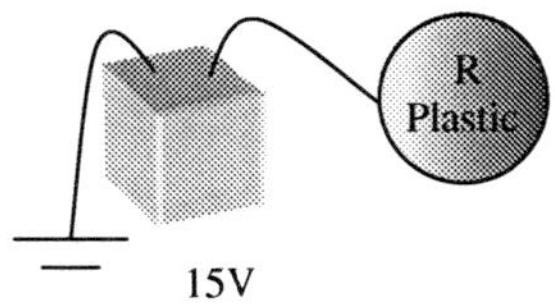

Q4. Represent on a graph how the potential (V) of a conductor varies during its charging process:

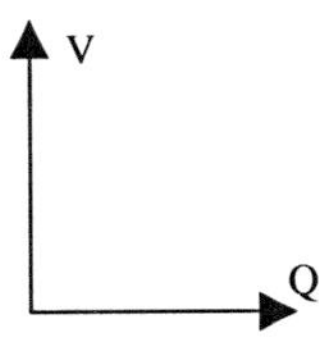

Q5. Explain in words, without using formulae, how the potential of a conductor varies whilst it is charging. What happens to the difference of potential of the battery-body system?

Q6. Explain what the electrical capacitance of a conductor means to you. Does it make sense to talk about the electrical capacitance of an insulator?

Q7. Two spheres with the same radius R, one made of metal and the other made of plastic, are separately connected (see diagrams) to a 15 volt generator. Which of them will have the greater electrical capacitance? Why?

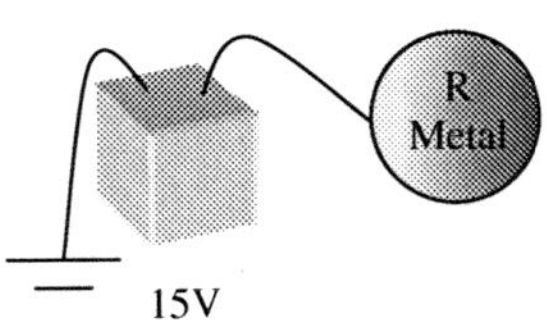

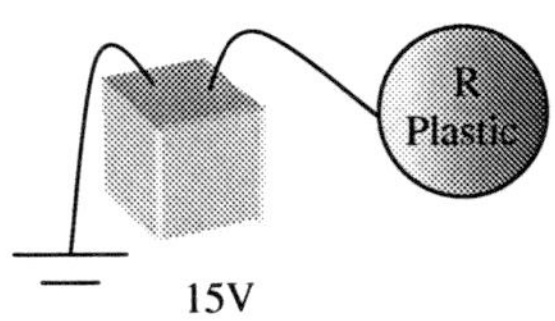

Q8. Consider the same conductor in two different situations (see Figures):

a) Its net charge is null. Does the conductor have electric capacitance?

b) The body is charged with a net positive charge +Q. Does the conductor have electric capacitance?

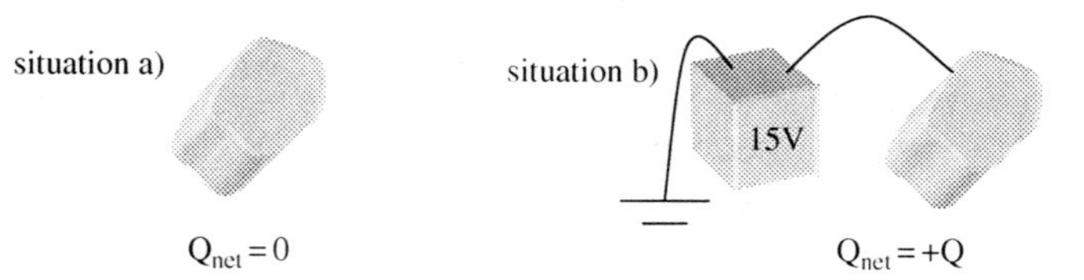

Q9. Consider the same conductor in two different situations:

a) surrounded by air.

b) submerged in oil.

Showing your reasoning for choosing which medium is better for electrically charging the conductor. Indicate if the capacitance of the conductor in air is greater, less or equal to the capacitance of the conductor in oil.

18. REPRESENTATION AND LEARNING ABOUT EVAPORATION

Abstract: While there has been considerable research on children's understanding of evaporation, the representational issues entailed in this understanding have not been investigated in depth. This study explored students' engagement with evaporation phenomena through various representational modes. Primary school classroom sequences and structured interviews shortly after, and a year later, indicated significant advances in learning flowing from negotiation of meaning around particle representations. A case study of one child's learning is used to demonstrate how a molecular distribution representation can offer the possibility of significant advances in children's thinking about evaporation. The findings suggest that teacher-mediated negotiation of representational issues can support enriched student learning

Keywords: Evaporation, Literacies of science, Representation and learning, Student learning in Science

1. INTRODUCTION

Factors affecting children's understanding of evaporation, and how and when learning challenges might be met, have attracted considerable research interest over the last twenty years (Osborne & Cosgrove, 1983; Russell et al., 1989; Bar & Galili, 1994). Following Piaget (1930, 1970) and others, much past research has drawn on a conceptual change or progression approach to children's understandings, where children are expected to pass through predictable stages of explanation. In this paper we consider the relationship between the conceptual and representational challenges for students in learning about evaporation.

There is also growing recognition in science education research that science as a discourse is a mix of languages entailing multi-modal forms of representation, where linguistic, numerical, and graphic and tabular modes are integrated to represent scientific explanations (Roth, 1995; Ainsworth, 1999; Kress et al., 2001; Russell and McGuigan, 2001; Stenning, 1998; Unsworth, 2001; Lemke, 2004). We draw on this literature to provide an interpretive lens for analysing an upper primary school child's engagement in a classroom program on evaporation. The research questions guiding the study were:

1. What representational challenges do children experience in engaging with particulate accounts of evaporation?

Russell Tytler, Professor of Science Education, Deakin University, Pigdons Road, Geelong Victoria 3217, Australia

R. Pintó and D. Couso (eds.), Contributions from Science Education Research, 237–248.

2. In what ways can this focus on representational issues contribute to current understandings of students' conceptual learning?
3. In what ways does an explicit focus on representation enable advances in learning?
4. In particular, what affordances does a molecular representation offer for understanding of evaporation by Grade 5 students?
5. What are the implications for teachers' understandings of children learning science, and for classroom practice?

2. LITERATURE REVIEW

Past research into children's understanding of evaporation, drawing on Piaget (1930, 1970), has tended to focus on explaining children's learning in terms of a developmentally-based progression in conceptions. Bar & Galili (1994) claimed to identify four distinct stages in children's progression in understanding evaporation and condensation: water disappears, water is absorbed into surfaces, water is transferred upwards, and water disperses into air. They associated these stages with particular ages, with the highest stage occurring at about age 13. Drawing broadly on a Piagetian orientation, researchers have sought to explain students' difficulties and misconceptions with particle explanations in terms of the abstract nature of the particle model (e.g. Driver, 1985) or ontological problems with the idea of particles (e.g. Scott, 1993).

However, the stability and sequential nature of evaporation conceptions have been questioned by various researchers, including Tytler and Peterson (2004), indicating both strengths and limitations to a Piagetian orientation. Other researchers, such as Johnson (2005) argued that particle theory is critical in supporting satisfactory representations of evaporation phenomena such as water boiling, because without particulate ideas, the notion of the liquid-gas transition cannot be successfully imagined.

At the same time, research in science education has also focused strongly on the representational demands of this subject as a crucial aspect of learning. From this perspective, students need to understand and conceptually integrate different representational modalities or forms in learning science and learning how to think and act scientifically (Karmiloff-Smith, 1992; Roth, 1995; Stenning, 1998; Ainsworth, 1999; Russell and McGuigan, 2001; Lemke, 2004). These researchers argue that to learn science effectively students must understand different representations of science concepts and processes, and be able to translate these into one another, as well as understand their co-ordinated use in representing scientific knowledge and explanation-building. We would follow Peirce (1978) and others, in distinguishing between a concept (for example, the idea of evaporation), its representation in a sign or signifier (verbal and diagrammatic accounts of evaporation), and its referent, or the phenomena to which both concept and signifier refer (changes to states of matter in the world). Each of these triadic elements are distinguishable from each other.

Drawing on these literatures, our paper reports on one student's responses within a classroom program that sought to engage students with molecular representations to support learning about evaporation. The students were expected to develop an understanding of this topic through interpreting, constructing, and integrating various representational modes such as diagrams, verbal accounts, gestures, and captioned drawings relevant to key inter-related concepts in the topic. This engagement entailed students a) clarifying their thinking through exploring representational resources, b) developing understanding of what these representations signify, and c) learning how to construct and interpret scientific representations about evaporation.

3. RESEARCH DESIGN

This study used a qualitative methodology (Denzin & Lincoln, 1995) to analyse one child's response, as part of a larger study of nine children, to a lesson sequence on evaporation with three grade 5 classes (age 11) in an Australian primary school. The sequence was followed by two interviews spaced a year apart, with nine of the children who have been involved in a longitudinal study (Tytler & Peterson, 2004, 2005). A sequence of classroom activities was devised to provide an instructional context in which children's ideas were probed and challenged. As part of this sequence, a molecular model was introduced as a potential representational resource and explanatory framework, and classroom discussion focused on using this to interpret aspects of phenomena such as boiling water, condensation, and the distribution of eucalyptus oil scent around a room. Students were asked to interpret these phenomena using diagrammatic representation, written captioning, and verbal explanation, and to integrate these representations. The sequence is described in more detail elsewhere (Tytler & Prain, 2005). In follow up interviews, the children were presented with a variation of these activities and asked to represent them in different modalities, and challenged by the interviewer to elaborate and refine these representations and their conceptual meaning.

3.1 Introducing the Molecular Representation

At the start of the sequence children filled in a questionnaire, using drawings and annotations, concerning their explanations of a number of evaporative phenomena. Following this, the researchers conducted a lesson in each of the three classrooms. This consisted of a whole class orientation activity in which children watched a frying pan of water boil, One of the researchers then presented the class with a model consisting of plastic beads in a perspex tray, which was used as the basis of a discussion on evaporation involving molecules of water breaking free of the surface and going into the air. A drawing was then constructed, during class discussion, which depicted water boiling in terms of molecules of water changing to a gas (spread out molecules) in the bubbles. The molecular model was presented as a scientific idea the children might find useful in thinking about these phenomena, and the language used was relatively open so children would not feel they were being presented with a 'one true perspective'. The discussion attempted to model the productive interplay that can occur between different representations: pictorial, physical, verbal and gestural.

Following this discussion children worked in groups on three successive activities representing different evaporative contexts (condensation on a cold can, a 'disappearing handprint', involving a wet hand placed on a towel, and a class activity in which a bottle of eucalyptus was opened in the front of the class and children asked to explain the patterns of scent that occurred round the room. For each of these activities children were asked to write on a worksheet an explanation of what happened. There were follow up activities run by the teachers in the subsequent week, including observation of a puddle drying up, for which children drew their interpretation. The children were interviewed the following week, and again a year later to explore whether their ideas were robust over this time.

The intention in introducing particle representations was to provide students with a way to a) make connections across different states of matter in the evaporative process, to b) imagine how water can exist in air, and to c) understand the various associated effects of the evaporative process. In the discussions surrounding the lesson activities, we noticed a tendency for the students to over determine the actions of molecules. In the subsequent interviews we focused specifically on representing molecular distribution.

4. CASE STUDY: CALUM

We have selected one case to report in some detail the connections between accessing and negotiating a molecular representation, and learning. The case was selected because Calum demonstrated strong and diverse connections across different representational modes, thus providing rich data for interpretation. The thrust of findings concerning learning pathways and gains are representative of most of the students (Tytler et al., 2006).

Calum is a child who has a history of interest in science from before he started school. During the course of the longitudinal study his ideas on evaporation shifted as he became able to construct coherent narratives that traced what happened to water when it evaporated, and related these to events from his personal life in ways that showed a commitment to ideas and interest in constructing deeper explanation. Calum tends to push for narrative accounts of phenomena that make sense of underlying mechanisms, and focus on details of phenomena to make sense of what is happening. His history of evaporative ideas is described in some detail in Tytler and Peterson (2004). By Grade 2 he consistently used water cycle imagery to talk about evaporation, and in Grade 3 made the advance of talking about water being taken up into the air.

"It's like when a puddle evaporates and it goes into the cloud but there's not clouds cos we're in a room here, so it just becomes a part of the air.... water that's evaporated. And it rises up."

Thus by Grade 3 Calum had made the advance over a water cycle notion of water going up through the air to the sun or clouds, to realise the local role of air in holding water, at least as a temporary station, while doors and windows are shut, on the way to clouds. He consistently demonstrates this. He seems not to have a detailed view on just how water can exist in air.

In the prior knowledge probe at the beginning of the Grade 5 sequence, his explanation of the water level in a fish tank dropping is consistent with this insight: "It slowly was evaporated from the heat and eventually will rise to the clouds". At no point up to this had he entertained the idea of a particle interpretation.

In his written accounts, during the first lesson, of condensation on a cold can, and the eucalyptus oil smell, he uses the idea of molecules, particularly to explain the 'gathering' involved with condensation: "the ice is cold on the inside and the monocles (sic) on the outside are warmer. The hot monocles are sucked to the cold and the monocles gather". This is the first time we are aware of Calum developing a mechanism for explaining condensation, and arguably is an advance made possible by the molecular representation. In explaining the disappearing puddle, presumably based on the classroom development of a molecular representation, Calum's written account overlays the water cycle imagery with a molecular representation: "The heat makes the monocles (sic) jump into the air the hot air rises and takes the monocles with it... to the clouds". Calum's verbal account in the interview confirmed and elaborated his pictorial and annotated representation.

While he was confident he could extend the water cycle image to include movement of molecules, he was still uncertain about the details of their spatial distribution. When probed about the relationship between the molecules and the definite edge in his diagram, and what was in between the molecules, he responded tentatively with 'Ah, maybe air'.

In drawing the evaporative process of a drop of alcohol during the interview, his representation of molecules clustered round the drop's edge closely matched his previous drawings for a puddle. When questioned about this, he explained that the evaporation would primarily occur at the edge since the water was deeper in the middle of the drop.

He accounted for what happened to the molecules in the following way:

"Ah, I have been saying the whole time the heat makes them kind of sizzle and they kind of are sucked or kind of jump a little bit into the air and when that happens the low... the temperature would um, kind of rise up and bring them up with it."

At this point in the interview a detailed discussion developed between the interviewer and Calum, based on his representation of 'where is the alcohol now?' (see Figure 1) in which he continued to draw and talk in response to the interviewer's questions. He was asked where the alcohol is now. He implied it slowly rises to the ceiling.

(C draws dots near the top of the frame)

I: Do you think it only goes up or would it spread out as well?

C: Ah, yeah, it would spread out and kind of...'cause otherwise if it just went up like, eventually when it was up near the roof it would be kind of be like just lumped together and that's not what really happens. So it kind of just all spreads out.

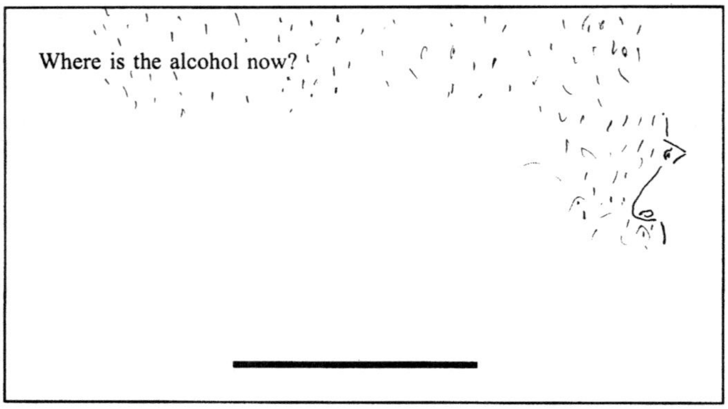

Figure 1. Calum's 1st interview drawing of the alcohol evaporating

I: OK. And the fact that you could smell it when we put the drop on . . . how would you explain that? What's happening with it that makes it possible to smell it?

C: Well I think probably each molecule has kind of like a little scent that's the same scent and sometimes maybe you can...they just let off a scent, each molecule . . .

I: So, the molecule, where would it be . . . when you were actually smelling . . . let's say I was to draw a nose (draws nose: see Figure 1) . . . what would you draw in order to show that person smelling the alcohol?

(Calum draws dots and lines around the face and nose)

C: Well all the molecules would go around and float around the room and saturate the room and when it comes past you, you can smell it and . . .

I: It wouldn't go up your nose?

C: Ah, well they might. A few might if you breathe in through your nose.

During this interaction, Calum was challenged to be more explicit about his understanding of the distribution of molecules. The drawing provides an explicit shared reference point for a collaborative process of clarifying and elaborating ideas. There is a complex interplay between the verbal account and the emerging drawing as a means to consolidate Calum's understanding. This interplay also provides insights for the teacher/interviewer into Calum's sense of the molecular distribution.

His responses tend to indicate an over-determined sense of the relationship between molecular movement and individual experience of smelling the alcohol ("when it comes past you, you can smell it"). Through this integration of discussion and pictorial production, Calum has reasoned his way from a simple picture of alcohol molecules moving upwards to the top of the room, to a more distributed view that explains why the alcohol can be smelt. This seems to be a shift from an upwards water cycle view of molecular movement, to a more complex appreciation of their distribution in a room. While this exchange between the student and interviewer does not necessarily demonstrate Calum's clear and durable knowledge of this distribution, his responses suggest that engagement with the representational task has deepened his understanding of (a) the topic, and (b) the challenges in attempting to represent this distribution pictorially.

A year later, in the interview dealing again with a drop of alcohol, Calum has retained his commitment to a molecular representation, voluntarily introducing the idea when prompted to say what you would see with a powerful microscope. He elaborated:

C: They're tiny drops of water that actually make up water and when heat is like over the water the hot water rises and it takes the top layer of molecules with it.

Calum draws little circles representing the edge of the drop (Figure 2). His focus on elaboration of a mechanism is consistent with the previous year, as is his view that heat and hot air rising takes the molecules up. He was in fact uncertain about the nature of the relationship between the substance, alcohol, and the molecules ("I am not sure if they would be alcohol molecules . . . after the molecules came up, the little particles would be left"). However, his concern for where the molecules

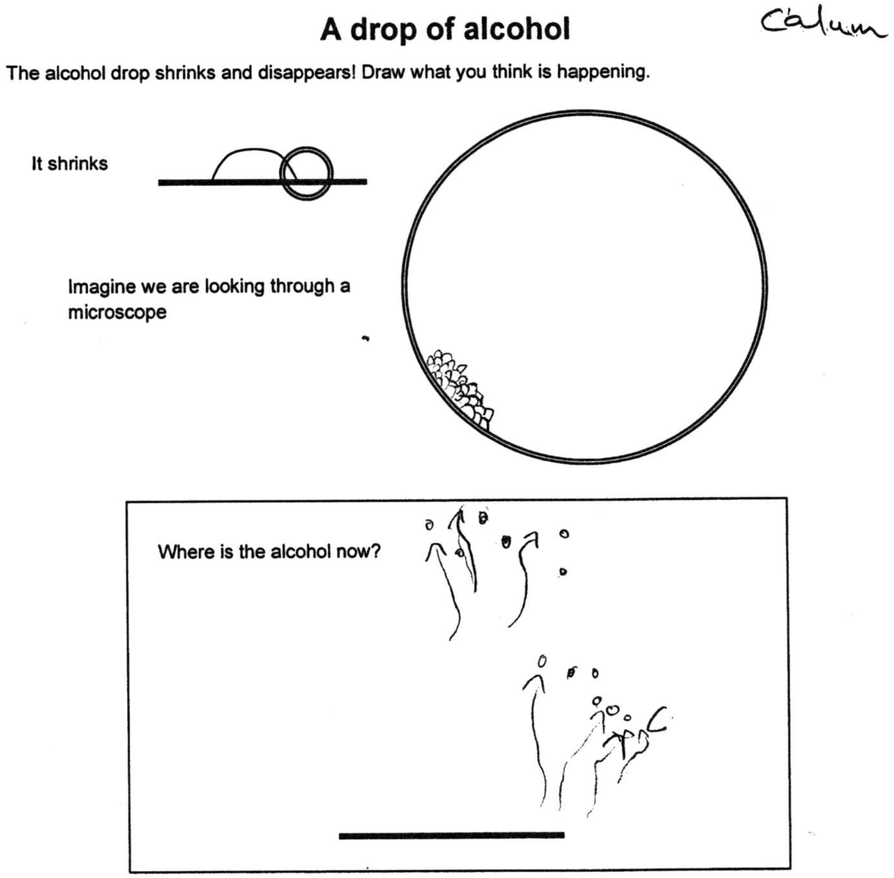

Figure 2. Calum's 2nd interview drawing of evaporating alcohol

go, prompted by the conversation with the interviewer centred on the representation sharpens his sense of how alcohol might exist in air, and the implications of the spatial distribution.

C: It would probably be gone up here like where the hot air rises to here and if you had it in here, because they can't go very high up and there's much space for it go it will just gather again.
I: So you are drawing the arrows here, is that the hot air rising?
C: Yes and taking the molecules up. (Those little dots)... they're the molecules.
I: You have only drawn two.
C: They are pretty much like spread out (he draws more)

Calum's drawing (Figure 2) separates out the molecules from the wafting mechanism of the air. He is combining two pictorial languages, one dealing with the distribution of material, the other with movement of air that carries the molecules.

C: You would smell their scent while they were being taken up by the hot air.

Again the interviewer inserts a nose into the drawing. Calum draws wafting and circulating patterns near the nose. He is using the drawing to co-construct with the interviewer a sense of spatial distribution and random movement. The confusion between air and alcohol is apparent in the drawing, since they are represented differently.

At this point Calum seems to harbour a few non-standard conceptions of molecules, being separate from the smell and possibly from the substances as such, and has a view that air may not be molecular. However, he is developing a sharper view of the spatial distribution, in coming to terms with the nature of the phenomenon.

5. REPRESENTATION AND LEARNING

Calum is one of nine children who were interviewed and for whom we had detailed information about their past history of evaporation explanations. These children generally had a history of development of their evaporation ideas to the point that they are all confident with a water cycle image, and in some cases could talk about water being held in the air, or coming out of the air in condensation phenomena (Tytler and Peterson, 2004). These ideas were reflected in the prior knowledge questionnaire in this sequence, and in most cases the written responses to the first lesson activities indicate a grasp of this understanding but little confidence with using the idea of molecules. What is added, in the conversations based round the interview drawings, is a closer sense of local molecular distribution that is used to explain details of the evaporation phenomenon, and encourages thinking about the mechanism of dispersal into air. We would argue that there is a shift in Calum's capacity to imagine the process whereby water can exist in air, involving the construction of a narrative of causation allied with the spatial representation.

With the children who were able to engage with the representation and use it flexibly in interview (there were others apart from Calum), there were interesting variations on how the movement (wafting, air currents, general directionality) was represented and explained. For Calum, heat continues to figure as an important causal mechanism (see Andersson, 1986). Other children drew wavy lines to represent the 'wafting' of the smell. Thus, children's understandings of this complex of features contains a range of pictorial elements dealing with different explanatory dimensions: descriptive (such as arrows showing the direction of movement of the vapour); causal (such as focusing on heat or hot air rising); and material (exploring the change in location and circumstance of the alcohol or water).

We do not argue for a promotion of standardised representations (such as happens in primary schools, for instance, with the water cycle) in supporting children's learning. Each child's representations were different in significant ways. Rather, our view is that we should treat representations as tools through which children can explore science ideas and clarify for themselves their understandings. The complexity of challenges these children faced in representing evaporative phenomena would argue against the possibility of achieving a resolved conceptual position. Rather, the different representations that comprise different aspects of the process should be seen as individually constructed and ultimately 'in process'. What matters, in the process of learning, is not a determined end point, but the quality of the process of exploring ideas itself.

6. IMPLICATIONS

Analysis of the lesson sequence and the case studies suggests that there is a range of learning challenges for children in engaging successfully with a particulate theory of evaporation. While our study may seem to confirm the research literature on the demanding conceptual challenges for children in understanding such a theory, we assert that such a conception can be generatively explored with 11 year-old children and readily implanted into a water cycle notion. However, we assert that the challenges for children in engaging with this topic are both conceptual and representational, and that these challenges are indivisible. One conceptual challenge lies in the notion of distribution and chance as causal. Children's understandings at this age are readily over-determined by a simple causal view of the movement of molecules, as noted by Piaget (1930, 1970). For example, the children's drawings of evaporation often echoed a highly determined pattern of movement associated with their prior understandings of the water cycle. In discussing the smell of eucalyptus oil, there was talk of molecules 'zig zagging' to explain why the pattern of people smelling seemed somewhat random. In the interviews the conceptual problem became apparent as a representational issue, with some children, in attempting to represent this pattern in a drawing, wanting to indicate a causal pattern based on localised interactions of particles with the nose.

Difficulties previously found in the literature, concerning imagining the relationship between molecules and the substance itself, or confusing categories of smell, or heat, or air, and substance, while present here, did not interfere with a productive engagement with spatial representations of the evaporative process. In fact, this connected use of visual and verbal modes to represent science understandings brought to the surface various 'misconceptions', in a way that would allow them to be fruitfully clarified and refined through further representational work.

It has been argued that molecular ideas are too difficult for young children because of this complexity of causation and distribution and properties that they embody. However, molecular ideas can be used at many levels, and in this study we have introduced them to help children visualise a specific problem with evaporation – how water might transform, and exist imperceptibly in air – without attempting to complicate the issue by focusing on surface mechanisms or molecular properties. Thus, it is a particular representation of particles, and a particular causal logic, that were introduced and which provided the representational challenge for these children. Johnson (2005) argues that the molecular model is necessary, at some level, for students to be able to imagine what is happening with evaporation. We would agree, and further argue that the complexity of envisaging particles beyond this is not relevant to examining the affordances of this representation for understanding evaporation.

There is evidence in Calum's case study that his engagement with a molecular representation enabled him to imagine how water could be contained in air, details of mechanisms for evaporation and condensation, and the implications for phenomena such as dispersion of smell, and the location of vapour in a room. The process of representational negotiation also provided insight into his thinking. The pedagogical implications of these assertions align with Vygotskyan, rather than Piagetian views of learning, given the focus on supporting children to develop and refine their representational resources. Our study suggests that learning about evaporation is not simply a story of cognitive readiness, but could entail extensive highly focused opportunities for teachers and students to negotiate appropriate representational tools for exploring this topic.

We would argue, arising out of our experience in this study, for a science program that fore-grounded teacher-mediated negotiation of representational issues as students construct different modal accounts of a topic. Such a program would need to focus on student engagement with a range of their own, and more authorised accounts of the topic, supporting research by Russell and McGuigan (2001) on the need for re-representation work across different modes. Students also need explicit instruction in the signifying functions of various elements within different representations. For example, arrows in a diagram can variously indicate direction, force, or chronological sequence. In primary school, children need to start to learn and integrate the languages of science. This study implies the need for an increased focus on this dimension of meaning making in learning science, and provides new insights into student understanding and student learning in terms of representational knowledge and choices in developing meaning.

REFERENCES

Ainsworth, S. (1999). The functions of multiple representations. *Computers & Education, 33*, 131–152.

Andersson, B. (1986). The experiential gestalt of causation: A common core to pupils' preconceptions in science. *European Journal of Science Education, 8*(2), 155–171.

Bar, V. & Galili, I. (1994). Stages of children's views about evaporation. *International Journal of Science Education, 16*(2), 157–174.

Denzin, N. & Lincoln, Y. (1995). *The handbook of qualitative research in education*. Newbury Park, CA: Sage.

Driver, R. (1985). *Children's ideas in science*. Philadelphia: Open University Press.

Johnson, P. (2005). The development of children's concept of a substance: A longitudinal study of interaction between curriculum and learning. *Research in Science Education, 35*(1), 41–61.

Karmiloff-Smith, A. (1992). *Beyond modularity: A developmental perspective on cognitive science*. Boston, MA: MIT press.

Kress, G., Jewitt, C., Ogborn, J. & Tsatsarelis, C. (2001). *Multimodal teaching and learning: The rhetorics of the science classroom*. London: Continuum.

Lemke, J. (2004). The literacies of science. In E.W. Saul (Ed.), *Crossing borders in literacy and science instruction: Perspectives on theory and practice*. Newark, DE: International Reading Association and National Science Teachers Association.

Osborne, R. & Cosgrove, M. (1983). Children's conceptions of the changes of state of water. *Journal of Research in Science Teaching, 20*(9), 825–838.

Peirce, C.S. (1978). *Collected papers of Charles Sanders Peirce*. 8 Volumes (Eds. Charles Hartshorne, Paul Weiss & Arthur W Burks, Vols 1–6), (Ed., Arthur W. Burks, Vols 7–8). Cambridge, MA: Harvard University Press.

Piaget, J. (1930, 1970). *The child's conception of physical causality*. New York: Harcourt Brace.

Roth, W.-M. (1995). *Authentic school science* (Vol. 1). Dordrecht, The Netherlands: Kluwer Academic Publishers.

Russell, T., Harlen, W. & Watt, D. (1989). Children's ideas about evaporation. *International Journal of Science Education, 11*, 566–576.

Russell, T. & McGuigan, L. (2001). Promoting understanding through representational redescription: An illustration referring to young pupils' ideas about gravity. In D. Psillos, P. Kariotoglou, V. Tselfes, G. Bisdikian, G. Fassoulopoulos, E. Hatzikraniotis & E. Kallery (Eds.), *Science education research in the knowledge-based society. Proceedings of the third international conference of the ESERA*, Aristotle University of Thessaloniki.

Scott, P. (1993). *Overtures and Obstacles: Teaching and learning about air pressure in a high school classroom*. Proceedings of the Third International Seminar on Misconceptions and Educational Strategies in Science and Mathematics, Ithaca, NY: Cornell University. Disk 18.

Stenning, K. (1998). Representation and Conceptualisation in Educational Communication. In M. W. van Someren, P. Reimann, H. P. A. Boshuizen & T. de Jong (Eds.), *Learning with multiple representations* (pp. 320–333). Amsterdam: Pergamon.

Tytler, R. & Peterson, S. (2005). A longitudinal study of children's developing knowledge and reasoning in science. *Research in Science Education, 35*(1), 63–98.

Tytler, R., Peterson, S. & Prain, V. (2006). Picturing evaporation: Learning science literacy through a particle representation. *Teaching Science, the Journal of the Australian Science Teachers Association*, 52(1), 12–17.

Tytler, R. & Prain, V. (2005, August). *Multiple representation in learning about evaporation*. (Paper presented at the European Association for Learning and Instruction Conference (EARLI), Nicosia, Cyprus).

Tytler, R. & Peterson, S. (2004) Young children learning about evaporation: Insights from a longitudinal study. *Canadian Journal of Science, Mathematics and Technology Education, 4*(1), 111–126.

Unsworth, L. (2001). Evaluating the language of different types of explanations in junior high school science texts. *International Journal of Science Education, 23*(6), 585–609.

M. CONSUELO DOMÍNGUEZ-SALES, CARLES FURIÓ-MÁS AND JENARO GUISASOLA

19. LEARNING FROM THE HISTORY AND PHILOSOPHY OF SCIENCE: DEFICIENCIES IN TEACHING THE MACROSCOPIC CONCEPTS OF SUBSTANCE AND CHEMICAL CHANGE

Abstract: This work analyses possible deficiencies in conventional teaching when the concepts of substance and chemical change are introduced. These deficiencies may be caused, in our opinion, by not taking into account the History and Epistemology of Chemistry. When teaching chemical reactions, the relationship between the two levels of representation – macroscopic, introduced with the empiricist model during the 16th and 18th centuries and microscopic from the 19th-century classical atomic model – are often forgotten. This study analyses chemistry teachers' opinions on these subjects and the way they are presented in textbooks. The results obtained confirm the existence of significant conceptual and epistemological teaching deficiencies, such as failing to introduce the macro and microscopic definitions of substance and the lack of references to both levels in the presentation of chemical reactions

Keywords: Chemical change, Compound, Historical evolution of concepts, Macro–micro relationships, Simple substance, Substance

1. INTRODUCTION, AIMS AND THEORETICAL BACKGROUND

Educational research in the field of Chemistry has highlighted the existence of numerous learning difficulties that later become mistakes made by students. The first concerns the operational definition of substance, which is essential to be able to understand other related chemical introductory concepts, such as the compound and chemical process (Vogelezand 1987; Domínguez-Sales & Furió-Más 2001).

If teaching fails to transmit the empirical definitions of substance and chemical reaction, learning problems can be expected. One of such problems would be students' difficulty to distinguish between a mixture and a compound (Sanmartí 1989) and not being able to distinguish between a physical and a chemical change would be another, as they lack the empirical criteria required to know whether the substances have been conserved during the interaction.

Moreover, research on alternative conceptions has highlighted the existence of a kind of isomorphism between certain students' difficulties and some of the problems

M. Consuelo Domínguez Sales, Escola Universitària de Magisteri Ausiàs March, C/Alcalde Reig, 8 València 46006

R. Pintó and D. Couso (eds.), Contributions from Science Education Research, 249–259.

that have arisen throughout the history of science (Furió et al. 1987; Wandersee et al. 1994). Consequently, by knowing the problems and their solutions provided by the elaboration of the two main historical models (the macroscopic and empiricist model from the 16th to the 18th century and the microscopic model in the 19th century) can help to both prevent students from running into possible conceptual and epistemological obstacles, and also to better organize the sequence of contents to be taught (Dominguez-Sales 2004).

Taking this into account, we believe that teachers must know the concepts and their evolution over history. In order to give a brief presentation of the macroscopic model, which we imagine is less used in teaching, a historical analysis is conducted. It will mainly focus on the qualitative advances that gave rise to the elaboration of the operational definition of substance, compound and chemical reaction in the macroscopic framework which made the birth of the classical atomic theory possible at the beginning of the 19th century:

a) Throughout the 16th and 17th centuries, the following evidence was proved: gases had weight (Torricelli, Pascal, von Guericke, etc.), they were elastic-like metals (Boyle) and some gases different to air were isolated (van Helmont). These discoveries overcame the differences in the Aristotelian paradigm between the corporeal matter (solids and liquids) and the rare matter (gases), which in turn led to gas materiality being accepted and this paved the way to search for a hypothesis to explain the behaviour of matter in a unitary way.
b) Advances in pharmacy and metallurgy over the 17th and 18th centuries made it necessary to introduce the operational concept of (pure) substance, as opposed to the idea of mixture, defining it as a body with a set of characteristic physical and chemical properties (Chalmers 1998). Within this historical framework, different empirical tables were elaborated, such as that by Geoffroy in 1718, where most of the compounds known at that time, together with their combination possibilities, were enumerated. Some science historians claimed that this implied a new concept of compound and chemical reaction (Klein 1994).
c) Mechanical philosophers in the 17th century offered the empirical notion of simple body or substance that cannot be separated into its constituent parts, as an alternative to the idea of Aristotelian element or principal constituent of bodies which marked a historical milestone in this macroscopic model. In the 18th century, Lavoisier supported this idea without giving it the ontological status of chemical element (Rocke 1986). Moreover, on considering the material participation of gases in closed systems, he propounded and confirmed the hypothesis of mass conservation in chemical reactions. These qualitative advances enabled the substantial change they experienced to be macroscopically understood, by connecting the reactants with the products through elements' conservation and, therefore, mass conservation. Nevertheless, this macroscopic model failed to solve some problems, such as that of compounds' constant composition (Proust and Berthollet controversy) which justifies the emergence of the new microscopic Daltonian model. At the beginning of the 19th century, this model created

a sound framework that made it possible to interpret the atomic definitions of compound, chemical reaction and atoms' mass conservation in the process.

This work presupposes that conventional Chemistry teaching will give little importance to the macroscopic empiricist model described above, which leads us to raise the following questions about Chemistry teaching:

- To what extent does conventional Chemistry teaching "forget" the operational definition of substance (contrary to mixture) created by the historical empirical model before presenting the classical atomic-molecular model?
- Does Chemistry teaching provide clear references about the existence of two macro-micro representation levels when explaining chemical reactions?

2. EXPERIMENTAL DESIGN AND SAMPLES

In order to answer these questions, we have elaborated a multiple, varied and convergent design, comprising two Sections. The first Section is divided into two parts: a questionnaire comprising five open questions aimed at obtaining teachers' ideas as regards the concepts of simple and compound substance and chemical reaction as well as determining which concepts they give importance to and which they consider necessary to be learnt by students to be able to understand chemical reactions.

The second part of Section 1 is a written essay which has been taken by a group of 42 teachers who were participating in an official exam in which they had to develop the topic "Material systems. Physical and chemical transformations. Separation procedures of the components of a mixture and a compound into its constituent parts". To evaluate the results, the whole test has been analysed, taking into account how the different aspects that we have been analysing regarding substances and chemical changes were approached.

The second Section of the design was elaborated after analysing teaching deficiencies. Taking these into account, a protocol for analysing textbooks was prepared in order to show how textbooks present these concepts. It comprises 27 questions, each related to only one aspect. A total of 128 textbooks have been analysed. The aim of this protocol was to relate the omissions shown in the teachers' answers to the presentation that textbooks made of them, detecting the presence or absence of different conceptual, historical and epistemological aspects, relating to the macro and microscopic concepts of substance and mixture, the macroscopic classification and the microscopic interpretation of simple and compound substance, the concept of chemical element and the study of chemical change in the macro and microscopic conceptualization.

2.1 Description of the Samples Used

All the teachers participating in the study have a degree in Chemical Sciences. The questionnaire has been answered by 88 secondary school teachers who have been organized into two different groups. The first one is made up of 39 teachers who were still doing their teaching training. The second group consisted of 49 teachers with different levels of experience. All of them were working in public secondary schools and were selected only for geographical reasons, on account of their proximity to the researchers' area.

The 42 people who completed the written test are a rich sample of the average teacher. They came from different Spanish regions. The importance of this group lies in the fact that the answers were part of an exam to prove their ability to teach in a secondary school with public financing, so they were supposedly doing their best.

Textbooks are the most commonly used curricular material in Chemistry teaching at every educational level. This is why our study analysed how they introduced the concepts of substance, simple substance, compound and chemical reaction. The sample of Chemistry textbooks used in the protocol of textbooks was made up of 128 books, which were written for secondary education (ages 14–18). These books are considered to be widely used for teaching chemistry in secondary education and the sample includes the most usual publishers in Spain for this subject. They were published between 1976 and 2003.

3. RESULTS

Due to space reasons, we will focus on the questions we consider to be the most representative and which provide the most significant data. The first test consisted of five questions, out of which two will be analysed. Next, we will go over some of the answers given in the written test. Finally we will show some of the items in the protocol for the textbook analysis. We will start with the first question on the teachers' questionnaire.

Before beginning to teach chemical changes, it is advisable for teachers to know whether the students master the prerequisites (basic knowledge) necessary to understand a complex concept such as chemical reactions. What prerequisites do you consider should be revised before beginning to study chemical changes?

The aim of this question is to determine the importance that teachers, in general, give to the conceptual prerequisites regarded fundamental to be able to understand chemical change as a substantial change that we intend to explain with the atomic-molecular theory of matter.

Table 1 shows that a minimal percentage of teachers refer to the definition of substance at both levels of study (macroscopic and microscopic). On the other hand, it is also meaningful that only a very small number of teachers touch upon the existence of two levels of study, to foresee the transpositions macro-micro that, on many occasions, the students make.

The second question refers to the study of chemical changes. Teachers must show what they consider to be the most important contents in order to study the topic.

Q2. *Tell us the sequence of contents you consider would be the best to explain the topic of chemical changes at introductory level.*

TABLE 1. Percentage of Teachers in Training ($N = 39$) and Working Teachers ($N = 49$) that do not Emphasize the Macro and Microscopic Definitions of Substance

Item	Prerequisite	% Teachers in training ($N = 39$)	% Working teachers ($N = 49$)
1	Places no emphasis on the macroscopic definition of substance.	89.7	85.7
2	Places no emphasis on the microscopic definition of substance.	89.7	91.8
3	Places no emphasis on the existence of two levels of representation (macroscopic and microscopic) that have to be linked to explain chemical changes.	100.0	98.0
4	Does not explain the meaning of simple and compound substance microscopically, emphasising the fact that having a fixed composition – macro – is explained by the hypothesis that the particles that form them always have the same number of atoms in each element – micro.	98.0	100.0

The results in Table 2 complement previous results, with a minimal percentage of teachers taking into account the macro and microscopic concepts of substance. Therefore, we must presume that, if the concept of substance is not taken into account as a prerequisite for studying of reactions (Table 1), nor during their study (Table 2), it is an aspect that on few occasions will be offered to the students. This idea is in accordance with the learning difficulties observed in this subject matter (Domínguez-Sales & Furió-Más 2001; Stavridou & Solomonidou 1989). It is also highly significant that such a low percentage of teachers consider it necessary to explain the macroscopic and microscopic concept of chemical reaction, even when

TABLE 2. Contents Regarded Important by Teachers in the Study of Chemical Changes

Item	Regarded concepts	% Training teachers ($N = 39$)	% Working teachers ($N = 49$)
1	To give a macroscopic definition of substance	2.6	4.1
2	To give a microscopic definition of substance	0.0	1.9
3	To explain the concepts of simple and compound substance and mixture at a macroscopic level	0.0	4.1
4	To explain the concepts of simple and compound substance and mixture at a microscopic level	5.1	4.1
5	To emphasize the macroscopic concept of chemical reaction as a substantial change	7.7	14.3
6	To explain the microscopic concept of chemical reaction	0.0	28.6
7	To study the law of mass conservation	41.0	63.3

TABLE 3. Conceptual Deficiencies Discovered in Teachers' Texts Relating to the Macroscopic Definition of Substance

Considered concept	**%Teachers (*N* = 42)**
Fails to explain the operational (macroscopic) definition of substance	88.1
Fails to present differences between mixture (material in general) and substance (a body with characteristic constant and well-defined properties) at the macroscopic level	92.9
Clearly identifies material and substance	26.2
Fails to explain that the words 'substance' and 'product' (natural, commercial, etc.) are not synonyms	100.0

introducing them. On the contrary, between 41% and 63% agree on studying the law of mass conservation as an important item, hence highlighting the importance given by teachers to the quantitative aspect of knowledge to the detriment of the qualitative aspect.

Next, we are going to show the results obtained from the written texts. We can appreciate that the quantitative results are completely convergent with previous results. Table 3 shows the results obtained in relation to the macroscopic definition of substance and its differentiation from mixtures.

In Table 3 we can appreciate the significant percentage of answers (88.1%) that do not clearly explain the macroscopic definition of substance or, most importantly (92.9%), which fail to explain the difference between this concept and that of mixture, which leads us to believe that they do not question it.

As we can see in Table 4, we have also noticed that most of the written answers do not make any reference to the microscopic analysis of the concept of substance nor those of simple and compound substance.

Once again we can appreciate that few teachers refer to the need to explain the microscopic definition of substance, maybe because it is supposed that it is a well-known concept. The problem that arises from the lack of a definition is the fact that teachers move from the macroscopic to the microscopic concept without making any reference and, in many cases (38.1%), superpose the models or make incorrect transpositions between them.

TABLE 4. Deficiencies Referring to the Microscopic Model of Substance and the Superposition of the Macro and Microscopic Model

Considered concept	**%Teachers (*N* = 42)**
Fails to clearly explain the microscopic definition of substance with all of its particles – atoms or molecules – being equal.	95.2
Fails to emphasize that simple substances are formed by the same kind of atoms and compound substances by different atoms in fixed proportions.	81.0
There are incorrect transpositions between the two conceptual macroscopic and microscopic models.	38.1

We will proceed to focus on the qualitative aspects that can be obtained from the written texts. It is worth remembering that Tables 1 and 3 show that more than 85% of teachers place no emphasis on the macroscopic definition of substance as basic knowledge to be able to understand chemical reactions. As we have seen in Table 2, 0.0% of teacher trainees and 4.1% of working teachers believe it is necessary to explain the concepts of simple and compound substance and mixture at a macroscopic level. Finally, Table 4 shows that a 95.2% of teachers fail to clearly explain the microscopic definition of substance and 38.1% of them make incorrect transpositions between the two conceptual macroscopic and microscopic models.

Therefore, if no context is provided when teaching these concepts, there is likely to be confusion among teachers. Two literal examples of the answers obtained in the written texts illustrating the way teachers reason are provided below. We can see that even teachers have some conceptual difficulties.

Paragraph 1: "Matter is any *substance* which owns a specific mass. Pure *substances* are those in which the same physical and chemical properties are present. [...] Mixtures are *substances* formed by simpler ones" (teacher 15)

This answer starts by identifying the concepts of matter and substance, in a statement made from a macroscopic point of view, relying on the properties. Next, there is a microscopic discourse in which *mixtures* are said to be *substances formed by simpler ones*, bringing to light ideas that remind us of the Aristotelian theory according to which all materials were made by the mixture of the four elements. The following text also shows a very common error:

Paragraph 2: "Mixture definition: it is a substance formed by more than one element. There can be homogeneous and heterogeneous mixtures" (teacher 27).

Here, the teacher touches upon the identification of matter and substance, but at the same time, on considering mixture as *a substance formed by more than one element*, the teacher shows the confusion between the concepts of matter and compound.

As a conclusion from the obtained results, we consider that teaching focuses on the microscopic level of representation, without placing much emphasis on the macroscopic level of the concept of substance and chemical changes. Thus, teaching is restricted to this perspective, not taking into account the macroscopic reference of the phenomena that they are trying to explain with the theory.

To finish the study of teaching deficiencies, we will show the results obtained from the analysis of 128 Chemistry textbooks. The percentages in Table 5 indicate deficiencies similar to those observed in the items answered by teachers. Next, we will mention the more significant ones, starting with those referring to the concept of substance.

As we can see, data are conclusive if we focus on the fact that most textbooks ignore the macroscopic and microscopic introduction of the concept of substance, and also fail to differentiate it from the mixtures at both levels.

To finish the study of textbooks, in Table 6 we will pay attention to the part of the study that refers to chemical reactions.

TABLE 5. Conceptual Deficiencies Making Reference to the Concepts of Substance and Mixture Found out in the Textbooks Analysis

Considered questions	15 years old ($N = 38$) %	16 years old ($N = 26$) %	17 years old ($N = 28$) %	18 years old ($N = 17$) %	TOTAL ($N = 109$) %
1. Fails to introduce the macroscopic notion of substance as a material system with a set of characteristic properties	65.8	69.2	28.6	76.5	58.7
2. Fails to present the concept of substance from a microscopic point of view as a system formed by many identical particles	94.7	88.5	85.7	94.1	90.8
3. Fails to present the concept of mixture microscopically by means of the atomic theory	92.1	96.2	82.1	94.1	90.8
4. Does not consider the material system macroscopic properties to make a difference between substance and mixture	63.2	65.4	71.4	94.1	70.6
5. Fails to mention that students identify the concept of substance with the broader idea of material (which is almost always a mixture)	100.0	100.0	100.0	100.0	100.0

It is worth recalling that only 14% of teachers mentioned the macroscopic definition of reactions and only 28% considered it necessary to explain their microscopic definition. As we can appreciate in the data shown in Table 6, the tendency displayed by textbooks is similar.

A summary of the data obtained is provided below:

a) More than 85% of teachers do not consider it necessary to explain the macroscopic definition of substance to approach the study of chemical changes and 58.7% of textbooks do not introduce these concepts. Moreover, about 90% of the teachers surveyed do not take into account the microscopic concept of substance, nor is this concept introduced in 90.8% of textbooks.

b) More than 90% of teachers do not explain the macroscopic differences between mixture (material) and substance on a macroscopic scale. In fact, 26% identify them explicitly. 70.6% of textbooks do not make the difference clear.

TABLE 6. Conceptual Deficiencies Revealed in the Textbook Analysis Referring to the Study of Chemical Reactions

Considered questions	15 years old ($N = 38$) %	16 years old ($N = 26$) %	17 years old ($N = 47$) %	18 years old ($N = 17$) %	TOTAL (N=128) %
1. Fails to present the macroscopic concept of chemical reaction as a change on which some substances disappear and some new ones appear, with different properties but related to the initial ones (substantial interaction).	97.4	92.3	89.4	100.0	92.2
2. Fails to establish the idea of chemical element as an answer to the question of what is conserved in a chemical change.	97.4	100.0	93.6	100.0	96.9
3. Fails to differentiate between a mixture (physical process) and a chemical reaction between substances nor does it explain that the mixture of reactants is necessary but not sufficient for them to interact.	89.5	96.2	95.7	100.0	94.5
4. Does not use the atomic theory to interpret simple chemical processes of interest to students.	42.1	76.9	70.2	82.4	64.8
5. Does not point out that the Dalton theory does not assume the possibility of metal transmutation in gold (atoms are immutable).	100.0	96.2	97.9	100.0	98.4

c) More than 98% of teachers do not consider it necessary to explain the meaning of simple substance and compound microscopically. They do not emphasize that having a fixed composition – macroscopic level – is explained by the hypothesis that the particles that formed the simple substance or the compound always have the same number of atoms of every element – microscopic level. Equally, 81.7%

do not mention the microscopic interpretation of these concepts from the point of view of atomic theory.

d) About 85% of teachers and 92.2% of textbooks do not explain the macroscopic concept of chemical reaction as a substantial change. Moreover, 80% of teachers and 64.8% of textbooks do not use the microscopic concept to do so.

e) Generally they do not make the element conservation explicit at the macroscopic level of representation of the chemical reactions as an idea that immediately explains the substantial diversity at the same time as it explains the relationship between the reactants and the products of the reaction. More specifically, less than 4% of teachers mention it as an important subject in the study of chemical changes and none of them take it into consideration as a possible difficulty for students nor do they express the need for it to be defined in the Daltonian context.

f) More than 98% of teachers and a 93.6% of textbooks do not refer to the existence of two different levels of representation (macroscopic and microscopic), which must be connected to explain the chemical changes.

4. CONCLUSIONS AND IMPLICATIONS

As a final conclusion, in this work we have verified that, historically, chemical changes were understood only after learning the macroscopic concepts of substance, compound and simple substance. Therefore, having knowledge of the History and Epistemology of Science can be a very useful tool in the learning and teaching process. Thus, we consider it essential for teachers to know the History of Science as a guide to organizing the sequence of contents.

The main educational implication owing to the lack of empirical conceptualizations in content sequencing is that the learning difficulties that arise, mainly at introductory levels, are not foreseen.

As a result, starting from the latter considerations, our proposal is that teaching must introduce the operational definition of substance as a body with a set of characteristic physical and chemical properties. The concept of chemical reaction as a substantial change must also be taken into account, in line with the macroscopic level of representation elaborated in a process that finished at the end of the 18th century.

It has also been shown that, preponderantly, neither the teachers nor the textbooks consulted establish the macro–micro relationships when these concepts are presented within the framework of the atomic-molecular model of matter. On the contrary, teaching focuses directly on the microscopic Daltonian model.

The main didactic implication of these conclusions that can be drawn is that prior comprehension of the chemical substance concept at both a macro and microscopic level is essential to be able to understand the concepts of compound and chemical process. Moreover, it would be convenient to study the concepts of substance and chemical reaction by relating them. At the same time atomic theory must be used to give a microscopic explanation of the chemical behaviour of substances.

Starting from the latter considerations, our proposal is that teaching must take the History and Epistemology of science into account as a guide to organizing the sequence of contents. In a future paper, we will prove that this may enhance students' understanding, help to prevent those who finish their studies from running into problems of distinguishing between a mixture and a compound or between a chemical change and a physical change and will lead to a better comprehension of Chemistry.

REFERENCES

Chalmers, A., 1998. Retracing the ancient steps to atomic theory. *Science & Education*, 7 (1), 69–84.

Domínguez-Sales, C., 2004. *Dificultats en la comprensió dels conceptes de substància química, substància simple i compost. Proposta de millora basada en estratègies d'ensenyament-aprenentatge per investigació orientada*. Tesi doctoral. Universitat de València.

Domínguez-Sales, C. & Furió-más, C., 2001. Knowing the history of science to understand students' difficulties with the concept of chemical substance. In D. Psillos et al. (eds), *Proceedings of the Third International Conference on Science Education Research in the Knowledge Based Society*. Vol I, 365–367 (ESERA-Aristotle University of Thessaloniki: Greece).

Furió, c., hernández, J. & Harris, H., 1987. Parallels between adolescents' conception of gases and the History of Chemistry. *Journal of Chemical Education*, 64 (7), 616–618.

Klein, U., 1994. Origin of the concept of chemical compound. *Science in Context*, 7, 163–204.

Rocke, A., 1986. *Chemical Atomism in the Nineteenth Century* (Columbus: Ohio State University Press).

Sanmartí, N., 1989. *Dificultats en la comprensió de la diferenciació entre els conceptes de mescla i compost*. Tesi doctoral. Universitat Autònoma de Barcelona.

Stavridou, H. & Solomonidou, C., 1989. Physical phenomena–chemical phenomena: do pupils make the distinction? *International Journal of Science Education*, 11 (1), 83–92.

Vogelezang, M. J., 1987. Development of the concept "chemical substance" – some thoughts and arguments. *International Journal of Science Education*, 9 (5), 519–528.

Wandersee, J. H., Mintzes, J. J. & Novak, J. D., 1994. Research on alternative conceptions in Science. In Gabel D. L. (ed), 1994, *Handbook of Research on Science Teaching and Learning* (McMillan Publishing Company: New York) 177–210.

PART 6

INNOVATIVE TEACHING–LEARNING ENVIRONMENTS IN SCIENCE EDUCATION

IVO ČÁP

20. NON-FORMAL SCIENCE TEACHING AND LEARNING

Abstract: The present contribution deals with the educational methods which would support education of young people and the whole society in science. After the historical overview the methods of formal, informal and non-formal education are described. The main attention is devoted to the non-formal methods which are important in the formation of the top level of talented young people, who represent the decisive element of the effective and sustainable development of the society and its competitiveness

Keywords: Competitiveness, Formal education, Informal education, Non-formal education, Science education, Science festival, Science Olympiad, Science and Society

1. INTRODUCTION

Nowadays, much effort is devoted to the problem of adaptation of the society to the changing conditions connected with the fast development of science, technology, social and political relations. The world at the beginning of the 21st century is qualitatively different from that of the previous period. At the threshold of the 21st century the society stands in front of the quite new problems arising from the rapid development of science and technology during and after the Second World War, resulting in an enormous progress in production, economy, worldwide trade and communication. This progress has started a worldwide competition influencing not only the race in the development of living level but also in the load of the living environment. The problem of the sustainable development is no more a problem of individuals or small communities – it is the problem of worldwide population. On the other hand the progress of technology has brought new demands not only on scientists and engineers but also on all users of the sophisticated products. It lays new demands on the qualification and education of the whole society. It is evident that subsequent development is impossible without adequate science and education. The coming period is characterized as the period of knowledge. We use to speak about "knowledge-based society", "information society", "learning society", etc. If we speak about education, we cannot ignore science, because education depends on the level of science and science depends on the level of the education.

Ivo Čáp, University of Žilina, Univerzitna 1, SK-010 26 Zilina, Slovakia

R. Pintó and D. Couso (eds.), Contributions from Science Education Research, 263–273.

2. HISTORY OF WORLDWIDE COMPETITION

In the second half of the 20th century the worldwide competition has obtained new dimension. The Second World War has divided the world into great competing world powers. If we follow historical aspects, we can see that world power was always connected with significant technical discoveries and their utilization. We can follow emergence and decay of great civilizations like Egyptian, Chinese, Greek and Roman. Flowering of the European civilization started with the scientific revolution based on the principal discoveries of Galileo, Newton and other scientists. The fundamental discoveries of modern science like theory of electromagnetism, theory of relativity, quantum theory, nuclear physics, modern chemistry and biology were initiated in Europe. Inventions of steam and electrical engines, new means of transport and communication influenced the whole society due to production, trade exchange and building of the modern military power. During the Second World War many scientists like Einstein, Fermi and others escaped to America and enforced there significantly development of science and technology. America became the leading power after the War. They sowed there seeds of the modern technology like electronic devices, nuclear, space, computer and communication technologies. It had reflection in increasing economic, military and political power. In the second half of the 20th century America has utilized the vast open market on the other side of the Pacific Ocean and has exported there its developed technologies. Japan and other countries have seized the opportunity and throughout the time they have invested much effort to the development of science and education. East Asia became the new economic power with significant role in the world competition. New rapid development is expected in China in the near future. We can follow the shift of the competitiveness westward around the globe.

One of the most important phenomena of today's development is the information revolution connected with invention of satellite communication and fast Internet connection, resulting in worldwide globalization. Knowledge and knowledge management is to be the serious problem of the near future.

This short historical excursion shows us that the main force, which forms every world power is science combined with education, followed by development of technology and corresponding social and political relations.

The progress of the society is no more the matter of individuals and isolated teams. Developed countries are aware of the necessity to integrate their effort to reach sustainable development. Western Europe initiated the integration process in the 50th year of the past century and it has resulted in the formation of the European Union. The problem of the competitiveness is always alive and it has its response in the Lisbon strategy. Therefore we speak more and more about science, research, education and their effectiveness and impact on the society.

3. SCIENCE LITERACY OF THE POPULATION

Progress of the present society depends on the quality of the population. Integration of countries into blocks is impossible without an adequate quality and "integration" of individuals. Activity of the society depends on its consciousness. It is tightly connected with knowledge and awareness. It has a lot of aspects – quality of a human has its physical and intellectual dimensions. Similarly the thinking of people has different features. The exact thinking results in science, technology, economy and management, the humanistic thinking influences the utilization of the technical inventions, social relations, political and environmental attitudes, culture, philosophical motivations, etc. Necessary condition of the harmonic development of the world is the adequate complex education of people.

Our topic consists in the exact education, especially in the "science education", which represents one of the poles of the natural development of the personality. Science education begins with the first sights of the child looking for the answers on the natural question "why?" After the first stage of the cognition it continues at the schools and proceeds throughout the whole life. Science education represents in a wide sense understanding of principles of the material world surrounding the human being and corresponding relations. Proper knowledge of principles of the material world (nature and technology) is important for creation of adequate humanistic, social and political thinking.

In connection with the level of "science education" of the society we speak about the "science literacy". What can be said about the needs of the society? Sometimes in history, genial individuals or high-level working teams were the most important, because the research was not so extensive and expensive. The top research of the present time is very heavy on the technical means and corresponding facilities and therefore very expensive. The present top research is the matter of the whole society, financing of science represents significant part of the GDP and it is an instrument of political decisive process involving the opinion of the whole society. The common opinion is important for building of the nuclear power plants, chemical factories, environment-protecting technologies, space research, etc. There are therefore two demands on the science education:

- to reach a proper standard level of the common people literacy
- to reach a top level of the scientific education of the limited part of the most talented people.

The first demand is necessary for the qualified decisions of the democratic society, including those dealing with support of science and education. The second one is necessary for the development of the innovative society – the most important condition of its competitiveness.

Development of both levels of the science literacy is vitally important for every society. The political wisdom of the public administration has its reflection in the care for the science and education quality and their support. We can see the direct relation between the support of the science and education and the advance of the national economy and living conditions.

4. SCIENCE AND EDUCATION – EDUCATION TO SCIENCE

One of the preferred problems of EU is the so-called "Science Education" or education to science. "Scientific" way of cognition is immanent to the human being but it must be developed throughout the life. The main stimulator of such a view of the world is essential curiosity, the quest to understand the unknown. History of the individual is similar to history of humanity. We can follow the historical way of discovering of the secrets of the world from antiquity until now, from primitive concepts and ideas up to the modern sophisticated theories. The same short history must pass every individual in the process of formation of his view of the world, from the elementary empirical experience up to the resulting knowledge and awareness of relations with the aid of educational system. The true education is a "research" as the method of the individual approach to the matter of education. The most important aspect of this process is the active participation of the individual in it. The role of books and teachers should be a stimulator of the proper curiosity and a supervisor of the cognitive process. The role of teacher is not only the delivering of information or reading of books, but to accompany the student on the way to discovering secrets of the world – the world of nature, people, technology and human relations.

4.1 Formal Education

It is evident that every student and every teacher are individuals with their conceptions, but the process of education has its own rules and in order to optimize the educational process it is necessary to follow at first the formal educational school system. It is given by the standard amount of knowledge to be taught and the standard methods with respect to the age and ability of students. The formal framework of the school education leads often to the formal approach of students and teachers. The result is that instead of research the educational process decays to the "delivering" of information and to memorization of encyclopaedic facts without notifying the relation to the daily life or to the near surroundings. The real "research" must cancel the boundary between the school classroom and the usual life. Because of shortage of such approach (mainly from the side of teachers) the "scientific view of world" decays to the routine, pragmatic and passive mode of life. It results in the resistance to the "educational activities" and to the science in the wide meaning of the word.

The main problem of the formal education consists in the formal aspects like curricula, text-books, teaching laboratories, specialized classrooms, organization, legislative rules, economic relations, etc. on one side and in teachers on the other side. Much effort is devoted to the improvement of the system of the formal education but it is impossible to avoid the main problem – the formal character of the mass education. The formal educational system is oriented preferably on the creation of a sufficient level of the common literacy. The standard formal system would form a proper basis for another development of the talented individuals. One of the aims of the formal educational system is to teach the students how to study, how to learn, how to manage the information – to create starting conditions for advanced and life-long learning.

If we remember our school years, there were not many teachers who have imprinted a significant sign into our mind, but most of them are put out of our mind. We can say "everyone thrown into a pool will remain wet, only a part will swim, but nobody is able to win in the swim-race without a good trainer". And a good trainer devotes his care only to the best individuals. The public instructor is not able to prepare a top swimmer during the public training.

4.2 Informal Education

Some possibility to improve the effectiveness of the standard education consists in utilization of informal methods; it means alternative methods based on a more friendly approach to the student. The used methods depend on the age of the pupil or the student. The primary school informal education would prefer the tuition by game. It copies the essential process of children's cognition and supports the pupil's creativity and curiosity. The supervised game is the elementary method of research and it leads the children to a scientific view of world. Significant space would be devoted to formulation of children's questions, typical for the pre-school period.

In case of older students the informal methods consist mainly in the project-oriented education. The interest of students is supported by their individual solutions of special problems related to the standard subject matter. The individual research-like work supports the interest of students in the work and development of their creativity. The effectiveness of the work can be increased by the competitiveness, which is connected with the presentation and evaluation of the results. Many educational systems have already included such methods into the regular curricula and organize local or national exhibitions of best projects. The main problem with introducing such methods into practice consists in teachers. This way of education is very demanding for the teacher because of creation of individual assignments, individual care for the working individual or group and evaluation of the individual projects. It is evident that this method is meant mainly for talented students or for the students of special schools or classes.

The mentioned methods are very effective in increasing the attractiveness of the science and scientific method of work. The realization of such education depends significantly on the ability of teachers. It is a challenge for the universities which are preparing new teachers. Progress in implementation of new methods is complicated because of conservative character of the educational system. There are many teachers prepared for their work 20 or 30 years ago and they are using the classical educational methods typical of the last century. The only way to improve the situation consists in organization of special training of the teaching staff and enforcement of the teaching staff with the young newly educated teachers.

In connection with new modern didactic tools, the multimedia and computer-aided tuition are taken as a way of attracting the interest of students. These means of support of the education can be useful but we must be very careful. One of the reasons for the decrease of the active interest in science consists in the expansion of the TV, Internet and other reproduction technology. All these things lead to an easy life and to the passive acceptance of information. Pupils and students grow in a comfortable world which offers a lot of interesting experiences without any significant effort. The preferred notion of a computer-aided virtual world weakens the ability of creative scientific work. It reduces the natural sense of reality. The modern electronic devices are a very powerful aid but it must be utilized very carefully. The computer-aided virtual world must not substitute the real world and the real experience.

4.3 Non-formal Education

Formal and informal educations are more or less the methods of the regular public education. The quality of this education has its reflection in the level of common literacy. On the other hand the innovative society must educate the small part of the population extremely talented in some field of knowledge to a topic level and prepare them for the scientific career. The number of such pupils or students is not very great and it is possible to take care of them individually. It is an opportunity for application of non-formal methods of education. Because of the existence of different target groups of population with special needs for education or training, there exist different forms of it. The main aim of non-formal methods is to support and stimulate the interest in top or special education and to devote special attention to such people, who wish to develop their knowledge and skills above the standard level of the formal education. Because of the special character of the non-formal education, specially qualified teachers or supervisors are needed. They are mainly the university teachers or researchers. The concrete methods have a great variety and demand individual effort of the organizers. In comparison with public formal education, which has its fixed curricula, textbooks and standard methods, the non-formal education has no strict regulations and has lots of supporting materials. The level of such education depends mainly on the individual organizers. It is therefore extremely useful to create a system of communication among organizers of such activities and offer them a platform for the knowledge and experience exchange. It should help the organizers, mainly volunteers, in what to do and how to do it.

4.3.1 Scientific projects The best way of support of the scientific view of the world is experimentation, real contact with the matter of research and experimental investigation of the object, formulation of conclusions and presentation of results. The level of the work depends on the age and ability of students. The same principle can be used for pre-school children up to the post-graduate university students. The system of scientific projects is perspective for the pupils and students attending the regular school education as the tool for preparation to a professional scientific or engineering career.

The system of scientific projects demands individual work with young researchers or their teams. It needs very professional and well-educated teachers. The small interesting projects solved by primary school pupils should have a character of observations of common phenomena from the nature or technical practice and can be supervised by their teachers. The more sophisticated projects of the students of secondary or high schools assume usually co-operation with institutions of higher education or research. The main problem consists usually in the formulation of proper problems and attraction of interest of their solvers. The publication of themes by means of some Internet portal should be a significant help to organizers of this kind of activities.

The important part of the projects is formulation of scientific results, elaboration of some report and its presentation. The stimulating form should be a presentation among schoolmates or other groups. The best projects are usually evaluated by means of contests or exhibitions. There exist lots of regional, national or international events. We have rich experience with
- regional and national festivals of science for the pupils up to the age of 15
- national contests of Tournament of Young Physicists (high schools)
- secondary school research activity (high schools)
- clubs of technical creativity (primary and secondary schools)
- clubs of young naturalists (primary and secondary schools)

There exist well-known international events, e.g.
- First step to Nobel price in Physics and Chemistry (Academy of Science of Poland – scientific projects of high school students)
- European Young Scientists Contest (European Commission – scientific projects of high school students)
- European or World science festivals (World competition and exhibition)

This form copies the effort of former scientists and inventors. The students work initiatively and slowly according to their pace and they can work on the solution freely in their free time. The students (or teams) then present and comment their resulting work. Solution of projects is a very popular and effective form in order to attract the students to science. On the other hand, it is very demanding on the time and effort of teachers – supervisors.

4.3.2 Special courses, correspondence seminars, summer (Winter) schools, etc.
Many schools and institutions of education organize for gifted pupils or students additional teaching activities exceeding the standard education in order to develop their abilities. These activities are voluntary and non-formal. The organizers prepare special lectures and training for the students in their free time, explain special and more sophisticated problems, train the students in methods of the solution of problems. Very popular forms of such education represent correspondence seminars.

Many governments support such activities. For example, ministries of education nominate national coordinators and coordinating committees which take care of these activities, mainly of scientific program of the courses, concrete problems, etc. Many such events have a national extent. The progress of attendants is permanently evaluated and the best ones according to the ranking list are invited to the usually one-week country summer or winter schools in an attractive ambience of mountain cottages or similar. These activities deepen significantly the knowledge and skills of gifted students and represent very precious preparation for the other forms of science education, mainly competitions. One of the very desirable effects consists in formation of something like national community of excellent young people interested in the special field of science or technology. After such events students start to communicate among them by means of Internet and create the positive environment for their talent development. The mutual contacts serve also as the platform for comparison of their level of knowledge. It stimulates significantly their effort to belong to the best ones. This form of non-formal education to science is very effective and its success has its reflection in the form of good results in scientific competitions and admission to universities.

The main problem consists in dependence on the willingness of volunteers who organize all these activities. The efficiency of these forms has grown up due to the modern communication technologies – Internet websites. This kind of non-formal education needs financial support for the final seminars and a lot of work on special lectures and problems preparation. Significant help could represent the sharing of materials among organizers in all cooperating countries by means of a proper website.

4.3.3 Science competitions – olympiads A very popular form of support of interest of young people in science is represented by different competitions. The most important competitions are science-oriented Olympiads. They have been founded about 50 years ago in the Middle Europe. The first and oldest is the Mathematical Olympiad and the Physics Olympiad. Throughout the years Olympiads have attracted a worldwide popularity and they have been extended to all natural sciences – biology, chemistry, environmental science, geology, astronomy, informatics, etc. Olympiads are the proper form of stimulation of interest of young people in deepening their knowledge and skills in corresponding fields of science. Organization of competitions and preparation for them has different forms in various countries.

The original idea actual today more than in the past is to organize an extensive survey among pupils and students of different ages in order to find those talented and to develop systematically their talent. We have experience with the organization of such activities from the age of 12 up to 18. It is organized in two or three rounds. The first one has a preparatory form; it represents a home work on central problems emitted by the national coordinator. It lasts about half year and it consists in preparation of students, mostly with the aid of their teachers or experts from universities and research institutions. The national coordinating body and the regional ones organize different supporting activities like lectures or short-term courses. The number of addressed and involved pupils and students of all levels are in thousands. Those who fulfil successfully this preparative period proceed to a competitive part, organized at the regional level. The most capable students are followed throughout their school years and motivated for the university studies of science. The top category for the oldest students finishes with a national competition. The team for the international competition is selected from the national winners. Popularity of such form of international competition increases from year to year. The Mathematical Olympiad attend about 90 countries, the Physics Olympiad about 80 countries and similar in the others. Asian countries organize, for example, the Asian competitions in advance.

The main problem consists in preparation of interesting and original problems for the competition every year. The most useful help for all organizers could be a systematic exchange of information among them.

At present we observe a new initiative in the field of natural science. New demands of the society consist in complex education of young people in natural science. There are new competitions in natural science like the European Union Science Olympiad (EUSO) and the World Competition of Young Scientists. Both competitions are oriented on physics, chemistry and biology.

The original goal of the Olympiads was to offer qualified free-time activity to the students. The present demand is not only entertainment. The top education of the most gifted students starts to be a necessary condition for the development of competitiveness in the society and science competitions are a good tool for it.

Science Olympiads represent an effective form of non-formal education along with other mentioned training activities like correspondence seminars and summer schools. It represents a complex care for the gifted students attending the regular school education.

4.3.4 Science exhibitions and festivals History of science and technology including the present time is rich with interesting and surprising things and moments. Following the progress of human knowledge can be very adventurous and inspiring for people of all ages. History of science and technology reminds us of our way of cognition. The most effective experience is that connected with a contact with real devices or processes. There are institutions that have assembled lots of historical devices, demonstrations of fundamental phenomena or function models of modern sophisticated devices and facilities.

There exist various forms of such public expositions. In many cities there exist science museums or museums of technological progress. There are exhibits of old devices of famous historical personalities, movable models of machines often controlled by the visitor, instructive films or multimedia presentations, etc. Such expositions exist often besides the big industrial or research institutions which offer explanation of the built-in equipment, processes and technologies, e.g. nuclear power plants, to the public. Other events are short term or occasional fairs or expos, e.g. science weeks. They should be arranged as presentation of projects of young scientists, presentation of universities, research and industrial institutions, etc.

These forms of non-formal public education are proper for all ages from young children up to their grandparents. Such activities enhance the public literacy and awareness of the real position of science and technology in the daily life. They present to the visitors the results of the modern science and technology or the impact of the human activities on the living surroundings. It is necessary in the era of sophisticated technology not only to use the miraculous technical means but also to understand their principles and the possible subsequence of their exploitation. It is also useful for the wide public to recognize the necessary effort of scientists, researchers and inventors who stand behind the comfort of the present civilized society. Such awareness of the value of the effort of top inventors should motivate the adult population to support science and education and to lead and orient children and youth to self-education and to the satisfaction of their essential curiosity.

4.3.5 Public forecasting, journals and computer-aided technologies The present period is typical of delivering all information to the consumer. Not only adults but also children are used to manage all things from the armchair. It is not very favourable but people spend too much time in front of the TV or computer screen. It is the reality and we are not able to change it. People do not read books, do not visit theatres and concerts. The only possibility of how to utilize this circumstance is to deliver the desired information into these media. The knowledge submitted by means of such media is accepted in a passive way but it is often the only possibility of how to offer the science-oriented programs to the common people. Because the competitiveness of criminal films, different shows and other entertaining programs is too great, the instructive programs on science and technology must be extremely professional, interesting and therefore attractive.

Another way of giving the desired information to the common people consists in popular scientific journals. Another information channel represents the computer network. Internet serves as a good tool for life-long electronic education. On the other hand the present generation utilize Internet more and more as a source of different interesting information. Because of the attractiveness of this tool of communication we must search for effective utilization of it for the desired information publication.

5. CONCLUSION

Development of science and technology has achieved such a level that the further sustainable development has to mobilize all abilities of the society. More than in the past the educational system must be optimized in order to achieve the maximum effectiveness with respect to the maintaining of desired competitiveness in the society. The attempt of the European Union in this field is formulated by the Lisbon strategy.

The maximum effectiveness of education desires high quality formal education on one side as well as a special non-formal one on the other side. Up to now the formal education was a matter of the central care of government and non-formal education has been a matter of volunteers willing to do it besides their ordinary occupation. Government more or less supports such activities, but the non-formal education is not the regular part of the educational system. It means that the results consisting in education of top experts are only accidental with the limited effectiveness.

One of the tasks of the Conference is to analyse the effective methods of education, formal and non-formal, and to suggest to the governments of European countries recommendations on how to transform the educational system in order to adapt it to the new demands of the 21st century. An important innovation should consist in involving the non-formal education into the standard methods of education, to accept corresponding legislative rules and financial support. It is necessary to integrate the whole educational system from primary schools up to universities and to deepen cooperation between teachers at all levels of the educational system.

It was shown in the contribution that there exist a great variety of different non-formal forms of education. The content and level depend on the individual organizers. There are not many top experts in the high level non-formal education and it would be very useful to utilize their experience, knowledge and skills for all similar activities. The second recommendation for the Conference is to discuss a possibility to create an international web portal of non-formal education in order to establish a commonly accepted platform for the best experience and educational materials exchange. Some isolated attempts to do it have been taken but similar initiatives are isolated and therefore not so effective. The web portal should be an effective solution for this problem. Some attempts to do it was made in the framework of activity of the expert group of the European Commission "Science Education", coordinated by the Slovak Republic. We have created the special website called Newton Network for this reason. We have had some first-hand experience in the framework of the ERA-NET project EUROSCENE 2003.[1] The idea of the web portal was approved but its realization and optimization demand much effort.

There are other successful countries all over the world like Europe, North America, South-East Asia, etc. It is interesting to search for the origin of their success. It is useful to study their systems of research and education and utilize their best experience. The worldwide competitiveness is the present reality and we all must do our best to contribute to the sustainable development of Europe and to contribute to the better state of the World.

NOTE

[1] 6th Framework Program of EU, ERA-NET Scheme, Project ERAS-CT-2003-510213 "EUROSCENE 2003".

21. ADULTS' UNDERSTANDING OF ANALOGY-BASED EXHIBITS IN AN INTERACTIVE SCIENCE MUSEUM

Abstract: This study analyses how adult visitors understood three different types of analogy-based exhibits in an interactive science museum and the factors affecting that understanding. 84 visitors were observed during their interaction and interviewed immediately afterwards. Results suggest that few visitors understood a target domain at a scientifically accepted level. The exhibit that "only shows similarities between relationships" was the most difficult to understand. Understanding occurred by triggering prior knowledge, by text comprehension, and by analogical reasoning. Analogical reasoning often failed because a source domain was unfamiliar, source and target were superficially dissimilar, some elements of the source were emphasised in relation to others in the exhibit design, or misconceptions about the target were retrieved. Finally, suggestions for the design of analogy-based exhibits are provided

Keywords: Adult visitors, Analogy-based exhibits, Cognitive strategies, Interactive science museum, Understanding

1. INTRODUCTION

Analogy-based exhibits are analogical representations of scientifically accepted models of phenomena (Gilbert & Stocklmayer, 2001). In these exhibits, different kind(s) of similarity (i.e., superficial and/or structural) can be shared between a source and a target domain (Gentner, 1983). Superficial similarity exists when the two domains share only a resemblance between entities; while a structural similarity refers to a sharing of a system of relational commonalities independently of any superficial similarity between source and target domain (Dunbar, 2001). Therefore, the exhibits can be one of three types: "only showing similarities between entities"; "showing similarities between both entities and relationships"; and "only showing similarities between relationships" (Stocklmayer & Gilbert, 2002). Examples of each subtype of exhibits are respectively: "Earthquake", "Black hole", and "Light harp" (Stocklmayer & Gilbert, 2002). In "Earthquake" a roller mechanism rocks a platform back and forth. The movement of the platform matches the movement of a tectonic plate but the relations between force and movements are not shown by the exhibit. In the exhibit "Black hole", balls can be rolled around the halls of a bowl which has a hole in the bottom. The bowl matches the gravitational field

Ana S. Afonso, Instituto de Educação e Psicologia, Universidade do Minho, Campus de Gualtar, 4710–057, Braga, Portugal

R. Pintó and D. Couso (eds.), Contributions from Science Education Research, 275–287.

and the consequences of rolling the balls around the halls of the bowl matches the vanishing of electromagnetic radiation in a region of intense gravitational field. Finally, "Light harp" shows the rectilinear propagation of light. It consists of several light sources whose beams can be interrupted by visitors' hand. Consequently, a sound is electronically generated. This exhibit shows a relationship between the location of a given light beam in a sequence and the sound generated (Stocklmayer & Gilbert, 2002).

Little is known on the outcomes of using analogy-based exhibits. One of the few known studies analysed how the type of analogy-based exhibit affected adults' mental models in the short term (Stocklmayer & Gilbert, 2002). In this study, adult visitors were observed interacting with an exhibit and were interviewed immediately after the interaction. Findings suggest that prior knowledge, the type of analogy-based exhibit, and its subject constrained the cognitive and affective outcomes. Exhibits related to everyday situations and of the types "only showing similarities between entities" and "showing similarities between both entities and relationships", enhanced the retrieval of relevant prior experiences and the building of scientifically accepted meanings. On the other hand, in exhibits "only showing similarities between relationships", the relationship between the experience and the target was frequently not clear to the visitors. As a consequence, aesthetical links with the exhibit and earlier experiences were made, but an understanding of the target analogue was not achieved.

In other areas of research, such as cognitive psychology or formal education, the usefulness of analogies in understanding a target domain has been studied. Empirical studies in cognitive psychology suggest that factors affecting the individuals' fluency in the mapping process are: task factors (e.g., time pressure or processing load), the characteristics of the individual (e.g., age and experience), and factors intrinsic to the analogical mapping (e.g., degree of transparency between domains, knowledge about source and target) (Gentner, 2003).

Studies in formal education have shown that although analogies are effective tools for developing knowledge about a not well known target (e.g., Glynn et al., 1995), they can result in undesirable outcomes because students may: learn the model rather than the target (Treagust, 1993); be unaware of the boundaries between a model and the reality it is representing (France, 2000); be unaware of the limitations of the analogy due to a lack of its evaluation (Duit et al., 2001); misunderstand the target due to inappropriate inferences (Newton, 2000); or have inappropriate support from the teacher (Dagher, 1995). Different strategies have been suggested in order to enhance the usefulness of analogies in formal education. For example, the use of "bridging analogies" (Brown & Clement, 1989) or the use of "multiple analogies" (Spiro et al., 1989). In both strategies, a series of intermediate analogies, related to each other, can gradually enhance the learning of a target. Furthermore, models for teaching with analogies have been proposed (e.g., Glynn, 1991; Zeitoun, 1984). Generally, they suggest that the source domain should be familiar; students and teachers should identify the similarities between source and target domains; and the limitations of the analogy should be identified.

The contexts of these empirical studies are different from those found in a museum. For example, the type of sample (e.g., usually individual students), the guidance given for analogical thinking, the type of sources used (e.g., usually structurally and superficially similar to the target), or the type of task posed (i.e. usually a well defined problem solving situation). Consequently, one cannot simply, base, on the results of these studies, an understanding of how visitors make sense of analogy-based exhibits.

In order to understand the potential relevance of analogy-based exhibits in engaging adult visitors' understanding of a target analogue, the research questions are: 1) What did visitors understand when an analogy-based exhibit was used? 2) Which factors affected that understanding?

2. METHODS AND SAMPLE

The study was framed in a "personal constructivist" perspective (Kelly, 1955). Data were collected in Lisbon, Portugal, at the "Pavilhão do Conhecimento" (Pavilion of Knowledge). 84 adults, having received no explanations from the explainers, were unobtrusively observed and informally interviewed (either alone or in a group) immediately after their interaction with an analogy-based exhibit. A characterisation of the participants is provided in Table 1. Visitors were asked about: how they operated the exhibit they were being asked about; how difficult it was to operate it; what was observed; whether the observations were familiar; the aim of the exhibit; what was understood from the experience; whether that understanding was novel; how difficult it was to understand the experience provided; and how was the knowledge achieved.

TABLE 1. Characterisation of the sample of visitors who participated in the study N = 84

Characteristics		First data collection	Second data collection	Total
Gender	Male	38	21	59
	Female	18	7	25
Age	18–29yrs	17	13	30
	30–39yrs	19	6	25
	40–49yrs	15	6	21
	50–59yrs	2	2	4
	60+yrs	3	1	4
Nationality	Portuguese	41	19	60
	Other	15	9	24

Three analogy-based exhibits of each subtype, all related to sound, were considered for this study (see appendix for a description and figures of the exhibits). The theme sound was chosen because little educational research has been carried out about this subject, which is of interest for citizens. All the interviews were recorded and fully transcribed. Data were collected in two phases: between 6th December 2003 and 5th January 2004 (excluding the period between 23rd December and 1st January) and five days during August 2004. The second phase aimed at providing an initial insight into the reasons for breakdown during the analogical reasoning process. During the second phase of data collection, data were not collected in the exhibit "showing similarities between both entities and relationships" because it was no longer in the museum. The length of the interviews was affected by the demands of the visitors' social group and by the way each visitor used the exhibit.

3. DATA ANALYSIS

The analysis was based on the model for data interpretation proposed by Gilbert & Watt (1983). Therefore, after analysing each interview independently of the others, an interpretation of multiple data was carried on and patterns in the data were identified. The categories reflect what visitors understood by using an analogical model and the factors affecting that understanding. The visitors' understandings were categorised into: target (in)completely understood; misconceptions about the target; behaviour of a source domain described; and not understood. The categories describing the factors affecting visitors' understanding of a target (once an analogy was perceived) are related to the (un)successful use of analogical reasoning; memory retrieval; and text comprehension. An analogy was considered to be perceived when a visitor was able to mention a target domain (intended or not) as the aim of the exhibit. Frequencies of each category were calculated.

4. RESULTS

4.1 Visitors' Understanding of an Analogical Model

Half of the visitors (11 out of 22) did not operate the analogical model "showing similarities between both entities and relationships". For some visitors this was because the operational instructions were not understood:

"That was the part that I didn't understand. Looking at this explanation [pointing to the instructions in the complementary text], I was expecting to see a graph, but I didn't see anything." (Portuguese male, 30–39, teacher)

For other visitors, the analogical model was not noticed or interest was not triggered. This may be a consequence of the exhibit design which integrates several interactive elements which are useless for operating the analogical model. Consequently, visitors may have been distracted from the main feature of the exhibit (Allen & Gutwill, 2004).

"No, I didn't [notice the analogical model]." (Portuguese male, 30–39, geologist)

"I noticed it...I don't know...I looked at it but, in fact, I neither understood it nor did I ask anyone about its purpose." (Portuguese female, 50–59, retired secretary)

Several understandings of the targets were reported by those visitors who operated the analogical model (Table 2). In the exhibit "only showing similarity between entities", the majority of visitors were aware of the target domain, but extending knowledge about it often resulted in misconceptions. For example, a visitor thought that the exhibit was showing the movement of the throat, rather than the behaviour of the vocal tract:

"It is clear that the shapes of those pipes symbolise our throat." (Portuguese female, 30–39, university teacher in engineering)

Misconceptions were rarely expressed in the exhibit "showing similarities between both entities and relationships", but several visitors were unable to provide a meaning to the experience:

"I don't know. I looked at the light but I didn't understand its purpose." (Portuguese female, 50–59, secretary)

In the same exhibit, some visitors, unaware of the target, focused on describing the behaviour of the source, which was sometimes helped by prior knowledge:

TABLE 2. Visitors' understanding of the experience as function of the type of exhibit

Visitors' understanding	Subtype of analogy-based exhibit		
	Only showing similarities between entities (n = 35)	Showing similarities between both entities and relationships (n = 11)	Only showing similarities between relationships (n = 27)
Target (in)completely understood	3	4	6
Misconceptions about the target	21	1	–
Behaviour of a source domain described	8	3	19
Not understood	3	4	2

"I was playing with the laser beam, with the reflection. I wanted to see how the reflection went through the oil." (American male, 40–49, managing director of an entertainment company)

Unawareness of a target domain was more common in the exhibit "only showing similarities between relationships", leading visitors to believe that the exhibit was demonstrating a phenomenon:

"In the exhibit there is a vibration of the dish which spreads the sand." (Portuguese male, 30–39, army officer)

Awareness of a target seems to be the first step to enhance learning about it. Unawareness of a target may be related to superficial dissimilarities between the source and visitors' representation of the target domain and to misunderstanding or disinterest of the complementary text. In addition, in the exhibit "only showing similarities between relationships" an intrinsic appeal of the source domain used might have also blocked the perception of the target (Stocklmayer & Gilbert, 2002).

In the situations in which the source was perceived, the results contrast with the findings reported by Stocklmayer & Gilbert (2002), who mentioned that when the target was related to an everyday world situation, visitors often forged strong links with the target. Consequently, one may ask: why were the analogies related to an everyday situation ineffective to some visitors and effective to others? Furthermore, some visitors achieved an understanding of the target in the exhibit "only showing similarities between relationships". So, one may also ask: Why was that understanding possible?

4.2 Factors Affecting Visitors' Understanding

When visitors were aware of a source domain, making inferences of a target was helped by analogical reasoning, prior knowledge, and text comprehension (Table 3). Table 3 suggests that achieving a consensual scientific understanding of the exhibit "only showing similarity between relationships" was possible either when visitors were familiar with the target or when they understood the complementary text. Examples of both cognitive strategies are respectively:

Visitor:	The sand shows how the plate vibrates.
Interviewer:	How do you know that?
Visitor:	How do I know? ... I am a physicist! (German male, 60–69, physics teacher)

Visitor:	The idea is that there are zones of the plate that vibrate and others that don't.
Interviewer:	How do you know that?
Visitor:	I read the text I was supposed to know but I read the text. (Spanish male, 18–29, aeronautic engineer)

On the other hand, the analogical reasoning was the main cognitive strategy activated as a consequence of using the exhibits "only showing similarities between entities" and "showing similarities between both entities and relationships". Successful analogical mapping was enhanced when the source domain was encoded at a structural level. For example, in the exhibit "only showing similarity between entities", some visitors matched the boxes with the vocal tract because:

"[they] are resonance structures that modify the acoustics conditions and so provide different sounds. For example an 'a' or an 'u' (visitor demonstrates the different openings of his mouth when producing these two sounds)." (Spanish male, 30–39, engineer)

Conversely, unsuccessful analogical mapping seems to be affected by the visitors' unfamiliarity with the source, which resulted in an encoding at a superficial level. For example, in the exhibit "showing similarities between both entities and relationships", the behaviour of the laser beam within the liquid was rarely mentioned. This may result from the visitors' unfamiliarity with optics in general. For example, in relation to the laser beam, a visitor mentioned that:

"I don't understand very well what they are for. Maybe, they are the waves. Those which are emitted by the sperm whale but I cannot associate sound with light beam. That's it. Maybe it's because of my ignorance or my lack of knowledge." (Portuguese male, 30–39, manager)

Superficial encoding of a source analogue also occurred when visitors selected only one part of the source domain for mapping. This may be the result of the characteristics of the source domain (such as the colours used, organization of the entities, or nature of interactivity), which may emphasise some elements of the source in relation to others. For example, in the exhibit "showing similarities between both entities and relationships", the laser beam pop-ups from the liquid. This may explain why visitors matched the laser beam with sound but neglected the mapping between the liquid and the SOFAR:

Visitor: May be the red rays are the sound waves that are emitted on the bottom of the sea.
Interviewer: What do you think this exhibit is about?
Visitor: I think it's about the communication between whales. (Portuguese female, 40–49, economist)

Similarly, in the exhibit "only showing similarities between entities", visitors neglected the elements of the source with low degree of interactivity. For example, a visitor explained that he did not identify the representation of the lungs in the source domain because:

Visitor: It's not obvious.
Interviewer: But the piston is red. Why do you say it is not obvious?
Visitor: Yes, but there are several exhibits here related to the movement of air. So the movement of the air is not so distinct. Besides, people are used to press things and to see air. (Portuguese male, 18–29, student)

Failures in the mapping process resulted in the generation of misconceptions. These mismatches were frequent in the exhibit "only showing similarities between entities". For example, visitors thought that the exhibit was representing the behaviour of the vocal chords:

TABLE 3. Cognitive strategies which helped visitors in understanding a target

Making sense helped by	Subtype of analogy-based exhibit		
	Only showing similarities between entities (n = 24)	Showing similarities between both entities and relationships (n = 4)	Only showing similarities between relationships (n = 6)
Analogical reasoning	20	3	–
Prior knowledge	2	1	2
Text comprehension	2	–	4

"I learnt that the way the vocal chords change is more complex than I imagined. We can see there the different positions for them. So, this part of our body, which is really small, has the capacity to form different shapes and I was surprised with that." (Canadian female, 40–49, art historian)

Interpreting the exhibit as representing the behaviour of the vocal chords rather than the vocal tract may have resulted from visitors' kinaesthetic experiences, such as the effort necessary in the vocal chords to produce a sound. These kinaesthetic experiences may have induced misconceptions about the mechanisms of vowel production:

Interviewer: How do you know that it is a representation of the vocal chords?

Visitor: Because I know that our vocal chords produce different sounds and so I made the connection. (Portuguese female, 18–29, student)

For other visitors, the same exhibit aimed at showing the behaviour of the pharynx and larynx during the production of vowels. In this situation, visitors did not notice the superficial dissimilarities between source and target domain, rather they looked for entities in the target that resembled those of the source. The mismatching suggests that visitors saw the model underlying the exhibit as a copy of the reality:

"[The mismatches resulted] probably because of my notions of the pharynx and larynx. For me they are represented on those boxes. They are neither the lips nor the teeth." (Portuguese male, 50–59, physical education teacher)

5. CONCLUSIONS

Findings suggest that for many visitors the analogies underlying the exhibits did not enhance a scientifically understanding of a target. Cognitive outcomes were constrained by: the type of analogy-based exhibits, the visitors' prior knowledge, the characteristics of a source, and the type of cognitive strategies used.

In the exhibits "only showing similarities between entities" and "showing similarities between both entities and relationships", many visitors activated analogical reasoning. However, its unsuccessful operation resulted in incomplete understanding or misconceptions about the targets. Some of the reasons for the (un)successful use of the analogy were similar to those already described in the literature. That is, familiarity with the source and its encoding at a structural level enhanced understanding (Novick, 1988), and superficial dissimilarity between source and target domains enhanced mismatching (Ross, 1989). In addition, misconceptions about a target and emphasis on some elements of the source in relation to others (for example, due to the choice of colours or organization of the elements of a source) seem to result in unsuccessful analogical reasoning.

Contrary to the study by Stocklmayer & Gilbert (2002), in the present study the exhibits related to an everyday situation were superficially dissimilar with the targets. This may partially explain why they were less effective in promoting a scientific understanding about a target.

Like in the study by Stocklmayer & Gilbert (2002), in the exhibit "only showing similarities between relationships", visitors were often unaware of the target domain because of the superficial dissimilarities between source and target. Consequently, visitors often describe the behaviour of the source domain. Nevertheless, there were some visitors who learnt about the target. That was possible when the complementary text was understood, rather than by operating analogical reasoning. Since understanding what is read is constrained by individuals' prior knowledge, this suggests that this type of exhibit is more appropriate for visitors with a level of expertise in science.

6. IMPLICATIONS FOR DESIGNERS

Exhibit designers need to be aware that exhibits "only showing similarities between relationships" are difficult to understand and do not often trigger analogical reasoning. Consequently, this type of exhibit needs to be designed having the more expert visitors in mind. In addition, complementary information should be clearly provided in order to guide visitors towards an understanding of the target analogue.

Analogical reasoning is often used for understanding exhibits "only showing similarities between entities" and "showing similarities between both entities and relationships". Consequently, in the design of these types of exhibits, source domains should be familiar to visitors (Gilbert & Stocklmayer, 2001; Glynn et al., 1991; Zeitoun, 1984); the relationships that hold in a source domain should be emphasised; there should be no emphasis on some entities of the source over others (e.g., by the organization of the elements of the source, level of interactivity, or colours used); and superficial dissimilarities between source and target should be avoided whenever possible. Furthermore, "bridging analogies" (Brown & Clement, 1989) or "multiple analogies" (Spiro et al., 1989) can be employed in order to overcome the reduced power of a single analogy.

ANNEX

Description, Classification and Reasons for the Classification of the Selected Exhibits

Vocal vowels

Description: Five boxes with different shapes opened at both ends. A piston is connected to one end of a tube, and a smaller tube, with two blades inside, is connected to the other end. By inserting the end of the smaller tube in one of the five boxes and by pressing the piston, air is released, the blades vibrate and the sound of an (English) vowel is produced. For each box there is a corresponding vowel.

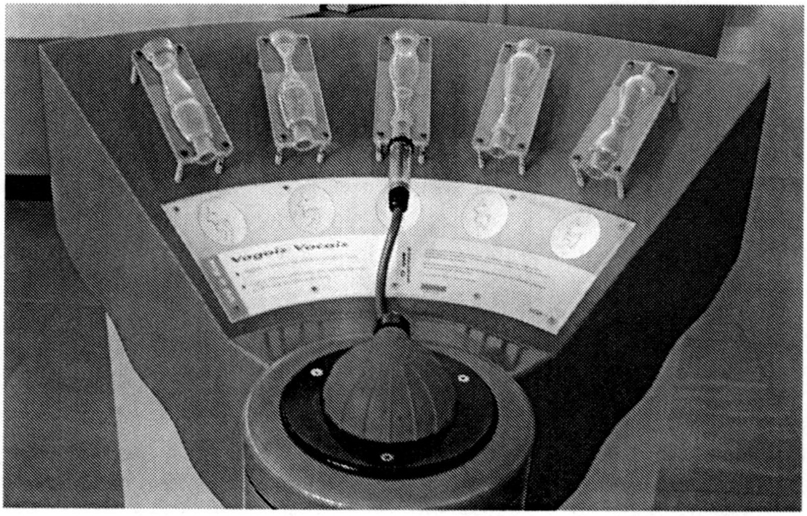

Figure 1. The exhibit "Vocal vowels"

Classification: "Exhibit only showing similarities between entities". The anatomy involved in the human's production of vowels is shown by the following mapping: the piston and the tube represent the subglottal system; the cylindrical tube matches the larynx; each of the boxes corresponds to the supralaryngeal vocal tract; the air released from the plastic piston represents the flow of air from the lungs; and the two blades match the vocal chords. The acoustical relations that hold on the target are not represented on the source because the flow of air released from the piston is not steady; the acoustic function of the vocal chords cannot be represented as a vibration of two blades; the acoustical relation between different shapes of the vocal tract and different vowels produced is not shown; and the acoustic energy that is radiated from the lips, influenced by a high-pass filter, is not demonstrated.

The global communication of the sperm whale

Description: A panel represents the ocean from the surface down to the depth of 1000m. In the panel there are images of different living beings placed in a kind of depth scale and a box with a liquid inside at the mark corresponding to the "depth of 1000m". By vertically moving a knob it is possible to hear the sound of the different animals at the "different depths". By placing the knob at the mark corresponding to the depth of 1000m it is also possible to activate a laser beam that propagates within the liquid.

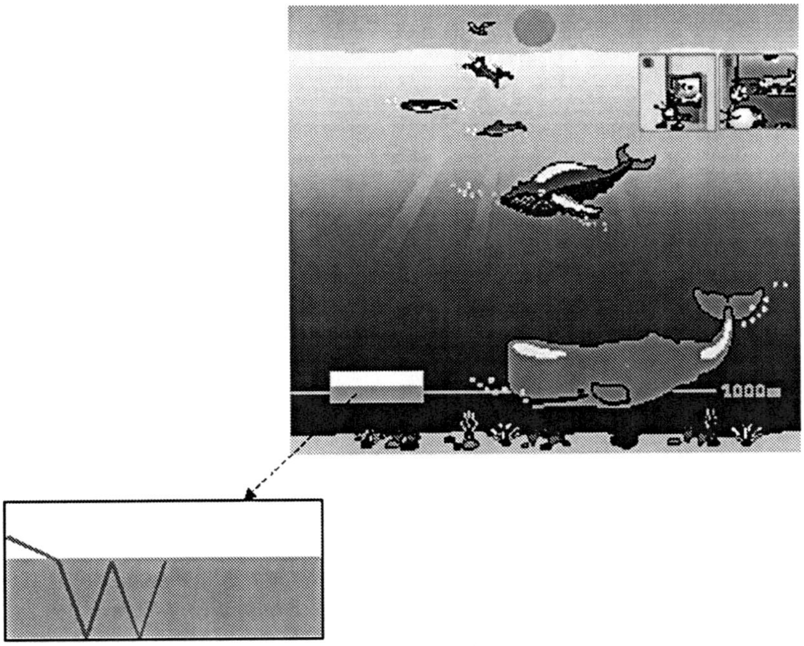

Figure 2. The exhibit "The global communication of the sperm whale"

Classification: "Exhibit showing similarities between both entities and relationships". The analogical mapping is shown in the following way: the liquid inside the transparent box is the entity analogue for the SOund Fixing And Ranging (SOFAR); the laser beam the entity analogue to the sound beam; and the "trapping" of the light beam inside the liquid is the relationship analogue for the "trapping" of the sound beam inside the SOFAR.

Bells

Description: Two metal plates with different shapes are clamped by their centres. By spreading sand on the top of the metal plates and by striking their edges with a violin bow, patterns of sand can be observed.

Figure 3. The exhibit "Bells"

Classification: "Exhibit only showing similarities between relationships". The patterns of sand match the nodal lines of vibration of the plate. They divide the plate in its various vibrating parts. By the location of the sand it is possible to visualize that Sections of the plate separated from each other by the nodal lines move in opposite directions. In the exhibit, the relationship analogue is drawn between the position of the lines of sand and the vibration of each Section of the plate.

REFERENCES

Allen, S. & Gutwill, J. (2004). Designing with multiple interactive: five common pitfalls. *Curator*, 47(2), 199–212

Brown, D. & Clement, J. (1989). Overcoming misconceptions via analogical reasoning: abstract transfer versus explanatory model construction. *Instructional Science*, 18, 237–261

Dagher, Z.R. (1995). Analysis of analogies used by science teachers. *Journal of Research in Science Teaching*, 32(3), 259–270

Duit, R., Roth, W.M., Komorek, M. & Wilbers, J. (2001). Fostering conceptual change by analogies-between Scylla and Charybdis. *Learning and Instruction*, 11, 283–303

Dunbar, K. (2001). The analogical paradox. Why analogies are so easy in naturalistic settings, yet so difficult in psychology laboratory? (In D. Gentner, J.K. Holyoak, & B.N. Kokinov (Eds.), *The Analogical Mind Perspectives from Cognitive Psychology* (pp. 313–334). Cambridge, MA: MIT Press.)

France, B. (2000). The role of models in biotechnology education. (In J.K. Gilbert, & C. Boulter (Eds.), *Developing Models in Science Education* (pp. 271–288). Dordrecht: Kluwer.)

Gentner, D. (1983). Structure-mapping: a theoretical framework for analogy. *Cognitive Science*, 7, 155–170

Gentner, D. (2003). Analogical reasoning, psychology of. (In L. Nadel (Ed.), *Encyclopedia of Cognitive Science* (pp. 106–111). London: Nature Publishing Group.)

Gilbert, J.K. & Stocklmayer, S. (2001). The design of interactive exhibits to promote the making of meaning. *Museum Management and Curatorship*, 19(1), 41–50

Gilbert, J.K. & Watt, D.M. (1983). Concepts, misconceptions, and alternative conceptions: changing perspectives in science education. *Studies in Science Education*, 10, 61–98

Glynn, S.M. (1991). Explaining science concepts: a teaching-with-analogies model. (In S.M. Glynn, R.H. Yeany, & B.K. Britton (Eds.), *The Psychology of Learning Science* (pp. 219–240). New Jersey: Lawrence Erlbaum Associates.)

Glynn, S.M., Duit, R. & Thiele, R.B. (1995). Teaching science with analogies: a strategy for constructing knowledge. (In S. Glynn, & R. Duit (Eds.), *Learning Science in the Schools: Research Reforming Practice* (pp. 247–273). NJ: Lawrence Erbaum.)

Kelly, G. (1955). *A Theory of Personality: The Psychology of Personal Constructs*. (New York: W.W. Norton & Company.)

Newton, D.P. (2000). *Teaching for Understanding*. (London: Falmer.)

Novick, L. (1988). Analogical transfer, problem similarity, and expertise. *Journal of Experimental Psychology: Learning Memory and Cognition*, 14(3), 510–520

Ross, H.B. (1989). Distinguishing types of superficial similarities: different effects on the access and use of early problems. *Journal of Experimental Psychology: Learning Memory and Cognition*, 15(3), 456–468

Spiro, R., Feltovick, P.J., Coulson, R.L. & Anderson, D.K. (1989). Multiple analogies for complex concepts: Antidotes for analogy-induced misconceptions in advanced knowledge acquisition. (In S. Vosniadou, & A. Ortony (Eds.), *Similarity and Analogical Reasoning* (pp. 498–531) Cambridge: Cambridge University Press.)

Stocklmayer, S. & Gilbert, J. (2002). New experiences and old knowledge: towards a model for the personal awareness of science. *International Journal of Science Education*, 24(8), 835–858

Treagust, D.F. (1993). The evolution of an approach for using analogies in teaching and learning science. *Research in Science Education*, 23, 293–301

Zeitoun, H.H. (1984). Teaching scientific analogies: a proposed model. *Research in Science and Technology Education*, 2(2), 107–125

STUART NAYLOR, BRENDA KEOGH, BRIGID DOWNING, JANE MALONEY AND SHIRLEY SIMON

22. THE PUPPETS PROJECT: USING PUPPETS TO PROMOTE ENGAGEMENT AND TALK IN SCIENCE

Abstract: The PUPPETS Project uses puppets as a stimulus for children to engage in conversations involving reasoning in primary science lessons. Data were gathered using a variety of methods, including audio and video recording, and analysed using an open coding approach based on grounded theory. The data indicate that puppets are engaging and motivating for children; that they promote talk involving reasoning; that they can be particularly effective with reluctant speakers; that they appear to be effective across the whole primary age range; and that they promote significant changes in teachers' professional practice

Keywords: Engagement, Motivation, Questioning, Puppets, Talk

1. INTRODUCTION

The PUPPETS Project is a research and development initiative funded by the Nuffield Foundation which aims to help teachers provide more opportunities for productive talk in science lessons, using puppets as a stimulus. The research focuses on the effectiveness of hand-held puppets to engage primary school children's attention, challenge their ideas and promote conversations involving reasoning in science.

The value of talk in children's learning is well-documented; talking about their ideas helps children to clarify their thinking and develops their capacity to reason (Kuhn et al., 1997; Mercer et al., 2004). This is particularly important in science, as the development of conceptual understanding and the ability to use reasoning are central goals of science education. Unfortunately the type of talk that promotes reasoning and thinking is frequently absent in science lessons (Newton et al., 1999). Reasons for this may include the uncertainty teachers have about the value of children's conversations, limited knowledge of appropriate teaching strategies, and insecure subject knowledge (Osborne and Simon, 1996; Solomon, 1998).

Focused and productive talk normally requires a suitable stimulus that engages children, enabling them to feel personally involved and committed to the conversation. The use of puppets is well-established in primary schools in subjects such as language and social education (Thorp, 2005). There is some evidence that puppets can have a valuable impact in motivating children and promoting language development (Low and Matthew, 2000). Puppets can engage the children's attention,

Stuart Naylor, Millgate House Publishing & Consultancy Ltd, 30 Mill Hill Lane, Sandbach, CW11 4PN, UK

R. Pintó and D. Couso (eds.), Contributions from Science Education Research, 289–296.

provide a context for conversation and promote purposeful activity. We believe that puppets could also act as an effective stimulus for generating talk involving reasoning in science lessons. This would build on previous work on group interactions (Maloney, 2002) and on using concept cartoons (Keogh and Naylor, 1999), which have proved to be a powerful means of generating discussion and argument (Downing, 2005). Puppets seem to provide a mechanism for 'bringing the concept cartoons alive', with dialogue between the puppets or between the teacher and the puppets acting as a stimulus for children to engage with that dialogue.

The main research questions to be investigated were:

1. In what ways can puppets be used in primary science lessons to enhance children's engagement and promote talk involving reasoning in science?
2. Is it possible to initiate changes in teachers' beliefs about the value of children's talk and in their management of talk in science lessons?

2. MODE OF INQUIRY

The project began with a pilot phase, involving eight teachers, to develop materials and guidance for teachers to use with puppets in their science lessons. These teachers attended a workshop to become familiar with the use of puppets, after which each teacher was audio and video-recorded during a science lesson where puppets were used. In each of these lessons sample groups of children were audio-recorded during small group conversations.

Preliminary analysis of the data from these puppets lessons led to the development of an analytical framework for classroom discourse using grounded theory (Glaser and Strauss, 1967). The focus for the analytical framework was on the nature of talk, including the use of reasoning and argument by the teacher, the puppets and the children. The pilot study also provided valuable feedback on the guidance provided for the teachers and on whether any classroom management issues were significant. The research team's preliminary judgments were tested out with the teachers at a follow up meeting, as a result of which the analytical framework and guidance were refined for the main study.

A further 16 teachers were involved in the main study, drawn from primary schools in the London and Manchester areas and covering a range of social, cultural and ethnic backgrounds. These teachers were observed and video-recorded teaching a typical science lesson. Two groups of children in each class were also audio and/or video-recorded during periods of pupil activity. Teachers were interviewed to obtain their views on the nature and value of children's talk and details about how they taught science. These initial observations and interviews provided a baseline against which future lessons could be compared. Transcripts of the teachers' and children's talk were scrutinised and the provisional analytical framework was further developed. This analytical framework was then available to apply to the data from lessons where puppets were used in science teaching.

These 16 teachers then attended a preparation meeting to provide guidance on how puppets might be used effectively, including issues such as role and characterisation of the puppets. Video clips and transcripts from the pilot phase were used to illustrate aspects of the teacher's role and how puppets might interact with the children.

After a period of several weeks, during which the teachers could get used to using puppets, a second science lesson was observed and recorded with the teacher using one or more puppets as part of their teaching. Teachers were asked to continue to teach lessons from their usual scheme of work, so that lessons without puppets and with puppets would be broadly comparable. Between them teachers were teaching a wide range of topics, both with and without puppets.

As before the teachers were observed and video-recorded, with two groups of children in each class also audio and/or video-recorded during periods of pupil activity. Teachers were interviewed to ascertain their views on the impact of the puppets; they also kept reflective diaries to enable them to capture their feelings about the use of puppets and record any significant classroom incidents at the time. Interviews with groups of children were video-recorded to provide data from the children's perspective. The range of data sources and collection methods allowed extensive triangulation and cross-referencing of the data. A follow-up meeting with the teachers provided feedback on the longer-term impact of using puppets and their response to issues emerging about the use of puppets in science lessons.

For various reasons not all the teachers were able to continue in the project. By the end of the main research phase, data were available for 13 of the original 16 teachers.

3. DATA ANALYSIS AND RESULTS

Using an open-coding approach (Strauss and Corbin, 1998), transcripts of all the conversations between teachers, puppets and children were coded and the coding framework discussed between researchers. An extensive range of codes was developed to identify episodes of discourse during the lesson which seemed significant in the development of children's engagement and thinking. Approximately 30 codes were generated in the initial data analysis, representing all the types of discourse which had occurred. These were reduced to a more manageable number (10) by conflating codes where possible. These 10 codes were related to either the teacher, the child or the puppet, giving us a total of 30 types of utterances.

The data were then scrutinised to identify where significant changes had occurred. Only some of the codes, where significant changes were evident, are reported on here.

Examples of the codes used included:

Q: question that requires reasoning

N: question that does not require a reasoned answer

A: use of argument to challenge or justify a point of view

R: recall of information from memory

P: gives information or asks question relating to the procedure to be followed

The analytical framework has enabled us to make comparisons between the time spent in various types of talk, to determine the nature of the discourse of the children, teachers and puppets, and to identify critical episodes of problem-solving and argumentation. The key themes emerging from this analysis include:

- puppets enhance engagement and motivation
- puppets provide a stimulus for talk involving reasoning
- puppets engage reluctant speakers
- the use of puppets is appropriate across the age range from 4 to 11 years
- puppets facilitate changes in the teachers' practice

3.1 Enchancing Engagement and Motivation

In each class nearly all the children showed high levels of engagement and motivation in response to the puppets. Evidence of engagement included children asking when puppets are to be used; physical interaction with the puppets (for example, holding their hands); following the puppet around the room; talking to the puppets. All of the teachers noted that puppets enhanced pupil engagement in their diaries and/or in their interviews; pupils in each group interviewed stated that they preferred lessons with puppets and that the puppets helped them to learn. Video evidence from lessons using puppets showed that pupils were focused on the puppets, that they had high levels of concentration, that they stayed on task and that they were eager to contribute to discussion.

Typical comments from teachers included:

"Children were keen to tell the puppet what they knew but also listened more attentively" (Teacher I)
"The children responded brilliantly. Thought they may be cynical, esp. Y5 boys, but they were especially motivated". (Teacher C)
"All children were keen to ask or answer a question and join in discussion" (Teacher C)

Analysis of interview transcripts with the children confirmed this high level of interest and engagement. Comments about the use of puppets included:

"Lessons are more fun"
"I understand better with the puppets"
"It inspired my imagination"
"You want to answer questions more"
"I am much more enthusiastic about learning now we have the puppets"

3.2 Providing a Stimulus for Talk Involving Reasoning

Analysis of interview transcripts showed that each of the teachers claimed that puppets are an effective stimulus for talk involving reasoning. They described how more children talk in science lessons with the puppets, how children give fuller explanations and listen more closely. They noted that children explained and justified their ideas more, and that this continued even when the puppet was not being used.

Video recording of children engaged in small group conversations allowed comparisons to be made of the nature of their conversations without (Lesson 1) and with (Lesson 2) the use of puppets. Table 1 shows comparisons of the amount of talk involving reasoning compared to talk about procedural matters. The table

TABLE 1. Comparison of Small Group Conversations without and with Puppets

	Talk involving reasoning	Procedural talk
Lesson 1 (without puppets)	48.5 mins	72.5 mins
Lesson 2 (with puppets)	128.5 mins	30 mins

shows the cumulative time spent in each type of talk across all the classes for which data was available.

Conversations were coded as talk involving reasoning when children appeared to use evidence, reasoning or justification in a concept-based conversation about a scientific question. For example, a child talking about what type of rock might be used for a rock climbing area replies:

"Waterproof, because if you go on them when they are not waterproof, like limestone, it could go, if they are still water in them you could slip, or because there is holes in it, it could just break away."

Conversations were coded as procedural talk when children focused on practical aspects of a lesson with no reasoning involved. For example:

"Use, oh, oh, (taps pencil on table, looks for an eraser and then carries on writing over the top of what she's done. Looks at another Child's work). Oh I've got it wrong! I need an eraser. (rubs out work and starts writing again). By using a measuring tape. How do you spell measure?"

There was a marked increase in the time that children were engaged in talk involving reasoning and a decrease in the time that they were talking about practical and procedural matters. A chi-squared test shows these changes to be statistically significant ($p<0.001$). The teachers' use of puppets appears to have had a significant impact on the amount of small group talk involving reasoning taking place in the majority of the classes. Other aspects of the data show that the total amount of time spent in small group conversations was broadly similar in lessons without and with puppets. In other words, the increase in the amount of small group talk involving reasoning represents a change in the **nature** of children's talk, not just an increase in the total amount of time made available by the teacher for talking. Children generally were highly focused and engaged in the whole class introduction to the lessons using puppets and this appeared to carry over into small group work.

3.3 Engaging Reluctant Speakers

The majority of the teachers noted in their interviews that the children were extremely keen to talk with the puppets. Several teachers noted that they were particularly effective with shy children who are normally reluctant to speak. This was confirmed by the children themselves in interview transcripts, where several children claimed that puppets help shy children to talk in class. Their comments included:

"I'm not nervous talking in front of the class (with the puppet) so it's helped me – because they all look at the puppet not me." (Child A, age 11)
"The puppet explains things using less scientific words – which is more understanding (Child B, age 10)
"It's more comfortable talking to the puppet. If you get it wrong with the teacher she says 'No'. The puppet says 'not really correct'." (Child C, age 10)
"The teacher already knows the answer anyway. So she's really just testing you. The puppet doesn't know the answer so we have to explain it in a way he will understand." (Child D, age 11)
"I put up my hand more with the puppet because I understand it more." (Child E, age 11)

The puppets appeared to create cognitive conflict for the children in a non-threatening manner, operating as peers with the children without the complications of the status attached to the teacher. By removing the status factor they enabled the children to speak more freely without fearing that judgments would be made about their conversations.

3.4 Puppets being used with a Wide Age Range of Pupils

The main research study was carried out with pupils in English schools in Years 3–6, with an age range from 7–11 years. No evidence was found of any significant differences in the children's responses across this age range. The age range was wider in the pilot study, with children from Years 1–6 (ages 5–11) being involved. Again no evidence was found of any differences in the children's responses according to their age.

In addition to this research study two additional small-scale research studies have been carried out with pre-service teachers. Both of these involved pre-service teachers working with all the classes across a whole school, from Reception to Year 6 (ages 4–11). Results in these two studies were more variable, with the student teacher's ability to manage the class successfully and the teacher's level of commitment to supporting the student teacher being significant factors in these studies. However in both cases the age of the children involved had no apparent bearing on the nature of their response.

The data are consistent in indicating that the use of puppets in science lessons is appropriate across the age range from 4 to 11 years.

3.5 Facilitating Changes in the Teacher's Practice

The teacher's use of questions which require thinking and reasoning rather than recall of information is a critical factor in promoting talk involving reasoning. Thinking and reasoning questions explicitly value justification and inference, act as a prompt to critical reasoning and begin to model the process of argumentation. Table 2 shows a comparison between the questions asked by teachers in typical science lessons (Lesson 1) and science lessons where puppets were used (Lesson 2). In Lesson 2 questions may be asked by the teacher or by the puppet. The numbers shown in the table are the cumulative total of all the questions asked by the teachers or puppets in the lessons.

TABLE 2. Comparison of Teacher Questions without and with Puppets

	Reasoning questions	Non-reasoning questions
Lesson 1 (without puppets)	73	388
Lesson 2 (with puppets)	263	291

Questions were coded as reasoning questions if they required thinking, reasoning or justification from children. For example:

"Well what do you think a nice rock climbing area should have, and what should it look like. What do you think?"

Questions were coded as non-reasoning questions when they did not require thinking, reasoning or justification (such as recall questions). For example:

"Think back to yesterday and why we eat, well what did we say? What were the two reasons why we eat food, William?"

In lessons where puppets were used there was a marked increase in the use of reasoning questions and a decrease in recall questions. A chi-squared test shows these changes to be statistically significant ($p<0.001$).

This increase in the use of reasoning questions is consistent with the way that puppets present problems to the class. Because the puppets present problems rather than instructions, whole class discussion with pupils is more likely to involve exploratory conversations about how the problem might be solved.

4. CONCLUSIONS AND IMPLICATIONS

The research indicates that puppets can provide a useful mechanism to enhance children's engagement and to promote talk involving reasoning in primary science lessons. There is evidence that when teachers use the puppets children talk more readily about scientific problems and higher-order thinking (such as explanation and justification) is promoted. They can be particularly significant in engaging reluctant speakers, such as shy children, and promoting greater involvement in scientific discussion with these children. The puppets appear to be effective in providing an interactive narrative which sets a context for learning and provides a purpose for children's talk and follow-up activity. It is notable that the increase in argument and talk involving reasoning is in relation to explicitly scientific problems, rather than in relation to the socio-scientific issues which are frequently used by researchers to promote argument in science lessons.

The research also indicates that teachers using puppets are likely to ask more questions involving reasoning and to ask fewer questions which only require recall.

This fairly fundamental change in professional practice was achieved with only a short (2 hour) preparation session and the provision of relatively inexpensive resources. The puppets were used within the teachers' existing schemes of work with no disruption to the usual curriculum. Little additional preparation was required for lessons involving puppets. They therefore appear to offer a valuable extension to the teaching/learning strategies typically used by primary school teachers in science and a potentially valuable mechanism for facilitating change in professional practice.

Further issues to be explored in the research will include analysis of the puppets' role at different parts of the lesson, how teachers create characters for their puppets and how teachers avoid stereotyping their roles.

ACKNOWLEDGEMENT

We are grateful to the Nuffield Foundation for funding the research.

REFERENCES

Downing, B. (2005) *Developing the nature and role of quality argument in primary science lessons through the use of concept cartoons*. Unpublished PhD thesis, Manchester Metropolitan University.

Glaser, B. and Strauss, A. (1967) *The discovery of grounded theory: strategies for qualitative research*. Chicago: Aldine.

Keogh, B. and Naylor, S. (1999) Concept cartoons, teaching and learning in science: an evaluation. *International Journal of Science Education*, 21, 4, 431–446.

Kuhn, D., Shaw, V. and Felton, M. (1997) Effects of dyadic interaction on argumentative reasoning. *Cognition and Instruction*, 15, 3, 287–315.

Low, J. and Matthew, K. (2000) Puppets and prose. *Science and Children*, 37, 8, 41–45.

Maloney, J. (2002) Effective group work in the classroom – the roles children play. *Paper presented at British Educational Research Association 2002 Annual Conference, University of Exeter.*

Mercer, N., Dawes, L., Wegerif, R. and Sams, C. (2004) Reasoning as a scientist: ways of helping children to use language to learn science. *British Educational Research Journal*, 30, 3, 359–377.

Newton, P., Driver, R. and Osborne, J. (1999) The place of argumentation in the pedagogy of school science. *International Journal of Science Education*, 21, 5, 553–576.

Osborne, J. and Simon, S. (1996) Primary science: past and future directions. *Studies in Science Education*, 27, 99–147.

Solomon, J. (1998) About argument and discussion. *School Science Review*, 80, 291, 57–62.

Strauss, A. and Corbin, J. (1998) *Basics of qualitative research*. London: Sage Publications.

Thorp, G. (2005) *The power of puppets*. Trowbridge: Positive Press.

23. INQUIRING THE INQUIRY LABORATORY IN HIGH SCHOOL

Abstract: In this research study, two groups of chemistry students were compared: A group of students who participated in an inquiry-type laboratory program implemented in the context of high school chemistry, with a group of students who were involved in a more traditional-type laboratory method. Three subjects were assessed: (1) the students' abilities to ask questions while conducting a chemistry experiment, (2) the students' perception of their classroom laboratory learning environment, and (3) the students' attitude toward the chemistry laboratory. In order to assess the development of inquiry skills, a practical test was administered. For the purpose of assessing the students' perceptions of the chemistry laboratory learning environment and attitude towards laboratory work, the Hebrew version of the *Science Laboratory Environment Inventory* (SLEI) and, the *'Attitude towards Science Laboratory'* questionnaires were administered respectively. In general, it was found that the inquiry group outperformed the control in these variables. In addition it was found that the students who conduct inquiry-type activities in the chemistry laboratory developed the ability to ask more and better inquiry-type questions

Keywords: Attitude towards science laboratory, Inquiry questions, Inquiry skills, Inquiry-type laboratory, Laboratory learning environment

1. INTRODUCTION

As the beginning of the twenty-first century, we are entering a new era of reform in science education. Both the content and pedagogy of science learning and teaching are being scrutinized, and new standards intended to shape and rejuvenate science education are emerging (National Research Council, 1996). Inquiry in the science laboratory is one of the important components of this reform (Bybee, 2000; Hofstein & Lunetta, 2004; Lunetta, 1998;). It should be mentioned that the issue regarding the teaching and learning in science laboratories was the focus of several surveys and studies that were conducted in recent years in Europe (e.g. Sere, 2002; Tiberghien et al., 2001). In the context of the science laboratory, inquiry includes the following components: asking relevant questions, hypothesizing, choosing a research question for further investigation, planning an experiment, conducting the experiment, and finally analyzing the findings and arriving at conclusions.

Laboratory activities have long had a distinctive and central role in the science curriculum, and science educators have suggested that many benefits accrue from

Mira Kipnis Department of Science Teaching, The Weizmann Institute of Science, Rehovot, 76100, Israel

R. Pintó and D. Couso (eds.), Contributions from Science Education Research, 297–306.

engaging students in science laboratory activities (Hofstein & Lunetta, 1982; Tobin, 1990; Hodson, 1990; Lazarowitz & Tamir, 1994; Garnett et al., 1995; Lunetta, 1998; Hofstein & Lunetta, 2004). More specifically, they suggest that, when properly developed, inquiry-centered laboratories have the potential to enhance students' meaningful learning, conceptual understanding, and their understanding of the nature of science. Hofstein and Walberg (1995) suggested that inquiry-type laboratories are central to learning science, since students are involved in the process of conceiving problems and scientific questions, formulating hypotheses, designing experiments, gathering and analyzing data, and drawing conclusions about scientific problems or phenomena.

The National Science Education Standards in the USA (National Research Council [NRC], 1996) use the term inquiry in two ways (Bybee, 2000; Lunetta, 1998): (a) inquiry as content understanding, in which students have opportunities to construct concepts and patterns, and to create meaning about an idea to explain what they experience; and (b) inquiry in terms of skills and abilities. Under the category of abilities or skills, Bybee included; identifying and posing scientifically oriented questions, forming hypotheses, designing and conducting scientific investigations, formulating and revising scientific explanations, and communicating and defending scientific arguments. It is suggested, that many of these abilities and skills are in alignment with those that characterize inquiry-type laboratory work, an activity that puts the student in the center of the learning process.

The main goal of this paper is to describe (and demonstrate) how providing chemistry students with opportunities to be involved in inquiry-type activities in the chemistry laboratory resulted a change in the students' questioning behavior, enhanced the students' attitude to and interest in chemistry laboratory, and improved their perceptions regarding the classroom laboratory learning environment.

2. THE INQUIRY LABORATORY PROGRAM

About 100 inquiry-type experiments were developed and implemented in 11th and 12th grade chemistry classes in Israel (for more details about the development procedure, assessment of students' achievement and progress, and the professional development of the chemistry teachers, see Hofstein et al., 2004). Almost all the experiments were integrated into the framework of the key concepts taught in high-school chemistry, namely: acids-bases, stoichiometry, oxidation-reduction, bonding, energy, chemical-equilibrium, and the rate of reactions. These experiments have been implemented in the school chemistry laboratory in Israel for the last five years (about 125 teachers in about 82 schools, and about 3500 students). Only teachers who underwent an intensive professional development were involved in teaching the inquiry laboratory program. In these professional development initiatives the teachers were given the opportunity to conduct all the stages of the inquiry experiments as will eventually be done by their own students.

Typically in the chemistry laboratory, the students perform the experiments in small groups (3–4), by following the instructions in the laboratory manual. Table 1 presents the various stages that each of the groups undergo in order to accomplish

TABLE 1. Phases in inquiry – type experiment

The phase	Abilities and skills
Phase 1: Pre-inquiry	
	• Conducting an experiment. • Observing and recording observations.
Phase 2: The inquiry phase of the experiment.	
1. Hypothesizing	
• Ask relevant questions. Choose one question for further investigations. • Formulate a hypothesis that is aligned with your chosen question.	• Asking questions and hypothesizing.
2. Planning an experiment	
• Plan an experiment to investigate the question. • Present a plan to conduct an experiment. • Ask the teacher to provide your equipment and material to conduct the experiment.	• Planning an experiment.
• Conduct the experiment that you proposed. • Observe and note clearly your observations.	• Conducting the planed experiment.
• Discuss with your group weather your hypothesis was accepted or you need to reject it.	• Analyzing the results, asking further questions, and presenting the results in a scientific way.

the inquiry task. In the first phase (the pre-inquiry phase), the students are asked to conduct the experiment based on specific instructions. This phase is largely 'close-ended', in which the students are asked to conduct the experiment based on specific instructions given in the laboratory manual. Thus, this phase provides the students with very limited inquiry-type experiences. The 'inquiry phase' (the second phase) is where the students are involved in more 'open-ended-type' activities such as asking relevant questions, hypothesizing, choosing a question for further investigation, planning an experiment, conducting the experiment (including observations), and finally analyzing the findings and arriving at conclusions. It is assumed that this phase allows the students to learn and experience science with greater understanding and to practice their metacognitive abilities. Moreover, it provides them with the opportunity to construct their knowledge by actually conducting authentic scientific work.

Of special interest regarding the issue of learning is part 1 of the 2nd phase in which students are asked to raise a hypothesis regarding a certain scientific phenomenon. This includes the following:

- Asking relevant questions concerning the phenomena that they have observed.
- Formulating a hypothesis that is in alignment with the suggested questions.

- Choosing an appropriate research question for further investigation, and
- Planning an experiment in order to investigate this question.

3. THE OBJECTIVES OF THREE STUDIES

This report is a summary of the research that was conducted over a period of 6 years (1999–2005). The common feature of these studies is that they are all based on a comparison of two groups of chemistry students; a group of students who participated in the inquiry-type laboratory program with a group of students who participated in a more traditional-type laboratory method. The students in the later group were exposed to more structured-type close-ended-type chemistry laboratory activities. The topics, textbooks that were used, and final examinations were similar in the two groups. Thus, the only difference was the approach to practical work in the chemistry laboratory.

Three topics related to students learning, perceptions, and attitudes were researched:

- The students' abilities to ask questions while conducting an inquiry-type chemistry experiment (Hofstein et al., 2004, 2005).
- The students' perception of their classroom laboratory learning environment (Hofstein et al., 2001).
- The students' attitudes toward and interest in the chemistry laboratory (Kipnis & Hofstein, 2004).

4. RESEARCH METHOD

4.1 Asking Questions Based on Inquiry-type Experiment (Hofstein et al., 2005)

In order to assess the development of inquiry skills, we administered a practical test in ten 12th grade chemistry classes. In six of the classes the students experienced the inquiry-type activities (N = 101) and in four classes (N = 84) they used a more traditional type approach.

The students were asked to perform a simple inquiry-type experiment, in which they had to mix sodium bicarbonate with citric acid and water in a small plastic bag. The students had to record their observations, ask questions about the experiment (as many as they can figure-out), choose an inquiry question for a further investigation, hypothesize (suggest) an answer to this question, and propose an experiment that can support (or reject) their hypothesis. The students in the control group, who had little (or no) experience regarding learning in inquiry-type laboratories, obtained in the pre-lab phase a short explanation about observations, the inquiry-question, the hypothesis, and planning a suitable experiment. The questions that were asked by the two groups of students were classified according to high-level questions and low-level questions (see Table 2). Low-level questions are related to the facts and explanations of the phenomena that were observed in the experiment performed by the students, e.g.: "*What is A*?" or "*Which reaction occurred?*" In general, the answers to these questions can be a single word, statement, or explanation. On the

TABLE 2. Criteria for classification of inquiry-type questions

Inquiry level	The questions raised by the students	Variables	Type of question
1	What happened?	No variables	
2	Why did it happen?		
3	What is the influence of A on B?	A **and** B cannot be measured.	Qualitative-type question
4		A **or** B can be measured.	Quantitative-type questions
5		A **and** B can be measured.	

other hand, high-level-type questions are questions that can be answered only by further investigation, such as conducting another experiment, *e.g. "Is the size of the bag influenced by the final temperature?"* or *"What would have occurred if we had used another liquid instead of water?"* These types of questions are more complicated, and the student has to think critically about the context of investigation in which he/she are able to pose them. The inquiry-type questions that were chosen by the students for further investigations were categorized as qualitative-type inquiry questions and quantitative-type inquiry questions (as summarized in Table 2).

In addition, the total number of questions asked by each of the students in the two groups was counted.

4.2 The Students' Perceptions of the Chemistry Classroom-laboratory Learning Environment (Hofstein et al., 2001)

For the purpose of assessing the students' perceptions of the chemistry laboratory learning environment, we used the Hebrew version of the *Science Laboratory Environment Inventory* (SLEI). This instrument was developed in Australia by Fraser et al. (1993), and was validated in Israel for use in its Hebrew version by Hofstein et al. (1996). The *actual* and *preferred* versions of SLEI were administered to the inquiry and control groups. The SLEI consists of 68 items in 8 scales. For the scales, its sample items and α Cronbach reliability coefficient see Table 3. The differences in the perceptions were compared using various variant and multivariate statistical methods. In addition, students and chemistry teachers were underwent oral interviews regarding their perceptions of the chemistry laboratory classroom learning environment.

4.3 The Students' Attitude towards and Interest in Chemistry Laboratory (Kipnis & Hofstein, 2004)

For assessing the students' attitude toward the laboratory activity, we used the *'Attitude towards Science Laboratory'* questionnaire that was developed and validated by Hofstein et al. (1976). The instrument was used in the past in other countries and was found to be sensitive to different laboratory approaches used in

TABLE 3. The SLEI questionnaire

Scale (No of items)	Sample item	α Cronbach reliability coefficient
Teacher supportiveness (16)	During the laboratory activity the teacher tries very much to help the students.	0.79
Involvement (10)	The students work hard during the laboratory activity.	0.60
Student cohesiveness (16)	In the laboratory class the students help each other.	0.65
Open-endedness (4)	The teacher tells us exactly how to conduct the practical work in the laboratory.	0.54
Integration (6)	The laboratory activity is based on the theoretical subjects that are taught in the classroom.	0.80
Organization (8)	The laboratory activity is accurately planned.	0.56
Rule clarity (4)	Every student knows the safety rules in the laboratory.	0.50
Material environment (4)	Our laboratory is well equipped.	0.68

the context of chemistry teaching and learning (e.g. in Nigeria by Okebukola, 1986). The questionnaire includes 62 items that were divided by factor analytic investigation into eight scales (see Table 4). The measure was administered among a cohort of 320 chemistry students at the beginning of the 11th grade and on a cohort of 833 chemistry students at the end of the 12th grade.

TABLE 4. Scales in the: attitude toward and interest in chemistry laboratory questionnaire

Scales	The interpretation of the factor	Sample items
I	Learning in the chemistry laboratory	A scientific theory discussed in a textbook is more useful than investigating that theory in the laboratory.
II	The 'amount' of laboratory work.	I would like more time to be devoted to practical work in the chemistry laboratory next year.
III	The 'value' of laboratory work.	Learning chemistry without doing experiments is uninteresting.
IV	The place of laboratory work within the framework of chemistry teaching and its value as a mean for learning chemistry.	Despite the fact that chemistry is called an experimental science it can be studied without doing actual experiments.
V	The student's enthusiasm for practical work and his enjoyment in working in the laboratory.	I enjoy making discoveries in the chemistry laboratory.
I	The student's assessment of his own experimentation versus teacher's demonstration.	I would rather perform an experiment myself than watch a teacher performs the same experiment.
VII	The immediate and future benefits a student gains from experimentation.	Students who perform investigations in the laboratory will be better research workers in the future.
VIII	The advantage of laboratory work.	I usually do not understand what I'm supposed to be investigating in a laboratory experiment.

5. RESULTS

The total number of questions asked by the inquiry-type students was found to be significantly higher compared to the control group. Comparison of the level of the questions asked by the inquiry group to those asked by the control group revealed a significant difference between the two groups regarding the level of the questions (for more details see Table 5).

When we compared the inquiry questions that were chosen for further investigation by the 'inquiry group students' to those that were chosen by 'non-inquiry group students' we found a highly and significant difference: $\chi^2_{(1)} = 20.1, P \leq 0.001$. (For more details see Table 6).

The 'inquiry students' asked more high-level and quantitative-type inquiry questions compared to the control (non-inquiry) group.

In regard to the learning environment, the statistical analysis revealed that the perceptions of the actual and preferred learning environments differed significantly (more pronounced in the actual form) in the inquiry and control groups. The most pronounced differences were with the following scales: *open-endedness, involvement*, and *material environment*, in which the inquiry group mean-scores were significantly higher than those of the control group. It was also found that the differences between the actual and preferred learning environments in the inquiry group were significantly lower than in the control group for *open-endedness, involvement*, and *integration* (Hofstein et al., 2001).

TABLE 5. The average number of questions asked by each of the students in the two groups

Questions	Mean number of questions per student (and standard deviation)		t	P ≤
	Inquiry group N = 101	Control group N = 84		
Total number of questions	5.19 (1.42)	3.05 (1.33)	10.55	0.0001
Low -order-type questions	2.96 (1.48)	2.75 (1.15)	1.06	NS
High-order-type questions	2.20 (1.26)	0.29 (0.61)	13.42	0.0001

TABLE 6. Classification of the inquiry questions

Inquiry level	Inquiry group, N = 101		Control group, N = 84	
	Number of questions	% of students asking a certain question	Number of questions	% of students asking a certain question
Qualitative-type question	44	44%	62	74%
Quantitative-type question	57	56%	19	23%

The statistical analysis of the *Attitude towards Science Laboratory* questionnaire revealed no significant difference between the inquiry group and the control groups in the 11th grade students. On the other hand, there was a significant difference between the attitudes of the two groups in 12th grade students regarding five dimensions of attitude (out of the eight dimensions) that are assessed by the instrument. In all these dimensions the inquiry group outperformed the control group (for more details see Tables 7 & 8).

From tables 7 & 8 it is seen that at the beginning of the 11th grade students' attitude toward and interest in chemistry laboratory was similar in the two groups. However, the comparison conducted at the end of the 12th grade the inquiry-group demonstrated a more positive attitude towards laboratory work compared to the control group. This is based on a MANOVA (multiple analyses of variance)

TABLE 7. The results of the attitude test for 12th grade students

Scale	Experimental group N = 378		Control group N = 455		t
	Mean	Standard deviation	Mean	Standard deviation	
I	3.98	0.57	3.92	0.53	NS
II	3.51	0.68	3.54	0.71	NS
III	3.61	0.50	3.49	0.48	3.42 ***
IV	3.54	0.85	3.17	0.88	6.22 ***
V	3.97	0.65	3.86	0.67	2.39 *
VI	3.85	0.65	3.65	0.64	4.49 ***
VII	3.50	0.77	3.16	0.81	6.11 ***
VIII	3.58	0.73	3.66	0.71	NS

* $0.05 \geq P \geq 0.01$, *** $P \leq 0.001$

TABLE 8. The results of the attitude questionnaire at the beginning of the 11th grade

Factor	Experimental group N = 164		Control group N = 156		t
	Mean	Standard deviation	Mean	Standard deviation	
I	3.88	0.63	3.89	0.61	NS
II	3.44	0.70	3.45	0.77	NS
III	3.57	0.52	3.46	0.55	NS
IV	3.52	0.80	3.39	0.89	NS
V	3.80	0.65	3.77	0.65	NS
VI	3.68	0.67	3.60	0.64	NS
VII	3.38	0.77	3.24	0.78	NS
VIII	3.70	0.70	3.45	0.79	2.96 **

** $0.01 \geq P \geq 0.001$

statistics that was conducted on all the variables (scales) of the questionnaire: Wilks' $\lambda = 0.87, F_{(8,823)} = 14.87, P < 0.0001$.

6. DISCUSSION AND SUMMARY

Based on these three studies, and on another study in which we found that the students' ability to plan an experiment in order to check their hypothesis improves during learning the inquiry laboratory program (Hofstein et al., 2004), we can conclude that the chemistry students who participated in this program obtained an opportunity to be involved in a worthwhile (student-centered) learning process that the chemistry laboratory provided. In addition, it appears that it also provided them with an opportunity to learn in a preferable and meaningful learning environment that resulted in a more positive attitude toward the chemistry laboratory.

In the Israeli case (described in this chapter), in order to make room for the inquiry laboratories the syllabus (content) was reduced by 20%. Well designed, inquiry-type laboratory activities can provide learning opportunities that help student develop high-level learning skills. They also provide important opportunities to help students to learn to investigate (e.g. ask questions), to construct scientific assertions, and to justify those assertions in classroom community of peer investigators in contact with more expert scientific community. There is no doubt that such activities are time consuming and thus, the education system must provide time and opportunities for teachers to interact with their students and also time for students to perform and reflect on such and similar complex inquiry and investigative tasks. Such experiences should be integrated with other science classroom learning experiences in order to enable the students to make connections between what is learned in the classroom and what is learned and investigated in the laboratory. This is highly based on the growing sense that learning is contextualized and that learners construct knowledge by solving genuine and meaningful problems (Brown et al., 1989).

REFERENCES

Brown, J. S., Collins, A., & Duguit, P. (1989). Situated cognition and the culture of learning. *Educational Researcher, 18*, 32–41.

Bybee, R. (2000). Teaching science as inquiry. (In J. Minstrel and E. H. Van Zee (Eds.), *Inquiring into inquiry learning and teaching in science*. Washington, DC: AAAS).

Fraser, B. J., McRobbie, C. J., & Giddings, G. J. (1993). Development and cross-national validation of laboratory classroom environment instrument for senior high school science. *Science Education, 77*, 1–24.

Garnett, P. J., Garnett, P. J., & Hacking, M. W. (1995). Refocusing the chemistry lab: a case for laboratory–based investigations. *Australian Science Teachers Journal, 41*(2), 26–32.

Hodson, D. (1990). A critical look at practical working school science. *School Science Review, 70*, 33–40.

Hofstein, A., Ben-Zvi, R., & Samuel, D. (1976). The measurement of the interest in and attitudes to, laboratory work amongst Israeli high school chemistry students. *International Science Education, 60*, 401—411.

Hofstein, A., Cohen, I., & Lazarowitz, R. (1996). The learning environment of high school students in chemistry and biology laboratories. *Research in Science & Technological Education, 14*, 103–115.

Hofstein, A., Levy Nahum, T., & Shore, R. (2001). Assessment of the learning environment of inquiry-type laboratories in high school chemistry. *Learning Environment Research, 4*, 193–207.

Hofstein, A., & Lunetta, V. N. (1982). The role of the laboratory in science teaching: Neglected aspects of research. *Review of Educational Research, 52*, 201–217.

Hofstein, A., & Lunetta, V. N. (2004). The laboratory in science education: foundation for the 21st century. *Science Education, 88*, 28–54.

Hofstein, A., Navon, O., Kipnis, M., & Mamlok-Naaman, R. (2005). Developing students' ability to ask more and better questions resulting from inquiry-type chemistry laboratories. *Journal of Research in Science Teaching, 42*, 791–806.

Hofstein, A., Shore, R., & Kipnis, M. (2004). Providing high school chemistry students with opportunities to develop learning skills in an inquiry-type laboratory: a case study. *International Journal of Science Education, 26*, 47–62.

Hofstein, A., & Walberg, H. J. (1995). Instructional strategies. (In B. J. Fraser and H. J. Walberg (Eds.), *Improving science education* (pp. 70–89). Chicago: The National Society for The Study of Education).

Kipnis, M., & Hofstein, A. (2004). Internal report, the Weizmann Institute of Science, Rehovot, Israel (in Hebrew).

Lazarowitz, R., & Tamir, P. (1994). Research on using laboratory instruction in science. (In: D. L. Gabel (Ed.), *Handbook of research on science teaching and learning* (pp. 94–128). New York: Macmillan Publication Company).

Lunetta, V. N. (1998). The school science laboratory: Historical perspectives and contexts for contemporary teaching. (In J. B. Fraser and K. G. Tobin (Eds.),*International handbook of science education*, Part One (pp. 249–262). Dordrecht: Kluwer).

National Research Council. (1996). *National science education standards*. (Washington, DC: Author).

Okebukola, P. A. O. (1986). An investigation of some factors affecting students' attitudes towards laboratory chemistry. *Journal of Chemical Education, 86*, 531–532.

Sere, G. M. (2002). Towards renewed research questions from outcomes of the European project lab-work in science education. *Science Education, 86*, 624–644.

Tiberghien, A., Veillard, L., LeMarechal, J. F., Buty, C., & Millar, R. (2001). An analysis of lab-work tasks used in science teaching at upper secondary school and university in seven European countries. *Science Education, 85*, 483–508.

Tobin, K. (1990). Research on science laboratory activities: In pursuit of better questions and answers to improve learning. *School Science and Mathematics, 90*, 403–418.

24. DEVELOPING STUDENTS' VIEWS ON THE NATURE OF SCIENCE THROUGH NON-TRADITIONAL WRITING-TO-LEARN EXPERIENCES IN THE SCIENCE CLASSROOM

Abstract: The chapter describes a theory-driven study, inspired and informed by the constructivist view of learning, science literacy theory, views on the NOS, and cognitive and social theories of writing. Relying on the epistemological effect of writing, it explores the effects that non-traditional forms of writing such as story, diary, and debate, have on students' views on the nature of science when used in combination with reflective writing on the writing process. Using a pre-post research design with intervention and comparative groups totaling 97 eighth graders from a homogenous, middle-class background, in a city in central Israel, the study demonstrates that the combined use of diversified types of nontraditional writing tasks and reflection on the writing process can enhance students' views on the nature of science, in a direction aligned with the type of scientific literacy sought in science education today. When structured as cognitively and attitudinally challenging assignments concerning scientific and societal issues, such writing genres can encourage students to perceive science as a more subjective, temporary, speculative, and interpretative endeavor that helps to solve problems and social issues. In contrast, no changes in views on the NOS were found for students in the comparative group

Keywords: Informal writing tasks, Nature of science (NOS), Science epistemology, Student views on NOS, Writing-in-science

1. INTRODUCTION

This chapter describes a study designed to investigate changes in students' views on the nature of science following writing-to-learn experiences in science classrooms. The study explores whether, how, and why non-traditional forms of writing in science can change students' views on the nature of science when used in combination with writing reflecting on the writing process.

Assisting students to develop informed views on the *nature of science* (NOS) or *epistemology of science* remains a central theme in science education and the development of scientific literacy (Millar and Osborne, 1998; OECD/PISA, 2003). In fact, one of the primary goals in current national standards for science education

Tili Wagner, Department of science teaching, The Weizmann Institute of science, Rehovot, 76100, Israel

R. Pintó and D. Couso (eds.), Contributions from Science Education Research, 307–318.

is for students to develop a sophisticated understanding of the nature of science (National Research Council, 1996). The nature of science is a complex, abstract construct blending the history, sociology, philosophy, and psychology of science. Yet, at its core, the nature of science refers to science epistemology and addresses the values and assumptions inherent to scientific knowledge (Lederman, 1992).

Although there is no single consensus regarding scientific epistemology that scientists, philosophers, and historians agree on, it is generally agreed that there are several issues pertaining to the nature of scientific knowledge and scientific work that students should understand(lederman, 2002; Osborne et al., 2003; Sandoval, 2003). There is also consensus regarding the desirable understandings of the nature of science in international science education standards documents (McComas and Olson, 1998). Nevertheless, despite the interest in developing students' understanding of the nature of science, research findings consistently show that students' views on the NOS are incompatible with current conceptions of the scientific enterprise (McComas, 1998) and that students leave secondary school with naive views on the nature of science. They believe in the absolute truth, fact, infallibility, logic and objectivity of science; that scientific knowledge is absolute (BouJaoude, 1996), and that theories can be proven by empirical evidence (Smith et al., 2000). Also, their views of science tend to ignore the socially constructed nature of science (Latour, 1987), and their perceptions appear essentially competitive.

Studies also show that students frequently waver between objectivist and evaluative constructivist perspectives. In other words, students' epistemological beliefs seem fragmented and contradictory depending on the context (Hammer, 1994; Sandoval and Morrison, 2003). More specifically Sandoval (2005), who distinguishes between a formal and a practical personal scientific epistemology, claims that student' beliefs regarding science and their views of themselves as science learners may be inconsistent with science education's goals for scientific inquiry and inquiry-based science learning.

To better understand and enhance students' views on NOS, research efforts have focused on the search for effective approaches to teaching the NOS. Studies of student epistemological beliefs that are not limited solely to their views on the epistemology of science, suggest that such beliefs are influenced by students' learning environment (Tolhurst, 2004). Moreover, the kind of tasks assigned to students, affect their epistemological development (Hofer and Pintrich, 1997). Laurillard (2002) for example suggests that when students are exposed to different opinions, when they are required to elaborate and support their own views, and when they are made responsible for finding information independently, they are likely to develop a more complex understanding of knowledge. However, it is still not quite clear which activities in fact support epistemological development for different students and what factors determine their success.

Of the general approaches reported in the research literature concerning students' views on NOS, the explicit-reflective approach has resulted in the most significant conceptual change in beliefs regarding science (Abd-El-Khalick and Akerson, 2004). The explicit-reflective approach emphasizes student awareness of certain NOS elements in their learning activities and focuses on student reflection on these activities from within a framework of these aspects. Similarly, Sandoval and Morrison (2003) maintain that inquiry itself cannot provide the basis for epistemological development unless it is accompanied by an explicit epistemological discourse that connects students' work to science in society. "Doing and talking science are not the same as talking about science" (Sandoval and Morrison, 2003, p. 384). In other words, investigating authentic problems is not necessarily enough to raise epistemological issues, even when epistemic criteria are made explicit. Students can apparently engage in sustained open-ended inquiry, appreciate its benefits for their personal construction of knowledge, but still maintain objectivist views on formal science.

This analysis suggests that *epistemic tools* can support meaningful epistemological discourse and students' epistemological understanding (Sandoval, 2003) and if students are to develop informed NOS conceptions, according to Lederman (1998), students should be engaged with structured opportunities to examine their understanding of key NOS concepts in different contexts. However, we still need to determine which teaching and learning opportunities can shape students' informed views on NOS – a lacuna addressed in the present study by treating writing as an epistemic tool.

There is great potential in writing to change the way we think and what we know (Millian and Camps, 2005; Schwonke and Renkl, 2005). We call this the epistemic effect of writing. it is therefore likely that if we use writing for its epistemic function, namely, for knowledge transformation rather than knowledge telling (Berieter, 1980), it can help students to reflect on their learning and encourage them to take a stance on their own writings (Marshall, 1987). Indeed, different studies have described writing as a tool for engaging in science and constructing science understandings (Yore et al., 2003). Moreover, writing it is claimed, can encourage NOS conceptualization (Carey and Smith, 1993; Ryder et al., 1999). This study therefore explores the pedagogical conditions that develop junior-high school students' views on NOS, and conceptualizes writing as a learning and thinking tool, able to generate educational experiences that promote science epistemology. Unlike most interventions, which mainly use scientific genres of writing, this study has used informal writing task assignments. Informal writing tasks are genres of writing not usually assigned in within the science discourse (Keys et al., 1999).

1.1 Theoretical Assumptions

The study makes two theoretical assumptions:

1. When writing-in-science requires the ability to explain support for claims, express doubts, question, suggest alternative views, and describe what is known (Yore et al., 2002), students who engage in writing not only deal with science contents, but also with the epistemic nature of scientific knowledge (Hand et al., 1999). Furthermore, since scientists' epistemological assumptions concerning science might be reflected in their choice of language and

construction of science claims (Yore et al., 2004), writing-to-learn experiences in science can both reflect and affect students' views on science.

2. Writing does not automatically translate into improved epistemological insights. It depends on the type of writing used (Hand and Treagust, 2004). Not every writing task supports or stimulates scientific epistemology. Controversial tasks, expressing complex and debatable questions (Mason and Boscolo, 2004), require explanation and justification (Sutton, 1996) and affect students' epistemological assumptions.

In light of the above, the present study explores the development of views on the NOS when using diversified modes of informal writing tasks and reflective writing on the writing process.

2. METHODS

2.1 Sample, Design and Procedure

We used an action research design for one and a half school years, applying both quantitative and qualitative methodologies. The sample consisted of 97 eighth graders from a homogenous, middle-class background, in a city in central Israel. The sample was divided into two intervention classes; each consists of 24 students, and two comparative classes, one with 23 students and the other with 29 students. Two teachers participated in the study, each teaching one-intervention class and one-comparative class.

Before and at the end of the study, we measured students' views on the NOS for both intervention and comparative groups. Following the pre-test, the intervention classes studied five science topics and selected one of three informal writing tasks at the end of each topic and one reflection writing task relating to their writing. Altogether, 15 (3 tasks X 5 units) writing tasks were developed for the first five units, to be used by students in the intervention group. The writing genres were: story, diary, and debate. The tasks engaged students cognitively and attitudinally in controversial discussion and creative and imaginary contexts. The comparative group studied the same syllabus with the same teachers, but they were not asked to write anything while studying the topics. Only after the final unit at the end of the study were two writing tasks and two reflection tasks never previously assigned to the intervention group assigned to both the intervention group and the comparative group. These novel tasks were "a plan" and "two related letters". By changing the type of writing task at the end of the study we were able to compare the differences in writing quality between the intervention group and the comparative group following the intervention and ruled out the risk of familiarity with a particular genre affecting the intervention group writing.

2.2 Examples of Writing Tasks (Unit on Heat and Temperature):

Story: *Tell the story of a group of water particles that were heated from a temperature of* −20°C *to* 150°C.

Diary: *An additional sun has appeared in the sky of our planet. The sun radiates continuous heat on earth. Write a diary of your own or the diary of someone else describing the effect on our world.*

Debate: *Two children, Dan and David, are arguing:*

Dan: In my computer game, the hero shoots at a tank with a small rifle and vaporizes the tank.

David: That's impossible. Tanks are made of iron. Iron is a metal and there is no heat that can make a tank evaporate. Continue the argument

Writing tasks for the Energy Unit (last learning unit)

The year is 2030. The Third World War, caused by international disputes over energy resources, is over. The United Nations decides on a different distribution of energy resources. According to its resolution, no matter where the energy resources are, they belong to all nations. An international committee is formed to discuss the distribution of resources.

The Plan writing task: *As a member of the committee, write a proposal for the new energy distribution.*

Two related letters writing task: *Write two letters to the committee, one from a boy in Saudi Arabia, the other from a boy in Israel.*

2.3 Data Sources

To explore students' views on the NOS, we adapted Fleener's (1996) questionnaire with some modifications. The questionnaire comprised 34 Likert type statements, statistically divided into six measures reflecting different dimensions of views on the NOS. Each dimension represented a continuum ranging from absolutist epistemology and realist ontology to relativist epistemology and idealist ontology. The six dimensions were (1) *The nature of the scientific process:* simple vs. complex; (2) *The nature of scientific knowledge:* absolute and fixed vs. temporary and speculative; (3) *The power and limitation of science:* emphasis on disciplinary setting vs. sensitivity to societal issues; (4) *The purpose of science*: utility and problem solving vs. interpretation of reality; (5) *Appreciating the value of science,* and (6) *Gender bias:* preference of men in science. The reliability for the NOS measures/dimensions ranged from 0.88 to 0.94.

2.4 Examples of Reflective Writing Tasks

Reflection on writing tasks was elicited through open-ended questions. Thirteen reflective writing tasks were developed. The following are examples of the reflective writing tasks assigned to the intervention group after each learning unit: (1) Describe what and how you felt while writing; (2) Why did you select the topic you chose? (3) What difficulties did you face when writing? (4) Is there a relationship between your writing and your thinking? If so, how is it reflected in your writing?

Content analysis of students' reflective writings explored how students' views on the NOS developed as they gained increasing writing experience and reflected more on writing. Two independent evaluators analyzed student writings: one researcher specializing in teaching and evaluation, and another researcher specializing in teaching in the sciences. A 95% agreement was found in the compared analysis of students' views on NOS as reflected in their writings.

3. RESULTS

3.1 Quantitative Analysis

Table 1 presents the means and standard deviations for each dimension of the NOS views for the two groups before and at the end of the study.

A repeated multivariate analysis of the data demonstrated a significant interaction between the "group and the measurement time", indicating that for four of the six NOS measures, the change in students' views in the intervention group was significantly higher than the change in students' views in the comparative group students. The direction of all changes was from positivist epistemological views toward more evaluative views. The change in means of the intervention and comparative groups measured in standard deviation units was: 1.19 vs. 0.16 for the *nature of scientific process* dimension ($F(1/95)=52.54$ $p<.001$); 0.97 vs. 0.20 for the *nature of scientific knowledge* dimension ($F(1/95)=25.01$ $p<.001$); 0.50 vs 0.12 for the *purpose of science* dimension ($F(1/95)=4.99$ $p<.05$); and, 0.73 vs. 0.33 for the *acknowledging the value of science* ($F(1/95)=9.48$ $p<.01$) dimension. Although no interaction was found for the *power and limitation of science* dimension, significant change was noted for this dimension between the beginning and end of the

TABLE 1. Means and standard deviations of NOS measures for the intervention and comparative groups – Pre- and Post-Test measures

NOS Dimensions	Beginning of study		End of study	
	Intervention group M sd	Comparison group M sd	Intervention group M sd	Comparison group M sd
1. Nature of Scientific Process	3.31 (1.01)	*3.26 (1.01)*	2.19 (0.87)	*3.11 (1.00)*
2. Nature of Scientific Knowledge	3.34 (0.96)	*3.21 (1.01)*	2.42 (0.85)	*3.00 (1.09)*
3. Power and Limitation of Science	3.20 (0.95)	*3.36 (1.05)*	3.43 (0.89)	*3.49 (1.06)*
4. Purpose of Science	2.98 (0.86)	*3.28 (0.90)*	3.45 (1.03)	*3.43 (1.01)*
5. Appreciating the Value of Science	3.46 (0.94)	*3.71 (0.85)*	4.08 (0.74)	*3.98 (0.76)*
6. Gender Bias in Science	1.33 (0.75)	*1.85 (0.99)*	1.26 (0.77)	*1.84 (1.18)*

study ($F(1/95) = 6.23$ $p < .05$) for both groups, indicating that science is socially and culturally embedded. Moreover, the only NOS dimension indicating no change was the *gender bias* dimension.

Table 1 present the interaction effect, showing no differences between the two groups with regard to student views on *the nature of scientific process* prior to the study, and a significant difference between the pre- and post-study measurements for the intervention group (1.19 sd). No significant difference was found for the comparative group (0.16 sd). The score for the intervention group at the conclusion of the study was also significantly lower than the score for the comparative group (about 0.92 sd units), indicating a view of the scientific process as a complex and creative empirically based process.

Table 1 shows a similar pattern of interaction between the "group" and the "measurement time", for the student views on the nature of scientific knowledge, with no significant differences between the measurements before the intervention or between the groups at the end of the study (0.55 sd). Significant interaction between the "group" and the "measurement time" was also found, indicating that the change in views in the intervention group was significantly different from the change in views of the comparative group. The direction of change shows that at the end of the study, intervention group students viewed science knowledge as more temporary, tentative, and subjective than the comparative group and more than they had thought at the outset.

The intervention group also showed significant development in their views on the "purpose of science" and the "appreciation of the value of science", whereas the comparative group showed no change (Table 1). In both cases, the results reveal that the intervention group students view science partly as the product of human inference, imagination, and creativity (inventing an explanation), in which a combination of observation and inferences and appreciation of the value of science to society are involved.

3.2 Qualitative Examples

Although so far the data have been based on students' self-reports using a Likert-type questionnaire, it is also important to explore how student views on the nature of science are expressed in their actual writing. Due to space limitations, however, we can only provide some illustrations of students' views on the NOS as reflected in their writing tasks and reflective writing on the writing process.

Student 1: In the first reflection task on the writing task assigned at the beginning of the study, this student expressed a view of science as a precise, objective discipline: "*There is no need to write creatively in science lessons because there is no connection between writing and science. We use writing to express ideas about world problems relating to economics and other non-precise and non-objective subjects.*" (after the second learning unit).

At the end of the study the same student wrote: "*It would be great if scientists wrote simply because we would finally be able to understand them and when they die we could follow their work and figure out what made them such geniuses – whether they were indeed geniuses or just diligent. I would love to know how people get insights which lead to new theories." (after the last learning unit).* Thus, we find that towards the end of the study, the student has acquired a more complex, subjective, and speculative view of science.

Concerning students' views on *the nature of scientific knowledge*, the following is an example of a student's writing at the beginning of the study: *Science deals with **facts** that we simply have to understand ...*" "*... If you are so confident, let's do an experiment to **prove** who is **right** and what is **really true!**...*" reflecting a belief in the absolute and fixed nature of scientific knowledge. However, at the end of the study, the same student wrote: *Writing about science is quite strange, though perhaps it hints that there are other ways of looking at physics* – reflecting on the idea that multiple conceptions might exist when interpreting scientific information.

In general, the qualitative examples in the students' writings reveal that as students acquired more writing experience, their reflective writings expressed an increased awareness of the role of the imagination and creativity in scientific work; the nature of the questions concerning scientific knowledge, thought and conclusions, and added concern for societal issues.

4. DISCUSSION AND IMPLICATIONS

The study demonstrates that the combined use of nontraditional writing tasks and reflection on the writing process can help shape students' views on the nature of science and make them consistent with the type of scientific literacy sought in science education today. Due to the intervention, the students exhibited a more complex view of science compared with their views at the beginning of the study. They showed greater awareness of the experimental and observational nature of science and of the fact that science is one of the ways of knowing the world, but not the only way. They were also more aware of the role of creativity and imagination in the scientific process. In contrast, the comparative group's views on the NOS remained the same. These results are most significant when viewing other studies showing that students at all grade levels express naive views on science epistemology, sometimes even after implementing different interventions designed to enrich their understanding of the NOS (Ryder et al., 1999; Sandoval and Morrison, 2003).

The results of the study confirm Tucknott and Yore's (1999) findings and suggest that exploration of student views on science through extended experiences in writing is a fruitful approach for the type of scientific literacy aspired to in students. The results of the study however do not match with Meyling's (1997) study showing that most students do not perceive intuition or speculation to be involved in the development of scientific thinking. They also oppose findings surveyed by Simpson's et al. (1994) noting that students' view on the nature of the scientific process is naïve, inductive and certain, rejecting the tentative nature of scientific knowledge.

Regarding the type of writing that can help change students' views on the NOS, this study suggests that informal writing genres shape students' views on science *differently* from the scientific genre of writing, as shown in other studies (Keys et al., 1999; Prain and Hand, 1999). More specifically this study demonstrates that when loosely defined problems expressing real dilemmas are presented in writing modes using different genres such as story, plan, or an argument, they provide a meaningful context for students to develop their epistemological beliefs by requiring them to interpret problems and identify the limitations of solutions (Kitchener, 1983). It also shows that argumentative writing on topics that differentiate pseudo-science from science and religion from science helps shape students' views on NOS, even if the writing assignments do not directly address epistemic views, thereby supporting Craven et al.'s (2002) claims. This contrasts with studies showing that the traditional genre of science writing with its typical focus on rational, objective, factual, and abstracted knowledge, can encourage specific understandings of the scientist's work, the nature of evidence, and collaboration in science but cannot generate significant changes in students' views of the nature of science (Keys et al., 1999) or influence their values, identities, beliefs, commitments, and science understandings (Hildebrand, 1999).

This probable distinction between the plausible effects of informal and formal writing on students' views of science is supported by Yore et al.'s findings, which show that many of the scientists who participated in their study used written language associated with the empiricist, positivist, or rationalist interpretations of science, ostensibly indicating traditional views of science. However, in informal discussions with a few of the subjects, Yore's study found that they actually held nontraditional views of the scientific enterprise. In terms of Sandoval's (2005) distinction between formal and practical scientific epistemology, this study encourages further exploration of whether formal epistemology is likely to be enhanced if students are exposed to diversified types of nontraditional writing tasks rather than the more traditional types of writing, which mainly influence the epistemology of their own knowledge production in school science.

It is likely that nontraditional writing tasks expressed in clear and natural language provide students with learning situations that spontaneously mediate both internal and external dialogues on the nature of science (Moscovici, 1993). According to Ryder et al. (1999), both kinds of dialogue can broaden students' insights and views on the nature of science. It is also likely that the non-traditional writing activities encourage conceptual awareness, which Vosniadou (1994) maintains is a precondition for conceptual change. One can also speculate that the writing activities combined with reflection on writing helps students to reorganize tacit and explicit knowledge into complex mental systems, which according to DiSessa (2002) involves a process of conceptual change. This explanation is also supported by studies emphasizing the gradual process of conceptual change (Vosniadou and Ioannides, 1998) and showing the importance of the reflection component within the explicit approach to enhance NOS (Abd-El-Khalick, 2001).

The study proposes an optimistic view of pedagogical opportunities for developing scientific literacy and student cognition, suggesting that using different informal writing tasks and reflective writing on the writing process, that do not explicitly refers to views on science, provide an effective epistemological tool for learning and self-discovery. At the same time, it challenges students' epistemological views by involving them in imaginary, creative, controversial, exploratory, open-ended, and personally-meaningful writing tasks that allow them to individually construct their views on science knowledge in a unique and meaningful way.

REFERENCES

Abd-El-Khalick, F. (2001). Embedding nature of science instruction in pre-service elementary science courses: Abandoning scientism, but... *Journal of Science Teacher Education*, 12(3), 215–233

Abd-El-Khalick, F., & Akerson, V. L. (2004). Learning about nature of science as conceptual change: Factors that mediate the development of pre-service elementary teachers' views of nature of science. *Science Education*, 88(5), 785–810

Bereiter, C. (1980). Development in writing. (In L. W. Gregg & E. R. Steinberg (Eds.), *Cognitive Processes in Writing* (pp. 73–93). Hillsdale, NJ: Laurence Erlbaum.)

BouJaoude, S. (1996). Lebanese students' and teachers' conceptions of the nature of science. (In: M. Debs (Ed.), *Proceedings of the Second Scientific Conference on the Future of Science and Mathematics Teaching and the Needs of Arab Society* (pp. 283– 303). Beirut: Arab Development Institute.)

Carey, S., & Smith, C. (1993). On understanding the nature of scientific knowledge. *Educational Psychologist*, 28, 235–251

Craven, J. A., Hand, B., & Prain, V. (2002). Assessing explicit and tacit conceptions of the nature of science among pre service elementary teachers. *International Journal of Science Education*, 24(8), 785–802

DiSessa, A. A. (2002). Why conceptual ecology is a good idea. (In M. Limon & L. Mason (Eds.), *Reconsidering Conceptual Change: Issues in Theory and Practice* (pp. 29–60). Dordrecht: Kluwer.)

Fleener, M. J. (1996). Scientific world building on the edge of chaos: High school students' beliefs about mathematics and science. *School Science and Mathematics*, 96(6), 312–319

Hammer, D. (1994). Epistemological beliefs in introductory physics. *Cognition and Instruction*, 12(2), 151–183

Hand, B., Prain, V., Lawrence, C., & Yore, L. D. (1999). A writing-in-science framework designed to improve science literacy. *International Journal of Science Education*, 21, 1021–1035

Hand, B., & Treagust, D. (2004, April) *Linking Pedagogical Requirements to Cognition and Literacy When Using Writing to Learn Strategies.* (Symposium on contemporary international issues facing science literacy: Classrooms and beyond. Presented at the Annual Meeting of the National Association for Research in Science Teaching, Vancouver, Canada)

Hildebrand, G. M. (1999). *Breaking the Pedagogic Contract: Teachers' and Students' Voices.* (Paper presented at the Annual Meeting of the National Association for Research in Science Teaching, Boston)

Hofer, B. K., & Pintrich, P. R. (1997). The development of epistemological theories: Beliefs about knowledge and knowing and their relation to learning. *Review of Educational Research*, 67(1), 88–140

Keys, C. W. (1999). Revitalizing instruction in scientific genres: Connecting knowledge production with writing to learn in science. *Science Education*, 83, 115–130

Kitchener, K. S. (1983). Cognition, metacognition and epistemic cognition. A three level model of cognitive processing. *Human Development*, 26, 222–232

Latour, B. (1987). *Science in Action.* Cambridge, MA: Harvard University Press.

Laurillard, D. (2002). *Rethinking University Teaching: A Conversational Framework for the Effective Use of Learning Technology*, 2nd edn. (London: Routlege Farmer)

Lederman, N. G. (1992). Students' and teachers' conceptions of the nature of science: A review of the research. *Journal of Research in Science Teaching*, 29(4), 331–359

Lederman, N. G. (1998). The state of science education: Subject matter without content. [Electronic version].*Science Education*, 3(2), 1–12

Lederman, N.G., Abol-El-Khalick, F., Bell, R.L., & Schwartz, R.S. (2002). Views of Nature of Science Questionnaire: Toward valid and meaningful Assessment of learners' Conceptions of Nature of Science. *Journal of Research in Science Teaching*, 39(6), 497–521

Lederman, N. G., Wade, P. D., & Bell, R. L. (1998). Assessing the nature of science: What is the nature of our assessment? *Science Education*, 7, 595–615

Marshall, J. D. (1987). The effects of writing on students' understanding of literary texts. *Research in the Teaching of English*, 21, 30–63

Mason, L., & Boscolo, P. (2004). Role of epistemological understanding and interest in interpreting a controversy and in topic-specific belief change. *Contemporary Educational Psychology*, 29(2), 103–128

McComas, W. F. (Ed.) (1998). *The Nature of Science in Science Education: Rationales and Strategies*. Dordrecht, The Netherlands: Kluwer Academic Publishers.

McComas, W. F., & Olson, J. K. (1998). The nature of science in international science education standards documents. (In W. F. McComas (Ed.), *The Nature of Science in Science Education: Rationals and Strategies* (pp. 41–52). Dordrecht, The Netherlands: Kluwer Academic Publishers.)

Meyling, H. (1997). How to change students' conceptions of the epistemology of science. *Science and Education,* 6, 397–416

Milian, M., & Camps, A. (2005). Writing and the making of meaning: An introduction. *L1 Educational Studies in Language and Literature*, 5(3), 241–249

Millar, R., & Osborne, J. (1998). *Beyond 2000: Science Education for the Future*. London: King's College London/Nuffield Foundation.

Moscovici, S. (1993). Towards a social psychology of science. *Journal for the Theory of Social Behavior*, 23, 343–374

National Research Council. (1996). *National Science Education Standards*. (Washington, DC: National Academy Press)

OECD/PISA. (2003, December). *Pisa Scientific Literacy Framework Doc: NPM (0132)18*. (Paper presented in the National Project Managers Meeting, Prague)

Osborne, J., Collins, S., Ratcliffe, M., Millar, R., & Duschl, R. (2003). What "ideas about science" should be taught in school? A Delphi study of the expert community. *Journal of Research in Science Teaching*, 40, 692–720

Prain, V., & Hand, B. (1999). Students' perceptions of writing for learning in secondary school science. *Science Education*, 83, 151–162

Ryder, J., Leach, J., & Driver, R. (1999). Undergraduate science students' images of science.*Journal of Research in Science Teaching*, 36(2), 201–219

Sandoval, W. A. (2003). The inquiry paradox: Why doing science doesn't necessarily change ideas about science. (In C. P. Constantinou & Z. C. Zacharia (Eds.), *Proceedings of the Sixth International Computer-Based Learning in Science Conference* (pp. 825–834). Nicosia: Cyprus.)

Sandoval, W. A. (2005). Understanding students' practical epistemologies and their influence on learning through inquiry [Electronic version]. *Science Education*, 89, 634–656. Published on line 27 May 2005 in Wiley InterScience (www.interscience.wiley.com)

Sandoval, W. A., & Morrison, K. (2003). High school students' ideas about theory change after a biological inquiry unit. *Journal of Research in Science Teaching*, 40(4), 692–720

Schwonke, A., & Renkl, A. (2005). The use of net-mediated environments for collaborative writing-to-learn. (In A. Méndez Vilas, B. G. Pereira, J. M. González & J. A. M. González (Eds.), *Recent Research Developments in Learning Technologies* (Vol. 3, pp. 1239–1243). Badajoz: Formatex 2005.)

Simpson, R.D., Kobala, T.R., Oliven, Y.S. and Crawley, F.E. (1994). Research on the affective dimension of science learning. In D. Gabel (Ed.), *Handbook of Research on Science Teaching and Learning*, chapter 6, New York: Macmillan

Smith, C. L., Maclin, D., Houghton, C., & Hennessey, M. G. (2000). Sixth-grade students' epistemologies of science: The impact of school science experiences on epistemological development. *Cognition and Instruction*, 18(3), 349–422

Sutton, C. (1996) Beliefs about science and beliefs about language. *International Journal of Science Education*, 18, 1–18

Tolhurst, D. (2004). The influence of web-supported independent activities and small group work on students' epistemological beliefs. *Proceedings of the Sixth Conference on Australian Computing Education*, 30, 311–316 (Dunedin, New Zealand)

Tucknott, J., & Yore, L. (1999, March). *The Effects of Writing Activities on Grade 4 Children's Understandings of Simple Machines, Inventions and Inventors.* (Paper presented at the National Association for Research in Science Teaching, Boston)

Vosniadou, S. (1994). Capturing and modeling the process of conceptual change. *Learning and Instruction*, 4, 45–69

Vosniadou, S., & Ioannides, C. (1998). From conceptual development to science education: A psychological point of view. *International Journal of Science Education*, 20(10), 1213–1230

Yore, L., Hand, B., & Prain, V. (2002). Scientists as writers. *Science Education*, 86(5), 672–692

Yore, L. D., Bisanz, G. L., & Hand, B. M. (2003). Examining the literacy component of science literacy: 25 years of language and science research. *International Journal of Science Education*, 25, 689–725

Yore, L. D., Hand, B. M., & Florence, .M .K (2004). Scientists' views of science, models of writing, and science writing practices. *Journal of Research in Science Teaching*, 41(4), 338–369

PART 7

MODELS AND MODELLING IN SCIENCE EDUCATION

JAN H. VAN DRIEL AND A.E. VAN DER VALK

25. TOWARDS A VALIDATED CONCEPTION OF SCIENTIFIC MODELS

Abstract: Teaching the use of models in scientific research requires a description, in general terms, of how scientists actually use models in their research activities. This paper aims to arrive at defining common characteristics of models that are used in present-day scientific research. Initially, a list of common features of models and modelling, based on the literature was compiled. Next, a questionnaire was developed, which consisted of ten statements with which the common features of models were probed. A sample of 77 research papers from 'hard science' journals was drawn. The questionnaire was sent to the first author of the selected articles. The useful response was 24 (31%). From an analysis of the results, it was concluded that although the respondents more or less agreed with all the common features of models, some characteristics were perceived differently, or formulated in different terms. This has led us to revise some of the features in our list, and suggest recommendations for the use of models in science curricula and science text books. In particular, more attention should be paid to modern uses of models and to aspects related to the nature of models

Keywords: Modelling; Nature of science; Scientific models; Teaching models

1. INTRODUCTION

In current attempts to reform science education, it is often argued that students should be given more opportunities to learn actively, in particular to be involved in research-like or inquiry activities (e.g. Millar et al. 1994; NRC 2000). Moreover, curriculum innovations focusing on the public understanding of science (e.g. NEAB 1998) have drawn attention to *reflection* on science, that is, increasing students' awareness of how scientific knowledge is constructed and applied, rather than focusing exclusively on the content of scientific ideas. In this context, as models rank among the main products of science (Gilbert et al. 2000) and science can be seen as a process of model building, "an understanding of the nature of models and model building is an integral component of science literacy" (Gilbert 1991, p. 78). Thus it is important that we, as teachers, textbook authors, science education researchers, and curriculum developers, know how models are actually used by present-day scientists in their research endeavours: How do scientists perceive the nature of the models they construct, test, and adapt? To answer this question, the literature on models and modelling in science was studied, resulting in a tentative

Jan H. van Driel, PhD, ICLON – Leiden University Graduate School of Teaching, PO Box 9555 NL-2300 RB Leiden, The Netherlands.

R. Pintó and D. Couso (eds.), Contributions from Science Education Research, 321–332.

description of common characteristics of models and their use in science. To validate this description for current scientific practice, a pragmatic approach was chosen. Scientists who recently published research papers in which models play a central role were asked to what extent the description applied to their work.

2. THEORETICAL BACKGROUND AND RESEARCH QUESTION

Models are used in research in all scientific disciplines (Gilbert 1991). In spite of differences between various models, it has been suggested that there are general features that are common to all models and to the ways they are used in scientific research. In particular, in some of our earlier work (Van Driel & Verloop 1999), we identified several common characteristics of scientific models on the basis of an analysis of the available literature. This literature consisted of publications from various domains, mainly the history and philosophy of science (e.g. Bertels & Nauta 1969; Black 1962; Giere 1991; Hesse 1966), and science education (e.g. Duit & Glynn 1996; Gilbert 1991; Gilbert & Boulter 1997). In empirical studies, we have used these common features to develop and test educational materials, aimed at understanding the role and the nature of models in chemistry (De Vos 1985; Van Hoeve-Brouwer 1996), and to probe teacher's understanding of models and modelling in science (Van Driel & Verloop 1999).

The first two general features of scientific models describe the nature and functions of a model.

1. A model is always related to a target (Duit & Glynn 1996; Gilbert 1991; Gilbert & Boulter 1997) *and is designed for a special purpose* (Bullock & Trombley 1999). The target is the actual object of research. The target may be an object, a phenomenon, an event, a process, a system, or an idea. A model is always a representation of the target, but the way in which the target is represented in the model (e.g. three dimensional model, mathematical equation) may be quite different, mostly depending on the purpose of the model. Nevertheless, it should always be possible to identify both the model and the target, and to distinguish between the two.

2. *A model serves as a research tool that is used to obtain information about the target which itself cannot be easily observed or measured directly* (Mayer 1992). A model is used to learn about the less known (i.e., the target) by comparing it to something that is more familiar (Bertels & Nauta 1969; cf. Kuhn 1970, on 'heuristic models'). The purpose of a model in scientific research is mostly to predict or to explain (Bullock & Trombley 1999).

The next two features refer to the criteria a model must fulfil:

3.a A model bears some analogies to the target. From analogies between the target and the model, certain aspects may be explained (cf. 'analogical models', Hempel 1965).

3.b These analogies enable the researcher to reach the purpose of the model; in particular to derive hypotheses from the model or to make predictions, which may be tested while studying the target (Hesse 1966). Regardless of its purpose, every model maps elements of the target (Bullock & Trombley 1999). Studying the model should enable the researcher to obtain information and to reformulate it into a hypothesis or a prediction that refers to the target under consideration. In this way the researcher knows what to look for and where and when to look for it (Black 1962).

4. A model differs in certain respects from the target. The differences make the model more accessible for research than the target (Woody 1995).In general, a model is kept as simple as possible (Bertels & Nauta 1969). Depending on the specific research interests, some elements of the target are deliberately excluded from the model (Black 1962). Other elements may be included although they are not similar to or shared with the target (cf. 'negative analogies', Hesse 1966). Usually, a model is much simpler than the target in order to make the target accessible for observation or other means of research. The target may be too small (an atom), too big (the universe) or too complex for direct observation, there may be ethical inhibitions (the human brain) or technical obstacles (the centre of the earth), or it may be difficult for other reasons to directly examine the target. The model is supposed to offer an alternative way of obtaining information about such an inaccessible target.

The last four features describe the selection and development of a model.

5. Since having analogies (3.a) and being different (4.) lead to contradictory demands on the model, a model will always be the result of a compromise between these demands (Black 1962; Van Hoeve-Brouwer 1996). The researcher is usually confronted with a choice between a complex model that resembles the target in many ways, and a simpler model that is easier to handle. The choice will depend on the nature of the research problem, on the facilities such as time and money that are available, and on the personal preference of the researcher.

6. *A model does not interact directly with the target it represents. Consequently, there is always an element of creativity involved in its design, related to its purpose* (Schwartz and Lederman 2005; Van Hoeve-Brouwer 1996). This means that a photograph, a spectrum or another source of information that does not exist independently of the target, does not count as a model, even if it is very helpful in reaching the purpose of obtaining information about that target (De Vos 1985). On the other hand, a model may contain elements that have been derived from the target, for instance through measurement, but it should also contain elements of interpretation, simplification, and the like.

7. Several consensus models may co-exist with respect to the same target (Van Oers 1988).This is an implication of the previous point: depending on the specific context and purpose of the research, different choices may be made, resulting in the development or selection of different models. For instance, biochemists and theoretical chemists may use quite different models for the molecular structure of water as they ask different kinds of research questions.

8. As part of the research activities, a model can evolve through an iterative process (Van Hoeve-Brouwer 1996). Empirical data obtained from the target will lead to a revision of the model, while the revised model will generate new hypotheses with respect to the target, thereby inducing new observations or experiments (Giere 1991). When, for example, the planet Neptune was found to deviate from its calculated orbit, the model of the solar system was revised so as to include another, as yet unknown planet. In 1930, Tombaugh found this ninth planet, Pluto, near its predicted position.

In the present study, we were interested in validating the above description by confronting it with the way models are actually used in present-day scientific enquiry. To this end a study was conducted, which was guided by the following general research question: *To what extent are the common features of scientific models, based on a study of the literature, recognised and acknowledged by practising scientists, who work with models in their research?*

3. METHOD

3.1 Instrument

Based on a series of pilot interviews, a questionnaire was developed consisting of ten statements each of which referred to one the features mentioned above. Statements were formulated in a way, which, hopefully, challenged the ideas of the respondents about the respective features. For instance, in the first statement, related to feature 1 (*A model is always related to a target),* the difference between model and target was emphasised: '... a model is a model of *something else*, something which it represents...'. Moreover, to ensure that respondents would react to the statements on the basis of experiences in their own research, all statements incorporated a phrase, such as 'in your research work', 'in your paper', and so on. Each statement had to be marked as 'correct' or 'incorrect', followed by an explanation or comment. The entire questionnaire can be found in the annex.

3.2 Sample

For the main study, a sample of 77 research articles published in 'hard science' journals was drawn from the natural sciences libraries of Utrecht University. Included in the sample were disciplines such as physics, astronomy, biology, chemistry, pharmacology, meteorology, geology, and interdisciplinary sciences such as biochemistry and geophysics. The first criterion to select articles from journal issues that had appeared between mid-1998 to mid-1999, was the appearance of the word 'model(s)' or 'model(l)ing' in the title. Next, the questionnaire was sent to the first author of the selected articles. In an accompanying letter the purpose of our study was explained. The overall useful response was 24 (31%). Although the response was rather low in quantitative terms, the response covered a large variety of disciplines, enabling us to explore the common model features across scientific disciplines.

3.3 Data Analysis

First, the response was analysed in terms of the numbers of respondents agreeing or disagreeing with the ten statements. Next, the explanations and comments of the respondents were analysed in a qualitative manner. The process of data analysis consisted of a multi-step procedure, in which a series of within-case analyses were followed by cross-case analyses (Miles and Huberman 1994). As a first step in this procedure, the written comments to the statements were analysed for each respondent by the first and second author individually. Next, by comparing and discussing our individual analyses (investigator triangulation; Janesick 2000), consensus was sought on the interpretation of the content of all answers and explanation of each individual respondent. In the light of our general research question (see above), the interpretation of these responses was focused on *falsification*: finding arguments whether or not to reformulate, nuance or even refute the provisional common features. After an individual attempt at finding such arguments, the first and second author compared and discussed the arguments they found for each feature before agreeing about the need to adapt a feature, or leave as it was formulated originally.

4. RESULTS AND DISCUSSION

4.1 Overview of Results

It appeared that, in total, 84% of the answers given were "agree with the statement", or the statement was evaluated as "correct". Eight respondents agreed with all ten statements. One respondent, however, disagreed with all but two statements. In total, 62% of the answers was illustrated or commented upon. Four respondents did not add any explanation, example, or comment to their answers; four other respondents illustrated all 10 items with a written comment. Moreover, four respondents provided a final comment or a comment in their accompanying letter. These respondents appeared to be very interested in the concept of models and/or in science education, e.g. "As a teacher, I feel that this kind of study is terribly missing in earth sciences and I am surprised of how your statements agree with my personal experience I have not yet so precisely felt.", or: "teaching the proper use of modelling is a primary objective within our education program".

4.2 Results and Discussion Per Model Feature

Although it is impossible in the context of this paper to do justice to all nuances made by our respondents, we will now discuss their responses in relation to the general features of models, described above. If necessary, we will propose changes of formulations of these features.

4.2.1 Feature 1 Feature 1 states that a model is always related to a target and is designed for a special purpose. In science education, feature 1 is commonly interpreted as 'there is a strict distinction between model and target'. This, however, has appeared not to be necessarily true for two reasons. To illustrate the first reason, we use the case of one respondent, a toxicologist, who disagreed with statement 1, however, without illustrating why. We were able to understand this disagreement because in the research of this respondent, 1-octanol, was used as a model to represent a class of chemical substances: non-polar toxics. As 1-octanol is a non-polar toxic itself, his model substance represents, among others, itself.

The second reason concerns the relationship between model and theory. One respondent suggested that an ideal model (of electrodynamics in physics) is almost identical to what it represents: Is the electron the target? Or (part of) a model? Or part of the theory? The suggestion of this respondent made us realise that a strict distinction between target and model may reflect a positivist view of the world in which the target exists independently of the observer and his theoretical preoccupations. In the past, molecules and atoms were not observable and only could be thought of by using models such as the Bohr model. But nowadays, we can 'see' atoms and even manipulate them individually with complicated equipment. In fact, we are observing 'theory-laden', with models and theories not only in our minds, but also made operational in our instruments.

We conclude that there are no objections against feature 1. However, there are objections against inferring from this feature that model and target can always be distinguished from each other.

4.2.2 Feature 2 This feature states: 'A model serves as a research tool that is used to obtain information which itself cannot be easily observed or measured directly'. This function of a model was recognised by our respondents. The information can be obtained through several methods: By computer simulation, by visualising, by experimenting with the model. Constructing such models can be perceived by scientists as an aim in itself. From the respondents' answers, we found that models in science can be used for another purpose as well: As a representation of scientific knowledge that facilitates making decisions in technology and in society. For instance, one respondent wrote: "Project goal was to develop a model (of toxicity of compounds in marine sediments) which could be used to set regulatory limits. In technology, medicine and society, models are used to make decisions based on scientific knowledge." To include this other way of using models, we suggest reformulating feature 2 as follows:

A model serves as:

- *a research tool that is used to obtain information about the target which itself cannot be easily observed or measured directly;*
- *a representation of scientific knowledge about the target, to be used to facilitate making decisions about issues (in technology, medicine, society, ...).*

4.2.3 Features 3.a and 3.b From our results (i.e., responses to statement 5), we have learnt that the criterion for bearing 'some analogies' (feature 3.a) should be explicitly related to the realm of the model's valid use. However, respondents working with statistical models pointed at a problem. For instance, a respondent investigating confounding in epidemiological models wrote: "Cannot tell: [...] without corroborative evidence, it is difficult to tell if the model has things in common with what it represents. Of course, then the model is useless and may be counterproductive. But we may not know it". We think that this is typically a problem of using statistical models, since the existence of statistical relationships does not necessarily indicate theoretical similarities between model and target.

As feature 3.b. was not disputed (see results concerning statement 7), we propose a revision for 3.a. only:

3.a. Within the realm of its valid use, a model bears some analogies to the target.

4.2.4 Feature 4 This feature states that a model differs in certain respects from the target and that these differences make the model more accessible for research than the target. From the respondents' answers to statement 4, it appeared that in most cases, but not in all, a model is more easily accessible than its target. For example, a respondent who did not agree with the idea of 'accessibility' wrote: "It is not only more accessible, it is simpler, it describes the fundamental properties rather than the total complexity of the object." Therefore, we propose to substitute the word 'accessible' in feature 4 by the word 'attractive':

A model differs in certain respects from the target. The differences make the model more attractive for research than the target.

4.2.5 Feature 5 The results, as indicated by respondents' answers to statement 6, did not provide arguments for a revision of this feature, which states that a model is always the result of a compromise between conflicting demands. One respondent explained his view as follows: "A 'good' model has the correct boundaries and level of detail to most effectively answer the question being asked about the real system." It could be added that the compromise is derived from (i) the aim of the study, as indicated by the research question or in design specifications, and (ii) the desired preciseness of predictions.

4.2.6 Feature 6 This feature states that a model does not interact directly with its target, and that there is always an element of creativity involved in its design so as to serve its purpose. The element of creativity was widely recognised by our respondents (i.e., by their responses to statement 8), except for those who 'borrowed' their model from other researchers.

We have found some examples of models, however, that did interact directly with their target. A clear example was already mentioned in the discussion of feature 1: The model used by a respondent concerned a substance (i.e., 1-octanol) that was representative of a group of chemical substances. Therefore, it could be said that, in this case, the model was part of the target. Nevertheless, in that case creativity was needed to formulate requirements for the selection of a suitable model-substance. Another example may be the broiler chickens model of another respondent: Chickens are the target and also make up part of the model. Therefore, we propose to skip the first part of the original feature 6 and reformulate this feature as follows:

The construction of a model requires creativity, among others in finding a compromise between 'having analogies with' and 'being different from' the target, so as to optimally serve its purpose.

4.2.7 Feature 7 From the responses to statement 3, we conclude that, in principle, two or even more different models can represent the same target. However, in practice, scientists may work with only one model. This may illustrated by the following response from a respondent: "The type of model has been chosen amongst a wide range of possible models [...] afterwards, the analysis was restricted to one model". This means that researchers may choose one specific model and have good reasons for doing so: Some models are better or more appropriate for a specific purpose than others. Therefore, characteristic 7 has to be extended as follows:

Several consensus models may co-exist with respect to the same target. However, depending on the precision requested (e.g. the precision of the predictions based on the model; the design specifications), one model can be the best, at least for the time being.

4.2.8 Feature 8 We did not find any objections against this feature, stating 'As part of the research activities, a model can evolve through an iterative process'. However, some respondents pointed at the fact that 'research' should not be regarded in the narrow sense of empirical research, for instance: "The model is not used to suggest new experiments, but to improve agreement between outcomes of theoretical calculations and measured values." Thus, advances in theoretical research as well as policy making and technological design activities can also add to the iterative process.

5. CONCLUSION AND IMPLICATIONS

We conclude that, on the whole, our initial list of common features of models, based on the literature, was recognised and acknowledged by the respondents in the empirical study. However, some features were perceived differently by some respondents, or formulated in different terms. This result aligns with the results of Schwartz and Lederman (2005), who suggested that "conceptions of scientific models and their use in science *may* differ with context of scientific practice." (p. 14)

The studies about models in the historical and philosophical literature we used to make the list of features, dated mainly from the nineteen sixties and early seventies, and focused on the use of models in fundamental research. From the nineteen eighties onwards, however, rapid changes have taken place in science, in strong interaction with new technologies (among others computers, DNA, microchemistry techniques). Moreover, models are increasingly used in technological design and in making decisions in society. Apparently, given some of the comments of our respondents, this has implications for the use of models and the way scientists perceive models and modelling.

Our results are relevant for gearing science education towards recent developments in science research. That is, if we want students to "learn science in a way that reflects how science actually works" (NRC 1996, p. 214), we can derive some implications from the findings of the present study for secondary science education.

Most models incorporated in secondary science education curricula to date aim at understanding fundamental scientific concepts. In the last decades, however, inquiry tasks including computer simulation models have entered the curricula and have become subject of educational research (De Jong & Van Joolingen 1998). In addition, because of the tendency to pay more attention to technology in science education (e.g. NRC 1996), models aimed at technological design are increasingly present in national curricula, for instance, in the UK (NEAB 1998) and in the Netherlands (OC & W 1998).

It has been suggested (Erduran 2001) that in textbooks for secondary science education, models are often presented as static facts, or as final versions of our knowledge. Features such as the relation between model and target (Feature 1), possible limitations of a model (Feature 5), or the way in which models are developed (Feature 8), are seldom addressed. Moreover, textbooks rarely include modeling assignments inviting secondary students to actively construct or test models (Erduran 2001). Research on teacher knowledge in the domain of models and modeling has repeatedly indicated that teachers have very limited knowledge of the nature of models and the act of modeling (e.g. Justi & Gilbert 2003; Van Driel & Verloop 1999). We assume that these shortcomings could be an important cause of the well-known problem in science education that many students consider a model to be a copy of the target (Grosslight et al. 1991). To broaden textbook authors' and science teachers' conceptions of models and modeling, we think it is important to make them aware of the fact that models do not always explain empirical observations, and that not all models refer to a simplified representation of an abstract concept. Moreover, the results of the present study provide arguments for paying attention to the role of models in technological design as well as in problem solving and decision-making in society.

Finally, the increasing attention for inquiry in science classrooms (NRC 2000) implies that not only the content of models, but also modelling activities should play a role in secondary science classes. In our view, in the context of such activities, teachers and educational materials should support students in recognising the similarities between the various models that are involved. We think the common features of models may assist teachers and textbooks to achieve this goal. In this way, such activities may contribute to students' awareness of the nature of scientific research, and their understanding of the development of scientific knowledge (cf. Gilbert & Boulter 1997).

ACKNOWLEDGEMENT

This project was started by Wobbe de Vos, who selected the articles and worked on the analysis of the responses until he became ill in the course of 2001. Sadly, Wobbe passed away in 2002. The authors of this paper decided to finalise his work.

REFERENCES

Bertels, K., & Nauta, D. (1969). *Inleiding Tot Het Modelbegrip [Introduction to the Model Concept.]* (Bussum: W. de Haan.)

Black, M. (1962). *Models and Metaphors. Studies in Language and Philosophy.* (Ithaca/London: Cornell University Press.)

Bullock, A., & Trombley, S. (1999). *The New Fontana Dictionary of Modern Thought* (3rd ed.). (London: Harper Collins.)

De Jong, T., & Van Joolingen, W.R. (1998). Scientific discovery learning with computer simulations of conceptual domains. *Review of Educational Research*, 68, 179–202.

De Vos, W. (1985). *Corpusculum Delicti.* (Utrecht: University of Utrecht (Unpublished PhD-dissertation)).

Duit, R., & Glynn, S. (1996). Mental modelling. (In G. Welford, J. Osborne & P. Scott (Eds.), *Research in Science Education in Europe: Current Issues and Themes* (pp. 166–176). Bristol, PA: Falmer Press.)

Erduran, S. (2001). Philosophy of chemistry: An emerging field with implications for chemistry education. *Science & Education*, 10, 581–593.

Giere, R. (1991). *Understanding Scientific Reasoning* (3rd ed.). (Fort Worth: Holt, Rinehart & Winston.)

Gilbert, S.W. (1991). Model building and a definition of science. *Journal of Research in Science Teaching*, 28, 73–79.

Gilbert, J.K., & Boulter, C.J. (1997). Learning science through models and modelling. (In B.J. Fraser & K. Tobin (Eds.), *The International Handbook of Science Education* (pp. 53–66). Dordrecht: Kluwer Academic Publishers.)

Gilbert, J.K., Boulter, C.J., & Elmer, R. (2000). Positioning models in science education and in design and technology education. (In J.K. Gilbert & C.J. Boulter (Eds.), *Developing Models in Science Education* (pp. 3–18). Dordrecht: Kluwer Academic Publishers.)

Grosslight, L., Unger, C., Jay, E., & Smith, C. L. (1991). Understanding models and their use in science: Conceptions of middle and high school students and experts.*Journal of Research in Science Teaching*, 28, 799–822.

Hempel, C.G. (1965). *Aspects of Scientific Explanation and Other Essays in the Philosophy of Science.* (New York: The Free Press.)

Hesse, M.B. (1966). *Models and Analogies in Science.* (London: Sheen & Ward.)

Janesick, V.J. (2000). The choreography of qualitative research design. (In N.K. Denzin & Y.S. Lincoln (Eds.), *Handbook of Qualitative Research* (2nd ed.) pp. 379–399. Thousand Oaks, CA: Sage.)

Justi, R., & Gilbert, J. (2003). Teachers' views on the nature of models. *International Journal of Science Education*, 25, 1369–1386.

Kuhn, T. (1970). *The Structure of Scientific Revolutions* (2nd ed.). (Chicago: The University of Chicago Press.)

Mayer, R.E. (1992). Knowledge and thought: Mental models that support scientific reasoning. (In R.A. Duschl & R.J. Hamilton (Eds.), *Philosophy of Science, Cognitive Psychology, and Educational Theory and Practice* (pp. 226–243). New York: SUNY Press.)

Miles, M.B., & Huberman, A.M. (1994). *Qualitative Data Analysis.* (Thousand Oaks, CA: Sage.)

Millar, R., Lubben, F., Gott, R., & Duggan, S. (1994). Investigating in the school science laboratory: Conceptual and procedural knowledge and their influence on performance. *Research Papers in Education*, 9, 207–249.

NEAB (Northern Examinations and Assessment Board) (1998). *Science for Public Understanding.* (Harrogate, UK: NEAB.)

NRC (National Research council) (1996). *National Science Education Standards.* (Washington, DC: National Academic Press.)

NRC (National Research council) (2000). *Inquiry and the National Science Education Standards.* (Washington, DC: National Academic Press.)

OC & W (Ministry of Education, Culture and Sciences) (1998). *Examenprogramma's Profielen vwo/havo: Natuurwetenschappelijke Vakken. [Syllabuses for the national Examinations vwo/havo: Natural Sciences].* (The Hague: SDU Servicecentrum.)

Schwartz, R., & Lederman, N.G. (2005, April). *What Scientists Say: Scientists' Views of Models.* (Paper presented at the annual meeting of the National Association for Research in Science Teaching, Dallas, TX.)

Van Driel, J.H., & Verloop, N. (1999). Teachers' knowledge of models and modelling in science. *International Journal of Science Education*, 21, 1141–1153.

Van Hoeve-Brouwer, G.M. (1996). *Teaching Structures in Chemistry: An Educational Structure for Chemical Bonding.* (Utrecht: CD- ß Press (PhD-dissertation)).

Van Oers, B. (1988). Modellen en de ontwikkeling van het natuurwetenschappelijk denken van leerlingen [Models and the development of scientific thinking among students].*Tijdschrift voor Didaktiek der ß-wetenschappen*, 6, 115–143.

Woody, A. (1995). The explanatory power of our models: A philosophical analysis with some implications for science education. (In F. Finley, D. Allchin, D. Rhees & S. Fifield (Eds.), *Proceedings of the Third International History, Philosophy and Science Teaching Conference* (Vol. 2., pp. 1295–1304). Minneapolis: University of Minneapolis.)

ANNEX

5.1 Questionnaire

(Each item refers to one of the general features of models discussed in Section 2 of this paper)

1. *(Feature 1)* In your research paper a model is a model of *something else*, something which it represents, e.g. the models simulate a process. In your research work a distinction was made between the model and what it represents.
   0 correct, or 0 incorrect
   Can you illustrate this, e.g. with an example?
2. *(Feature 2)* That which is represented by the model (or models) in your paper, is the actual *object of research*. The model is, ultimately, a tool in that research, not an aim in itself.
   0 correct, or 0 incorrect
   Could you identify the object of your research?
3. *(Feature 7)* In your research work, *two or even more* different models can represent – and often do represent – the same object.
   0 correct, or 0 incorrect
   Comment:
4. *(Feature 4)* The model in your paper differs from whatever it represents in that it is *more accessible*. If it was not more accessible, it would not be a useful model.
   0 agree, or 0 disagree
   Can you describe the difference in accessibility, or any other difference you think is essential?
5. *(Feature 3.a)* At the same time, the model in your paper has things *in common* with what it represents. If it had nothing in common with it, it would not be a useful model.
   0 correct, or 0 incorrect
   Can you identify the essential common properties?

6. *(Feature 5)* The model in your paper is a *compromise* between (a) accessibility, and (b) correspondence with the object it represents. Models can therefore be situated in a kind o spectrum, with simple, easily accessible models on one side and more advanced, complicated models on the other side.
   0 correct, or 0 incorrect
   Could you illustrate or refute this with examples?
7. *(Feature 3.b)* From the model in your paper, one can derive *hypotheses or predictions* with respect to that which the model represents. Such a hypothesis or prediction requires testing before it can be accepted as correct.
   0 correct, or 0 incorrect
   Comment:
8. *(Feature 6)* The model in your paper did not follow directly and automatically from measurements or other observations, or from other models. Instead, there was an element of *creativity* involved.
   0 correct, or 0 incorrect
   Could you throw light on the role of creativity in your research work?
9. *(Feature 8)* As part of a research cycle, the model in your paper suggests new experiments or observations to be carried out. These, in turn, may lead to newer and more advanced models to be developed, et cetera.
   0 correct, or 0 incorrect
   Could you explain your answer?
10. *(*) Elegance* is, in your experience, somehow a consideration in designing or selecting a model.
   0 correct, or 0 incorrect
   Could you describe a relevant experience?

(*) NB This feature was added because the aspect of 'elegance', which was not included in our tentative description of features, was mentioned in one of the pilot interviews.

MARIOS PAPAEVRIPIDOU, CONSTANTINOS P. CONSTANTINOU AND ZACHARIAS C. ZACHARIA

26. THE DEVELOPMENT OF ELEMENTARY STUDENTS' UNDERSTANDING OF COMPLEX ECOSYSTEMS THROUGH A MODEL-BASED APPROACH

Abstract: We report on a study in which we designed and implemented an activity sequence within the context of ecosystems aimed at fostering scientific modelling skills and enhancing conceptual understanding among fifth graders through the use of Stagecast Creator™. We describe how activities supported students to improve and expand their models by adding new information collected from observations and personal knowledge about real-life ecosystems. Paper-and-pencil tests were used both before and after the study to evaluate students' modelling skills and understanding of concepts related to ecosystems. The data analysis followed both qualitative and quantitative methods. The results indicated the importance of the synergy of the study's goals: to learn about marine ecosystems and develop modelling skills. We found significant differences between students' pre- and post-test scores, suggesting that our approach facilitated the development of both modelling skills and enhanced conceptual understanding of marine ecosystems of fifth graders

Keywords: Computer-based modelling tools, Ecosystems understanding, Modelling skills

1. INTRODUCTION

A number of researchers underline the necessity of integrating technological tools in science teaching and learning (Barab et al. 2000; Hogan and Thomas 2001). The research outlined in this paper has been triggered from these calls for reforming the science curriculum. Specifically, we report on a study in which we designed and implemented a curriculum within the context of ecosystems aimed at fostering scientific modelling skills and enhancing conceptual understanding among fifth graders through the use of Stagecast Creator (SC). In addition, we discuss how the SC contributed towards the enhancement of conceptual understanding and modelling skills.

2. THEORETICAL BACKGROUND

A plethora of curriculum units in science involve the study of dynamic systems, e.g. biotopes in ecology, barometric phenomena in climatology, circulatory system in anthropology, etc. The complexity of the structure of these systems as well as the

Learning in Science, Group, Department of Educational Sciences, University of Cyprus, P. O. Box 20537, Nicosia 1678, Cyprus

R. Pintó and D. Couso (eds.), Contributions from Science Education Research, 333–345.

simultaneous multivariate processes being executed within them impede learners' understanding of concepts and relationships related to the system. Therefore, the acquired knowledge is isolated, fragmented and inert. Many researchers (Louca et al. 2003; Stratford et al. 1998; Wilensky and Resnick 1999) have argued that systems' exploration and understanding would be facilitated, if students are given the opportunity to explore these systems through creating dynamic computational models. Through the creation of dynamic models students are engaged in a process of combining isolated knowledge about poorly understood concepts and relationships into larger, more clearly understood constructs in a way that allows them to represent, reconstruct and explore that knowledge within a computational model (Stratford et al. 1998).

Consequently, meaningful learning in science can be thought of as a dynamic process of building, organizing and elaborating knowledge of the natural world. According to Constantinou (1999), physical science can be characterized as a complex network of models interrelated by a system of theoretical principles. Models are units of structured knowledge used to represent observable patterns in physical phenomena. Accordingly, physical understanding can be perceived as a complex network of modelling skills, that is cognitive skills for making and using models (Booch et al. 1999; Stern 2000). The development of modelling skills provides the opportunity for students to make sense of their own physical experiences and to evaluate information reported by others.

The modelling approach can potentially provide a backbone structure for constructing meaning in physical science. In this approach, students are guided to develop a set of generic modelling skills in one domain and to transfer those same skills in other domains, further elaborating and developing them with experience and practice. The modelling approach to learning is iterative in that it involves continuous comparison of the model with the reference physical system with the express purpose of gaining feedback for improving the model so that it accurately represents as many aspects of the system as possible. It is also cyclical in that it involves the generation of models of various forms until one can be found that successfully emulates the observable behaviour of the system (Papadouris and Constantinou 2001).

The study of ecosystems can be used as a multidisciplinary platform for the development of modelling skills, because ecosystems are physical entities consisting of organized units of biotic and abiotic elements, as well as relationships and interactions among its consisting units. The constituent components of an ecosystem determine, control and modify ecosystem's operations. Each part affects the behaviour of the whole, depending on part's interaction with other parts of the ecosystem. Understanding and reasoning effectively about ecosystems involves understanding a number of different types of causal patterns. Particulary, while tracing food effects perturbations, one might apply two kinds of causality patterns: simple linear causality or domino-like and interactive causality (Bell-Basca et al. 2000). Linear causality refers to a causal pattern where an initial cause triggers a chain of consecutive effects, and every new effect becomes a new cause (e.g., the disappearance of the primary consumers within an ecosystem may result in the disappearance of secondary consumers and in turn the disappearance of tertiary consumers and so on). Interactive causality refers to the case where a change in a species population can be a cause and effect at the same time. For example, an increase to the population of primary consumers within a food web could be the cause of a decrease in the population of the producers and the effect of the increase in the population of secondary consumers.

A number of researchers that aimed at exploring students' conceptual understanding of various concepts related to ecosystems came to a congruity that students hold numerous misconceptions about ecosystems and the nature of interrelatedness of the components of ecosystems (Adeniyi 1985; Barman et al. 1995; Bell-Basca et al. 2000; Hogan and Thomas 2001). For example, Strommen (1995), while examining first graders' conceptions of a forest habitat, found that students' ideas consisted of mostly concrete facts that missed the broader conceptual relationships among the rest of the organisms of the ecosystem. This weakness was associated with students' deficiency of systemic thinking. Additionally, Barman et al. (1995) noticed that students believed that a change in one population will not be passed along several different pathways of a food web and that a change in one population will only affect another population, if the two are related in a predator–prey relationship. Similarly, the results of Hogan's (2000) study point to limitations in awareness of patterns of systems interactions as constraining students' systems reasoning in ecology.

The advent of age-appropriate computer-based modelling tools (e.g. Stagecast Creator, Model-It, Logo) enabled new ways of overcoming some of the conceptual and reasoning difficulties associated with the representation and interpretation of complex systems. An important aspect of computational modelling is that it allows students to visualize abstract concepts (Barab et al. 2000), complex relationships (Singer et al. 2000) and multiple processes by creating structures through which they can explore and experiment. Although computational tools seem to be promising in facilitating individual's learning, a few studies have focused on student's development of modelling skills (Carney 2002; Talsma 2000). Van Driel and Verloop (2002) stated that 'not much is known about the variety of teaching activities concerning models and modelling that are actually applied by science teachers…' (p. 1258). Our study is focusing on this exact issue. More specifically, our study aims at answering the following questions:

- Is it feasible to foster conceptual understanding in the domain of ecosystems among 11-year-olds through a modelling-based approach?
- Can a computer-based environment be used to support, in an efficient way, students' development of modelling skills and conceptual understanding in the context of ecosystems?

3. METHODS

3.1 Participants and Setting

The sample was comprised of 16 fifth graders (8 boys and 8 girls) who were studying in a mid-sized suburban school in Cyprus. Participants were randomly assigned in groups of four and were asked to work for the purposes of the study in the school's computer lab. Each group had access to a computer. Despite the fact that all participants were computer literate, special attention was given to orienting students to the use of SC. The intervention lasted about two months.

3.2 Experimental Design

The research was carried out in two stages (see Figure 1).

Stage 1

Stage 1 involved the development of the curriculum materials that were implemented during the intervention. Hence, we undertook epistemological analysis of both modelling skills and the concepts related to ecosystems in order to identify prerequisite concepts and skills for the development of curriculum focusing on the enhancement of modelling skills and conceptual understanding.

Based on the epistemological analysis of each of our objectives, we developed a series of pre-tests to evaluate students' modelling skills and conceptual understanding related to ecosystems prior to the intervention. Descriptions of each of the pre-tests are provided in Table 1.

Data collected through the various pre-tests were analysed using phenomenography (Marton and Booth 1997). Both the types of responses and the difficulties (e.g. conceptual, reasoning, epistemological) that emerged from the analysis guided our effort to design a structured intervention to help participants overcome specific difficulties and gain conceptual understanding concerning concepts related to ecosystems.

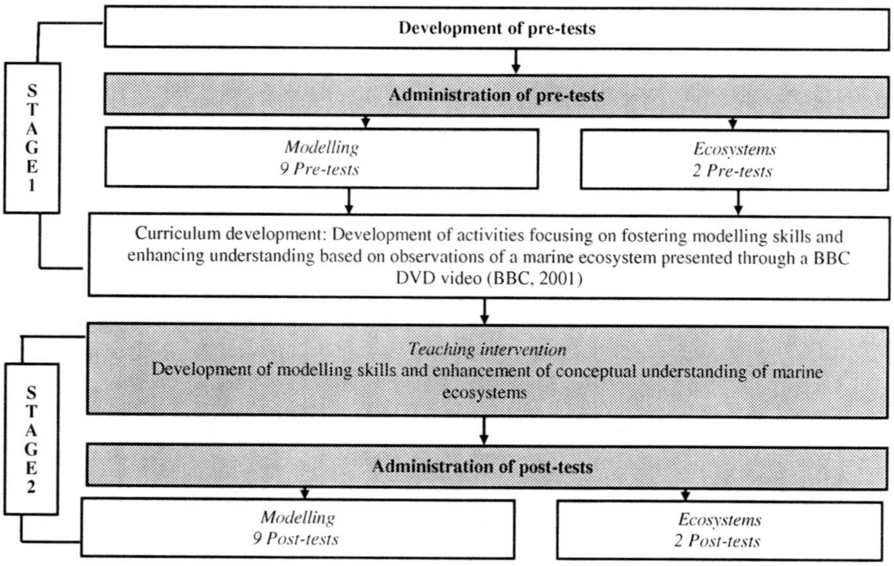

Figure 1. Experimental design

TABLE 1. Descriptions of Pre-Tests that Evaluate Modelling Skills and Conceptual Understanding of the Conceptual Aspects of Ecosystem Dynamics

Skill/Conceptual Understanding	Pre-Test	Context	Aspects
Modelling	*Pre-test 1*	Traffic system	Model formulation
	Pre-test 2	Traffic system	Extraction of information from a given model
	Pre-test 3 & *Pre-test 4*	Bicycle	Comparing models and real phenomena
	Pre-test 5 & *Pre-test 6*	Elbow	Comparing one model to another
	Pre-test 7 & *Pre-test 8*	Respiratory system	Appreciation of a model as a representation of a physical phenomenon
	Pre-test 9	Ant colony	Consistency of a model with all relevant phenomena and ideas for improvement
Ecology	*Pre-test 1*	Forest ecosystem	• Understanding of the basic needs of living organisms • Formulation of food chains • Populations relationships • Ecological balance
	Pre-test 2	Forest ecosystem	• Understanding and formulation of food chains • Food relationships • Population concept • Reproduction rate • Species interactions • Understanding and formulation of food pyramids • Ecological balance

The development of modelling skills begins by observing a physical phenomenon, whose complex procedures and operations between its features make it difficult for individuals to gain a clear understanding of how this phenomenon is executed in real life. In our case, the complex phenomenon that students were expected to model was a marine ecosystem. Students were given a DVD titled "The Blue Planet" (BBC 2001) and had the opportunity to access and study it in several stages of the intervention. The idea was to allow students to make observations of the physical phenomenon, collect information that was needed for their model, make comparisons between the physical phenomenon and their model, etc.

The next decision we had to consider concerned the modelling medium which students had to use to build their models. We decided to utilize SC in our study as it has been used effectively for the development of modelling skills among middle-school students (Constantinou 1999). SC is a programming environment that enables the design of microworlds (Smith and Cypher 1999), and hence the building of models, although it has not been designed to be used as a modelling tool. Programming in SC is done by demonstration (the user can create a script which is monitored and modelled by the program, and this script can be performed by the program when the criteria of its original design are met), without writing a code or syntax as is needed in traditional programming languages, and hence it becomes accessible for use with younger children.

The development of the curriculum was based on the philosophy and principles of the *Physics by Inquiry* pedagogy (McDermott 1996). This teaching approach rests on the premise that students are more likely to learn physics when they are engaged in both hands-on and minds-on activities. Students always work in groups of three or four and interact with materials in order to gain experiences and develop practical and scientific skills, as well as the ability to process and manipulate multiple representations. Throughout the instruction, students are expected to perform observations, formulate hypotheses, construct operational definitions for the concepts they are inventing, develop interpretative models accounting for their observations and evaluate them on the basis of specific epistemological criteria.

3.2.1 Overview of Activity Sequence In the first activity, students observed a marine ecosystem as presented in the DVD (BBC 2001) and at the end they answered specific questions (e.g. "What is going to happen if the phytoplankton vanishes in the sea?", "Which of the following organisms (shark, phytoplankton, herring) has the largest population in the sea?", etc.) based on what they already knew about marine ecosystems and what they observed through watching the DVD. They discussed their answers in their groups and no answer was given to the students by the instructor until the end of the intervention. Right after, they were prompted to make a drawing of the observed phenomenon, compare it to their peers' drawings, and discuss about similarities and differences between their drawings. Through the curriculum, students were informed that the drawing they sketched to represent the observed physical phenomenon was called a *model*, and that individuals often construct models to improve their understanding of a physical phenomenon.

Afterwards, students watched preselected segments of the DVD and they were asked to identify the three basic needs (nutrition, reproduction, shelter) of the organisms that lived in a marine ecosystem. The next step of the activity sequence was related to model improvement; students had to go back to their initial paper-and-pencil model and add further information to represent the three previously identified needs of the marine organisms. After completing this task, a classroom discussion was organized by the instructor to summarize the steps that were followed for model improvement [(i) compare the model to the physical phenomenon, (ii) identify new features from the physical phenomenon that have not been included in the model, (iii) represent the new identified features to the model, (iv) the new product is the improved model of the physical phenomenon].

By the end of this discussion, students were challenged to state whether the drawing was the most appropriate tool to formulate a "good" model for the observed phenomenon, and hence they identified possible disadvantages of the paper-and-pencil models (e.g. they are static representations, they do not inform us what is happening as time passes, etc). After identifying the weakness of formulating paper-and-pencil models, students were encouraged to suggest other tools that would enable them to develop dynamic models to represent in a better way the physical phenomenon. Students recalled their previous experience[1] with SC that enabled them to build better models, as it allows to make the characters of the model move around in the screen. Therefore, they improved the paper-and-pencil model by building a dynamic model of the observed marine ecosystem in SC. At this stage we did not expect that students would build a complete model of the physical phenomenon by programming all characters to perform all the observed behaviours, but what was anticipated was to include some of the organisms and setting, through programming some of their behaviours (e.g. movement) to their models.

In order to reduce the complexity of the ecosystem due to the multiplicity of its inhabitants, we decided to isolate nine organisms of the marine ecosystem (phytoplankton, kelp, herring, crab, mussel, urchin, sea otter, sea-gull, shark) to help students focus on organisms' behaviour and interactions within the ecosystem. Students collected relevant information about their nutrition habits for each of these organisms from particular segments of the DVD, as well as from additional informative leaflets specially designed for the purpose of this activity. As soon as students organized the collected data in a two-column table referring to "Who eats who" and "Who is eaten by whom", they were introduced to new concepts related to the previous activity. The new concepts were *nutrition, producers, consumers, food relationships, predator and prey*, and *energy*, and they were prompted to apply the new concepts to the case of the organisms being studied (e.g., name the producers/consumers, write sentences to show which organism is the predator for whom, etc). This activity served as a prerequisite to the formulation of food chains; students were called to represent a food relationship between two organisms of the marine ecosystem in a diagram by linking the corresponding organisms with an arrow according to their food relationship. At the end of this task, they got informed that this was called a "food chain" and they were encouraged to discuss why the term "food chain" was appropriate for this type of diagram. What we aimed through this activity was to guide students to identify the food chain as a *food relationship model* and, also, to appreciate the importance of the food chain model as a means to represent a relationship between two organisms in a minimal manner. An extended activity came next, challenging students to produce a diagram in which every single food chain was linked to another through energy transfer arrows, based on common food relationships among the organisms, and therefore the concept of a food web was introduced as an extended version of the food relationship model.

Based on what they had learned so far, students were encouraged to go back to their original SC model and make improvements in such a way that their refined model included the nine organisms being studied, and also their food behaviour for each of them. In order to build the food behaviour for a character, students were expected to use the concept of energy for this character. Then, they created a rule so that a character's level of energy was associated to its eating habits and movement rules (e.g. the character would gain some amount of energy when eating a particular prey and loose an amount of energy as it moved in the sea). Finally, they created a rule for removing a character from the screen (it represents its death), when its energy level dropped below a specified number. During this activity, students could access the real marine ecosystem through the DVD, make comparisons between their evolving model to the physical phenomenon, and improve their model so that it represents more aspects of the corresponding phenomenon. At the end of this activity, students were introduced to the concept of *ecosystem*, and discussed issues regarding its components and the processes that occurred within it.

The next activity aimed at introducing two new concepts: *population* and *reproduction*. The new concepts were discussed with reference to the concept of energy, and right after students were prompted to compare their SC model to the physical phenomenon by focusing on the population and reproduction issues. Finding out that their model did not represent the process of reproduction, and that the population size for each of the species had been verified in a random manner, students went through a new model revision phase in order to enrich their model. What we expected students to do at this stage was to create rules for reproduction, decide the energy level of the parents that would be able to mate and specify the energy level of the offspring.

At this stage, the instructor organized a classroom discussion on whether the food web model that had been developed in the previous sessions enabled students to collect information of the population size of each of the marine organisms. After identifying the weakness of this particular model concerning the representation of the population size, students were told to make a drawing (a new model) to represent the relative size of the population of the nine organisms. Through guiding instructions and prompts, students worked collaboratively to draw a food pyramid, and hence the curriculum introduced the concept of *food pyramid*, as well as relevant concepts such as *producers, first class consumers, second class consumers*, etc. As soon as students got familiar with these concepts, the instructor organized a classroom discussion through which students identified the pyramid as a model to represent population and food relationships.

At this point, students would have already formulated three different models based on their understanding of how the marine ecosystem being studied operates: these are the*SC model,* the *food web model* and the *food pyramid model*. In order to differentiate the three models based on their role and their scope, students made comparisons among these models and specified the different kinds of information one could collect through studying each one of the models. We considered this activity as fundamental to the development of modelling skills, because it set the boundaries to the development of epistemological awareness concerning models and modelling in science. In other words, we expected that students after the study would be able to reason why it was important to build models of a physical phenomenon.

In order to complete the development of the concept of the ecosystem, students were introduced to the concept of *ecological balance* through studying two food pyramids: one being at the state on ecological balance and the other at the state on ecological imbalance. Students were prompted to predict what was going to happen over time to the ecosystem that each pyramid represented, and in order to test their prediction, they were asked to decide which of the previous models they developed would help them in this way. Hence, they tested their prediction using the SC model (the population size of each of the species was reduced or increased according to the food pyramid), and observed what happened as time passed. Based on this activity, students were expected to name the SC model as the "ecological balance model", and investigated different scenarios related to the ecological balance of the ecosystem (e.g. what was going to happen if the phytoplankton vanished? Etc.).

As a final activity, the instructor organized a classroom discussion where students were guided to summarize the various steps that should be followed when modelling a physical phenomenon and discussed the features that should be included in a model: (*objects, variables, interactions, procedures*). Students were encouraged to identify and applied each of these features to the ecological balance model.

3.2.2 Stage 2 Stage 2 involved the implementation of the study's curriculum. The curriculum included checkpoints at specific stages where students were asked to discuss their findings and reasoning with the instructor. The role of the instructor is to help students articulate their thoughts, identify inconsistencies regarding their models and also negotiate epistemological, conceptual, practical or any other difficulties they might encounter.

After the completion of the intervention, the same pre-tests that were administered at the beginning of Stage 1 were administered again to the participants to obtain comparable pre-test and post-test results concerning students' mastery and transfer of modelling skills and conceptual understanding in previously unfamiliar domain. The context of each of the pre- and post-test differed with the context of the curriculum that was implemented during the intervention in order to ensure students' ability of transferring skills and concepts in new contexts.

3.3 Data Analysis

The data obtained through the various tests were analysed in two ways. First, open-ended questions were analysed using phenomenography (Marton and Booth 1997), and responses subsequently coded and transformed for analysis by MANOVA, Repeated Measures, to test whether there were any significant differences in the responses prior to and after instruction.

4. RESULTS

The results that emerged from the statistical analysis concerning the comparison of students' conceptual understanding of various aspects related to ecosystems between pre- and post-test, are shown in Table 2.

As is shown in Table 2, the results indicate significant differences between students' scores in pre- and post-test scores for all aspects of the concept of ecosystem. The multivariate η^2 based on Wilks' λ for all aspects of the ecosystem was quite strong (ranging between 0.5 and 0.92), indicating that a great percentage

TABLE 2. Summary of Quantitative Results Obtained Through the Pre- and Post-Tests Concerning the Development of Conceptual Understanding in the Context of Ecosystems

Conceptual aspects of ecosystem dynamics*	Wilks' λ	F	Hypothesis df/error df	Sig.	Partial η^2
Understanding of the basic needs of living organisms	0.11	125.69	1/15	0.0001	0.89
Organizing a set of given species on a food pyramid	0.44	19.129	1/15	0.001	0.56
Understanding of population relationships among given species	0.11	120.65	1/15	0.0001	0.89
Prediction of the consequences within an ecosystem due to a dramatic change in the population of a species	0.076	183.54	1/15	0.0001	0.92
Evaluating the change within an ecosystem with the most serious consequences	0.12	140.55	1/15	0.0001	0.88
Identifying a food pyramid that is at the state on ecological balance	0.50	15.00	1/15	0.002	0.50
Drawing the progress of two food pyramids, one that is at the state on ecological balance and one that is at the state on ecological imbalance, over time	0.17	73.54	1/15	0.0001	0.83

* *For each aspect, students were required to explain the reasoning behind their answer.*

of multivariate variance of the dependent variable (conceptual understanding performance) of the MANOVA model is associated with the group factor (intervention).

In order to highlight the contribution of SC to the development of modelling skills and conceptual understanding in the context of ecosystems, we provide the qualitative results of two questions related to students' ability to predict the consequences to an ecosystem, when a particular change in the population of one species occurs. Those results are shown in Table 3.

The results provided in Table 3 indicate that, after instruction, participants were able to provide correct predictions regarding the consequences to the population of each species after the occurrence of a particular change. Based on the quotes of the correct responses with correct reasoning, it is evident that students appreciate SC contribution as a modelling tool (e.g. "I would like to have a model of ecological balance in SC") as well as a conceptual tool (e.g. "I need a model in SC to run and see if my answer is correct") to guide their learning in the context of ecosystems. Their responses also emphasize the development of epistemological awareness towards models and the modelling process; this is inferred when students end their response by highlighting the importance of hypothesis or prediction testing through the use of an SC model.

TABLE 3. Types of Responses, Quotes and Frequencies of Students' Performance in Two Items of Pre-Post-Test 1 Prior to and After Instruction

Item	Type of response	Quote	Frequency (n)	
			Before	After
Item 4.1 (pre-test 1)	Correct response with correct reasoning	*Yes, it will change. The crows will be on the verge of extinction and probably the insects will be increased and eat more plants, so the plants will be reduced and the monkeys will probably be reduced. The crows will be on the verge of extinction, because they will not be able to find food, e.g. frogs. The insects might increase, because their predators – the frogs – will vanish and the monkeys might also disappear, because they feed on plants and the insects will eat more plants now. To be more confident with my response, I would like to have a model of ecological balance in SC to observe what is going to happen when this change occurs.*	0	15
	Partial response answer with correct reasoning	*The crows will disappear, because they will not have any food to feed in.*	12	1
	Wrong response or irrelevant response or failure to respond	*It will change, because if the frogs that are green, then there will be only soil on the ground.*	4	–
Item 4.2 (pre-test 1)	Correct response with correct reasoning	*The plants will decrease and since the plants are on the basis of the pyramid, every other organism will decrease. When the plants decrease, the insects will not be able to find food and they will decrease. Hence the frogs won't find food and they will decrease, the crows will decrease, because they feed on frogs. But in order to be surer of what is going to happen, I need a model in SC to run and see if my answer is correct.*	0	13
	Partial response answer with correct reasoning	*The plants will decrease, because they will be consumed by the monkeys.*	9	3
	Wrong response or irrelevant response or failure to respond	*It won't change anything, the monkeys feed on birds, if they left (the monkeys), it would be better.*	7	–

5. DISCUSSION

The study reported in this chapter was intended to answer two research questions. According to the first question, which refers to whether it is feasible to enhance students' conceptual understanding in the domain of ecosystems through a modelling-based approach, the results indicate that students' conceptual understanding concerning marine ecosystems was improved. This finding is aligned to the notion of Halloun and Hestenes (1987) that modelling, as an instructional approach, may serve as a medium for the development of conceptual understanding in the domain of ecosystems and that it enables students to transform their preconceptions to scientific ideas, as well as to overcome any conceptual, reasoning and epistemological difficulties that hinder their learning.

Moreover, the results concerning students' appreciation of SC as both a modelling and conceptual medium, which enables prediction making and hypothesis testing, indicate that SC has been substantiated as an effective tool to scaffold students' development of modelling skills and conceptual understanding in the context of ecosystems. The impact of SC on students' learning has also been evident through the implementation of the activity sequence that was described in a previous section of the chapter. Particularly, students used SC to create dynamic models of the observed marine ecosystem, they applied new concepts to their models through building and debugging rules to set a required behaviour to a character, and validated their model by repeatedly contrasting it to the physical phenomenon and adding missing information.

A potential factor that accounts for the effectiveness of the present study is associated with the activity sequence that was specially designed for the purpose of our research. It is important to state that, from our perspective, curriculum design is an essential empirical process that integrates research and development. Effective curriculum development emerges through a rigorous methodological approach that integrates the results of the research into the curriculum design process. In the instance of integrating computer-based tools with instruction, there is an emerging need to supplement these tools with appropriately designed, carefully structured and properly validated curriculum material.

NOTE

[1] The study reported in this chapter is part of an extended study that aimed to develop programming skills and modelling skills of fifth graders through the use of SC in the context of marine ecosystems. Prior to the reported study, the same participants have been working with SC for a two-month period and they were guided through a specially developed curriculum to develop several aspects of the programming skill.

REFERENCES

Adeniyi, E. O. (1985). Misconceptions of selected ecological concepts held by some Nigerian students. *Journal of Biological Education, 19(4)*, 311–316

Barab, S. A., Kenneth, H. E., Barnett, M. & Keating, T. (2000). Virtual solar system project: Building understanding through model building. *Journal of Research in Science Teaching, 37(7)*, 719–756

Barman, C. R., Griffiths, C. R. & Okekukola, P. A. O. (1995). High school students' concepts regarding food chains and food webs: A multinational study. *International Journal of Science Education, 17(6)*, 775–782

BBC (2001). *The Blue Planet: A Natural History of the Oceans (DVD)*. BBC productions

Bell-Basca, B. S., Grotzer, T. A., Donis, K. & Shaw, S. (2000, April). *Using Domino and Relational Causality to Analyze Ecosystems: Realizing What Goes Around Comes Around.* (Paper presented at the National Association of Research in Science Teaching, New Orleans)

Booch, Gr., Rumbaugh, J. & Jacobson, I. (1999). *The Unified Modelling Language.* Addison Wesley

Carney, K. (2002). When is a Tree a Process? Influences on Student Representations of Process in "Low Floor" Qualitative Modelling Tasks. (In P. Bell & T. Satwicz (Eds.), *Keeping Learning Complex: The Proceedings of the Fifth Annual International Conference of the Learning Science* (pp. 49–56). Mahwah, NJ: Lawrence Erlbaum)

Constantinou, C. P. (1999). The Cocoa Microworld as an environment for modelling physical phenomena. *International Journal of Continuing Education and Life-Long Learning, 9(2)*, 201–213

Halloun, I. & Hestenes, D. (1985). Modelling instruction in mechanics. *American Journal of Physics, 55(5)*, 455–462

Hogan, K. & Thomas, D. (2001). Cognitive Comparisons of Students' Systems Modelling in Ecology. *Journal of Science Education and Technology, 10(4)*, 319–345

Louca, L., Druin, A., Hammer, D. & Dreher, D. (2003). Students' collaborative use of computer-based programming tools in science: A descriptive study. (In B. Wasson, S. Ludvigsen & U. Hoppe (Eds.), *Designing for Change* (pp. 109–118). Netherlands: Kluwer Academic Publishers)

Marton, F. & Booth, W. (1997). *Learning and Awareness.* (New York: Lawrence Erlbaum)

McDermott, L. C. (1996). *Physics by Inquiry (Volumes I & II).* (New York: John Wiley & Sons, Inc)

Papadouris, N. & Constantinou, C. P. (2001). The design and development of research-based, technology-enhanced science curriculum: What support can we expect from communication and information tools? (In A. Gagatsis (Ed.), *Learning and Assessment in Mathematics and Science* (Vol. II, pp. 129–155). Nicosia: University of Cyprus)

Singer, J., Krajcik, J. & Marx, R. (2000). The design and evaluation of classroom supports for seamless integration of a dynamic modelling tool. (In B. Fishman & S. O'Connor-Divelbiss (Eds.), *Fourth International Conference of the Learning Sciences* (pp. 62–69). Mahwah, NJ: Lawrence Erlbaum)

Smith, D. C. & Cypher, A. (1999). Making Programming Easier for Children. (In A. Druin (Ed.), *The Design of Children's Technology.* San Francisco, CA, USA: Morgan Kaufmann Publishers)

Stern, J. (2000). The design of learning software: Principles learned from the computer as learning partner project. *Journal of Science Education and Technology, 9(1)*, 49–65

Stratford, S. J., Krajcik, J. & Soloway, E. (1998). Secondary students' dynamic modelling processes: Analyzing, reasoning about, synthesizing, and testing models of stream ecosystems. *Journal of Science Education and Technology, 7(3)*, 215–234

Strommen, E. (1995). Lions and tigers and bears, oh my! Children's conceptions of forests and their inhabitants. *Journal of Research in Science Teaching, 32(7)*, 683–698

Talsma, V. (2000). Scientific understandings revealed by students' computer models of a stream: a trickle or a flood? (In B. Fishman & S. O'Connor-Divelbiss (Eds.), *Fourth International Conference of the Learning Sciences* (pp. 62–69). Mahwah, NJ: Erlbaum)

Van Driel, J. H. & Verloop, N. (2002). Experienced teachers' knowledge of teaching and learning of models and modelling in science education. *International Journal of Science Education, 24(12)*, 1255–1272

Wilensky, U. & Resnick, M. (1999). Thinking in levels: A dynamic systems approach to making sense of the world. *Journal of Science Education and Technology, 8(1)*, 3–19

27. EFFECTS OF MODEL-BASED TEACHING ON THE DURABILITY OF PRE-SERVICE PHYSICS TEACHERS' CONCEPTIONS OF LUNAR EVENTS

Abstract: This longitudinal study was conducted to determine the durability of pre-service physics teachers' scientific understanding of the Moon and some lunar events over an extended period after the model-based teaching. This research was a continuation of the previous study (Ogan-Bekiroglu, 2007) including determination of pre-service physics teachers' existing conceptions and their reasoning of the Moon and some lunar events, and examination of the changes in their conceptions after the model-based teaching. Participants of the previous study were given delayed post-questionnaire six months after their model presentations. Results of the previous study showed that nearly all participants developed conceptions consistent with the scientifically accepted perspective of many moon-related events after the model-based teaching. This study indicated that while a majority of the pre-service physics teachers continued to hold scientific conceptions, some of them had conceptual decay in their understanding of some lunar events

Keywords: Astronomy, Conception, Longitudinal, Mental model, Model-based, Moon, Pre-service Teachers

1. INTRODUCTION

Learning about science, about science teaching, and about students' learning of science requires effective professional development (Dana et al., 1997). Therefore, teacher education programmes should include strategies to improve pre-service teachers' content knowledge as well as pedagogic skills. This study investigated the long-term effects of model-based teaching on pre-service physics teachers' conceptions of some lunar events.

2. LITERATURE REVIEW ABOUT CONCEPTIONS OF THE MOON

Various studies have been conducted to determine students' and pre-service teachers' understanding of the Moon. A considerable amount of research has focused on both children's and adults' conceptions of Moon phases (Callison and Wright, 1993; Dai and Capie, 1990; Rider, 2002; Stahly et al., 1999; Taylor, 1996; Trundle et al., 2002). "There is evidence that people with varied levels of schooling

Feral Ogan-Bekiroglu, Sengul Sk. Sarikonak, Apt. 5/12 Bostanci, Istanbul 34744, Turkey

R. Pintó and D. Couso (eds.), Contributions from Science Education Research, 347–359.

and training from elementary school through in-service teachers do not understand the cause of the Moon phases" (Trundle et al., 2002, p. 634) and "some of them still hold misconceptions even after they were provided a scientifically correct explanation" (Hermann and Lewis, 2003, p. 51). Research indicates that the most common misconception of the cause of Moon phases held by the participants is "Earth's shadow falling on the Moon".

Besides conceptions of lunar phases, researchers have been interested in students' and pre-service teachers' conceptions of some other moon-related events. Barnett and Morran (2002), for instance, conducted a study in the United States to investigate elementary school students' confusion between a full Moon and a lunar eclipse after instruction. In addition, Trumper (2001) worked with high school students in Israel. The most common misconception he discovered was that the Moon neither revolved around the Sun nor rotated on its axis. Dove (2002) found that the reason why the Moon always presented the same face to the Earth was less well understood by secondary students in the United Kingdom even after instruction. He added that estimations of the time in earth days from sunrise to sunrise on the Moon were also varied among the students. Furthermore, Suzuki (2003) revealed that Japanese pre-service teachers assumed that the Moon could only be seen at night. Analysis of the research shows that no matter which country they are from and which grade they are in, learners have some common non-scientific understandings of the Moon and lunar events.

Researchers highlight some reasons for the difficulty in learners' conceptions of lunar phenomena. Samarapungavan et al. (1996) state that "the initial construction of children's cosmological understanding is constrained by core principles originating from naïve physics as well as by certain ontological and epistemological assumptions about cosmological entities" (p. 520). According to Parker and Heywood (1998) and Pena and Quilez (2001), students' unfamiliarity with optics, geometry and light poses a difficulty in learning the moon-related concepts. Treagust and Smith (1989), in addition, relate students' misconceptions of Moon phases with their limited understanding of relative size, motion and distance. Thus, learners' conceptions of lunar events might be related to their naïve knowledge based on their casual observations and experiences and/or their understanding of some physics concepts. Regarding this interpretation, the question would be: "How can the difficulties in learners' conceptions be eliminated so that learners can construct scientific knowledge about the Moon and moon-related events?"

3. MODEL-BASED TEACHING

Baxter (1991) suggests that by placing students in a situation where they confront their original ideas, much of the astronomical ignorance that appears to pass into adulthood can be challenged, and students will be encouraged to emerge from their pre-Copernican worldview. Barnett and Morran (2002), furthermore, advocate that students develop more scientifically sophisticated understanding of natural phenomena when instruction is designed to actively engage students in activities that afford them opportunities to not only become aware of their own understanding but also to reflect and discuss their understanding. From Trumper's (2001) point of view, some possible activities including a process of prediction, observation, discussion and conclusion are necessary in order to teach astronomy. These suggestions can be implemented under model-based teaching. Boulter et al. (2001) reveal that a model-building sequence begins with students expressing initial models. Then, students face challenges to their existing models in the field along with the need to negotiate new group models, and, finally, they report models in their presentations and endure more challenges (Boulter et al., 2001).

Researchers (Harrison and Treagust, 2000; Samarapungavan et al., 1996; Trumper, 2001; Trundle et al., 2002) also emphasize the importance of using and construction of a model to help children simulate the relative motions between the Sun, Earth and Moon and mentally test a simulation of their reconfigured cosmology.

4. PURPOSE OF THE STUDY

When instruction does lead to desirable conceptual change, the durability of the effects becomes a concern. Assessment to determine the durability of conceptual change after longer delays is needed (Georghiades, 2000). Therefore, the purpose of this longitudinal study was to investigate the effects of the model-based teaching on durability of the pre-service physics teachers' conceptions of the Moon and some lunar events. In order to achieve this aim, the pre-service physics teachers' conceptions were determined six months after the model-based teaching and compared to their conceptions examined in the previous research (Ogan-Bekiroglu, 2007).

5. THEORETICAL FRAMEWORK

Three forms of conceptual knowledge were mentioned in the literature according to their functional characteristics (Ozdemir, 2004). One of these forms is models. Conceptions are proposed as mental models in this study. Mental models are internal, cognitive representations used in generating external representations and in reasoning of many kinds in response to phenomena (Buckley, 2000).

According to Chi and Roscoe (2002), mental model is a representation of knowledge as a set of interrelated propositions that are embedded in a structure and it can be categorized for its coherence and completeness. They define conceptual change as processes of repairing misconceptions or the process of reassigning or shifting a miscategorized concept from one ontological category to another ontological category, from flawed or incomplete mental model to correct mental model. Furthermore, Chi and Roscoe claim accumulation of two learning processes to remove misconceptions and repair flawed mental models. These two processes are assimilation, which means adding new pieces of knowledge, and revision, which means correcting pieces of knowledge.

Because the aim of this study was to explore how model-based teaching was effective in promoting conceptual change, it was more reasonable to categorize the pre-service physics teachers' mental models as correct or incorrect according to the scientific point of view.

6. METHODOLOGY

The combination of three questionnaires developed during the previous study was used as a delayed post-questionnaire in this study. The delayed post-questionnaire included 20 factual, explanation and generative questions (see Annex). The types of the questions were determined as open-ended to be able to evaluate the participants' reasoning behind their answers. The following three factors were considered while preparing the questions: the national curriculum, phases of the model-based teaching and participants' background, which was physics. The National Curriculum in Turkey does not start from an observational base. Instead, it directly presents the phenomenon and continues with the explanation. On the other hand, the first phase of the model-based teaching was observation. Thus, the questions directed the pre-service teachers to do observations. In addition, the content of the questions enable the pre-service physics teachers to apply their knowledge of physics concepts to the lunar phenomena.

6.1 Model-Based Teaching Context

The whole implementation lasted 14 weeks. The pre-service physics teachers started the sequence with observations. In the tenth week, they were asked to divide into six equal groups to work on construction of their models as groups. Between the tenth and thirteenth weeks, some class time was allocated for the groups to compare their observations, look for patterns, discuss their views and construct a three-dimensional model. The groups discussed and decided how their model would be. In the last day, the groups showed their mental models of some moon-related events on their constructed models. During their oral presentations, the pre-service physics teachers were also asked to express the limitations of their models and the events that their models failed to represent.

6.2 Participants and Setting

The questions were answered by the same participants six months later to investigate their mental models of the moon-related events and reasoning. The participants were 36 pre-service teachers attending the physics teacher education program. Their ages ranged from 22 to 27; of whom, 15 (42%) were males. They graduated from eleven different universities and all had obtained a degree in physics. All of them studied reflection, scattering, gravitation, Kepler's laws, and angular momentum as parts of both high school and undergraduate physics curriculum. They also studied Moon phases and lunar eclipse as parts of elementary science curriculum.

6.3 Data Collection and Analysis

The participants' written responses to the delayed post-questionnaire were the data source. Participants had enough time to respond all the questions. The responses indicated that all the questions were assessed what they had been intended to assess. The fact that the researcher being the instructor of the participants might have kept them answering the questions in a careless way.

Their responses were analyzed inductively to identify themes and patterns that described their conceptual knowledge of the moon-related events. For the purpose of this research, if both the response and the reasoning behind it were consistent with the scientifically accepted perspective, it was coded as "correct mental model". On the other hand, if the reasoning behind the answer was non-scientific or incomplete, or there was no response, it was coded under the "rest of mental model" category. The data were re-examined a few times to detect any response that did not fit into one of the mental model categories.

In order to evaluate the significance of the results, the McNemar test was performed to compare delayed post-, post- and pre-test responses. For this reason, correct mental models were characterized as "correct" and the rest of the mental models were characterized as "incorrect".

7. RESULTS AND DISCUSSION

The pre-service physics teachers' mental models of some moon-related events before, just after and six months after the model-based teaching are presented in Table 1. Table 1 also illustrates the groups that the frequency values belong to. The three-dimensional models constructed by the groups at the end of the model-based sequence were explained in the previous research (Ogan-Bekiroglu, 2007) but it might be useful to mention here again. Six groups of participants constructed six models. The first model constructed by Group A was a three dimensional illustration of the figure of the Sun-Earth-Moon system that existed in most of the textbooks. Eight moons were shown in their phase positions in a circular orbit of the Earth and each moon was half illuminated by the Sun's rays. The model constructed by Group B consisted of a small basketball representing the Earth, a tennis ball representing the Moon and a torch representing the Sun. The members of the group were holding the objects. The model constructed by Group C consisted of a steady globe on a panel and a table tennis ball connected to the globe. The table tennis ball represented the Moon which could revolve around the Earth. Both the Earth and the Moon could rotate on their axes. One group member was holding a torch. The fourth model constructed by Group D was made up of a light source, a globe, and a table tennis ball representing the Moon. The globe was connected to the light source and the whole system was on a panel. The Moon was connected to the globe and could be revolved around it. Both the Moon and the Earth could rotate on their axes. The models constructed by Group E and Group F were very similar to each other by means of their functionality. They consisted of a light source, a globe and a table tennis ball representing the Moon. The Moon and the Earth could rotate on their axis from west to east. The Moon could revolve around the Earth from west to east and the Moon-Earth system could revolve around the Sun with the help of wheels. The only difference between two models was that Group F connected the Moon to the Earth with a small angle.

During the oral model presentations, the participants did not illustrate the following events on their models and used gestural modes of representation to explain them: the daily difference in moonrise time, Moon's elliptic orbit shape, the changes in the orbital velocity of the Moon, the changes in the colour of the Moon when it is near the horizon and when it is near the zenith, magnitude change in the Moon's appearance as it rises, a halo around the Moon and the reddish colour in the lunar eclipse. Moreover, only Group F could illustrate why the lunar eclipse occurred twice a year.

Results showed that while a majority of the pre-service physics teachers still had scientific understanding of the lunar events, some of them went back to their initial mental models. After six months, all of the participants could still explain why the Moon was seen in different phases, express that the Moon could be seen in daytime and locate the full Moon when the Sun was on the eastern horizon. Almost all of the pre-service physics teachers (97%) saved their correct mental models of the reason for moonrise and moonset, sequence of the phases of the Moon, and the reason for lunar eclipse after six months, but this 3% decrease in their correct responses was not found statistically significant for these phenomena. These results indicate that periodic observations, construction of models and using the constructed models to explain the phenomena might provide assimilation and revision of knowledge, and constant conceptual change.

The conceptual decay in the participants' mental models after six months occurred mostly in the following events: seeing the same face of the Moon (from 100% to 64%), daily time difference in moonrise (from 50% to 6%), the reason of tides (from 100% to 48%) and tidal effect (from 100% to 67%). The McNemar test revealed that the decreases in the frequency values of correct mental models from post- to delayed post-test were statistically significant for these events ($p=0.00$). Although all the participants could express scientifically the reason for always seeing one face of the Moon from the Earth just after the model-based teaching, 36% of them left their correct mental model six months later. Additionally, the reason for daily time difference in moonrise was the event that the pre-service physics teachers had most difficulty in generating the scientific model in the beginning. That is, none of the participants had explained the reason correctly before the model-based teaching. They had not even noticed the time difference. Even though half of the participants had correct mental model of daily time difference in moonrise just after the model-based teaching, only two of them could hold it after six months. None of the groups illustrated this phenomenon on their constructed models. Therefore, the reason for their difficulty might be not having enough experience to internalize the time difference and/or the rote learning they developed during the model-based teaching. Furthermore, the reason for the tides on the Earth was the event that the pre-service teachers had the biggest conceptual decay. Though all the participants

TABLE 1. Pre-service physics teachers' mental models before, just after and six months after the model-based teaching

Events	Before the model-based		Just after the model-based			Six months after the model-based		
	Category	Frequency (%) n=36	Category	Frequency (%) n=36	Groups	Category	Frequency (%) n=36	Groups
Moonrise and moonset	Correct	14	Correct	100	A+B+C+D+E+F	Correct	97	A+B+C+D+E+F
	Rest of	86	Rest of	0		Rest of	3	A
Same face of the Moon	Correct	6	Correct	100	A+B+C+D+E+F	Correct	64	A+B+C+D+E+F
	Rest of	94	Rest of	0		Rest of	36	A+B+C+E
Moonlight and phases of the Moon	Correct	97	Correct	100	A+B+C+D+E+F	Correct	100	A+B+C+D+E+F
	Rest of	3	Rest of	0		Rest of	0	
Sequence of the phases of the Moon	Correct	94	Correct	100	A+B+C+D+E+F	Correct	97	A+B+C+D+E+F
	Rest of	6	Rest of	0		Rest of	3	A
Seeing the daytime Moon	Correct	69	Correct	100	A+B+C+D+E+F	Correct	100	A+B+C+D+E+F
	Rest of	31	Rest of	0		Rest of	0	
Lunar eclipse	Correct	72	Correct	100	A+B+C+D+E+F	Correct	97	A+B+C+D+E+F
	Rest of	28	Rest of	0		Rest of	3	C
Day to day time difference in moonrise	Correct	0	Correct	50	D+E+F	Correct	6	B+F
	Rest of	100	Rest of	50	A+B+C	Rest of	94	A+B+C+D+E+F
Location of the full Moon	Correct	72	Correct	100	A+B+C+D+E+F	Correct	100	A+B+C+D+E+F
	Rest of	28	Rest of	0		Rest of	0	
Moon's orbit shape	Correct	36	Correct	83	B+C+D+E+F	Correct	81	B+C+D+E+F
	Rest of	64	Rest of	17	A	Rest of	19	A+C
Moon's orbit direction	Correct	55	Correct	100	A+B+C+D+E+F	Correct	86	A+B+C+D+E+F
	Rest of	45	Rest of	0		Rest of	14	A+B+C+D
Moon's rotation	Correct	64	Correct	100	A+B+C+D+E+F	Correct	81	A+B+C+D+E+F
	Rest of	36	Rest of	0		Rest of	19	A+B+C+D

(Continued)

TABLE 1. (Continued)

Events	Before the Model-Based		Just After the Model-Based			Six Months After the Model-Based		
	Category	Frequency (%) n=36	Category	Frequency (%) n=36	Groups	Category	Frequency (%) n=36	Groups
Changes in the orbital velocity of the Moon	Correct	33	Correct	67	B+D+E+F	Correct	61	B+D+E+F
	Rest of	67	Rest of	33	A+C	Rest of	39	A+B+C
Tides	Correct	19	Correct	100	A+B+C+D+E+F	Correct	48	A+B+C+D+E+F
	Rest of	81	Rest of	0		Rest of	52	A+B+C+D+E+F
Tidal effect	Correct	0	Correct	100	A+B+C+D+E+F	Correct	67	A+B+C+D+E+F
	Rest of	100	Rest of	0		Rest of	33	A+B+C+D
Changes in the colour of the Moon	Correct	28	Correct	67	A+D+E+F	Correct	61	A+D+E+F
	Rest of	72	Rest of	33	B+C	Rest of	39	A+B+C+D+E+F
Same full Moon in America	Correct	0	Correct	100	A+B+C+D+E+F	Correct	94	A+B+C+D+E+F
	Rest of	100	Rest of	0		Rest of	6	B+D
Magnitude change in the appearance of full Moon	Correct	3	Correct	83	A+B+D+E+F	Correct	72	A+B+D+E+F
	Rest of	97	Rest of	17	C	Rest of	28	A+B+C+D+E
Halo	Correct	0	Correct	67	A+D+E+F	Correct	58	A+D+E+F
	Rest of	100	Rest of	33	B+C	Rest of	42	A+B+C+D+E+F
The number of the lunar eclipses in a year	Correct	6	Correct	83	A+B+D+E+F	Correct	67	A+B+D+E+F
	Rest of	94	Rest of	17	C	Rest of	33	A+B+C+D
The colour of the lunar eclipse	Correct	0	Correct	67	A+D+E+F	Correct	55	A+D+E+F
	Rest of	100	Rest of	33	B+C	Rest of	45	A+B+C+D+E+F

Correct: Correct mental model, Rest of: Total of the other mental models in the categorization, i.e. complete flawed, incomplete correct, flawed, and incoherent mental models and no responses.

developed correct mental model i.e., both the Sun's and the Moon's gravitational forces cause tides on the Earth, just after the model-based implementation, 52% of them ignored the Sun's effect and went back to their initial reasoning after six months. Even though the most positive effect of the model-based teaching was found on the participants' mental models of the relationship between tidal effect on the Earth and the appearance of the Moon so that all the pre-service teachers' conceptions shifted to correct mental model in the previous research (Ogan-Bekiroglu, 2007), this conceptual change was durable for only 67% of them sixth months later. The McNemar test also showed a significant decrease from post- to delayed post-test in the frequency values of the correct mental model of the number of lunar eclipses in a year (from 83% to 67%, $p=0.00$). The ways learners represent the given information depend on the organization of their existing knowledge. Thus, the significant conceptual decays taking place six months later in the pre-service physics teachers' mental models of seeing the same face of the Moon, daily time difference in moonrise, tides, tidal effect, and the number of the lunar eclipses showed that some pre-service physics teachers could not develop durable scientific knowledge related to reflection of sunlight, gravitational force, and relative positions of the Sun, Earth and Moon. On the other hand, the McNemar test between pre-test and delayed post-test presented significant increases in the participants' correct mental models of these lunar events ($p \leq 0.05$). In other words, the number of pre-service teachers having correct mental models of seeing the same face of the Moon, daily time difference in moonrise, tides, tidal effect, and the number of the lunar eclipses was higher six months after the model-based teaching than the number before the model-based teaching.

The changes in the orbital velocity of the Moon, the changes in the colour of the Moon, a halo around the Moon and the reddish colour in the lunar eclipse were the events that the participants needed to have knowledge related to some concepts of mechanics and optics such as reflection and Kepler's laws to generate correct mental models. Table 1 illustrates that in the beginning of the model-based teaching, the number of pre-service physics teachers having correct mental models of these four events was very low. After the model-based teaching, this number increased and not all of them but 67% of the participants had correct mental models. This raise almost resisted to time. That is, most of these pre-service teachers kept their correct mental models after six months. Nevertheless, the McNemar test results were not significant. As mentioned above, the participants did not illustrate the changes in the orbital velocity of the Moon, the changes in the colour of the Moon, a halo around the Moon and the reddish colour in the lunar eclipse on their constructed models. However, the gestural modes of representation they used might help them revise their pieces of physics knowledge and enhance the durability of conceptual change.

According to Table 1, some participants (14%) left their correct mental model of the Moon's orbit direction around the Earth six months later ($p \leqslant 0.05$). However, there was an increase in the frequency value of correct responses from the beginning of the model-based teaching (55%) to six months after the model-based teaching

(86%) ($p=0.00$). Although, some pre-service teachers (19%) left their correct mental models of the Moon's rotation on its axis six months later ($p \leq 0.05$), the number of participants who responded consistently with the scientific model increased from pre-test to delayed post-test (from 64% to 81%, $p \leq 0.05$). Generating correct mental models in response to the Moon's orbit direction and its rotation requires some declarative knowledge. These results present that the model-based teaching had positive effects on the participants' skills of recalling theoretical facts.

Apart from one participant, the pre-service physics teachers saved their correct mental model of the Moon's elliptical orbit shape. Possibly because of the assumption used in solving many physics problems, 64% of the pre-service physics teachers had thought that the Moon revolved the Earth in a curricular orbit before the model-based implementation. Even though the participants could neither directly observe the elliptic orbit nor show it on their models, discussion of this event during the model construction might cause to repair their flawed mental model and permanent conceptual change.

Regarding the magnitude change in the appearance of the full Moon, 33% of the participants had showed linear reasoning and expressed that the full Moon looked larger than usual as it rose because it came closer to the Earth in its elliptical orbit before the model-based implementation. After six months, 72% of the pre-service teachers (the decrease from 83% to 72% from post- to delayed post-test was not significant) still had the correct mental model that this event was a psychological phenomenon and the full Moon actually was not larger than usual as it rose.

A direct proportion between the pre-service physics teachers' mental models and their three-dimensional models was appeared in the previous study (Ogan-Bekiroglu, 2007). That is, the closer their models were to the real situations, the more scientific conceptions they gained. In this study, the least conceptual decay occurred in the members of Group F's mental models, whose three-dimensional model was the closest to the real situation. On the other hand, the most conceptual decay took place in the members of Group A's mental models, whose model was a three dimensional illustration of the figure of the Sun-Earth-Moon system that existed in most of the textbooks and had many limitations. The images in the textbooks may effect students' conceptions (Pena and Quilez, 2001). Further study might be needed to explore this relationship.

8. CONCLUSION AND SUGGESTION

This study was conducted to examine the durability of scientific conceptions of lunar events that the pre-service physics teachers gained after the model-based teaching. In order to do that, the participants' mental models were determined six months later. Everyday experience as well as misleading illustrations can lead to false perceptions of astronomical events (Vosniadou, 1992). Dove (2002) suggests four strategies to prevent students from going back to their intuitive ideas. These strategies are observations, classroom models, questions requiring students to apply their knowledge and pictures of moons in impossible positions. Therefore, the model-based implementation was composed of daily Moon observations, and construction and use of three-dimensional models as groups.

The pre-service teachers in this study had neither done periodic observations nor constructed a model before. Some participants' understanding reverted to their initial positions and some of them lost pieces in their propositions six months after the model-based teaching. These results are consistent with the results of Tytler's (1998) study where some participants' conceptions were regressed. However, even though there was conceptual decay in the pre-service physics teachers' correct mental models of some lunar events, they had more scientific conceptions of the moon-related events than they had before the model-based teaching. Hence, it can be concluded from the overall results of this research that model-based teaching helps learners assimilate new knowledge, organize their existing knowledge and assign propositions to correct ontological categories so that durable conceptual change may occur.

This study suggests that model-based implementation can be promoted in teacher education programs to facilitate pre-service teachers' subject matter knowledge.

REFERENCES

Barnet, M. & Morran, J. (2002). Addressing children's alternative frameworks of the Moon's phases and eclipses. *International Journal of Science Education*, 24(8), *859–879.*

Baxter, J. (1991). A constructivist approach to astronomy in the national curriculum. *Physics Education*, 26, *38–45.*

Boulter, C., Buckley, B. & Walkington, H. (2001, April). Model-based teaching and learning during ecological inquiry. (*Paper presented at the Annual Meeting of the American Educational Research Association, Seattle, WA.) (ERIC Document Reproduction Service No. ED454048*)

Buckley, B. C. (2000). Interactive multimedia and model-based learning in biology. *International Journal of Science Education*, 22(9), *895–935.*

Callison, P. L. & Wright, E. L. (1993, April). The effect of teaching strategies using models on pre-service elementary teachers' conceptions about relationships. *(Paper presented at the Annual Meeting of the National Association for Research in Science Teaching, Atlanta, GA.) (ERIC Document Reproduction Service No. ED360171)*

Chi, M. T. H. & Roscoe, R. D. (2002). The process and challenges of conceptual change. (In M. Limon & L. Mason (Eds.),*Reconsidering Conceptual Change: Issues in Theory and Practice* (pp. 3–27). Dordrecht, The Netherlands: Kluwer Academic Publishers)

Dai, M. F. & Capie, W. (1990, April). Misconceptions about the moon held by pre-service teachers in Taiwan. *(Paper presented at the Annual Meeting of the National Association for Research in Science Teaching, Atlanta, GA.) (ERIC Document Reproduction Service No. ED325327)*

Dana, T. M., Campbell, L. M. & Lunetta, V. N. (1997). Theoretical bases for reform of science education. *The Elementary School Journal*, 97(4), *419–432.*

Dove, J. (2002). Does the man in the moon ever sleep? An analysis of student answers about simple astronomical events: a case study. *International Journal of Science Education*, 24(8), 823–834.

Georghiades, P. (2000). Beyond conceptual change learning in science education: Focusing on transfer, durability, and metacognition. *Educational Research*, 42, 119–139.

Harrison, A. G. & Treagust, D. F. (2000). Learning about atoms, molecules and chemical bonds: A case study of multiple-model use in grade 11 chemistry. *Science Education*, 84, 352–381.

Hermann, R. & Lewis, B. F. (2003). Moon misconceptions. *The Science Teacher*, 11, 51–55.

Ogan-Bekiroglu, F. (2007). Effects of model-based teaching on pre-service physics teachers' conceptions of the Moon, Moon phases, and other lunar phenomena. *International Journal of Science Education*, 29(5), *555–593*.

Ozdemir, O. F. (2004). *The Coexistence of Alternative and Scientific Conceptions in Physics*. Dissertation, The Ohio State University.

Parker, J. & Heywood, D. (1998). The earth and beyond: Developing Primary teachers' understanding of basic astronomical events. *International Journal of Science Education*, 20(5), *503–520*.

Pena, B. M. & Quilez, M. J. G. (2001). The importance of images in astronomy education *International Journal of Science Education*, 23(11), *1125–1135*.

Rider, S. (2002). Perceptions about Moon phases. *Science Scope*, 6, 48–51.

Samarapungavan, A., Vosniadou, S. & Brewer, W. F. (1996). Mental models of the earth, sun, and moon: Indian children's cosmologies. *Cognitive Development*, 11, *491–521*.

Stahly, L. L., Krockover, G. H. & Shepardson, D. P. (1999). Third grade students' ideas about the lunar phases. *Journal of Research in Science Teaching*, 36(2), *159–177*.

Suzuki, M. (2003). Conversations about the moon with prospective teachers in Japan. *Science Education*, 87, *892–910*.

Taylor, J. I. (1996). Lunar phases: Students construct fundamental knowledge of moon phases. *The Science Teacher*, 11, 39–41.

Treagust, D. F. & Smith, C. L. (1989). Secondary students' understanding of gravity and the motion of planets. *School Science and Mathematics*, 89, *380–391*.

Trumper, R. (2001). A cross-age study of junior high school students' conceptions of basic astronomy concepts. *International Journal of Science Education*, 23(11), *1111–1123*.

Trundle, K. C., Atwood, R. K. & Christopher, J. E. (2002). Pre-service elementary teachers' conceptions of moon phases before and after instruction. *Journal of Research in Science Teaching*, 38(7), 633–658.

Tytler, R. (1998). Children's conceptions of air pressure: Exploring the nature of conceptual change. *International Journal of Science Education*, 20, 929–958.

Vosniadou, S. (1992). Designing curricula for conceptual restructuring: Lessons from the study of knowledge acquisition in astronomy. *(ERIC Document Reproduction Service No. ED325327)*.

ANNEX

QUESTIONNAIRE

1. Does the moon rise and set? Where? Why?
2. Do we always see the same face of the moon? Why?
3. Why do we see the different phases of the moon?
4. Please draw the shapes of the moon phases in sequence.
5. Can we see the moon in the daytime?
6. How does a lunar eclipse occur?
7. Does the Moon rise everyday at the same time? Why?
8. Assume that the Sun was just rising on the eastern horizon. The Moon was full and it could be seen someplace in the sky. Where would the Moon be?
9. What is the shape of the orbit of the Moon during its revolution around the Earth?
10. When the Moon revolves around the Earth, what direction does it follow?
11. When the Moon rotates on its axis, what direction does it follow?

12. Does orbital velocity of the Moon change during its revolution around the Earth? Why?
13. What is the reason for tides on the Earth?
14. Please put in order the tidal effect on the Earth from biggest to smallest for these phases: new moon, full moon, first quarter and last quarter. Explain your answer.
15. The Moon appears yellow or orange, when it is near the horizon; and almost white, when it is near the zenith. Why?
16. If there were a full moon in Turkey today, would it be a full moon in America today? Why?
17. Why does the full moon appear larger when it is near the horizon than when it is near the zenith?
18. Sometimes we see a halo around the Moon. Why?
19. Lunar eclipse occurs when the earth gets between the Sun and the Moon, why does this phenomenon occur two times a year, instead of every month?
20. Why does the Moon have a reddish-copper glow during a total lunar eclipse?

RENÉ WESTRA, KERST BOERSMA, AREND J. WAARLO AND ELWIN SAVELSBERGH

28. LEARNING AND TEACHING ABOUT ECOSYSTEMS BASED ON SYSTEMS THINKING AND MODELLING IN AN AUTHENTIC PRACTICE

Abstract: This paper is a report on educational design research concerning learning and teaching contemporary ecology. To be able to understand ecosystem behaviour as derived from a complex and dynamic view, learning and teaching systems thinking and modelling skills is essential. To accomplish context-based ecology education, a cultural-historical approach was chosen, using three authentic social practices in which ecology is involved. A sequence of learning and teaching activities was thought out, elaborated and tested in classrooms. Throughout the field test the learning and teaching process was monitored in detail using various data sources.

The results show that the students acquired basic systems thinking; they were able to articulate similarities and differences between the levels of biological organization (individual, population, and ecosystem). In addition, they understood which factors are crucial in an ecosystem and how they work, in particular how they impact quantitatively on each other. Most students were able to explore the required computer models. However, for most of them it remained problematic to build models themselves

Keywords: Authentic practices, Complexity, Cultural historical approach, Developmental research, Dynamics, Ecosystem, Modelling, Systems thinking

1. INTRODUCTION

Our biological environment can be regarded as a complex adaptive system (Gell-Mann, 1995). Such a system behaves according to three key principles: order is emergent as opposed to predetermined, the system's history is irreversible, and the system's future is often unpredictable. These features result from the interaction of various 'building blocks' and processes at different levels of biological organization (individual, population, and ecosystem) (Holling, 1987). The dynamic behaviour of such a system proves hard to understand for secondary school students studying ecology (Barman et al., 1995; Magntorn & Helldén, 2003; Munson, 1994).

In the traditional approach to ecology teaching, dependencies between populations tend to be represented through 'food webs'. Although this format conveys the idea of a network, it does not contribute to students' insights into the dynamic

René Westra, Centre for Science and Mathematics Education, P.O. Box 80000, 3508 TA Utrecht, The Netherlands

R. Pintó and D. Couso (eds.), Contributions from Science Education Research, 361–374.

interdependencies of populations in the web (Hogan, 2000). Moreover, in many cases the food web representation tends to be transmitted, rather than constructed by the students themselves, which may lead to another misunderstanding: that the food web is a fact of nature, rather than a way of getting hold on the complexity of an ecosystem.

Several solutions have been proposed to prevent these misunderstandings. A recent study (De Ruiter et al., 2005) suggests to replace the metaphor of a *static* structure like the food web by the metaphor of the structures to be built in the game of Jenga (see Figure 1). This could contribute to visualizing the *dynamics* of a complex food web.

Each block in the game could be considered a potential keystone. It is hard to foresee which blocks will be essential for stability in a constantly changing Jenga structure. In a comparable way, the importance of extinct or imported populations for stability of the food web in an ecosystem can vary over time. In such a view, food webs are open and dynamic systems. This new idea could help in the management of ecosystems that are being changed by human influence, although the exact relation between the structure of a food web and the stability of an ecosystem appears to be very complicated (Neutel, 2001).

It has also been claimed that explicit systems thinking and modelling could improve students' understanding of ecosystems (Boersma, 1997; Hogan & Thomas, 2001; Schaefer, 1989; Zaraza, 1995). In systems thinking, an ecosystem is considered to be an open and complex adaptive whole, in which the parts (popula-

Figure 1. In the game of Jenga, each player successively takes away a block and places it on top of the tower, until the structure becomes unstable and crashes

tions or functional groups of populations, and all kinds of abiotic factors) are influencing themselves and each other in nonlinear ways, giving rise to dynamic patterns over time. This kind of thinking may be helpful in directing attention at particular features of the ecosystem like the distinction of hierarchical levels, feedback and temporal delay which cause dynamic, often cyclic but sometimes chaotic patterns.

Students should develop a competence to relate different levels of organization. In addition, students should come to regard each level of organization both as a conceptual unit, which can have properties of its own, and as an assembly of smaller interacting units at a lower level. This could lead to more coherence in their understanding (Verhoeff, 2003).

Concerning the dynamics of a system, computer models, in comparison to the Jenga game, show three advantages. They enable to follow *quantitative* aspects of processes in time, introduce various interacting factors, and study changes on the level of the individual, the population, and the ecosystem.

Research suggests that students learn more about systems behaviour by building or using dynamic (computer) models than by creating static depictions of systems relationships (Kurtz dos Santos & Ogborn, 1994; Louca et al., 2003).

In our view, modelling is an essential part of systems thinking. Ecosystem behaviour is far too complex to be explained and predicted by a single, unified theory. Models draw on a number of theories to help understand a specific problem. They never contain all the features of reality, but only those that are essential to the specific problem. Ecological models facilitate on particular factors, on a specific level of organization, depending on the goals of the model. Such models allow us to encompass our knowledge about the components that interact in a system, the way they interact, and how crucial these interactions are in light of the specific problem (Jørgensen & Bendoricchio, 2001). A model never contains all the features of reality; it only contains those features that are essential in the context of the problem to be solved. An ecological model could be compared to a geographical map, which in fact is a model. Different types of maps serve different purposes, i.e. they focus on different objects. They also differ in scale. In a similar way, an ecological model enables us to focus on particular objects, on a specific level of organization, depending on the goals of the model. For example, in a marine ecosystem the modeller concentrates on cod and its density (population level), because he wants to know what causes the imminent extermination of this species. He does not care about the weight of an individual cod, or about a complete survey of all species of fish in the area under study.

However, both systems thinking and computer modelling are demanding approaches for most students, and it is not self-evident that these approaches can be successfully taught in secondary education. It appears to be a complicated task for students to apply systems thinking to concrete biological instances (Verhoeff, 2003). Moreover, students do not fully distinguish the ideas and/or purposes underlying models, the content of the models, and the experimental data which support or refute the validity or usefulness of models (Grosslight et al., 1991;

Westra et al., 2002). They expect a model to represent the full richness of the real world (Hogan & Thomas, 2001).

Notwithstanding these problems, we stick to introducing systems thinking and modelling in secondary biology education. In our view, these competences are essential to ecological literacy. Perceptions of nature and management of ecosystems are strongly determined by the level of biological organization that a person has in mind. Therefore, seemingly irreconcilable positions in a debate may arise from the participants building their arguments on different levels of organization. In addition, many management measures are based on modelling to predict their impact.

So, our challenge will be to identify ecology-related problems that are simple and transparent enough for students to develop and test their own models, and yet sophisticated enough to 'understand nature', in the sense that students understand what is actually happening when changes in an ecosystem (in a number of cases caused by human intervention) take place.

The central research question in this study is:

How could upper secondary school students acquire an adequate understanding of an ecosystem, emphasizing its complexity and dynamics?

The research question entails the following sub-questions:

- What ecology-related practices seem appropriate for enabling students to grasp and value the role of modelling and systems thinking?
- What pedagogical approach seems helpful in acquiring the skills of modelling and systems thinking?
- How practicable and effective is the resulting learning and teaching strategy?

2. METHOD

In our study, a learning and teaching strategy (LT strategy) has been developed by means of a 'developmental research' or 'design research' approach. In developmental research, theory-driven, creative and practicable solutions to learning and teaching problems are designed in iterative consultation with experienced teachers. Researchers and teachers also co-operate in testing the developed LT activities in classroom settings (Lijnse, 1995). To ensure ecological fidelity, mussel breeders, forest rangers, as well as professional ecologists, were consulted in the design process.

The first version of the LT strategy underlying this series of lessons was field-tested in four 5VWO (A-level, 16–17 years old) classes in two different schools in December 2004 and March 2005. The students worked in dyads. A revised strategy will be tested again in 2006.

Throughout the field-test, the learning and teaching process has been monitored in detail, using video and audio recording, classroom observations, notes, sketches, computer models of the students, and interviews with the teachers and students.

Below, we will first present our theoretical framework. Then, we will describe and justify how we proceeded in designing the teaching sequence. Finally, we will report on the field test and come to conclusions.

3. THEORETICAL FRAMEWORK

In our view, based on the Vygotskyan cultural-historical approach (Blanck, 1990; Hedegaard, 2001), learning requires a practice that invites students to perform all kinds of activities in a social context. Students work together, talk, discuss, and reflect on their activities. According to this approach the teacher makes a 'double move' (Hedegaard, 2001): he steps down to the actual zone of development of the students, but also challenges them to move to their proximal zone.

Practitioners use knowledge in activities that are relevant in their practice. Not all students have an interest in the field of ecology as such. But every student is a member of society. And ecology matters in society, because man influences ecosystems, and is being influenced by ecosystems. We expect that, by starting from an authentic social practice in which ecology is involved, learning activities could become meaningful for students (Boersma, 2004; Bulte et al., 2004; Kattmann, 1977).

Our cultural-historical approach leads to reinterpreting the use of contexts in science education, i.e. a context is a social practice in which a number of *activities* are carried out to meet specific objectives (Van Oers, 1987; Van Oers, 1998). So, learning will not be meaningful without carrying out activities. These activities are essential to cognitive development in terms of changing students' prior knowledge and skills.

However, to be meaningful, students do not only need a 'broad motive' from the start to act and acquire knowledge and skills they need to answer a central problem. To keep the learning process going, they also need 'local motives' to find answers to partial problems which connect already existing knowledge and skills with the goals that have to be attained during their learning process (Lijnse & Klaassen, 2004).

Since knowledge and skills are often strongly situated, students have to adapt their cognitions when it is required to use them in another non-familiar social practice. This process of adaptation is called re-contextualisation (Van Oers, 1998). In this process students have to infer an abstraction of a concept as it is used in the social practice and to adapt (re-contextualize) it to be useful in the new practice.

4. DESIGN OF THE TEACHING SEQUENCE

The first issue was to identify a suitable pedagogical approach to introduce systems thinking and modelling. There have been several attempts to implement a context-concept approach in teaching and learning. It proved hard to tune the chosen contexts with the conceptual requirements of the curricula (Bennett & Holman, 2002). Recently, a new attempt has been made in the Netherlands to develop and implement a context-concept approach in the renewal of upper secondary biology education (Boersma et al., 2005). By relating the context-concept approach to the cultural-historical approach, we might have a solution for the problem mentioned above.

The second issue was to choose as a context a social practice in which ecological key concepts, like complexity and dynamics of an ecosystem and relations between individuals and populations, play an important role in activities. In a densely populated country like the Netherlands, there are many examples of human activities interfering with ecosystems. However, in many cases human control is so dominant that the dynamic behaviour of the system becomes rather predictable even without a model. To the students, such systems would not provide the required need to build a model. By contrast, the context of the mussel culture in an estuarine ecosystem (Easter Scheldt) seems promising because of its economics, the human impact and its manageable complexity from a student's perspective. In comparison to a 'natural' ecosystem, the complexity of the system is reduced by the breeders introducing the mussels as young animals on special locations, in desired quantities, and by the mussels being harvested when they are fully grown, which brings ecological factors like birth rate, density, and death rate under control.

Mussel breeders want to achieve an optimal and sustainable mussel culture. They have requested scientists from NIOO (Netherlands Institute of Ecology) to study ways of optimizing mussel culture in this dynamic ecosystem. In other words, what density of mussels on a bank results in a maximum yield of full-grown animals, without damaging the environment? In the practice of these scientists, working in the so-called MABENE-project funded by the EU, they carry out activities like studying the anatomy and physiology of the mussel, collecting data on biotic and abiotic factors that influence mussels, building apparatus to collect corresponding data, and modelling systems with mussels and their environment (Herman, 2004). So, the practice of the mussel breeders and the practice of the NIOO-scientists do exhibit a certain amount of overlap.

For use in a series of lessons in classrooms both practices have to be separated. In addition, they need an educational transformation, a.o. by identifying the essential activities. For example, to the scientists the relation between their knowledge and skills and the required activities is clear. To the students it is not. They know mussels as organisms, but not their anatomy and physiology, neither how they are part of the population and the ecosystem respectively. But they need to understand how an individual mussel (representing *the* mussel) influences, and is being influenced by his environment, that this individual mussel is part of a population with special emergent characteristics, and that this population is part of the ecosystem, again with emergent characteristics but of a different nature.

For re-contextualisation of the acquired concepts, an authentic practice of nature management (especially rabbits) in a water resource area, the PWN Dune Reserve in Northern Holland has been chosen. The students are confronted with a non-recovering rabbit population in the dune area after a VHS-virus epidemic, in a more complex ecosystem.

For computer modelling Powersim Constructor Lite, a graphic modelling tool, has been used.[1]

The series of lessons concludes with a test. The test items deal with an 'ecological practice' to test the acquired knowledge and skills of the students in context. This is a context of nature conservation dealing with the decision whether or not to shoot elephants in overpopulated areas in Southern Africa. Figure 2 presents the outline of the teaching module. Table 1 describes the sequence of learning activities.

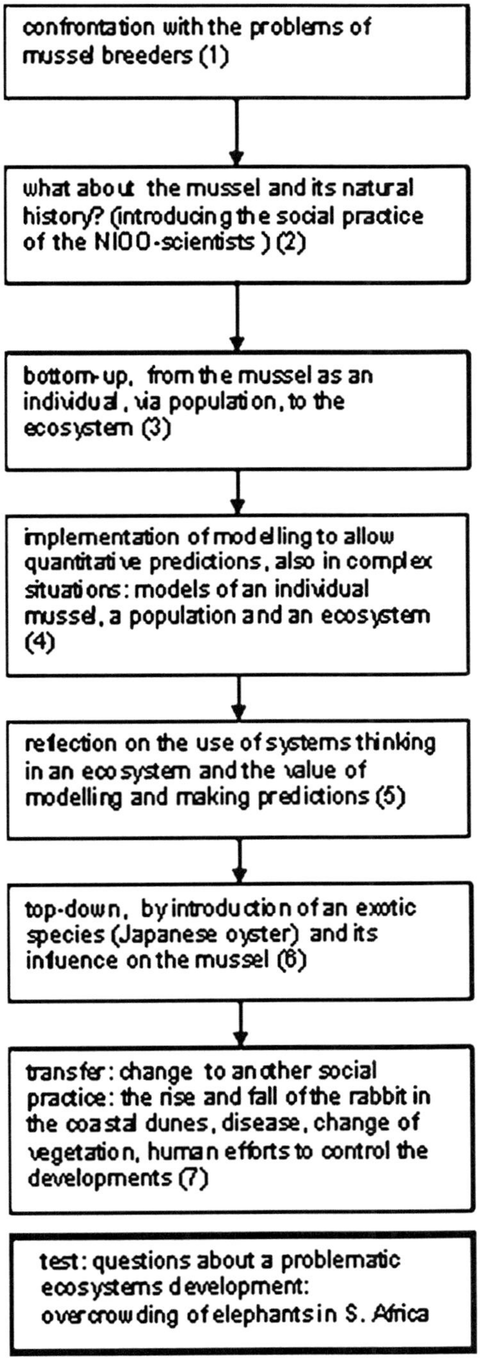

Figure 2. Topics of lessons

TABLE 1. The Sequence of Learning Activities

Step	Learning activity
1	Viewing a video
2	Posing the central problem by means of a group discussion
3	Investigating mussel anatomy, especially about feeding structure; Determining flesh-weight and dry-weight of a mussel; Depicting models on paper of: an individual mussel and its environment/the population of mussels and its environment/the ecosystem and its environment (bringing in systems thinking)
4	Modelling with Powersim modelling tool on a computer: a mussel/a population/an ecosystem
5	Reflection by comparing the models/group discussion
6	Exploring the effects on the population and on individual mussels of the introduction of a new species in the computer model
7	Re-contextualising, using acquired knowledge and skills when building a model in another ecology-related practice.

5. FIELD TEST

The collected data from different sources have been inspected and interrelated to find answers to the research questions.

Authentic citations or other results being used are coded with a number, marking the actual lesson (45–50 minutes) out of a series of nine. Besides, a symbol is added (see Table 2) to show the data source. For example, 3C means citations that were recorded in the third lesson.

The students could imagine the problems of the mussel breeders and the dune forest rangers. They also empathized with the role of ecological scientists. However, after the first part of the series of lessons, their enthusiasm declined. Their interest in the mussels was not enduring, and in retrospect, they would have preferred to move on to the rabbits' context earlier. However, when the reasons for the chosen practices and their sequence were explained, they accepted them as a logical choice.

They produced a long list of factors influencing a mussel. Most students were able to depict a simple model including the necessary factors and their interrelations (see Figure 3).

TABLE 2. References to Data Sources

Data sources	Video- or Audio-recorded citations	Interviews with students or teachers	Modelling activities	Answers on the questions in the final test	Text or drawings on worksheets
References	C	I	M	T	W

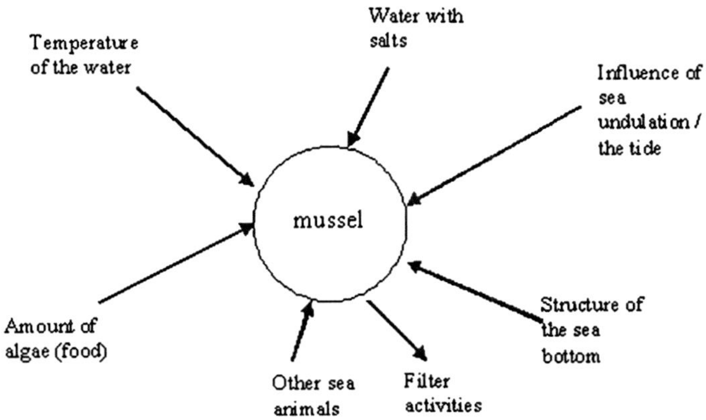

Figure 3. Factors influencing a mussel according to Corinne & Nienke (2W)

Teacher: *"What actors did you enter into your scheme with the mussel?"*
Marjeleine: *"Sea currents."*
T: *"Do they influence the mussel?"*
Marjeleine: *"Yes."*
T: *"What else?"*
Jessie: *"A sufficient amount of algae."*
T: *"Anymore?"*
Edwin: *"The presence of other marine animals."*
T: *"Why?"*
Edwin: *"Well, like crabs and starfish, they eat mussels."*
Lisanne: *"Food, temperature of the water, they cannot live in cold water, I think. Also depth."*
T: *"Why depth, does a mussel need light?"*
Jop: *"No, but phytoplankton does."*
Edwin: *"The soil, the currents."*
Jop: *"Currents do influence the soil, I think? And I think salt-content, especially near the coast."*
Lisa: *"The concentration of mussels." (2C)*

They were well aware of the similarities and differences between the level of organization of the individual, the population, and the ecosystem. For example, they realized that it is not an individual mussel, but the population of mussels that influences the concentration of plankton (food).

"So, when the density is raised, the mussel competes, and the individual does not grow so well, he is so meagre that he cannot be harvested." (4W)

Or that foraging birds like the Common Goldeneye or the Eider Duck exert influence on population level (by decreasing the number of mussels) as well as on individual level (by increasing the weight of the mussels that are not consumed).

They could tell that when birds forage on mussels, this has a negative effect on the density (population level), but a positive effect on dry-weight of the remaining mussels (individual level).

But not everybody agrees on this being positive!

Fabian: *"The more mussels are eaten by the birds, the better for the remaining individuals. Their dry weight increases."*

Josine: *"Oh, but what is the benefit of that for the eaten ones?" (7C)*

The students were able to build a simple computer model of the growth of an individual mussel based on the daily increase of dry weight of the animal with the help of a worksheet. However, most of them had severe difficulties in formalizing the relations, when they had to build a more complex model themselves. They did not know how to describe the sort of relation between two factors (like multiplication or addition) or how to quantify a relation by using some constant. They had also problems with validating the outcomes of their computer models when a population of mussels was involved and some variables were manipulated (see also Table 3). Many of them were engaged, but got disconnected from the biological reality of the mussels and also lost much of their motivation. When they were invited to explore a complete model, they showed understanding of what was happening and linked their biological knowledge to the model. They discovered that such a complete model could expand their biological understanding. The teachers agreed with this.

Hamid: *"I understand a model when it is explained, but I am not able to build it myself."*

Josine: *"I work so hard modelling that I tend to lose contact with real world. But after all, I think modelling helps me understanding complex situations in nature."*

Teacher: *"I think it a surplus value that complexity and dynamics become clear in these lessons." (I)*

In accordance with the results above, in the final test, most students proved to be able to discriminate between the three levels (individual, population, and

TABLE 3. The Performance of Dyads in Carrying out Various Types of Computer Modelling Activities

Type of activity	% of dyads, capable of performing activity	
	School 1	School 2
Building a model with the help of a work sheet	√	√
Independently sketching a model	x	x
Introducing the correct relations into a model	x	x
Quantifying relations in a model	x	x
Validating a model, by comparing with real world	√	x
Exploring an existing model	√	√
Extending an existing model	x	x

√ more than 50% capable, x = less than 50% capable

TABLE 4. Percentages of Students' Correct Matches of Scientists and their Level on Focus (T)

Scientist	Nr 1 (individual)	Nr 2 (ecosystem)	Nr 3 (population)
School 1 (n = 22)	100	48	52
School 2 (n = 34)	95	74	63

ecosystem) in the ecology-related and practice-oriented text about how to deal with overpopulation of elephants (see Table 4).

They were also able to draw models using information from the texts, but most of them did not discriminate well enough between the character (stock, constant or variable, see Figure 4 and Table 5) of the factors that they had used in their Powersim computer models.

A population of elephants is recognized as a stock, but a population of trees is not. That sunshine is a constant factor, not influenced by other factors that have to be used in the model, seems to be understood, but that the factors poaching & hunting and anti-conception, which can be only influenced from outside the model (by human interference) are therefore also constants, seems not be understood.

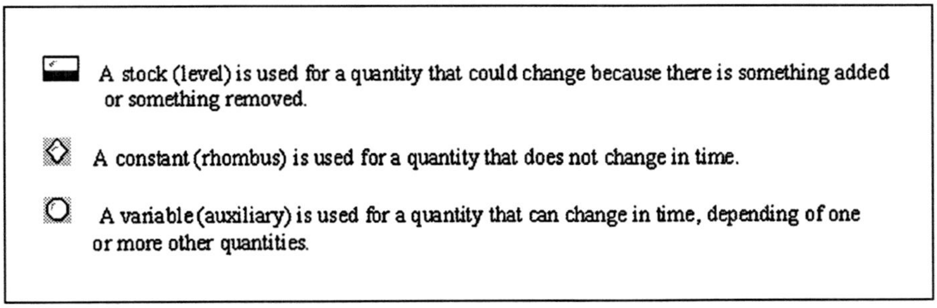

Figure 4. The characters of factors used in Powersim models

TABLE 5. Percentages of Students' Correct Matches with the Types of Factor in Powersim

School	Elephants			Trees			Poaching & Hunting			Anti-Conception			Sunshine		
	S*	V	C	S*	V	C	S	V	C*	S	V	C*	S	V	C*
1	62	38	0	21	58	21	5	95	0	14	43	43	5	10	86
2	81	19	0	27	70	3	0	52	48	0	47	53	7	30	63

S = stock; V = variable; C = constant, where the ones with * are the correct ones (T)

6. CONCLUSIONS

In the former section, the third sub-question has been answered.

As to the first sub question, our choice of the ecological practices turned out to offer good opportunities to introduce systems thinking as well as modeling. The mussel context appeared to be simple enough for students to build initial models, yet complex enough to make the models useful.

The rabbits' context, although far more complex, was also feasible; students were able to apply systems thinking skills acquired in the first context to the second context.

However, embedding ecological problems in a social practice by itself does not necessarily provide sufficient motivation to the students. In addition, the students need to understand why just these practices were selected to deal with the issue. Students' involvement in the issue may fade away at various points during the lessons. These critical moments require explicit attention and reflection in order to keep the students motivated.

As to the second sub-question, we found that students were able to explore models and to derive new biological implications from their models. Students were able to express ideas about effects on individual, population, or ecosystem level. They seemed to be aware of (quantitative) effects from, for example, population level on individual level. However, when it came to designing and implementing their own models, students still experienced severe difficulties in formalizing and quantifying relations, and in evaluating model outcome by applying their biological knowledge. Part of the difficulty may arise from the students being focused at creating a running model; once this goal has been established, the students are satisfied. However, a deeper obstacle may be that the students do not perceive the biological world in terms of numbers; for instance, they do not have any expectation on a plausible range for the dry weight (that is the biomass) of a mussel. It seems to be necessary to: 1. spend time on teaching formalizing and quantifying, for example by group discussion of different solutions and deciding what is the most logical one; 2. stimulate the students to compare their result with 'real world' data'; 3. clarifying the concept of dry weight by let them measure the dry weight of a mussel.

The problems in the correct use of the Powersim symbols for specific factors, found in the test, can be due to problems in understanding the character of a factor and the relation of this factor with others. But another possibility is that the students still have problems with the syntax of the modeling tool.

Further investigations with a revised series of lessons have to be done, with special attention to modeling, to find a teaching and learning strategy which leads to a better development of modeling skills, and in the end to a more distinguished insight into complexity and dynamics of an ecosystem.

NOTE

[1] This tool is functionally equivalent to the more well-known STELLA-software.

REFERENCES

Barman, C., Griffiths, A. K. & Okebukola, P. A. O. (1995). High school students' concepts regarding food chains and food webs: A multinational study. *International Journal of Science Education*, 17, 775–782.

Bennett, J. & Holman, J. (2002). Context-based approaches to the teaching of chemistry: What are they and what are their effects? (In J. K. Gilbert, R. Justi, D. F. Treagust & J. H. Van Driel (Eds.), *Chemical Education: Towards Research-Based Practice* (pp. 165–184). Dordrecht: Kluwer Academic publishers.)

Blanck, G. (1990). Vygotsky: The man and his cause. (In L. C. Moll (Ed.), *Vygotsky and Education* (pp. 31–58). Cambridge: Cambridge University Press.)

Boersma, K. Th. (1997). *Systeemdenken en zelfsturing in het biologieonderwijs*. Utrecht: University (inaugural lecture).

Boersma, K. Th. (2004, June). *Sleutelbegrippen en handelingspraktijken in biologieonderwijs van 4 tot 18*. (Paper presented at the ORD 9, Utrecht)

Boersma, K. Th., van Graft, M., Harteveld, A., de Hullu, E., van den Oever, L. & van der Zande, P. A. M. (2005). *Basisdocument van de Commissie Vernieuwing Biologieonderwijs over ontwikkeling en invoering van samenhangend biologieonderwijs*. Utrecht: CVBO.

Bulte, A. M. W., Westbroek, H. B., Van Rens, L. & Pilot, A. (2004). Involving students in meaningful chemistry education by adapting authentic practices. (In B. Ralle & I. Eilks (Eds.), *Quality in Practice-Oriented Research in Science Education* (pp. 105–116), Aachen: Shaker)

De Ruiter, P. C., Wolters, V., Moore, J. C. & Winemiller, K. O. (2005). Food web ecology: Playing Jenga and beyond. *Science*, 309, 68–70.

Gell-Mann, M. (1995). *The Quark and the Jaguar. [Adventures in the Simple and the Complex]*. London: Little, Brown and Company.

Grosslight, L., Unger, C., Jay, E. & Smith, C. (1991). Understanding models and their use in science: Conceptions of middle and high school students and experts. *Journal of Research in Science Teaching*, 28(9), 799–822.

Hedegaard, M. (2001). *Learning and Child Development*. Aarhus: Aarhus Universitetsforlag.

Herman, P. M. J. (2004). MABENE-veldcampagne in de Oosterschelde. Retrieved from http://www.nioo.knaw.nl/news/pdf/mabene2004.pdf

Hogan, K. (2000). Assessing students' systems reasoning in ecology. *Journal of Biological Education*, 35(1), 22–28.

Hogan, K. & Thomas, D. (2001). Cognitive comparisons of students' systems modeling in Ecology. *Journal of Science Education and Technology*, 10(4), 319–345.

Holling, C. S. (1987). Simplifying the complex; the paradigms of ecological function and structure. *European Journal of Operational Research*, 30, 139–146.

Jørgensen, S. E. & Bendoricchio, G. (2001). *Fundamentals of Ecological Modelling*. Amsterdam: Elsevier.

Kattmann, U. (1977). *Bezugspunkt Mensch [Grundlegung einer humanzentrierten Strukturierung des Biologieunterrichts]*. Köln: Aulis Verlag Deubner & Co KG.

Kurtz dos Santos, A. C. & Ogborn, J. (1994). Sixth form students' ability to engage in computational modeling. *Journal of Computer Assisted Learning*, 10, 182–200.

Lijnse, P. L. (1995). 'Developmental research' as a way to an empirically based 'didactical structure' of science. *Science Education*, 79(2), 189–199.

Lijnse, P. L. & Klaassen, C. J. W. M. (2004). Didactical structures as an outcome of research on teaching-learning sequences? *International Journal of Science Education*, 26(5), 537–554.

Louca, L., Druin, A., Hammer, D. & Dreher, D. (2003). Students' collaborative use of computer-based programming tools in science: A descriptive study. (In B. Wassen, S. Ludvigsen & U. Hoppe (Eds.), *Designing for Change* (pp. 109–118). Dordrecht: Kluwer Academic publishers.)

Magntorn, O. & Helldén, G. (2003, August). *The Development of Student Teachers'* Understanding *of Ecosystems During a Course-Their Views*. (Paper presented at the ESERA, Noordwijkerhout)

Munson, B. H. (1994). Ecological misconceptions. *Journal of Environmental Education*, 25(4), 30–34.

Neutel, A. M. (2001). *Stability of Complex Food Webs [Pyramids of Biomass, Interaction of Strengths and the Weight of Trophic Loops]*. Utrecht (dissertation).

Schaefer, G. (1989). *Systems Thinking in Biology Education* (vol. 33). Paris: Unesco.

Van Oers, B. (1987). *Activiteit en begrip. Proeve van een handelingspsychologische didactiek.* Amsterdam: VU-uitgeverij (dissertation).

Van Oers, B. (1998). From context to contextualizing. *Learning and Instruction*, 8(6), 473–488.

Van Weelie, D. (2001). Contextualizing biodiversity. (In O. de Jong, E. R. Savelsbergh & A. Alblas (Eds.), *Teaching for Scientific Literacy. Context, Competency, and Curriculum* (pp. 99–116). Utrecht: CDß-Press.)

Verhoeff, R. P. (2003). *Towards Systems Thinking in Cell Biology Education.* Utrecht: CDß-Press (dissertation).

Westra, R. H. V., Savelsbergh, E. R., Kortland, K., Prins, G. T. & Mooldijk, A. H. (2002). Leren door zelf modelleren: Constructief en uitdagend onderwijs. *NVOX*, 27(7), 331–334.

Zaraza, R. (1995). Systems thinking in the classroom. *Curriculum Technology Quarterly*. Retrieved from http://www.ascd.org/publications/ctq/1995fall/zaraza.html

PART 8

DISCOURSE AND ARGUMENTATION IN SCIENCE EDUCATION

CLAUDIA VON AUFSCHNAITER, SIBEL ERDURAN, JONATHAN OSBORNE AND SHIRLEY SIMON

29. ARGUMENTATION AND THE LEARNING OF SCIENCE

Abstract: Within the last years a consensus about the importance of argumentation in school science has developed. Students should not only be able to follow and judge scientific debates in public, it is also assumed that argumentation contributes directly to science learning. However, detailed studies on the interrelationship between argumentation and the development of science knowledge are rare in research on students' learning. In the study reported in this paper, Junior High School students' processes of argumentation and their cognitive development occurring in science lessons based on argument were investigated. Using video and audio documents of small group and classroom discussions, students' performance of argumentation was analysed using a schema based on the work of Toulmin (1958). In parallel, students' development and usage of scientific knowledge was investigated drawing on a schema for determining the content and area of abstraction of students' meaning making. Results show that when engaging in argumentation students draw on their prior experiences and knowledge. Activities based on argumentation enabled students to consolidate and elaborate their existing knowledge but did mainly not result in new (conceptual) understanding. However, students were able to develop high level arguments with relatively little knowledge and vice verca.

Keywords: Argument, Argumentation, Conceptual change, Conceptual development, Learning about science, Learning of science, Scientific knowledge

1. INTRODUCTION

Within the last years, an increasing number of research projects have focussed on argumentation in school science teaching (e.g., Osborne et al., 2004). The theoretical underpinnings of these projects refer to three different arguments for promoting argumentation in science education which briefly summarized are:

(1) Scientists engage in argumentation to develop and improve science (e.g., Lawson, 2003).

(2) The public has to use argumentation to engage in scientific debates (e.g., Simon et al., 2003).

(3) Students' learning of science requires argumentation (e.g., Osborne et al., 2004).

All three rationales emphasize that the coordination of argumentation and scientific knowledge plays an important role but it is especially the third argument which stresses that argumentation can also serve as a heuristic to develop an understanding

Claudia von Aufschnaiter, Leibniz University of Hannover, Physics Education, Bismarckstrasse 2, 30173 Hannover – Germany

R. Pintó and D. Couso (eds.), Contributions from Science Education Research, 377–388.

of scientific concepts. Students not only have to learn to develop valid arguments but will also learn science while arguing. Whereas several researchers share the idea that both argument and content specific knowledge play an important role in scientific communities and for the public understanding of science, their interrelationship for science teaching and learning has not yet been explored in great detail. That is, how much scientific knowledge is needed to perform a valid argument? Furthermore, how does the performance of argumentation influence conceptual understanding? And, conversely, how does the lack of content knowledge limit performance on argumentation? Overall objectives of this research, therefore, were (see also v. V. Aufschnaiter et al., 2008):

a) How do students incorporate their content specific knowledge in lessons based on argumentation?
b) How do students develop and improve their science understanding in these lessons?
c) How are quantity and quality of students' arguments influenced by students' content specific knowledge?

For the research reported in this paper, mainly the first two research questions are addressed whereas the results to question c) are only briefly summarized. These questions were explored through a set of case studies of video-documented lessons on science and socio-scientific issues for which students were asked to engage in argumentation. By means of an analysis of the transcripts both the quantity and quality of students' arguments and students' cognitive development were considered and their interrelationship examined.

2. THEORETICAL FRAMEWORK

2.1 Argument in Science Teaching

Research with a focus on classroom discourse during the teaching and learning of science emerged during the 90s (e.g., Driver et al., 2000; Lemke, 1990; Martins et al., 2001; Newton et al., 1999). Within this area, research on argumentation has concentrated on the process of argument rather than on the specific content and validity of an argument. Mainly, researchers have focussed on argumentation in scientific communities, either in classrooms or amongst scientists themselves. Using video, or audio, and by transcribing the dialogues, teachers' and students' arguments have been investigated both in terms of their quality and quantity, often using a Toulmin-like schema to address the nature of students' arguments (Toulmin, 1958). Results indicate that teaching of argument can increase the quality of students' arguments and the frequency with which students use arguments (e.g., Kelly et al., 1998; Osborne et al., 2004; Zohar & Nemet, 2002). In addition, some investigations have also focussed on students' understanding of science. Basically, these studies use either a descriptive analysis of the content that students deploy during argumentation or a pre-post-test design in order to examine the increase in students' knowledge (e.g., Jiménez-Aleixandre et al., 2000; Jiménez-Aleixandre & Pereiro-Munhoz, 2002; Leach, 1999; Mason, 1996; Zohar & Nemet, 2002). Results in these approaches indicate that both students' conceptual understanding and their argumentation improve. Nevertheless, as valuable as this work is, it should be noted that a descriptive analysis of students' knowledge usually does not give any hint about the general interrelationship between argument and cognition, and a pre-post-test design does not answer in detail *how* students incorporate and develop conceptual understanding while arguing.

In summary, this research provides a body of evidence that the teaching of argument can increase students' engagement in science, their ability to perform argument, and their understanding of science. However, research also indicates that a lack of scientific knowledge may result in poor performance of argumentation (e.g., Hogan & Maglienti, 2001; Koslowski, 1996). This finding would suggest that we need a better understanding of how a lack of specific knowledge limits engaging in meaningful argumentation and how argumentation enhances conceptual understanding.

2.2 Development of Scientific Knowledge

For the last three decades a large body of research has concentrated on students' development of scientific knowledge (e.g., Duit & Treagust, 1998, 2003). Basically, these studies draw on a constructivist epistemology and aim to determine students' conceptual change as well as to identify teaching strategies that promote such a change. However, although theoretical frameworks of these projects often address students' learning processes, the majority of the empirical procedures investigate the products of students' learning rather than the processes. Using pre-post-test designs with interviews or questionnaires, students' conceptions prior to instruction and afterwards are analysed. The main criteria as to whether or not students have increased their scientific knowledge is how near, or how far, students' conceptions are from those of scientists. However, it can be argued that increasing competence is not only a result of more appropriate knowledge but can also be recognised in students' ability to solve problems more quickly or to reach a higher level in their mental activities.

Examining the relationship between research on argumentation and research on students' learning of science, it is evident that the former has developed procedures and methods to investigate processes of classroom discourse but has mainly not investigated how processes of argument interrelate with the development of scientific understanding. In contrast, research on science learning has developed methods to investigate individual understanding but its interest is usually focussed on the start and end points of learning instead of the processes of gaining new knowledge. The combination of both procedures in this study aims to shed new light on the interrelation between processes of argumentation and (situated) scientific understanding.

In order to obtain more insight into students' development of scientific understanding we have used a theoretical framework and methodology that has been developed by a group of researchers at the Universities of Bremen and Hannover (Germany). From several research projects aiming at a general description of the development of physics competencies C. von Aufschnaiter and S. von Aufschnaiter (2003) have concluded that at least three dimensions are needed in order to describe students' situated understanding: content, level of abstraction, and time (see also e.g., V. Aufschnaiter, 2001; V. Aufschnaiter & Welzel, 1999). The first two dimensions were used within this study and are, therefore, briefly described below.

2.2.1 Content Although the description of scientific appropriateness of students' (situated) understanding has predominated in research in science education (e.g., Duit, 2006), we would argue that it is the ability of the student to integrate separate aspects of such knowledge which is also an important cognitive development. In addition to the integration of aspects of knowledge students should also be able to differentiate which aspects are important within a specific situation and which are not. Students need to be able to discern those aspects which are important in order to describe and (later on) conceptualise specific phenomena (see also Marton & Booth, 1997). Even if neither all relevant elements of content are recognised nor integrated in a complete (correct) physics description, those that are enable students to move towards an improved understanding of physics.

2.2.2 Abstraction For the study reported in this paper we refer to a model which has been shown to be useful in describing students' development of physics knowledge (for details see V. Aufschnaiter & V. Aufschnaiter, 2003). The distinction between this model and more general Piagetian ones is that content knowledge is seen as a specific feature of higher level reasoning. At its core, the model distinguishes four areas of abstraction (each containing two to three levels), of which only the first two are frequently seen in high school students' learning:

Area I Students grapple and (mentally) deal with *concrete* situations and objects, that is, they consider single phenomena, describe observations and experiences, link physics expressions to particular objects/phenomena but not to physics concepts or theories.

Area II Students develop a more *general* understanding within which they summarize classes of phenomena and objects (establish a conceptual understanding). Furthermore, students combine different general properties into a rule-based understanding. However, the relationship between the properties is seen as *static*, that is, students cannot mentally create functional variation besides connecting values (e.g., small x/small y versus big x/big y).

The dimensions presented above, content and level of abstraction can be used to describe and investigate students' situated development of knowledge. Furthermore, comparing how students develop their knowledge at different points of instruction can result in a description of the learning that has occurred.

3. DATA AND METHODS

Research was conducted in two phases. The first phase took place from September 1999 till September 2000. Here, material for use in argument-based lessons was developed and 12 teachers were trained on skills of promoting argumentation in science lessons (e.g., Erduran et al., 2004; Osborne et al., 2004; Simon et al., 2003). The teachers involved also incorporated a series of nine argument-based lessons in grade 8 (age 12–13) classes of London Junior High Schools. Whereas the first and the final lesson were based on a socio-scientific issue the lessons in-between had a science focus. In the second year (September 2000–2001), a subset of six teachers who had made progress in their ability to promote argumentation were chosen to repeat their teaching on argument. The additional training was reduced and the teachers were provided with in situ feedback about their teaching. Again, nine lessons were taught. In each class, two groups mainly consisting of four students were documented with video and audio for two science and two socio-scientific lessons (for examples see Table 1). Transcripts were made for all groups and in all lessons recorded.

Data were taken from the second phase of the programme in order to use those lessons which were taught by teachers who had already been trained on how to promote argumentation in the classroom. In order to explore the nature of the interrelationship of argumentation and cognitive development in detail, we focused on four of the teachers each with two scientific and one socio-scientific lesson (in total 24 student videos plus transcripts).

Data analysis was performed in two steps. In attempting to assess the nature of argumentation the focus of our analysis has been on the process of arguing –

TABLE 1. Examples of Teaching Material Used to Promote Argumentation. Statements Include Scientific Justifications as Well as Typical Student Misconceptions (More Examples in Osborne et al., 2003; V. Aufschnaiter et al., 2008)

Content	**Description**	**Structure of argument**
Funding a zoo (socio-scientific)	Students are asked to generate arguments for and against the funding of a new zoo in their community. Their arguments will determine if the zoo gets funded by a fictitious international funding agency.	Arguments for and against an issue
Phases of the Moon (scientific)	Students are asked to use the following four different statements to decide which one explains best why we have at least four different shapes (phases) of the Moon: (a) The Moon gives out light, spins around, and when the light side faces the other way it looks dark. (b) The Moon shrinks and then gets bigger during each month. (c) The rest of the Moon is blocked out by a cloud. (d) You can't always see the part that is lit by the Sun, so when we look at it we can only see the part that is lit by the Sun's rays (e) The Moon moves in and out of the Earth's shadow, so the light doesn't always get to the Sun.	Constructing an argument

argumentation – rather than the content of an argument itself. To develop a framework to measure the quality of argumentation, we have made two major distinctions. The first is to determine whether an argument contains any reasons and grounds i.e. data, warrants or backing to substantiate its claim, as transcending mere opinion and developing rational thought is reliant on the ability to justify and defend one's beliefs with evidence. In addition, teachers need to be able to identify such discourse moves and expose their limitations – the lack of justification – to their students. Hence, our second distinction is between second level arguments which are accompanied by grounds containing data or warrants and arguments of the third level consisting of claims, data, warrants *and* rebuttals. Episodes with rebuttals are, however, of better quality than those without. As Kuhn (1991, p. 145) argues the ability to use rebuttals is 'the most complex skill' as an individual must 'integrate an original and alternative theory, arguing that the original theory is more correct.' Our schema differentiates three levels of argument with rebuttals: arguments with weak or incomplete rebuttals; arguments with clear rebuttals; and arguments with multiple rebuttals. These we see as a hierarchy of increasing quality. This analysis has led us to define quality in terms of a set of five levels of argumentation (e.g., Erduran et al., 2004; Osborne et al., 2004).

For analysing students' development of knowledge both students' verbal and nonverbal activities were considered. However, we basically refer in this paper to those parts in which students verbally interacted as otherwise a reconstruction of their situated understanding becomes even more difficult. In order to analyse the development of students' understanding of science during the lessons based on argument we try to hold a second-order perspective on students' ideas:

> *[...] in the sciences, as well as in daily life, statements are made about world, about phenomena, about situations. These statements are made from what we call a first-order perspective. The ways of experiencing the world, the phenomena, the situations, are usually taken for granted by the experiencer; they do not see them, they are not aware of them. In phenomenography, where a second-order perspective is taken, it is these underlying ways of experiencing the world, phenomena, and situations that are made the object or research.*
> *(Marton & Booth, 1997, p. 118)*

For the analysis of students' knowledge the dimensions of content and abstraction (e.g., V. Aufschnaiter & V. Aufschnaiter, 2003), that is, to which content elements students referred within their argumentation and which areas of abstraction they reached (concrete, area I, v abstract (conceptual) knowledge, area II). Furthermore, content and abstraction were also used to investigate the complexity of the tasks to argue about. In cases where we could not agree on one area of abstraction, two possible areas are given.

In this work, emphasis was given to those parts in which students' constructed an argument. Every student's contribution was investigated both for its content and level of abstraction. In addition, for each argument a level was ascribed. The outcome was a description of the argument itself based on the Toulmin analysis with its assigned levels as well as a content specific description of each individual student's contribution to the argument. Furthermore, tracing students' cognitive development during the whole lesson allowed us to analyse how students changed or at least improved their understanding during, or as a result of, argumentation.

4. RESULTS

Comparing students' performance and frequency of argumentation at the beginning of the year (all six teachers of the sample included) to the lessons at the end of the year an increase in the quality of the arguments can be noticed (Osborne et al., 2004). Thus, we conclude that the teaching of argumentation had an impact on students' use of dialogic discourse (similar results are given in Zohar & Nemet, 2002). In our investigation of students' scientific understanding we analysed what students initially employed when being exposed to the teaching material (see examples in Table 1) and how students' understanding develops throughout the lessons on a specific topic. Finally, we analysed what kind of knowledge was incorporated in high and low level arguments.

4.1 Students' Situated Sense Making

When presented with a task, students often immediately constructed an understanding of the given situation. In such situations, whether or not they were able to engage successfully with the tasks was explicable by their previous experiences with the task's content. Those tasks with familiar content were much easier to engage with as the following two examples demonstrate (for the task see Table 1).

Example 1a: group 1

S3 *I think it's D.*

S4 *Why do you think it's D?*

S1 *I think it's D or E.*

S3 *What it is, the Moon, the Moon orbits the Earth. [area II] This is what it's like, yeah? [area II]*

S1 *The Moon orbits the Earth. [area II] When the Earth gets in front of the Sun... [area I or II]*

S3 *So this is the Sun, the big thing (draws). [area I] Right? That's the Earth (draws) [area I]*

S1 *The Moon moves around the Earth. [area I]*

S3 *And that's the Moon yeah? (points to his sketch about the positions of the Sun, Moon and Earth) [area I]*

S4 *No, the Sun, yeah . . . [area I]*

S3 *Yeah, but listen that's the Sun, that's the Earth, the Moon goes like that, and the Earth goes like that (points to sketch), so now you've got that, and that, going on at the same time. [area I] And what it does, look, as it goes there, you can only see the back of it and it only shows like parts of it. [area I] The bit you are going to see is just that bit. [area I]*

S3 *As it moves, it's a round shape... [area I] If it's there (points to sketch) you can see that bit. [area I]*

S4 *What happens when you can see the whole thing? [area I]*

S1 *The whole thing, it would be about there wouldn't it? [area I] It would be there, the whole thing. [area I]*

S3 *INDISTINCT REMARK*

S1 *When it's behind here (points to sketch), that's when I think it might be E. [area I] That's when I think it might be . . . [area I]*

S4 *I think it's E. [–]*

S3 *I think it's D or E. It might be E, but I think it's D. [–]*

Example 1b: group 2

S2 *Miss, which one of these questions says that the Moon spins around, the Earth goes around and . . . [area II]*

T *None of those actually say that. [area II]*

S2 *No.*

T *So you have to decide which one of these is true [supports argument] Now, what you know is true.... you need to decide which one of these is supporting what you know? [supports argument]*

[. . .]

S2 *You can't always see the Moon, which is there by the Sun, because as we turn around on our own axis the Moon spins as well, around us. [area II]*

S3 *Not every day. [area I or II]*

S2 *To see it again. [area I] Because the Moon spins on it's own axis and if you spin around the Moon is still there and we are on the other side. [area I or II] So we can't see it, because of the Earth. [area II]*

Both groups referred to the movement of the Moon and the Earth. Furthermore, in example 1a the positions of the Sun, the Moon, and the Earth were also considered. Here, the students created and used a sketch showing the Sun, the Moon and the Earth in order to engage with statements D and E. With this picture, students started to refer to a concrete situation (as presented in the drawing). Therefore, mainly area I is ascribed. The second group (example 1b) could not find their initial understanding of the general features of the Moon's and Earth's movement in one of the statements presented. After the discussion with their teacher they mainly repeated their ideas and then most members of the group stopped working on the task. Later in the lesson they mainly copied results from one group member and talked about private issues.

From this and several similar results amongst the different groups and tasks given, we conclude that students can only engage in the tasks at content and areas of abstraction that are familiar to them. Even though the tasks may give additional or extending information, students did not make use of this information during their discourse. Students were also not able to use additional information given by other students or the teacher and thus, could not rebut data or warrants about a specific context which were not already familiar to them. Also, teachers aiming to scaffold students' argumentation were not successful as long as students were struggling with content specific aspects. Even though this result may sound trivial, taking the well known results on the impact of prior knowledge on students' learning into account, it is striking that already a relatively small mismatch between students' knowledge and the tasks' demands can result in students skipping the task or at least not making sense to additional information. On the other hand, if students can incorporate their knowledge while discussing the tasks, the data indicate that they often reach (almost) initially relatively high areas of abstraction. That is, they reach area II more often than reported in other studies with the same age group (e.g., V. Aufschnaiter & V. Aufschnaiter, 2003). By itself this is a significant achievement of engaging in argumentation.

4.2 Students' Learning

Comparing students' ideas presented at the beginning of the lesson and their development throughout the lesson we can infer about students' content specific learning. For instance, student S3 from Example 1a concludes at the end of the session:

S3 *Sometimes you can see the Moon when it's light outside [area I], and it's coming across and it's blocking the Sun, [area I] but where the Sun's a lot bigger and the Earth's a lot bigger than the Moon [area I], it doesn't block it, the Sun's rays are still coming to the Earth because it's not hitting the Moon. [area I] But you can still see the rays coming from the Sun [area I], that's why I think the Moon actually blocks the Sun and not the Earth blocks the sunlight from to get to the Moon. [area I] I think it's the Moon that puts a shade over the Earth and not the Earth that puts a shade over the Moon. [area I]*

In example 1a the students started to create an understanding of the setting of the Sun, Moon, and Earth. Here they already started to use a sketch of the relative positions of the Sun, the Moon, and the Earth in order to create meaning for statements D and E. Throughout the lesson, they developed a more precise description of how statements D and E can be explained by the relative positions of the Sun, the Moon, and the Earth and discussed further how these bodies move. Even though the way S3 expressed his idea at the end of the lesson looks fairly fluent and is in some sense more integrated, he still only partially understood statement D.

The majority of our results show that the elaboration of students' ideas increased but higher areas of abstraction were rarely reached for the first time (rarely exceeded the initial understanding within a single lesson). There is almost no evidence in our data that students develop a "new" (conceptual) understanding of the topic throughout their discussions and argumentations. Rather, their understanding becomes more elaborated and more fluent but does mainly contain ideas that were already raised initially. We conclude that argumentation is essential to support students to develop a stable understanding as basically it provides the opportunity to use similar ideas in differing circumstances. Such a process leads to consolidation as ideas which are initially tentative are confirmed and elaborated. However, lessons based on argumentation seem to have no direct impact on students developing a new understanding in a sense that it emerges within the discourse directly. But argumentation seems to have an important function: it supports students' improvement of thinking as students' discourse leads to a quicker development of specific ideas and helps to make connections across (familiar) contexts. It is such an improvement that is the basis of further learning.

4.3 Interrelationship between the Quality of Students' Arguments and their Understanding of Science Content

An analysis of the content and the area of abstraction of students' high and low level arguments revealed that students can develop high level arguments with little knowledge (narrow content elements at area I). Conversely, some low level arguments consist of more expanded knowledge (several content elements connected and/or developed at area II). Overall, the data support the assumption that it is not the quality of students' knowledge (described by its content and level of abstraction) but students' experiences with the content, about which to argue, that has an impact on the level of argumentation reached. Students' ability to undertake high level argumentation will then depend on whether or not the content of the argument relates to students' experiences, especially for the data on which the argument is based. That is, students can undertake high level argumentation with relatively low level knowledge but will probably not do so when they lack experiences with the corresponding data. Therefore, we would expect that even scientists may not be able to engage in high level argumentation when confronted with an unfamiliar task.

5. CONCLUSIONS AND IMPLICATIONS

The aim of this study was to explore the interrelationship between processes of students' argumentation and their development of scientific knowledge. Two main results were reported briefly (see also V. Aufschnaiter et al., 2008):

(1) Although argumentation seems to have a strong potential for students' engagement in science contexts, the learning of new science concepts via lessons based on argumentation seems to be limited. This result does not derogate the importance of argumentation in school science but may indicate possible limitation to the aims that are associated with such approaches.

(2) Results indicate that high level arguments can be developed with low level knowledge and, conversely, high level knowledge may not result in high level argumentations. This result may indicate that it is important to understand better what limits and enables students to argue so that instruction on argumentation can be designed accordingly.

Maybe even more important than the particular outcomes of this study is the attempt to integrate two different frameworks in order to enhance our understanding of the details of the teaching and learning of science. We expect that this and similar attempts are promising approaches to science education research in the future.

ACKNOWLEDGEMENTS

The study was funded by the UK Economic and Social Science Research Council (R000237915). The German Research Foundation supported the data analysis during an academic visit (AU 155/3-2) and has been funding a project on the interrelationship between argumentation and conceptual development in physics (AU 155/5-1,2).

REFERENCES

Driver, R., Newton, P. & Osborne, J. (2000). Establishing the norms of scientific argumentation in classrooms. *Science Education, 84(3)*, 287–312.

Duit, R. (2006). *Bibliography – STCSE: Students' and Teachers' Conceptions and Science Education.* Retreived June 2006, from http://www.ipn.uni-kiel.de/aktuell/stcse/stcse.html

Duit, R. & Treagust, D. F. (1998). Learning in science – from behaviorism towards social constructivism and beyond. (In B. J. Fraser & K. G. Tobin (Eds.), *International handbook of science education* (pp. 3–25). Dordrecht, Boston, London: Kluwer.)

Duit, R. & Treagust, D. F. (2003). Conceptual change: A powerful framework for improving science teaching and learning. *International Journal of Science Education, 25(6)*, 671–688.

Erduran, S., Simon, S. & Osborne, J. (2004). TAPping into argumentation: Developments in the application of Toulmin's argument pattern for studying science discourse. *Science Education, 88*, 915–933.

Hogan, K. & Maglienti, M. (2001). Comparing the epistemological underpinnings of students' and scientists' reasoning about conclusions. *Journal of Research in Science Teaching, 38(6)*, 663–687.

Jiménez-Aleixandre, M. P., Bugallo, A. & Duschl, R. (2000). "Doing the lesson" or "Doing science": Argument in high school genetics. *Science Education, 84(6)*, 757–792.

Jiménez-Aleixandre, M. P. & Pereiro-Munhoz, C. (2002). Knowledge producers or knowledge consumers? Argumentation and decision making about environmental management. *International Journal of Science Education, 24(11)*, 1171–1190.

Kelly, G. J., Druker, S. & Chen, C. (1998). Students' reasoning about electricity: Combining performance assessments with argumentation analysis. *International Journal of Science Education, 20(7)*, 849–871.

Koslowski, B. (1996). *Theory and evidence: The development of scientific reasoning.* (Cambridge, MA: MIT Press)

Kuhn, D. (1991). *The skills of argument.* (Cambridge: Cambridge University Press)

Lawson, A. E. (2003). The nature and development of hypothetico-predictive argumentation with implications for science teaching. *International Journal of Science Education, 25(11)*, 1387–1408.

Leach, J. (1999). Students' understanding of the co-ordination of theory and evidence in science. *International Journal of Science Education, 21(8)*, 789–806.

Lemke, J. L. (1990). *Talking science: Language, learning, and values.* (Norwood, NJ: Ablex Publishing)

Martins, I., Mortimer, E., Osborne, J., Tsatsarelis, C. & Jiménez-Aleixandre, M. P. (2001). Rhetoric and science education. (In H. Behrendt, H. Dahncke, R. Duit, W. Gräber, M. Komorek, A. Kross & P. Reiska (Eds.), *Research in science education – past, present, and future* (pp. 189–198). Dordrecht: Kluwer.)

Marton, F. & Booth, S. (1997). *Learning and awareness.* (Mahwah, NJ: Lawrence Erlbaum)

Mason, L. (1996). An analysis of children's construction of new knowledge through their use of reasoning and arguing in classroom discussions. *Qualitative Studies in Education, 9(4)*, 411–433.

Newton, P., Driver, R. & Osborne, J. (1999). The place of argumentation in the pedagogy of school science. *International Journal of Science Education, 21(5)*, 553–576.

Osborne, J., Erduran, S. & Simon, S. (2003). *Ideas, evidence and arguement in science. Teacher training pack.* (London: Nuffield Foundation)

Osborne, J., Erduran, S. & Simon, S. (2004). Enhancing the quality of argument in school science. *Journal of Research in Science Teaching, 41(10)*, 994–1020.

Simon, S., Osborne, J. & Erduran, S. (2003). Systemic teacher development to enhance the use of argumentation in school science activities. (In J. Wallace & J. Loughran (Eds.), *Leadership and professional development in science education* (pp. 198–217). London: RoutledgeFalmer.)

Toulmin, S. (1958). *The uses of argument.* (Cambridge: Cambridge University Press)

von Aufschnaiter, S. V. (2001). Development of complexity through dealing with physical qualities: One type of conceptual change? (In H. Behrendt, H. Dahncke, R. Duit, W. Gräber, M. Komorek, A. Kross & P. Reiska (Eds.), *Research in science education – past, present, and future* (pp. 199–204). Dordrecht: Kluwer.)

von Aufschnaiter, C. & von Aufschnaiter, S. (2003). Theoretical framework and empirical evidence on students' cognitive processes in three dimensions of content, complexity, and time. *Journal of Research in Science Teaching, 40(7)*, 616–648.

von Aufschnaiter, S. & Welzel, M. (1999). Individual learning processes - a research programme with focus on the complexity of situated cognition. (In M. Bandiera, S. Caravita, E. Torracca & M. Vicentini, (Eds.), *Research in science education in Europe* (pp. 209–215). Dordrecht: Kluwer.)

von Aufschnaiter, C., Erduran, S., Osborne, J. & Simon, S. (2008). Arguing to learn and learning to argue: Case studies of how students' argumentation relates to their scientific knowledge. *Journal of Research in Science Teaching,* 45(1).

Zohar, A. & Nemet, F. (2002). Fostering students' knowledge and argumentation skills through dilemmas in human genetics. *Journal of Research in Science Teaching, 39(1)*, 35–62.

30. STUDENTS' ARGUMENTATION IN GROUP DISCUSSIONS ON A SOCIO-SCIENTIFIC ISSUE

Abstract: Socio-scientific issues in class have been proposed in an effort to democratise science in society. They offer opportunities to develop students' argumentation skills. We have explored how students elaborate arguments in the context of small group discussions on a socio-scientific controversy about mobile phones. The analysis of the discursive practices allowed to identify various argumentation processes within students' group discussions. Students' arguments were elaborated from scientific data, common ideas and epistemological and strategic considerations. Students' social interactions influenced the patterns of argumentation elaborated within the group discussions. Implications of this study for the teaching of socio-scientific issues in class are discussed

Keywords: Argumentation, Group discussions, Science democratisation, Socio-scientific issues

1. FRAMEWORK AND PROBLEM

For several years now, researchers have been proposing the debating of socio-scientific issues in the classroom (for a literature review, see Sadler, 2004). This can lead students to debate energy choices, examine global climate change (Schweizer & Kelly, 2001; Sadler et al., 2004), or evaluate the dangers of mobile phones and base stations (Albe, 2005), among other issues, in an effort to democratise science in society. Some authors have suggested training citizens on how to be critical for the purpose of social reconstruction and political action (Pedretti & Hodson, 1995; Roth & Désautels, 2002; Bader, 2003). As Bader (2003) emphasised, the objective is to give young citizens the means to be able to participate in socio-technical controversies and to negotiate with specialists. According to Driver, Newton and Osborne (2000: 297), "... in our democratic society it is critical that young people receive an education that helps them to both construct and analyse arguments relating to the social applications and implications of science". For other authors, it is necessary to emphasise the nature of science itself. According to Oulton et al. (2004: 419), "society would benefit if science education encouraged pupils, who are both today's and tomorrow's citizens, to adopt a more positive and realistic view of science and its potential for resolving conflicts [...], to recognize the tentative nature of scientific knowledge".

Virginie ALBE, Toulouse EducAgro, Ecole Nationale de Formation Agronomique, B.P. 87, 31326 Castanet-Tolosan, France

R. Pintó and D. Couso (eds.), Contributions from Science Education Research, 389–401.

When confronted with socio-scientific issues, students refer to a variety of arguments. Kolstø (2001) emphasised that attitudes with respect to risk are of primordial importance in students' decisions and argumentation concerning the issue of burying very high-tension power lines. Students also discussed the credibility of scientists, the degree of consensus among scientists, the epistemic nature of scientific proof and other considerations related to the scientists' honesty. The significance attached by students to scientific knowledge varied, with students declaring that such knowledge was desirable but not necessary for making decisions.

For others, the social and epistemological considerations raised by students are of crucial importance. Students may engage in discussions to look for scientific proof as in an empiricist and realistic point of view of science, it is considered to be the key element for resolving controversies (Driver et al., 1996; Bader, 2003; Sadler, 2004). According to Leach and Lewis (2002: 204) "many science students tend to over-attribute significance to empirical processes in suggesting how scientific disputes might be resolved and in justifying viewpoints on scientific issues". Sadler et al. (2004) analysed how high-school students had interpreted and assessed contradictory expertise on global warming. They demonstrated that the students' interpretation and evaluation of this contradictory information was influenced by assumptions about the nature of science, such as, for instance, how data are interpreted and the interaction of science with society, but also by the personal opinions of students and their scientific knowledge. The use of scientific knowledge in argumentation regarding socio-scientific issues has also been questioned. During a debate in the form of a citizen's conference on climate warming, Schweizer and Kelly (2001) showed that opponents of the thesis that human activities had contributed to an increase in the greenhouse effect developed more arguments, whose structure was more complex and for some of them had no scientific basis. Furthermore, Levinson (2003) identified some confusion about scientific concepts which had not been clarified during a students' discussion on the diagnosis of illnesses in foetuses. Moreover, new scientific confusion appeared as students reinforced their positions on abortion during the debate. For Aikenhead (2003), knowledge of science and scientists is more relevant than scientific knowledge itself in the context of contradictory scientific expertise. For Fensham (2002), students can be prompted to emphasise values.

According to the literature review of Sadler (2004: 523), "the most fruitful interventions would be those which encourage personal connections between students and the issues discussed, explicitly address the value of justifying claims and expose the importance of attending to contradictory opinions". It often implies debates or group discussions. But some authors warned that the social demand of group discussions may be too high. Dawes (2004) underlined that "group talk can help learners to exchange ideas, to have access to different perspectives and to make meaning together. However, this may not happen if groups of children remain unaware of talk as a tool for thinking together." Differences in discussion groups can also lead to question equity towards learning through this type of activity (Kelly et al., 2001). According to Alexopoulou and Driver (1997), "there is a complex interrelation of conceptual, contextual and social factors which influence the collaborative sharing and building of meanings in science classrooms". Some authors have shown that students' group discussions were focused on the procedural aspects of the activities (Bianchini, 1997; Kittleson & Southerland, 2004; O'Neil & Polman, 2004). Others have stressed that students have difficulty in forming arguments (Zeidler, 1997; Chinn & Brewer, 1998), in presenting opposing arguments (for and against), or in presenting diverse points of view on a question (Driver et al., 2000). For Kuhn (1992) an "epistemological naïveté may be an important factor in the limited argumentative reasoning ability that people display". Others have shown that the way students construct arguments is different with unknown data (Kelly et al., 1998; Jiménez-Aleixandre et al., 2000). While confronted with socio-scientific issues, how do students elaborate arguments? The purpose of this chapter is to document the argumentation students develop in small group discussions on a controversial socio-scientific issue. A study was undertaken to identify the content of the arguments which were elaborated by students to decide whether or not mobile telephones are potentially dangerous and to explore factors that influence argumentation. This took place in small group discussions dedicated to the preparation of a role-playing on this socio-scientific issue to enable students to learn how to evaluate the quality of scientific data (Hind et al., 2001).

2. MODE OF INQUIRY

A role play on the controversial issue of the danger of mobile telephones was undertaken in a science class of 12 secondary education students specialising in sciences and technologies for agronomy and the environment. They were from 16- to 18-years-old. Students were divided into two groups to defend the case that mobile phones either are dangerous (group A) or are not dangerous (group B) for human health. Students played the roles of expert witnesses in a trial in which an employee was taking his employer to court over his poor state of health, on the grounds that this was caused by using a mobile telephone and was the reason he had to leave his job. Students studied seven research extracts on the occurrence of illnesses in animals which had been subjected to experiments but also on epidemiological surveys and memory tests. These extracts were supplied by Hind et al. (2001) in the module on which we based this study. Each group was asked to establish for each research extract if the relationship between the results obtained on animal tissue and health effects in humans was valid; if other research groups obtained similar results; if the size of the sample used meant that it was statistically relevant or sufficiently reliable. Then, each group wrote down the arguments they intended to develop to convince the other group during the role play. Each group discussion was audio-taped and students' discourse fully transcribed. The discussions lasted approximately 50 min in each group.

With a micro-ethnographic perspective (Green & Bloome, 1997; Kelly et al., 2001), discourse events in group discussions were examined. This perspective inspired by ethnography and discourse analysis is related to the hypothesis that interactions and negotiations between individuals in a social group shape local, contingent and group dependant "discourse objects". Following Garfinkel, interactional linguistics invite to analyse discursive practices without separating cognition from social. As Kelly et al. (2001) underlined, meaning is considered as one of a group, not an individual, and therefore the substance of cognition is viewed as social. This perspective is then focused on the content of arguments without separating knowledge and social aspects. It differed from studies using Toulmin's argument pattern which laid more emphasis on argumentation "to determine whether that process can be facilitated and its quality assessed" (Osborne et al., 2004: 1015), to identify the structure of arguments (Kelly et al., 1998) or to analyse argumentation discourse within different learning environments (Jiménez et al., 2000). Focus was on the analysis of students' discourse to examine how arguments were constructed by students within and across the two discussion groups. The nature and content of each student's intervention are identified. It can be questions, claims with or without justifications, oppositions, expressions of agreement or disagreement, repetitions, reformulations. Students can bring into discussion elements from research extracts, strategic, social, epistemological considerations, scientific or technological knowledge, everyday knowledge, points of view on the activity. These methodological procedures allowed to identify the processes of elaboration and the content of the arguments within the two discussion groups.

3. FINDINGS

Several processes of group argumentation have been identified. The simplest process for collective development of knowledge was based on acceptance. More complex methods were found in collaborative processes, where the knowledge objects were collectively built. In some cases, students articulated their disagreements during processes of collective argument elaboration. Disagreements, on the other hand, could lead to the destabilisation of the object under discussion, sometimes resulting in its fragmentation. Processes of group argumentation can be summarised as follows:

- Acceptance: one student proposes an argument and the others agree. This can occur through authority or not.
- Collaborative argumentation: students co-construct arguments i.e. one student proposes something, it is developed by another, and others continue, etc. This can occur within the group, in pairs, or with role-playing.
- Contradictory confrontations: one student raises objections, questions or proposes an alternative. Others disagree. Confrontations can lead students to abandon or to reach an agreement by resolving disagreements, articulating opposing argumentative positions, or voting.

Students' interactions involved different processes of argumentation for the two groups. Within group A episodes of collaborative argumentation were more numerous than in group B. Moreover, students in group A articulated opposing positions through votes or collaborative argumentation, such as shown in the following two episodes.

After studying a research extract (n°1) students expressed contradictory viewpoints. Arnaud firstly proposed a strategic argument ("Anyway, it is our interest to say that there is a link between the results obtained from animal tissue and the results on humans", line 7). The debate involved determining whether the relation between the results obtained on animal tissue and the effects on humans was valid, which led them to define the nature of humans: *a mammal? made up of animal cells? an animal?* An extract from the transcript of the exchanges of students in group A is given below. Students' names have been replaced with fictitious ones.

15.	*Sébastien: Well, [Arnaud], for the first one there. In my opinion it's not the same thing because we're mammals and those are animal cells.*
16.	*Katia: Yes, we're mammals and those are animal cells.*
17.	*Arnaud: Yes, but it's still the animal kingdom.*
18.	*Sébastien: Yes but it's big, the animal kingdom.*
19.	*Arnaud: You're not going to say to the others: "Yes, well you're right, the mobile..."*
20.	*Sébastien: We could say yes, besides, there are lots...*
21.	*Arnaud: You have to say yes to everything.*
22.	*William: That way we can be sure.*
23.	*Arnaud: They're going to say no to everything, so...*
24.	*Sébastien: OK, why don't you give your opinion now, so there're a few of us and then we can vote. [Agnès], what do you think? You're going to say yes, just say yes or no.*

Sébastien disagreed and argued the point. Katia accepted this and through this agreement supported the argument set out by Sébastien. William supported Arnaud's strategic proposals (at turns 19 and 21), thus disagreeing with Sébastien. Sébastien suggested a vote. Here, we can observe disagreements on the discourse objects (strategy and nature of humans) and social roles structured in the discussion and shaping of the discussion. Two students expressed opposing positions (Sébastien and Arnaud), each supported by another student (Katia and William) and two other students (Agnès and Marianne) did not participate in the group discussion. Students elaborated a collective agreement by voting on whether the relation between results on animals and effects on humans was valid. The question of the validity of the transposition on humans of the results of research carried out on animals, of an epistemological nature, was not directly dealt with by the students, as the emphasis was shifted to a question about the relative ontological nature of the status of humans. This could mean that the need to deal with the epistemological question was too challenging for the students (with reference to the concept of "learning demand" defined by Leach and Scott, 2002) and that the discussion steered them towards another debate, in which they also had difficulty in collectively elaborating a point of view. Other studies (for example by Hill, 1989, quoted by Larochelle, 2002) have also shown the difficulty (which, in addition, is not simply cognitive) students have in comparing human to animals in categorisations. However, in this extract it also seems that strategic considerations guided the discussions and shaped the discourse objects, the objective of the activity for some students clearly being to win the case. At this point, students did not resolve the disagreements through discussion but everyone expressed his own personal point of view when voting. During the discussion about which extracts to choose in order to justify their thesis, the students came back to the question about the relation between the results obtained on animal tissue and effects on humans.

518. Arnaud: Those, are they nematodes?
519. Sébastien: You could say if it acts on worms, then it could have an effect on us.
520. Arnaud: There you go ... After all they are animals, well...
521. Sébastien: Yes, they're living organisms, they're ...

At turn 519, Sébastien put forward an argument establishing a relationship between observed effects on animals and possible effects on humans. During his next turn, Arnaud evaluated Sébastien's proposal ("There you go") and provided a justification for this argument ("After all they are animals") in reference to a discourse object from the beginning of the activity. Previously (turns 15 to 18), students discussed the nature of humans, here Arnaud put worms and human beings in the same category of animals with a formulation that made possible to reach an agreement. He qualified his formulation with "well", which enabled Sébastien to state his agreement and to finish what he was trying to say by reformulating the category using the term "living organisms". Disagreements of an ontological nature were thus avoided and the argument put forward by Sébastien was accepted.

Students' disagreements can also lead to the fragmentation of the discourse objects when they concerned epistemological considerations. Extract n°5, in which "no increase in the rate of breast tumours in mice was observed" led to disagreements. The discussion between Arnaud and Sébastien related to the exposure time of mice in the research. It can be noted that the formulations "yes but" which initiated each speech indicated the opposition between the two participants. Their disagreement is due to the fact that 18 months was or was not considered too short a time and indicated that the reference of the discussion is human carcinogenesis. Students' tendency to use anthropomorphic justifications has also been observed during students' argumentation (Jimenez-Aleixandre et al., 2000; Grace, 2005). Here, in reference to human carcinogenesis, the integration into the argumentation of common ideas, social accepted knowledge or 'cultural truisms' (Billig, 1987) was observed. Students did not manage to overcome their disagreements and this resulted in a fragmentation of the discourse objects and to a vote. Even though students' individual expressions indicated that for three of them 18 months was considered too short a time for cancer to develop, one disagreed and two were uncertain, it did not allow them to reach an agreement.

Students' arguments were elaborated from research extracts studied, common ideas and epistemological and strategic considerations. They also integrated elements into their argumentation such as additional data from research extracts. For instance, students co-elaborated an argument using research extract n°4 and with reference to extract n°5. Students established, through a process of collaborative argumentation, that in research extract n°4 "twice as many subjects were affected" and by adding elements that "mice developed lymphomas even with a daily exposure time of less than the 20 hours reported in extract n°5".

Another argument about the danger of using cellular telephones was collectively developed by the students discussing a research extract (n°7). Their collaborative argumentation led to disagreements of an epistemological nature which could not be overcome by discussion, so collective elaboration concentrated on formulating their argument as efficiently as possible. The students established that "as the percentage is higher in affected people, we can say that this illness may be linked to the telephone".

Using another research extract (n°3) and with the integration into the discussion of social common knowledge, the students established through a process of collaborative argumentation that "the beneficial effect observed on the memory is a temporary response, the long-term effects may be harmful, as with drugs in sport".

Extract n°2 concerned an observed change in the electrical activity of the brains of rats and a reduction in responses to stimuli, but the text indicated that "the hippocampus would be buried too deeply within the brain to be affected by cellular telephones". This point was raised three times without leading to collective discussion, then was resumed with elements introduced by the students based on their knowledge of the development of cancer, through collaborative argumentation. The students established that "once it reaches a part of the brain [...] afterwards, it spreads everywhere". Finally, the students elaborated, by a process of collaborative argumentation on the basis of general knowledge, that "waves from mobile telephones disturb the functioning of the brain".

The students in group B co-elaborated arguments most of the time with processes of acceptance. It was noted that this happened most often between two students, rather than in the discussion group. In addition, it was noted that one instance of collaborative argumentation took place between two students, whereas the others occurred collectively on the basis of research extracts. Parallel discussions took place and on three occasions, the authoritarian personal expression of a student acting as the 'leader' was used in order to reach agreement on a object discussed within the group as shown in the following extract.

88. *Fabien: But it is enough reliable, no? honestly! But it is reliable.*
89. *Caroline: Yes*
90. *Sandra: Did you write << we don't know >>?*
91. *Max: But isn't it from the sample size?*
92. *Sandra: Yes, here.*
93. *Max: we need to see in function of the sample size*
94. *Sandra: then we don't know?*
95. *Max: No, we don't know.*
96. *Fabien: eh, at the first one, Miss, we have written no.*
97. *Caroline: we can't say.*
98. *Fabien: we all put no, why do you put that, you?*
99. *Nathan: you, Sir, make a serious mistake, because, it is you that run…*
100. *Fabien: for this, it is no, it is not that we don't know.*
101. *Sandra: I have put no.*

Fabien questioned the agreement set up by Caroline, Max and Sandra to answer "no" to the question of the limited value of the extract research because it was not reliable enough. Nathan taking the role of a lawyer then accused Fabien of being incompetent in the role of the "leader" of the group. Fabien repeated with authority his point of view: the research was reliable and the answer to the question is "no". Sandra accepted Fabien's view point declaring that she wrote "no" to answer the question. It has the effect of reaching an agreement and Fabien initiated the study of another research extract. It appeared that the social interactions between students shaped the discourse objects under discussion. Moreover, conflicting exchanges occurred on several occasions. By adopting the roles of lawyers students were on one occasion able to reach agreement.

104. *Fabien: << genetically modified mice have been used to increase their… >>*
105. *Max: they have been genetically modified, it is not good.*
106. *Nathan: No, but, they say that they have been genetically modified.*
107. *Cécile: Yes*
108. *Nathan: it then distorts everything. It is not good.*
109. *Fabien: Yes, yes, exactly, you are right.*
110. *Nathan: Do you agree with me Dear Mr [name]?*
111. *Fabien: I totally agree with your proposition which is acceptable.*
112. *Nathan: I am glad that we reach an agreement.*
113. *Cécile: I am strongly delighted*
114. *Nathan: Mister [name] what is your opinion please?*
115. *Max: here, oh yes*

Nathan considered that the use of GMOs was a source of bias in the research results. Fabien agreed with a double affirmation and a sign of support to Nathan.

Nathan then asked for a confirmation of the agreement taking the role of lawyer. Fabien confirmed the agreement with a formulation playing the role of a discussion between lawyers. Nathan, in the same role play, mentioned that he accepted the agreement with delight. Then, Cécile entered the play and Nathan asked Max his opinion. He confirmed the agreement. Elsewhere, students in group B did not manage to articulate their disagreements in processes of collective discussion as reported below.

145. *Nathan: again a distorted research, I think*
146. *Caroline: genetically modified!*
147. *Cécile: Yes.*
148. *Nathan: We can directly cancel these two experiments because ... they are distorted, so ...*
149. *Fabien: No, no, they are not. This one, it will help us to defend ourselves because the mice exposed 20 hours a day for 18 months: << ... There was no increase in the rate of tumours in these mice. ... >>, therefore it means no increase of cancer*
150. *Cécile: Yes, but, previously, it was not the same thing, here, it is breast tumours, and there it is [inaudible] observed. So, maybe with the different stuff, it is not the same.*
151. *Fabien: Well, it is not so ...*
152. *Nathan: Do mice have breast?*
153. *Nathan: is not necessarily valid, yes.*
154. *Max: But yes, it's valid.*
155. *Nathan: But no.*
156. *Max: Yes, because they show that ...*
157. *Fabien: Maybe ... How many years does it have to be done?*
158. *Nathan: anyway, it's not valid, they are genetically modified.*
159. *Fabien: Yes, this, this will help us ... therefore it will be valid. Well, the research ... the second, yes or no ? I don't know.*

Here, the disagreements on the reliability of the research extract could not be overcome by discussion or an authoritarian personal expression based on strategic considerations. Previously students reached an agreement to consider that the research was not reliable as it involved genetically modified mice. We noted that this agreement was reached with students taking the roles of lawyers and that in this phase of activity, confrontations took place. Strategic considerations also appeared to orient the group interpretations of the research extracts and argumentation.

On other occasions contradictory confrontations occurred and students abandoned their discourse object, i.e. changed the topic under discussion. From a research extract (n°3), students firstly established, by means of collaborative argumentation, that this research indicated a beneficial effect from the use of cellular telephones.

Then students wondered about the permanence of this effect over time, with references to doping in sport ('It is as with EPO, someone that makes EPO injections, he will run quickly, but after, at 40 years old, he will die [...] Hey so yes! On short term effects, it can be positive, but after, long term effects') and the importance of percentages ("response times are improved by 4%"). The emergence of this new discourse object had the effect of destabilising the co-elaborated argument and later students' discussion led to the rejection of this previous argument collaboratively constructed but non stabilised.

333.	*Nathan: these texts for me, It doesn't mean nothing*
334.	*Sandra: We found something there, the research as he said, it improves, it increases response times, here we are, but there is doping ...*
335.	*Fabien: we say that it is positive, but it is not reliable because it can after a long time bring some diseases to brain*

This episode ended the group discussion on the validity of research extract n°3. Two students within the group co-rejected the use of this research extract to argument on mobile phones' positive effects, as students considered that long-term effects could be harmful. As students could not overcome their disagreements, it raised some questions. Was the social demand too high as Dawes (2004) warned? As students' disagreement was related to the nature and duration of observed effects in a research where response times of the recall of words and pictures were measured, the epistemological demand may also be too high. It has also been observed that students integrated common ideas, socially accepted knowledge or 'cultural truisms' (Billig, 1987) into their argumentation when referring to long-term harmful effects of doping in sport. Collaborative argumentation on epistemological considerations was also observed elsewhere. Considering that contradictory results were obtained from extracts n°6 and 7, students argued collaboratively that the researches were not reliable. This may be linked to students' representations of science. Later, during a discussion about which extracts to choose in order to justify their thesis, the students came back to the question about the reliability of research extracts. Through a process of collaborative argumentation, three students of group B established that the research extracts were not reliable because the mice have been genetically modified. Strategic preoccupations were, moreover, expressed by one student but not discussed collectively, which evidenced a fragmentation of the discourse objects. In a process of argument co-elaboration between two students, it was also established that a research extract (n°2) was not reliable using considerations about the way that researchers establish knowledge ("researchers base their work on hypotheses which are not necessarily correct"). Other epistemological considerations were raised which were not discussed collectively. These related to the question about whether the results were reliable ("the results are not 100% reliable").

4. CONCLUSIONS AND IMPLICATIONS

As this was a case study, it is impossible to generalise, but the analysis does show that role play based on a socio-scientific controversy is an activity in which group discussions need to be examined carefully. Some authors have shown that students' group discussions were focused on the procedural aspects of the activities (Bianchini, 1997; Kittleson & Southerland, 2004; O'Neil & Polman, 2004). Here, students' discussions focused on the study of research extracts and on argumentation. According to Osborne et al. (2004), socio-scientific issues offer opportunities to develop students' argumentation skills. It seems that the issue discussed was very motivating for the students and stressed the educational interest of dealing with socio-scientific issues in class. The analysis of the discursive practices allowed to identify various argumentation processes within students' group discussions. Qualitative categories were generated from a micro-ethnographic analysis and can serve as a framework for analysing students' discussions in future studies.

Students elaborated a variety of arguments from which scientific knowledge was rarely implied. As other authors have shown, this study underlined the importance of considerations of an epistemological nature and the general knowledge used by students arguing about a socio-scientific controversy. Students in group A did not manage to resolve disagreements through discussion when they concerned epistemological requirements. In group B, students had difficulties in elaborating collective discourse objects and did not manage to articulate their disagreements. On some occasions they showed difficulties in regulating their interactions on the social plane. Peer's modes of interaction influenced the patterns of argumentation that were employed in the groups. Through role-playing, students on one occasion build an argument through a process of acceptance. In this case, students' role-playing functioned as a means to avoid exploring their disagreements. This emphasised that the relationships between the students had an influence on the collective discourse objects elaborated and structured the discussion.

It can then be considered that the activity of reviewing research results and elaborating arguments in a discussion group, with a view to participating in a role play on the controversial issue of the danger of using cellular telephones, is very demanding in social and epistemological terms. This is not to say that small group discussions should not be organised on socio-scientific issues. On the contrary, considering the educational interest of dealing with socio-scientific issues in class, it suggests that teachers' mediation and students epistemological training in this context become crucial.

Teachers should attempt to arrange group discussions to avoid students' conflicts. Dawes (2004) proposed language tools to initiate and sustain exploratory talk (Mercer, 1995) in group discussions. In the same line, Grace (2005) considered that guidance on appropriate ground rules for collaborative discussion may be valuable in helping pupils organise group discussion. A discussion at the end of the activity may also be valuable to enable students to reflect on their views and appreciate the value of group discussion. It may also be proposed that students' work on socio-scientific controversies should be accompanied by an examination of the way in which scientific knowledge is produced within a community and, in particular, the role of controversy in this process. For instance, the inclusion of considerations of the nature and limitations of science, the status, role and limits of evidence, the interests involved and the ways science communities operate to establish knowledge are very important when dealing with contemporary science controversies.

REFERENCES

Aikenhead, G.S. (2003). *Review of research on humanistic perspectives in science curricula.* (Paper presented at the ESERA 2003 Conference, The Netherlands).

Albe, V. (2005). Positions d'étudiants et d'étudiantes sur une question techno-scientifique controversée: La dangerosité des téléphones cellulaires. *Canadian Journal of Science, Mathematics & Technology Education*, 5, 361–376.

Alexopoulou, E. & Driver, R. (1997). Small group discussions in physics: Peer interaction modes in pairs and fours. *Journal of Research in Science Teaching*, 33, 1099–1114.

Bader, B. (2003). Interprétation d'une controverse scientifique: Stratégies argumentatives d'adolescentes et d'adolescents québécois. *Canadian Journal of Science, Mathematics & Technology Education*, 3, 231–250.

Bianchini, J. (1997). Where knowledge construction, equity, and context intersect: Student learning of science in small groups.*Journal of Research in Science Teaching*, 43, 1039–1065.

Billig, M. (1987). *Arguing and thinking: A rhetorical approach to social psychology.* (Cambridge: Cambridge University Press).

Chinn, C.A. & Brewer, W.F. (1998). An empirical test of a taxonomy of responses to anomalous data in science. *Journal of Research in Science Teaching*, 35, 623–654.

Dawes, L. (2004). Talk and learning in classroom science. *International Journal of Science Education*, 26, 677–695.

Driver, R., Leach, J., Millar, R. & Scott, P. (1996). *Young people's image of science.* (Buckingham: Open University Press).

Driver, R., Newton, P. & Osborne, J. (2000). Establishing the norms of scientific argumentation in classrooms. *Science Education*, 84, 287–312.

Fensham, P.J. (2002). De nouveaux guides pour l'alphabétisation scientifique. *Canadian Journal of Science, Mathematics & Technology Education*, 2, 133–150.

Grace, M.M. (2005). *Developing personal values and argumentation skills through decision-making discussions about biological conservation.* (Paper presented at the ESERA 2005 Conference, Barcelona).

Green, J. & Bloome, D. (1997). Ethnography and ethnographers of and in education: A situated perspective. (In J. Flood, S.B. Heath & D. Lapp (Eds.), *Handbook for literacy educators: Research in the communicative and visual arts*. New York: Macmillan).

Hill, G.L.C. (1989). Students' 'untutored' beliefs about natural phenomena: Primitive science or commonsense? *Science Education*, 73, 155–186.

Hind, A., Leach, J., Ryder, J. & Prideaux, N. (2001). *Teaching about the nature of scientific knowledge and investigation on AS/A level science courses*. Leeds: CSSME.

Jiménez-Aleixandre, M.P., Bugallo Rodríguez, A. & Duschl, R.A. (2000). "Doing the lesson" or "Doing science": Argument in high school genetics. *Science Education*, 84, 757–792.

Kelly, G., Crawford, T. & Green, J. (2001). Common task and uncommon knowledge: Dissenting voices in the discursive construction of physics across small laboratory groups. *Linguistics and Education*, 12, 135–174.

Kelly, G.J., Druker, S. & Chen, C. (1998). Students' reasoning about electricity: Combining performance assessments with argumentation analysis. *International Journal of Science Education*, 20, 849–871.

Kittleson, J. & Southerland, S. (2004). The role of discourse in group knowledge construction: A case study of engineering students. *Journal of Research in Science Teaching*, 41, 267–293.

Kolstø, S.D. (2001). To trust or not to trust ... pupils' ways of judging information encountered in a socio-scientific issue. *International Journal of Science Education*, 23, 877–901.

Kuhn, D. (1992). Thinking as argument. *Harvard Educational Review*, 62, 155–178.

Larochelle, M. (2002). Rapport au savoir et socialisation à la cité scientifique. (In Actes des 3èmes journées d'études franco-québécoises (pp. 58–76). Paris: La Sorbonne).

Leach, J. & Lewis, J. (2002). The role of students' epistemological knowledge in the process of conceptual change in science. (In M. Limón & L. Mason (Eds.), *Reconsidering conceptual change. Issues in theory and practice* (pp. 201–216). The Netherlands: Kluwer academic publishers).

Leach, J.T. & Scott, P.H. (2002). Designing and evaluating science teaching sequences: An approach drawing upon the concept of learning demand and a social constructivist perspective on learning. *Studies in Science Education*, 38, 115–142.

Levinson, R. (2003). *Teaching bioethics in science: Crossing a bridge too far?* (Paper presented at the annual meeting of the National association for Research in Science Teaching, Philadelphia).

Mercer, N. (1995). *The guided construction of knowledge: Talk amongst teachers and learners.* (Clevedon: Multilingual Matters).

O'Neil, D.K. & Polman, J.L. (2004). Why educate "Little Scientists?": examining the potential of practice-based scientific literacy. *Journal of Research in Science Teaching*, 41, 234–266.

Osborne, J., Erduran, S. & Simon, S. (2004). Enhancing the quality of argumentation in school science. *Journal of Research in Science Teaching*, 41, 994–1020.

Oulton, C., Dillon, J. & Grace, M. (2004). Reconceptualizing the teaching of controversial issues. *International Journal of Science Education*, 26, 411–424.

Pedretti, E. & Hodson, D. (1995). From rhetoric to action: Implementing STS education through action research. *Journal of Research in Science Teaching*, 32, 463–485.

Roth, W.-M. & Désautels, J. (Eds.) (2002). *Science education as/for social action.* (New York: Peter Lang).

Sadler, T.D. (2004). Informal reasoning regarding socio-scientific issues: A critical review of research. *Journal of Research in Science Teaching*, 41, 513–536.

Sadler, T.D., Chambers, F.W. & Zeidler, D.L. (2004). Student conceptualisations of the nature of science in response to a socioscientific issue. *International Journal of Science Education*, 26, 387–410.

Schweizer, D.M. & Kelly, G.J. (2001). *An investigation of student engagement in a global warming debate.* (Paper presented at the NARST Annual Meeting, St Louis).

Zeidler, D.L. (1997). The central role of fallacious thinking in science education. *Science Education*, 81, 483–496.

31. EXEMPLARY TEACHING OF ARGUMENTATION: A CASE STUDY OF TWO SCIENCE TEACHERS

Abstract: The teaching of argumentation has been advocated as a significant goal for science education worldwide. Argumentation involves the coordination of evidence and theory to support or refute an explanatory conclusion, model or prediction. Even though argumentation has gained popularity as a pedagogical strategy, there is limited understanding of how enculturation into pedagogical practices around argumentation influences science teachers. The main objective of this chapter is to present a case study of two middle-school science teachers who participated, over 5 years, in various school-based research projects on argumentation ranging from basic research in teaching and learning to the development of professional development programs for training teachers in argumentation. The projects took place between 1999 and 2004 in the United Kingdom. The teachers were asked to reflect as a pair on various aspects of teaching and learning of argumentation. The results address the teachers' views and knowledge of argumentation, their perceptions of the goals, constraints and successes in their teaching of argumentation, their perceptions of themselves as learners and teachers, and their reflections on the professional development that they received. Implications for professional development of pre-service and in-service teachers are discussed

Keywords: Argumentation, Case studies, Professional development, Science teaching

1. INTRODUCTION

In recent years, the teaching of argumentation has emerged as a significant educational goal (e.g. Driver et al., 2000; Erduran & Jimenez-Aleixandre, in press; Kelly & Chen, 1999; Zohar & Nemet, 2002) Argumentation involves the coordination of evidence and theory to support or refute an explanatory conclusion, model or prediction (Toulmin, 1958). The case made is that argumentation is a critically important discourse process in science (Lemke, 1990), and that it should be taught and learned in the science classroom (Jimenez-Aleixandre et al., 2000). Argumentation is also perceived to facilitate the appropriation of community practices that provide the structure, communication and motivation required to sustain scientific inquiry (Gee, 1994). In this regard, it becomes imperative to make argumentation

Sibel Erduran, University of Bristol, Graduate School of Education, 35 Berkeley Square, Bristol BS8 1JA, United Kingdom

R. Pintó and D. Couso (eds.), Contributions from Science Education Research, 403–415.

a focus of professional development of science teachers (Erduran et al., 2006; Erduran, 2006a, b; Erduran & Osborne, 2005; Simon et al., 2006).

The main objective of this chapter is to present a case study of two middle-school science teachers who participated, over 5 years, in various school-based research projects on argumentation ranging from basic research in teaching and learning to the development of professional development programs for training teachers to teach argumentation. The projects that took place between 1999 and 2004 were supported by the Economic and Social Research Council, the Nuffield Foundation and the Gatsby Foundation in the United Kingdom (e.g. Erduran et al., 2004; Osborne et al., 2004a). The primary source of data in this chapter is an interview conducted with two of the teachers together as a pair after 5 years of involvement in the projects. The teachers were asked to reflect on the various projects, their developments across these projects, and their ideas on themes such as the transfer of effective teaching skills in argumentation to other aspects of science teaching. The interview themes are consistent with current literature on how teachers' knowledge and practices relate (Cochran-Smith & Lytle, 1999) and they aim to help investigate the nature and development of teachers' pedagogical content knowledge (Magnusson et al., 1999) in argumentation. For example, the teachers' reflections about themselves as well as their students in the context of argumentation provides one facet of professional development where a new body of knowledge – argumentation – was appropriated by the teachers into a theoretical knowledge base as well as classroom practices. The interview was analyzed and supplemented with some data from the teachers' classroom practice.

2. THEORETICAL BACKGROUND

The philosophical and cognitive foundations of argumentation have played a central role in the justification of research in argumentation in science education (e.g. Duschl & Osborne, 2002). Contemporary perspectives in philosophy of science (e.g. Giere, 1991; Kitcher, 1988) emphasize that science is not simply the accumulation of facts about how the world is. Science involves the construction of theories that provide explanations for how the world may be. Science often progresses through dispute, conflict and argumentation rather than through general agreement (e.g, Kuhn, 1970; Latour & Woolgar, 1986). Thus, arguments concerning the appropriateness of experimental design, the interpretation of evidence and the validity of knowledge claims are at the heart of science, and are central to the everyday discourse of scientists. Scientists engage in argumentation and it is through this process of argumentation within the scientific community that quality control in science is maintained (Kuhn, 1992).

Beyond coherence with current philosophies of science, there are cognitive values of argumentation in science education. From the cognitive perspective, to the extent that argument involves the public exercise of reasoning (Billig, 1996; Kuhn, 1992), lessons involving argument will require children to externalise their thinking. Such externalisation requires a move from the *intra*-psychological plane, and rhetorical argument, to the *inter*-psychological and dialogic argument (Vygotsky, 1978). When children engage in such a process, and support each other in high quality argument, the interaction between the personal and the social dimensions promotes reflexivity, appropriation and the development of knowledge, beliefs and values. Furthermore, to grasp the connection between evidence and claim is to understand the relationship between claims and warrants and to sharpen children's ability to think critically in a scientific context, preventing them from becoming blinded by unwarranted commitments (Quinn, 1997).

From the sociocultural perspectives on cognition, argumentation is a critical tool for science learning since it enables within learners the appropriation of community practices including scientific discourse (Kelly & Chen, 1999). If enculturation into scientific discourse is significant to science learning, then it becomes imperative to study such discourse to understand how the teaching and learning of argumentation can be traced, assessed and supported (Duschl & Osborne, 2002; Erduran, 2007). In this sense, the improvement and development of tools for capturing implementation of significant features of argumentation becomes a major concern for science education research.

Through his well-known book titled *The Uses of Argument*, Stephen Toulmin has made a significant impact on how science educators have defined and used argument. Toulmin's definition of argument has been applied to the study of a wide range of school subjects including science (e.g. Jimenez et al., 2000), history (Pontecorvo & Girardet, 1993) and English (Mitchell, 1996). It has also been used for supporting student learning. For example, Mitchell (1996) has successfully adapted Toulmin's framework as a heuristic to scaffold university students' writing.

Toulmin's definition of argument embraces concepts such as claim, data, warrant, backing, rebuttal and qualifier which are detailed elsewhere with examples (Erduran et al., 2004). It has framed the way that we have promoted argument in all of the projects that the teachers participated. This approach has included the design of resources for professional development as well as materials for students. Toulmin's framework has also been modified as an analytical tool in the coding of transcripts from classroom data (Erduran et al., 2004).

3. DATA SOURCES AND METHOD

The two teachers, one male and the other female (referred to with pseudonyms of Martha and Brian in the rest of the chapter) in their early 30s, both with physics specialisation were interviewed together by the first author about their experiences in the various projects. The interview lasted for an hour and a half. The teachers were selected on the basis of their performance in the range of projects that they participated in relative to the following criteria: the improved quality in argumentation in classroom verbal data, their ability in coordinating group discussions, promoting counter-argumentation (playing devil's advocate), giving quality feedback, engaging students in higher level questioning and using meta-language of argument in the classroom.

Both teachers' practices had improved significantly during the first two years of involvement in the initial project (Simon et al., 2006) and their students' group discussions traced during a school year had also indicated an increase in higher quality argumentation (Osborne et al., 2004a). Coded verbal data from these teachers' lessons (Table 1) illustrate the increased use of higher quality arguments in terms of more complex arguments.

Typically, arguments in terms of Toulmin's framework or Toulmin's Argument Pattern (TAP) were generated between students and teachers whereby for instance, a student would provide a claim or data and a teacher would provide a warrant for this claim-data pair. Once the transcripts were coded, the trends in the distribution of TAP in each lesson were traced in the following fashion. Each teacher implemented the same activity one year apart with comparable students. In other words, the students in each school across the two years came from the same neighborhood with similar ethnic, linguistic and racial backgrounds. The lessons were similar in structure. That is there was an introduction, group discussions, group presentations and finally assignment of homework in either case for both years. From analysis of argument frequencies in the classroom (1), we observed that Martha and Brian had significantly higher quality of arguments in the second year implementation of their lessons (Osborne et al., 2004a).

The goal of the research underlying the present study was to explore how effective teachers who have participated in argumentation projects for about 5 years reflect on their experiences as learners and teachers of argumentation. More specifically, we were interested in shedding some light on the following questions:

- How do experienced argumentation teachers articulate their views and knowledge of teaching and learning of argumentation?
- What do teachers perceive as goals, constraints and successes of teaching argumentation?
- What shifts do the teachers perceive in themselves as learners and teachers?
- What recommendations do they have for professional development?

TABLE 1. Frequency of arguments generated in teachers' whole class teaching (from Osborne et al., 2004a)

Teacher	Year	TAP/2	TAP/3	TAP/4	TAP/5
Brian	Year 1	48	47	5	0
	Year 2	59	27	14	0
Martha	Year 1	48	32	16	4
	Year 2	5	85	10	0

The teachers were asked to respond to a set of questions in a semi-structured interview format. The interview questions are included in the Annex. The pair interview which lasted for about an hour and a half, was intended to provide teachers with the opportunity to reflect on each others' ideas and to provide support for an extended discussion. The teachers were asked to reflect on the various projects, their developments across these projects, and their ideas on themes such including their views on and recommendations for teaching and learning of argumentation.

4. RESULTS

Qualitative data analysis of the interview data resulted in the generation of categories on the basis of some emerging themes. The teachers' responses were categorised relative to the following constructs: views and knowledge of teaching and learning of argumentation, perceptions of goals, constraints and successes in teaching argumentation; shifts in teachers' perceptions of themselves as learners and teachers, and recommendations for professional development. In the following sections, these categories will be detailed with some selected excerpts from the data.

4.1 Views and Knowledge of Teaching and Learning Argumentation

4.1.1 Knowledge of argument Teachers demonstrated a highly sophisticated understanding of argument as well as nuances of the nature of evidence in construction of argument. Brian, for instance, was versed in talking about not only hard evidence but also interpretations of evidence where models could act as evidence in an argument:

"If you had a model of water flowing down pipes as the electrons flowing around the circuit, you can use that model to support your argument, because the model itself can withstand the argument a little bit. But you can tear the model apart as well, it depends what you want to do. You could argue about the quality of a piece of evidence, but if you break everything down enough, at some point you end up with something you can see, and it does this, therefore it's fact."

Martha contributed to this conversation by relating the discerning qualities of models and arguments:

Martha: *Isn't evidence repeated facts?*
Brian: *It's there, so you can't argue about it.*
Martha: *Whereas a model is just an interpretation of something and therefore will have limitations.*

Here we see an instance where the teachers' voices converge and where they seem to be completing each other's ideas in agreement. Both teachers talked at great length about the nature of evidence and how evidence relates to argument. The interviewer, Sibel, asked the teachers to consider a scenario where a claim (i.e. I feel that there is a God) is substantiated by appeal to a piece of information (i.e. My conviction tells me that) which, in strict Toulmin terms could act as a claim-data pair.

Sibel: *Yeah, I mean, the example that they raised was this contrast of religion and science. In religion you can say – there is a God, because I feel that there is a God, or something. My conviction tells me that.*

Martha: *That's not a piece of evidence. I feel it is not a piece of evidence.*

Brian: *No, your evidence becomes that there are a million people that have a feeling that this thing exists. It does become an evidence then, to a point, doesn't it? It doesn't fit the scientific part of evidence which is that it's only hard evidence is you see it and it happens.*

The teachers concluded this discussion by agreeing that there are different types of evidence:

"There are definitely different types of evidence. Models are a collection of other bits of information to make evidence isn't it? The model itself has to be based on its own evidence. Otherwise it's not a model. A model is based on some things that happen. A collection of evidence. You don't just get a model out of thin air, do you? Even the religious models are a collection peoples' view of something that happened. And hence that's the evidence." (Martha)

The excepts on how TAP applies to an example of religion highlights the importance of considering the content of an argument as well as its structure if a claim is to be made about a scientific argument. Nevertheless, such nuances in teachers' understandings of the nature of evidence exceed the expectations of the professional development agenda. In other words, even though Toulmin's framework was emphasised throughout the duration of the projects, detailed analysis of the nature of evidence was not an explicitly promoted outcome. However, both Brian and Martha (along with the other project teachers) still expressed difficulty at the completion of the projects with Toulmin's framework in differentiating between claim, data, and warrant.

4.1.2 Role of content In discussing the role of content in the teaching of argumentation, the teachers made a distinction between socio-scientific topics and science topics they taught. This was not a coincidence as within the previous project that they participated, the researchers had made such a distinction and data were collected from their classrooms where they concentrated on the teaching of either kind of lessons. Both teachers indicated that they particularly enjoyed teaching argumentation in the context of socio-scientific topics:

Martha: *I do like the social science ones because I think it's really important that children learn to engage in debate that's going on. Because no one is expecting them to listen to radio or Newsnight and have that debate, but to understand there are ethical decisions about stem cells, and that is important, because they don't understand that is happening.*

Brian: *And it's a real nice link between why we should discuss this little thing today and get this conclusion. And these people are making huge decisions that will effect your future.*

In terms of the role of teachers' own disciplinary background and their teaching of argumentation, there was agreement that teaching argumentation outside of one's own science background was more comfortable as Martha illustrates below:

I suppose, in some respects, I found it easier to do things that were outside of physics because I am less obsessed with the correct answer. I didn't really know, in biology or chemistry, whereas in physics...

However in terms of the impact of argumentation on the teachers' content knowledge in the subject matter, there was no particular agreement. Brian thought that as a physicist, his biology knowledge might have improved whereas Martha did not feel that doing argumentation work has made a significant impact on her subject matter knowledge.

4.2 Teachers' Perceptions of Goals, Constraints and Successes in Teaching Argumentation

4.2.1 Goals of argumentation Both teachers tended to emphasise democratic and social goals in using argumentation in their teaching. For example, they indicated that argumentation can be used to inspire students, not to create scientists. Even though the teachers possessed a good understanding of argument in science, they did not envision the use of argumentation in science lessons towards such a goal but rather for broader goals of citizenship. Furthermore, the teachers agreed that argumentation skills were valuable because of their transferability to contexts other than science, as Brian illustrates below:

All of the skills are very transferable. If they learn how to argue successfully, with science, probably more so, they can probably argue with everything. I suspect it's easier to transfer... trying to argue about science concepts is actually quite tricky because their evidence bank isn't huge. They need to recognise they must understand more science in order to argue the point well.

The issue of argumentation as a transferable skill had been raised as part of the professional development programme. Indeed one of the earlier workshops had concentrated on the uses of argumentation skills in everyday life. Hence it is possible that the teachers have embraced this idea from the workshop input.

4.2.2 Constraints and successes Both teachers indicated that the availability of resources, time, good classroom management skills and incentives were essential for the effective uptake of the teaching of argumentation by other teachers. They were pleased with their participation in the projects which they perceived as having contributed to their personal as well as professional development. They both stated that the teaching of argumentation helped build their confidence and being more open-minded. Some other successes that they listed included the generation of an access to a bank of lessons, the opportunity to share resources with other project teachers and improved management skills in coordinating group work. Seeking and presenting evidence, reflecting on the lesson outside of the classroom and thinking from the point of view of students were given as some further examples of how the argumentation projects improved their skills as teachers. However some dreaded situations still remained that bothered them despite their long involvement in argumentation. These included students arriving at multiple explanations with the same evidence and students accepting wrong explanations without questioning them. Some of these issues had already surfaced during the professional development programmes and persisted at this stage as well.

4.3 Shifts in Teachers' Perceptions of Themselves as Learners and Teachers

In reflecting on how they became very good at teaching argumentation, the teachers mentioned persistence and being adventurous as two important attributes that helped them adopt the new approach. Open-mindedness enabled Brian to overcome his initial skepticism:

... I think even when you came to observe us at round about Christmas... I still sat at the end of that and said – yeah, this is great for the transference of skills across the curriculum, but I don't really trust it for science.

Even though the teachers believed that good teachers would normally engage their students in discussions and would play the devil's advocate to push their students' thinking, Brian believed that "what was new and refreshing about this [project] was taking it into areas where we hadn't previously had a discussion." The teachers acknowledged the role of understanding the terminology (like claim and evidence) involved in discussing the features of an argument in helping them appropriate those terms and making them part of their classroom discourse.

Both teachers mentioned how participation in the project has helped them manage the pupils through assigning a specific number of students to a groups and ensuring that each of the students have roles. Those skills have become part of the teachers' management strategies whether they are engaging the students with argumentation or not. Another skill that the teachers gained was that of immediately appealing to evidence any time a debate or disagreement came up in class, consulting with "scientists in the other subject areas" to ensure they are on the right track, and considering how "a 13 year old would see as a point of conflict."

4.4 Recommendations for Professional Development

Teachers were asked to reflect on the professional development opportunities provided by the research and development projects. The video exemplars produced following the research project (Osborne et al., 2004b) were perceived as being insufficient for new teachers who had had no sustained involvement in research projects. The video shows particular clips focusing on key strategies (e.g. managing group discussions, modeling) and even though there is one full edited argument lesson provided, the teachers felt that the use of focused clips might not be very useful for teachers who are not familiar with this strategy at all.

I always think that's the trouble with every bit of training I've seen. And I like doing varied things. But the most useful thing for me to see is someone teach a good lesson from start to finish. And I know that's never ideal because you always have one set of kids in front of that person. And really, what you need to see is that lesson taught to weak, able and very able classes, the same lesson. And watch what they do. And you can actually, genuinely see – that's how you approach that, that's how you deliver that. It's still very bitty. It does provide food for thought but I don't think it teaches them how to deliver the lesson. (Brian)

Furthermore, the teachers felt that they would have benefited from seeing the same lesson taught to a different student populations. Fitting an argumentation lesson into the school's program of study was also an issue to consider in expecting teachers to implement a new strategy:

... but again it would have to fit with what your scheme of work. Because if you've got five days arguing about the ethics of genetic engineering and you've only spent half a lesson on it, that's four and a half lessons that you've lost. So it's getting that balance right. And I see more of it coming down to, actually, much shorter things within lessons. And actually more regular. As opposed to a whole lesson. We were told to do a whole lesson, but sometimes it would have been a bit snappier if they were shorter sometimes. Because you'd just focus on a point. (Martha)

The professional development program had the expectation that the teachers would teach 9 argument lessons across a school year. Each lesson would be based on one period. The choice on a period-length lesson was intentional to make the less of a demand on the teachers and to enable a more realistic uptake in the long term. The teachers' emphasis on short teaching episodes is consistent with this agenda.

5. CONCLUSIONS

In this study we have discussed a case study of two science teachers with long term involvement in research and development projects on argumentation. Both teachers displayed sophisticated understanding of argument as well as its teaching and learning. Their recommendations centred around effective professional development to take into account a holistic presentation of teaching scenarios and a range of student abilities. Both teachers indicated that their own success with the project was due to their persistence in learning something new and the nature of the workshops conducted with them and other teachers – which have been summarized, trialled and published subsequently (Osborne et al., 2004a, b). They also indicated that among many teaching strategies, they are now more conscious of doing group work and they view the ability to conduct and coordinate group discussions as a significant skill that can be transferred to other aspect of teaching. When asked to reflect on what kinds of developmental and cognitive skills they would expect students to undergo in the learning of argumentation, both teachers referred to a scheme used in the research project to analyze the quality of student argumentation in group discussions. The scheme derived from a theoretical account of argument based on Toulmin's work (1958) focussed on the use of rebuttals and the use of data and warrants to support one's claim while another person is in opposition to an original claim (2). Both teachers, whose classroom practices included meta-level language with students about the nature of rebuttals (Table 1) indicated that a development in argumentation skills would necessitate the presence of improved skills with rebutting an argument.

6. IMPLICATIONS

The teachers' responses in the interview along with the data from earlier studies provide a rich source of data not only to trace these teachers and their students' development in argumentation but also to illustrate how the community of practice (Lave & Wenger, 1991) in argumentation has evolved across different projects over 5 years. The case studies of these teachers underscore the need to strike a balance between encouraging teachers to take the risks needed for development, and providing them with sufficiently supportive strategies to try something innovative. Establishing the capabilities and inclinations of teachers towards such risks requires teacher professional development courses to take into account teachers' existing knowledge (Fullan, 2001; Loucks-Horsley et al., 1998) and conceptualisation of science teaching. Teachers will only take risks, however, if they are offered a clear vision of why this approach is important, how it might be implemented effectively, and how it builds on their existing skills. Articulating this vision through a case study such as the one described here will, we believe, contribute to sustainable educational reform. Future areas of research will include situating the role of argumentation in influencing teachers' other pedagogical skills, and tracing the developmental stages in the learning to teach argumentation from novice to expert teaching. Establishing argumentation as a goal in pre-service science teacher education (Erduran et al., 2006), we believe, will help teachers in adapting and sustaining argumentation as a long-term pedagogical strategy.

ACKNOWLEDGMENTS

We would like to thank the teachers for participating in the study. Table 1 is from data collected as part of the ESRC project (UK Economic and Social Science Research Council grant number R000237915) which had contributions from Jonathan Osborne and Shirley Simon.

NOTES

[1] Results from TAP coding of each transcript were further summarized in the following fashion. First, the TAPs were grouped in terms of the occurrence of double, triple, quadruple and quintuple combinations. We called these permutations of TAP features 'clusters'. For example, the 'claim-data-warrant' and 'claim-data-rebuttal' were grouped as an instance of cluster 3. It is implicit in our grouping that with increasing number for a cluster, the argument becomes more complex in nature. In other words, we are assuming that a 'claim-data' argument is a less sophisticated form of an argument than a 'claim-data-warrant' argument where there is an added feature of justification (in terms of a warrant) in the latter scenario. Furthermore, for our coding purposes, we concentrated on identifying arguments in terms of the quantity of TAP features in arguments, not qualitative differences across different permutations of TAP. That is to say, by collapsing different arguments into clusters, we did not differentiate between arguments that might have a different qualitative composition despite a quantitative equivalence in terms of TAP. That is 'claim-data-warrant-rebuttal' and 'claim-data-warrant-backing', both instances of cluster 4, are grouped together since each has 4 features of TAP even though qualitatively there is a difference between the arguments in terms of presence/absence of rebuttals and backings.

[2] The scheme based on levels of argument is detailed in Erduran et al. 2004.

REFERENCES

Billig, M. (1987). *Arguing and thinking: A rhetorical approach to social psychology Cambridge*: Cambridge University Press.

Cochran-Smith, M., & Lytle, S. L. (1999). Relationships of knowledge and practice: Teacher learning in communities. In, A. Iran-Nejad, & C. D. Pearson (Eds.), *Review of research in education*, vol. 24, pp. 249–305. Washington, DC: American Educational Research Association.

Driver, R., Newton, P., & Osborne, J. (2000). Establishing the norms of argumentation in classrooms. *Science Education*, 84(3), 287–312.

Duschl, R., & Osborne, J. (2002). Supporting and promoting argumentation discourse. *Studies in Science Education*, 38, 39–72.

Erduran, S. (Ed.) (2007). Argument, discourse and interactivity. Special Issue. School Science Review, 88(324).

Erduran, S. (2006a). Argumentation in initial teacher training: Empowering prospective chemistry teachers with evidence and justification. In, I. Eilks, & B. Ralle (Eds.), *Towards research-based science teacher education*, pp. 41–52. Aachen: Shaker Verlag.

Erduran, S. (2006b). Promoting ideas, evidence and argument ininitial teacher training. *School Science Review*, 87(321), 45–50.

Erduran, S., & Jimenez-Aleixandre, M. P. (Eds.) (in press). *Argumentation in science education: Perspectives from Classroom-Based Research*. Dordrecht: Springer Academic Publishers.

Erduran, S., & Osborne, J. (2005). Developing arguments. In, S. Alsop, L. Bencze, & E. Pedretti (Eds.), *Analysing exemplary science teaching: Theoretical lenses and a spectrum of possibilities for practice*. Buckingham: Open University Press.

Erduran, S., Simon, S., & Osborne, J. (2004). TAPping into argumentation: developments in the use of Toulmin's Argument Pattern for studying science discourse. *Science Education*, 88(6), 915–933.

Erduran, S., Osborne, J., & Simon, S. (2005). The role of argument in developing scientific literacy. In, K. Boersma, O. de Jong, H. Eijkelhof, & M. Goedhart (Eds.), *Research and the quality of science education*. Dordrecht: Kluwer Academic Publishers.

Erduran, S., Ardac, D., & Yakmaci-Guzel, B. (2006). Learning to teach argumentation: case studies of preservice secondary science teachers. *Eurasia Journal of Mathematics, Science and Technology Education*, 2(2), 1–14.

Fullan, M. (1991). *The new meaning of educational change* (2nd ed.). London: Cassell.

Gee, J. P. (1994, April). *Science talk: How do you start to do what you don't know how to do?* Paper presented at the Annual Meeting of the American Educaitonal Research Association, New Orleans, LA.

Giere, R. (1991). *Understanding scientific reasoning* (3rd ed.). Fort Worth, TX: Holt, Rinehart, and Winston.

Jimenez-Aleixandre, M., Rodrigues, A., & Duschl, R. (2000). "Doing the Lesson" or "Doing Science": Argument in high school genetics. *Science Education*, 84(6), 757–792.

Kelly, G. J., & Chen, C. (1999). The sound of music: Constructing science as sociocultural practices through oral and written discourse. *Journal of Research in Science Teaching*, 36(8), 883–915.

Kuhn, D. (1992). Thinking as argument. *Harvard Educational Review*, 62, 155–178.

Kuhn, T. E. (1962). *The structure of scientific revolutions*. Chicago: University of Chicago Press.

Latour, B., & Woolgar, S. (1986). *Laboratory life: The construction of scientific facts* (2nd ed.). Princetown, NJ: Princetown University Press.

Lave, J., & Wenger, E. (1991). *Situated learning: Legitimate peripheral participation*. Cambridge: Cambridge University Press.

Lemke, J. (1990). *Talking science: Language, learning and values*. Norwood, NJ: Ablex.

Loucks-Horsley, S., Hewson, P., Love, N., & Stiles, K. E. (1998). *Designing professional* development *for teachers of science and mathematics*. Thousand Oaks, CA: Corwin Press.

Magnusson, S., Krajcik, J., & Borko, H. (1999). Nature, sources and development of pedagogical content knowledge. In, J. Gess-Newsome, & N. G. Lederman (Eds.), *Examining pedaogical content knowledge*, pp. 95–132. Dordrecht: Kluwer Academic Publishers.

Mitchell, S. (1996). *Improving the quality of argument in higher education interim report*. London: Middlesex University, School of Education.

Osborne, J., Erduran, S., & Simon, S. (2004a). Enhancing the quality of argumentation in school science. *Journal of Research in Science Teaching*, 41(10), 994–1020.
Osborne, J., Erduran, S., & Simon, S. (2004b). *Ideas, evidence and argument in science*. In-service Training Pack, Resource Pack and Video. London: King's College London.
Pontecorvo C., & Girardet H. (1993). Arguing and reasoning in understanding historical topics. *Cognition and Instruction*, 11(3 & 4), 365–395.
Quinn, V. (1997). *Critical thinking in young minds*. London: David Fulton.
Simon, S., Erduran, S., & Osborne, J. (2006). Learning to teach argumentation: Research and development in the science classroom.*International Journal of Science Education*, 28(2–3), pp. 235–260.
Toulmin, S. (1958). *The uses of argument*. Cambridge: Cambridge University Press.
Vygotsky, L. (1978). *Mind in society*. London: Harvard University Press.
Zohar, A., & Nemet, F. (2002). Fostering students' knowledge and argumentation skills through dilemmas in human genetics. *Journal of Research in Science Teaching*, 39(1), 35–62.

ANNEX

Theme	**Question**
Transfer	Do you think that your skills with teaching argumentation makes any difference to other aspects of your teaching? How?
	How has the teaching of argumentation contributed to your understanding of the (a) nature of science (b) scientific reasoning (c) social aspects of science?
	Do you think that students gain something else than argumentation skills by doing argumentation activities?
	Do you think that getting better at argumentation means that students' conceptual understanding also improves?
	In what ways (if any) do you think students' skills with argumentation help them with other aspects of science /subjects?
	How does the science curriculum benefit from inclusion of argumentation? What are some benefits/disadvantages with respect to other subjects?
Teacher development	What are some major issues in initial teacher development and argumentation?
	If you were to convince a new teacher about the value of teaching argumentation, what would you say? What would you show them?
	What are some main issues in continuing professional development and argumentation?
Science domain	Having done argumentation lessons in different subjects and topics, do you think that there are any particular issues about teaching argumentation in say physics versus chemistry?
	Do you think that there are certain features of argument that are similar across subject areas?

Evaluation of self effectiveness	Why do you think have you been very good at teaching argumentation?
	What would you say are the best features of your teaching of argumentation?
	In what ways has your teaching of argumentation contributed to other aspects of your teaching?
Misconceptions	One main criticism of argumentation teaching is its opening up of space to include students' misconceptions in the lesson. How would you react to these concerns?
Role of evidence in argument	One of the main differences between science and religion is their emphasis on evidence. Do you think that we have delineated the nature of evidence effectively? Do you think that models could count as evidence?
Toulmin's Argument Pattern	Some teachers had concerns about using TAP. What are you thoughts on the use of TAP in teaching argumentation?
	If you could use another definition of argument, what would it be?
Good teaching	There are some researchers who say that good teaching requires a sound understanding of not only the subject matter but the nature of knowledge in that subject. What do you think?
Materials/ strategies	What are you most favourite sets of materials/strategies in teaching argumentation?
	As you know, most of the lessons you have taught in our projects were single lesson based. If you were to teach a sequence of lessons, what would be some major issues to consider for you as a teacher? How would you structure the sequence of lessons? How would you evaluate student work along the way and also in a summative assessment?
Developmental aspects	Do you think that students of any age can learn to argue?
	Do you think that students will retain their knowledge of argumentation beyond middle school? What makes you think that?
	If we were to design activities for primary/secondary/tertiary science education, how would the curriculum be best structured?
Action research	There has been some resistance to this idea in one of our projects. Yet this is an idea that's very familiar to argumentation: it's about arguing about your own teaching through evidence from your teaching. Given your effectiveness and understanding of argumentation, how would you convince new teachers to deal with evidence of their own teaching?

MARINA CASTELLS, JUAN ENCISO, JOSEP M. CERVERÓ, PERE LÓPEZ AND MODESTO CABELLOS

32. WHAT CAN WE LEARN FROM A STUDY OF ARGUMENTATION IN THE STUDENTS ANSWERS AND GROUP DISCUSSION TO OPEN PHYSICS PROBLEMS?

Abstract: The chapter focuses on students' arguments with a framework based on the Theory of Argumentation of Perelman and Olbrechts-Tyteca (1958). From this perspective, which joints dialectics and rhetoric, argumentation is understood as the set of discursive techniques that will convince the audience of the validity of the theses presented. The general aim is to find a new way to interpret students' misconceptions and reasoning, with particular reference to the role of premises and argumentative schemes they use, which we consider part of common sense. The chapter is based on two studies. The first study was carried out with teacher trainees' students in the university. Written answers to three qualitative kinematics problems were compiled. In the second study, held in a secondary school class, small group discussions answering a question about free fall were video recorded. The results provide new insights into students' scientific misconceptions and into the reworking of knowledge that emerges in the students' discussions in small groups. The implication for science education is that focusing explicitly on argumentation in science classes and using the same kind of arguments students use, and discussing them, is likely to improve not only students' argumentative skills, but also their conceptual understanding of science itself

Keywords: Argumentation, Problem solving, Physics education, Secondary education, Teacher training

1. INTRODUCTION AND BACKGROUND

This study focuses on the argumentations students use in their answers and in the group discussions when trying to solve informal problems in physics.

We assume that teaching and learning sciences is fundamentally a communicative process. Scientific meaning is constructed through language and other semiotic tools that mediate the building of scientific entities and models, which represent new pieces for thinking with and will facilitate the production of arguments and explanations about real or laboratory phenomena. Today it is widely agreed that a fundamental aim of science teaching in school is to develop competences that will permit students to modify their knowledge and to acquire new knowledge

Marina Castells, Dept. DCEM, Universitat de Barcelona, Passeig de la Vall d'Hebron, 171 (08035) Barcelona, Spain.

R. Pintó and D. Couso (eds.), Contributions from Science Education Research, 417–431.

throughout their lives. This means that students must learn to obtain information by themselves, to distinguish between good and deficient information; to do so, students must learn to reason and to argue. In general, argumentation is not part of the science curriculum, it is assumed that students learn mainly to argue by imitating the examples of argumentations used by their teachers, by reading texts or through social interaction, and it is unusual to find practices of meta thinking about arguments in textbooks or in students' discourse.

In recent years agreement has been reached on the importance of argumentation and rhetoric in the construction of scientific knowledge, and on its status as a central feature of scientific activity (Gross, 1990; Kuhn, 1993; Pera & Ahea, 1991; Prelli, 1989; Kelly et al., 1998). This central role of argumentation is supported by psychologists, philosophers and sociologists of science, and, therefore, cannot be ignored in any science education (Duschl & Osborne, 2002). Parallel to these considerations, for many years in mathematics there has been a line of thinking called 'heuristic mathematics' which detaches the value of 'plausible reasoning'[1] in the construction of mathematical knowledge (Polya, 1945). Mathematical theories are held to appear in two stages: a first stage, in which the plausible reasoning is fundamental, followed by a second stage of formalization in which the discoveries are re-examined and given a formal structure. This structure is normally deductive. Polya (1953) asserts that the evidence used in all subjects (including Physics) and in every day context as well can be defined as plausible reasoning; and he advocates the teaching of plausible reasoning in mathematics classes (and by extension, in physics classes as well).

As far as science education is concerned, argumentation and rhetoric are considered fundamental in the learning of scientific knowledge (Kitcher, 1988; Khun, 1992, 1993). Numerous studies have focused on the analysis of argumentation discourse in educational contexts (Kelly et al., 1998; Duschl et al., 1999; Driver et al., 2000; Jiménez-Aleixandre et al., 2000; Erduran et al., 2004). More recent research has sought to elucidate the dynamics of classroom interactions that initiate and sustain argumentative discourse by collaboration between researchers and middle-school science teachers and to develop models of educational activities in an effort to make argumentation a component of instruction (Zohar, 2002; Osborne et al., 2004). An example is the development of the IDEAS project, whose instructional model is based on the use of Toulmin's Argument Pattern (TAP) (1958).

The TAP is the model of argumentation mainly used in researches done in this field. Our hypothesis is that, for research, the TAP is suboptimal for analysing the spontaneous arguments students use to persuade their peers or the teachers of their ideas. The TAP is fundamentally discourse-centred; if the aim is to study the argumentation used to convince others (an audience), a model centred on the speaker and on the audience, such as Perelman and Olbrechts-Tyteca's Theory of Argumentation (TA) (1958), must be used (Hoogaert, 1995). As the TA has not been used as a framework for science education research, we will pause at this point to explain why we chose this theory for our study.

What do we understand by the term 'argumentation'? Argumentation has two main complementary meanings: (1) a social and collaborative (or individual) process necessary for solving problems and for advancing one's knowledge (Duschl & Osborne, 2002) and (2) the discursive techniques that will persuade the audience of the validity of the theses presented (Perelman and Olbrechts-Tyteca, 1958). In these perspectives, notions as justifying, reasoning and arguing are all submitted under the heading of argumentation.

2. INTEREST OF THE STUDIES AND JUSTIFICATION OF THE FRAMEWORK

Solving open problems, especially in everyday contexts, offers a favourable learning context for argumentation that can be carried out either in groups or individually. Much research have been done on problem-solving in science (e.g. Heller et al., 1992; Heller & Hollabough, 1992) but not from a rhetorical and argumentative perspective. Studies centred on argumentation present to students questions that are related more to values and opinions than to scientific concepts. Here we study argumentation in relation to problems with physics content, and so our research should be contextualized inside the construction of scientific knowledge.

In fact, we are interested in identifying students' argumentations in problem-solving activities (individually or in small group) in order to obtain arguments to be used in science teaching. Our hypothesis is that arguments will convince students about scientific models if these arguments are not very far from common sense.

These ideas suggest that an analytical framework based on rhetoric and argumentation may be well suited to the study of the students' argumentations. We choose Perelman and Olbrechts-Tyteca's TA (1958) because it focuses on plausible reasoning and joints dialectics and rhetoric.

3. A FRAMEWORK FOR ANALYSING ARGUMENTATION: PERELMAN AND OLBRECHTS-TYTECA'S THEORY OF ARGUMENTATION

Perelman and Olbrechts-Tyteca's theory consists mainly of three related essential aspects of an argumentation,[2] in accordance with their impact on the audience: premises, argumentative techniques and theses. Perelman and Olbrechts-Tyteca (1958) deals in depth with many other elements that have traditionally been considered as belonging to rhetoric, but as the title of his book indicates,""[3] he includes them under argumentation.

The 'theses' or conclusions are statements about which the speaker wants to convince the audience (oneself or others) through argumentation. The 'premises' are the data and the agreements on which the argumentation is built. Much of Perelman and Olbrechts-Tyteca's book is devoted to premises and their degree of adaptation to the argument, reflecting the importance he attributes to them in argumentation. According to Perelman and Olbrechts-Tyteca the premises can be of several kinds.

From the real: facts,[4] truths or theories and presumptions. From the preferable: values, hierarchies of values and "places".[5]

The 'argumentative techniques' include 'argumentative schemes' of isolated arguments, as well as their interaction, amplitude of argumentation and the order of arguments, which Perelman and Olbrechts-Tyteca (1958) relates with the convincing force of the argumentative discourse. The 'argumentative schemes' are the discursive structures that allow the transfer of agreements from the premises to the theses. The argumentative schemes are grouped into two main categories: 'schemes by association' or connection that joint separated elements in a new structure, and schemes 'by dissociation' or separation that separate elements considered related or part of a whole, thus changing systems and notions. Inside these broad categories many other subcategories can be identified.[6]

4. METHODS, PARTICIPANTS AND EDUCATIONAL CONTEXT

4.1 Research Questions

The research is based on the following questions:

- Are students without specific training in argumentation able to elaborate arguments? What kinds of arguments can be found in the argumentations used by students to solve informal problems in physics?
- What kind of premises are students' arguments based on? Where do these premises originate? Do they accord with scientific ideas or theories?
- Can the analysis based on the Theory of Argumentation (TA) help us to understand misconceptions and strategies of reasoning identified previously by researchers?
- Do peer group discussions contain relevant interventions that help to modify students' initial theses? Can the argumentative process in students' discussions be considered as a collaborative process?[7]
- What are the implications for science education of using the TA as a tool for the analysis of students' argumentations?

4.2 Mode of Inquiry and Materials

The research is basically qualitative. Our data are taken from two different studies carried out in Barcelona. The first study was carried out with teacher trainees' students in the university. Written answers to three informal problems of kinematics were compiled. The number of students answering the problems ranged between 25 and 40 students, depending on the problem. The three problems analysed represent the same physics problem in three different situations: a river, a train and conveyor belt. [8]

The common question of the problems can be summarized as follows:

"A living being is situated in the middle point between two objects; s/he/it and the objects are being dragged by a moving support. This living being has to decide which of the two objects, the one ahead or the one behind, s/he/it would reach first if he/she moves in one direction or the other? Why?"

The second study is situated in a secondary school class where scientific investigations in small group take place. The subject matter was "Scientific Investigations", in which the students had to answer some scientific problems by discussion and experiments. The problems were to be proposed by the teacher or the students.

One of the proposed problems is about free fall:

"Does the time of fall from a specific height depend on the mass of the objects that fall?"

5. DATA ANALYSIS AND RESULTS

The analysis in both studies involves mainly identifying the arguments used by the students and characterizing them by thesis, premises and argumentative schemes. The analytical categories are based on those presented in Perelman and Olbrechts-Tyteca's book. In the second study, we also analysed the specific contribution of certain students or the teacher in the argumentation, which can be considered especially relevant in relation to how a premise, initially shared by all the members of the small group, may be changed.

Two individual written answers and fragments of the students' discussion are presented and analysed to illustrate the collected data and the analysis procedure. Along with the written answers and transcriptions, we also present our interpretation based on the TA.

5.1 Illustration from the First Study

The river problem, answers and our interpretations (only of question b)

> p. 1 – Imagine that you are floating on a tyre in the middle of a river and not swimming. Some metres upstream there is a wooden box floating and some metres downstream is another wooden box.
> (a) If you are floating, not swimming, will the distance between you and each box change with the time? Why?
> (b) Imagine the tyre no longer holds you up. Which of the two boxes would you reach first? Why?

Student S_1's answer:

"Even if the box upstream is nearer to us, swimming against the current requires a great deal of effort and would be a waste of time, so even if the box downstream is further away, we would reach it first, because the speed we acquire by swimming is complemented by the impulse of the current".

The argument used is based on the effort necessary to reach the boxes. We can see here the use of the premise: *swimming against current requires great effort.* Could this be considered a 'place'? Everybody will agree with this idea even if they did not have the particular experience. The argument used is an 'argument of comparison'. What is compared? The *effort* the swimmer has to make to reach one or the other box. Regarding the box upstream, the student says that the effort of swimming must be very great because the person has to *swim against the current*, regarding the box downstream the student says that the distance that he has to swim will be greater than the distance to the box upstream, but the distances do not seem to be relevant and he makes his decision on the basis of the effort that he has to make to reach the boxes. In the case of the box downstream, the current will help the swimmer, through an "impulse" that it gives him, and so (this is implicit) the effort needed is less. In reality, the magnitude considered is the *effort by unit of time* and not the total quantity of effort. It is a typical 'quasi logic argument'[9] because the students cannot measure the total effort, and so the comparison is not between two measurable quantities. This kind of argument also has a very important consequence: Indeed, the student's answer depends on intuition. We can also be aware that the student does not answer the original question of the problem which was about elapsed time and not about effort, but this does not matter to the student.

The argument of this student could also be considered as an 'argument of the easiest option',[10] included among the 'arguments based on the structure of reality'. Underlying this is the place of 'the preference of what is easy above what is difficult'[11] (into the 'place of quantity', according to TA).

Student S_2's answer:

"I would go towards the box situated some metres upstream because although I have to swim against the current, the same current will bring the box towards me; in contrast, if I swam towards the box situated downstream I would take longer to reach it because at the same time as I swim towards it, the box would move downstream by the force of the current".

A premise is: *Swimming against the current is difficult, it requires effort.* Another premise seems to be*: the current moves floating bodies in the water.*

The argument is an 'argument of comparison' of magnitudes: what is compared is *the time taken* to reach one or the other box. The time is calculated via the *measure of distance to swim.* We can say that the argument is a 'quasi logic argument by mathematical relationship'. The student compares distances and makes conclusions about times, arguing by a 'schema of more implies more', more distance implies more time (like a direct proportionality). Perelman and Olbrechts-Tyteca does not put explicitly this kind of argument into the 'quasi logic arguments'; instead, we can find it as an argument of 'double hierarchy'[12] (TA, p. 517), which is one of the groups of 'arguments based on the structure of reality'. This is a very common schema that everybody would agree with: has practically a status of 'place'. But beyond this apparently "more formal" argument, we also see a kind of 'pragmatic argument'[13] (TA, p. 409): 'I will go to the box that is easier to reach', included among the 'arguments based on the structure of reality'. We can consider that a premise of 'place of quantity' is also the substratum of this argument, the choice of the most favourable option is the manifestation of this place here.

What is the problem with this answer? It is that the students do not consider that the current will carry all the objects that float, both person and boxes. *The carrying of the person is ignored* in this reasoning but not the carrying of the boxes. Why? We hypothesize that it is because the student sees the person in the water as an active body while the boxes are seen as inactive, only floating. Focusing on the action of swimming, the student ignores that the person will be carried by the current.

Answers to the train problem and our interpretations:

> p. 2 – Imagine you are in a train, and you are exactly in the middle of the carriage. If you want to leave the train as quickly as possible, would you go to the door in front of you or to the one behind you?

We compared the argumentations given by students in the river problem with the same problem in the train and in the belt situations in order to assess the influence of context. Here we present only one example of the answers to the train problem.

Student S_3's answer:

"I would decide to go out by the rear door. If you go to the door at the front, there are forces that push you back and so you have to make much more force in order to walk. But, if go to the rear part of the train, you will almost be running because it is as if somebody is pushing you from behind".

This reasoning is similar to the one presented in the river problem, but here the premise changes completely. The premise is: *the train exerts a force on the person from behind*. We can see that this force is in the opposite direction to the one the current exerts on the swimmer. The argument is an 'argument by comparison', the magnitudes to be compared are *the force that the person has to exert to walk in one direction or the other*. What is the origin of the premise? Is it acquired by experience? For us it could be related to the force of inertia that we teach in Physics. However, another interpretation is possible. The students may be thinking that they are inside a train that is stopping and that the braking force can be regarded as a force in the direction opposite to the motion. In summary, the student interprets the problem in a direction that contributes to the premise used in the reasoning.

5.1.1 Some results These and other analysed cases suggest that in the situation of individual written work the considered premises of the students change with the context of the problems, which indicates that in the reading of the problem the students are making an interpretation of it. The same problem (from the perspective of Physics) presented in three different contexts can be converted into three different problems by the students. These interpretations of the problems can depend on several factors, any one of which could be 'places'. This result agrees with the importance that Perelman and Olbrechts-Tyteca (1958) attributes to the interpretation of the data used in an argument.

The argumentative structures detected in the students' answers to the problems mentioned above mainly come under the heading 'arguments of association'. Inside this large group of arguments we find any 'quasilogical argument' in almost all the individual students' answers; we also find 'arguments based on the structure of reality' and 'arguments that build the structure of reality'. The use of quasi logic arguments in science classes is interesting that means students try to argument like "formal physics" and although they are not able to do these kind of formal arguments correctly, they use similar arguments to these. This inability may be due to their limited knowledge of physics and mathematics.

In the group of 'arguments based on the structure of reality', in the studied cases, we find 'arguments of *gaspillage*'[14] (TA, p. 430), specifically, arguments similar to 'arguments by sacrifice'[15] (TA, p. 432), also 'pragmatic arguments' (TA, p. 409) such as the 'argument of what is possible', the 'argument of the easiest option' and the 'argument based on an opportunity'. They are all 'arguments by succession'[16] (TA, p. 404). We find also 'arguments by causal nexus' (TA, p. 405), a kind of 'arguments by succession'.

In the 'arguments that establish reality' (TA, p. 536) our students present 'arguments based on a particular case' (TA, p. 536), such as 'argument by illustration' (TA, p. 546) and 'arguments based on analogy' (TA, p. 569). This last kind of argument is presented in almost all the answers analysed, though the analogous case is not always explicit.

5.2 Illustration of the Second Study

The question:

Does the time of fall from a specific height depend on the mass of the objects that fall?

Here we present a fragment at the beginning of students' discussions:

14. S2: *Yes, the one that weighs more, falls first, that is normal.* (Theses)
15. S4: *But…*
16. S2: *Then the hypothesis is…*
17. S4: *We have already said that. Does the heaviest body fall first?*
18. S1: *But the hypothesis cannot be a question.*
19. S2: *No, the hypothesis was an imagined answer.*

....

23. S3: *We have to check that.*
24. S4: *How stupid! And how can we check this?*
25. S1: *Well, then our hypothesis is that the heavier body lands first.* (Thesis)

In this group discussion, the initial thesis of all the students participating is: *The greater the mass, the shorter the time the object takes to fall,* or perhaps: *The greater the mass, the higher the speed of the object.* This thesis, which corresponds to the answer of the question, is taken by the students as a premise supported by the place of 'what is normal' (TA, p. 151), and so does not need to be justified. This can be illustrated by comments made by two students: "Yes, the one that weights more, falls first, it is normal", "How stupid!..." (i.e. that there is no point in proving the hypothesis). Perelman and Olbrechts-Tyteca (1958) considers the place of 'what is normal' (TA, p. 151) a 'place of quantity' (TA, p. 148). We can also consider this common place as an argument of 'double hierarchy' (TA, p. 517), which means the schema: more → more (like a direct proportionality).

How should we interpret this initial fragment of the discussion? Students have a common schema, which is a universal agreement or place. They see it as corresponding to the structure of reality: this is how things are and there is no more to be said. To use the language of science education, this is a common conception; indeed, a misconception. Based on TA, we interpret this as a universal argumentative scheme or perhaps a 'place'.

But as the teacher asked the students to demonstrate their hypothesis by experiments, the premise will function as a thesis to be proved. The teacher's intervention is fundamental in this transformation of a premise into a thesis. In other discussions, a student may also play this role questioning the premise.

The discussion continues as follows:

156. S4: *Repeat, repeat*
157. S1: *Wait*!
158. S3: *Ok, the one, two, three*
159. S2: *Do they fall at the same time, or not?*
160. S1: *Let's do it again...*
161. S3: *Get on here* (on the chair)
162. S3: *One, two, and three.*
163. S2: *They fall the same.*
164. S4: *They fall the same.*
165. S3: *This one falls faster at the beginning, but later they fall the same.*
166. S2: *Perhaps it's like parachutists who fall very fast at the beginning and more slowly at the end?*
167. S1: *Wait; let's do it with the keys and the rubber.*
168. S3: *They fall equally...*
169. S2 (he/she writes) *Mass doesn't influence the...*
170. S1: (he/she takes the pen from S_2 and crosses it out)
171. S1: *What I think is that it depends on the distance.*
172. S4. *We have reached a point where we break all the outlines of the investigation.*
173. S3: *Let's see, this box weighs more than this other one.*
174. S4: (he/she picks up the box and drops them) *They fall equally.* (They call for the teacher)
175. Teacher: *We drop two objects of different mass, see what happens.*
176. S4: *They fall equally.*
177. S1: *I believe that the distance also, partly ...*
178. S2: *That's all* (he/she writes) *the mass doesn't influence.*
179. S4: *Proved* (he/she hits the table). *The mass doesn't influence.*
180. S1: *I believe it is the distance*

181. S2: *You're very stubborn; you can see that it doesn't affect it.*

By the end of the experiments some students are convinced that their initial hypothesis was wrong. We can see here a process of reworking knowledge (Ogborn et al., 1996). But, for some students, this transformation is very hard to accept; they try to defend their conceptions. The resistance to change may lie in the fact that the interpretation according to Physics clashes with a very common argument. What strategy do students use to maintain his initial hypothesis? We find that students introduce new factors or variables that they think could affect the result of the experiment. For example, one student (S_3) proposes the following argument to maintain the initial hypothesis: "at first, the heavier body falls more quickly, then both bodies fall at the same speed". This argument seems a superposition of the scientific idea and the common idea regarding the free fall. Although another student (S_2) gives some comments that suggest where this idea arises from, establishing an analogy with the parachute, this new factor is not taken into account by the other students. However the distance introduced by student S_1 is taken into consideration in the group. His argument is: "two bodies of different weight land on the floor at the same time, but if the distance of the fall is not great enough, it cannot be observed". The argument used is an 'argument relative to the difference of degree' (TA, p. 527), an argument 'based on the structure of reality', according to Perelman and Olbrechts-Tyteca. Although the other students seem readier to change the initial hypothesis than S_1, they agree to do the experiments to test S_1's idea. In the following fragment of the discussion, students are carrying out new experiments, taking into account S_1's ideas:

207. S2: *Well, now, what are we doing?*
208. S1: *It is necessary to make it in a bigger space.*
209. S2: *He is obsessed with the idea that it influences.*
210. S1: *Let's throw two cases up and let see which one will fall first.*
211. S2: *Let's go*
212. S1: *One, two, three.*
213. S3: *They fall at the same time.*
214. S2: *At the same time.*
215. S1: *Let's repeat it.*
216. S4: *I'll get the camera*
217. S3: *Now let's see if we throw two objects of different mass up, if they will fall at the same time.*
218. S4: *Wait, Alex you can't see it well.*
219. S1: *Come on, whenever you want.*
220. S2: *The same, exactly the same, nailed!*
221. S3: *Wait, this box and that one, can you see it? This one weights more.*
222. S4: *Do they fall nailed, eh? Do they fall nailed.*
223. S1: *Wait, repeat it, come on!*
224. S4: *Yes, they fall nailed, how strong!*
225. S2: *The same, the same*
226. S1: *But I think we could do again in a bigger space*

227. S4: *But why? Why do you think that they won't fall at the same rate?*
228. S1: *Juan, can we check it from somewhere higher?*
229. Teacher: *Why?*
230. S1: *If the height is great, the one with the bigger mass will build up more speed.*
231. Teacher: *Do you believe that the speed of the fall depends on the mass?*
232. S1: *Yes, I believe it does.*
233. Teacher: *Try this: you stay in the classroom and the others go down to the playground, and you drop some objects out of the window.*
234. S3 & S4: (From the class they drop objects of different masses down onto the playground).
235. S1 & S2: (They watch the objects fall in the playground and then come back to the laboratory)
236. S1: *I still think that the height has an influence.*
237. S2: *Mass doesn't influence the falling of the bodies.*
238. S1: *Well, I will try it at home; we've got a very high balcony.*

Despite the experiments, S_1 has not changed his mind. His teacher told us later that he only accepted the new idea when mathematical formulas were used to calculate the time of fall (in the next class, not presented here). This episode can be interpreted as an indication of the authority of mathematics (the formulas) to persuade the student who is not convinced by experiments.

5.2.1 Some results The premises on which students base their reasoning are interpreted 'facts', and 'places' underlying these interpretations. The free fall of two bodies of different mass is a fact that students have never experienced, but it is a fact in Perelman and Olbrechts-Tyteca's meaning of the term, which means a fact of observation or an idea, conception, etc. that is commonly accepted in a society. But what is accepted here? It is an interpretation, we never see two different masses falling together, but we agree that they take different times to land from the same height. We have some experiences that are related to this; for example, we have seen a piece of paper or any other lightweight body landing very slowly, which reinforces the concept we learn from others, i.e. that two bodies of different mass take different time to land. This conception maybe related to some common argumentative schemes underlying it. These very common schemes, the schema of 'more → more' (like a direct proportionality) and the schema of 'cause and effect', increase the convincing force of the conception. Those schemes could be considered as common places in our culture.

In the group discussion, the teacher helped to re-direct the students' experimental work by asking them why they think their premises were true and thus contributed indirectly to the reworking of the knowledge. Argumentation is a dynamic process, in which there are motions from premises to thesis and from thesis to premises.

6. CONCLUSIONS

The results of the analysis stress the importance of the premises considered in the arguments, which are linked to the students' personal interpretation of the situation of the problem or of the data. This interpretation must depend on experiences, ideas and conceptions that students possess, which are mainly based on common places or very common schemes in a specific culture. In consequence, our results suggest that the incorrect answers to the problems or questions proposed are related less to the kind of argumentative schemes used than to the premises on which the arguments are based.

The premises on which students base their reasoning are mainly facts that are interpreted, often by means of common places, or truths agreed and accepted in a society – underlying some common argumentative schemes – which may differ from the premises of the scientific world. In the situation of individual work, they change with the context of the problems, indicating that as students read the problem they also make an interpretation of it. These interpretations of the problems influence their answers. In the small group discussion, the initial interpretation of the problem is the same for all the students, and corresponds to a common-sense interpretation that is difficult to change.

Especially relevant in science classes are the premises of 'places'. These premises belong to what we termed "common sense" and are the substrate of all informal reasoning. Many large obstacles to students' comprehensions of scientific issues may arise from these 'places'. They are rarely made explicit in the explanations of the teachers or in the textbooks students read, and so they are never discussed; nor is there any comment on their inadequate use by particular students.

In the group discussion reworking of knowledge is made possible by means of students' argumentation and occasional intervention from teachers. When the teacher (or any student) asked why they think their premises are true, his/her intervention contributes to the reworking of knowledge. Questioning premises, especially common places on which they are based, is fundamental to the construction of new scientific knowledge.

The argumentative structures detected in the students' individual answers or in the group discussion can be characterized according to the arguments identified in the Perelman and Olbrechts-Tyteca's book, that is they argue using structures commonly used in other contexts, such as legal context, politics or everyday contexts, but here in science classes, they are of no use for justifying or obtaining the correct answer to the problems. The terms Perelman and Olbrechts-Tyteca uses to categorize arguments help us to understand the thinking of the students and to appreciate from his perspective the most common science misconceptions.

The analysis stresses the importance given by students to some kinds of argumentative schemes in their argumentations; these schemes can be used by the teachers in science classes. They can contribute to persuading students of the validity of scientific models because they are consistent with their own arguments. Among others, the 'more -> more' schema, the 'cause and effect' argument and pragmatic arguments will be effective in persuading students.

7. DISCUSSION AND IMPLICATIONS

Our results agree with those of other research on students' science misconceptions, but the theoretical framework used here offers a new interpretation. The theoretical perspective relates misconceptions to the characteristics of common sense in our culture, in fact many of the argumentative schemas that underlie common sense were identified in the ancient times (for example in Aristotle's books) and collected in the TA. So, we can establish a link between the misconceptions and incorrect answers of our students with a set of deeply held beliefs.

An important implication for science education is that focusing explicitly on argumentation and discussing the specific arguments students use can help improve not only their argumentative skills but their conceptual understanding of science as well.

Finally, some comments about the theoretical frame of the research: the Theory of the Argumentation is a good basis for studying argumentative aspects of students' answers to qualitative and open problems, and also for the study of their argumentations in the context of a group discussion. Although it is a very rich theory that is highly relevant in the teaching and learning of scientific knowledge, it is not sufficiently structured and some of its concepts not very well defined (for example, the concept of place), nor are the categories of arguments well delimited. More work needs to be done in the construction of a specific analytical framework for science education research based on this TA.

ACKNOWLEDGMENTS

We would like to acknowledge the support of the DURSI (Generalitat de Catalunya) for the 'Xarxa de Referència: Recerca per a l'educació matemàtica i científica (REMIC)', n° U2201-U9UAB2005, coordinated by Roser Pintó (UAB).

NOTES

[1] Polya identifies non-deductive reasoning and plausible reasoning, this kind of reasoning coincides with what Perelman and Olbrechts-Tyteca names not "compelling" argumentation and, in general, argumentation in opposition to demonstration (with the meaning this word has in formal logic). In this chapter we use mainly the term 'argumentation' and also the term "plausible reasoning" in the meaning of "not deductive reasoning".

[2] Id note 1.

[3] Perelman and Olbrechts-Tyteca, L. (1958) *Traité de l'Argumentation. La Nouvelle Rhetorique. Bruxelles:* Editions de l'Universite de Bruxelles

[4] Facts are universal agreements, and therefore unquestionable. An argument gains convincing force if it based on facts (Perelman and Olbrechts-Tyteca, 1958).

[5] "Places" are premises of a general character that are useful for all sciences and all kind of discursive genres. (Perelman and Olbrechts-Tyteca, 1958).

[6] We do not detail in this chapter all the kind of argumentative schemes of the TA, but in Data analysis and results we will explain some of the categories found in our studied cases.

[7] The places are agreements of a general character, they are also the substratum of the argumentative schemes, because only agreement about their argumentative value can justify their application to

particular cases (Perelman and Olbrechts-Tyteca, 1958, p. 255).
Collaborative argumentation: Students listen ones to the others and collaborate in finding an argued solution to the proposed problem or question.

[8] p. 3 – A conveyor belt is used to carry plants from the store to the shopfloor. The flowerpots are placed on the belt, one metre apart. The belt is moving at a constant speed. An ant is exactly in between two of the pots, 50 cm from each, so it sees flowerpot A ahead of it (according to the direction of the belt) and flowerpot B behind.
Which pot – flowerpot A or flowerpot B – would the ant reach first if it decided to move towards one or the other?

[9] Quasilogic argument, according the TA, is an argumentative scheme whose convincing force lies in the fact that it is comparable to any formal argument of logics or of mathematics.

[10] Argument of the easiest option: Between two options, we choose the most easier to do.

[11] 'The preference for the easy over the difficult' is a kind of place of quantity: an option is preferable to the other one because the "quantity" of the facility to obtain is bigger.

[12] Argument of double hierarchy: "The double hierarchy expresses currently an idea of proportionality, direct or inverse, or, as a minimum, a nexus term by term" (TA, p. 517).

[13] Pragmatic argument: We appreciate an act or an event in function of its favourable or unfavourable consequences. It is an argument into the arguments based on the structure of reality. (TA, p. 409).

[14] Arguments of the *gaspillage* (it is the term used by Perelman and Olbrechts-Tyteca in his French book): Because of an action (task) has begun, accepted the sacrifices and the wastes made, the agent is reluctant to abandon the action (TA, p. 430).

[15] Argument by the sacrifice: This refers to the not useful sacrifice (TA, p. 432).

[16] Arguments by succession: They join a phenomenon with its consequences or its causes (TA, p. 404).

REFERENCES

Driver, R., Newton, P. & Osborne, J. (2000) Establishing the norms of scientific argumentation in classrooms. *Science Education, 84*, 287–312

Duschl, R. & Osborne, J. (2002) Supporting and Promoting argumentation discourse. *Studies in Science Education, 38*, 39–72</REF

Duschl, R. et al. (1999) *Middle school science students' dialogic argumentation* (Paper presented at European Science Education Research Association Conference. Kiel 1999)

Erduran, S., Simon, S. & Osborne, J. (2004) TAPping into argumentations: Developments in the application of Toulmin's argument pattern for studying science discourse. *Science Education, 88*(6), 915–933

Gross, A. (1990) *The rhetoric of science* (USA: Harvard University Press)

Heller, P. & Hollabough, M. (1992) Teaching problem solving through cooperative grouping. Part 2: Designing problems and structuring groups. *American Journal of Physics, 60*(7), 637–644

Heller, P. et al. (1992) Teaching problem solving through cooperative grouping. Part 1: Group versus individual problem solving. *American Journal of Physics, 60*(7), 627–636

Hoogaert, C. (1995) Perelman and Olbrechts-Tyteca et Toulmin. Pour une rhétorique néo-dialectique. *A : Hermès,* 15, 155–169

Jiménez-Aleixandre, M. P., Burgalló, A. B. & Duschl, R. A. (2000) "Doing the lesson" or "Doing science": Argument in high school genetics. *Science Education, 84*(6), 1–36

Kelly, G. J., Druker, S. & Chen, C. (1998) Students' reasoning about electricity: Combining performance assessments with argumentation analysis. *International Journal of Science Education, 20*(7), 849–871

Kitcher, P. (1988) The child as parent of the scientist. *Mind and Language, 3*(3), 215–222

Kuhn, D. (1992) Thinking as argument. *Harvard Educational Review, 62*, 155–178

Kuhn, D. (1993) Science as argument: Implications for teaching and learning scientific thinking. *Science Education, 77*(3), 319–337

Ogborn, J. et al. (1996) *Explaining science in the classroom* (Buckingham: Open University Press)

Osborne, J., Erduran, S. & Simon, S. (2004) *Ideas, evidence and argument in science.* [In service Training Pack, Resource Pack and Video]. (London: Kings' College, London)
Pera, M. & Ahea, W.R. (1991) *Persuading Science. The Art of Scientific Rhetoric* (USA: Science History Publications)
Perelman, G. H. & Olbrechts-Tyteca, L. (1989) *Tratado de la Argumentación. La Nueva retórica* (Madrid: Gredos). Translation of *Traité de l'Argumentatio. La nouvelle rhétorique* (Bruxelles: Editions de l'Université de Bruxelle) (1st edition, 1958)
Polya, G. (1945) *How to solve it (mathematics problems)* (New Jersey: Princeton University Press)
Polya, G. (1953) *Mathematics and plausible reasoning* (New Jersey: Princeton University Press)
Prelli, L. (1989) A Rhetonic of Science: Inventing Scientific Discourse (Columbia, SC: University of South Carolina Press)
Toulmin, S. (1958) *The uses of argument* (Cambridge: Cambridge University Press)
Zohar, A. & Nemet, F. (2002) Fostering students' knowledge and argumentation skills through dilemmas in human genetics. *Journal of Research in Science Teaching, 39*(1), 35–62

PART 9

TEACHING AND LEARNING SCIENCE USING MULTIMEDIA AND COMPUTER TOOLS

YASMIN B. KAFAI AND KATHERINE A. MUIR WELSH

33. EVALUATING STUDENTS' MULTIMEDIA SCIENCE DESIGN PROJECTS IN THE ELEMENTARY CLASSROOM

Abstract: Science standards have emphasized the need for authentic activities and assessments of students' science inquiry. Project-based classroom activities promote such inquiry in which students are often asked to create multimedia reports or other computer-based artifacts. One issue has been how to evaluate the educational quality of such student-generated artifacts. This paper reports on a three month long project in which single gender teams of elementary students (grades 4 and 5) worked on designing and implementing instructional multimedia products to teach younger students (grade 3) in their school about human physiology. Our analyses focus on evaluations of final multimedia science software conducted in class by the teacher and the students. We present and discuss the evaluation rubrics developed by the teacher and by the students. In our results we found that both rubrics presented viable efforts in assessing the quality of science content and other pedagogically relevant aspects. We then discuss the process through which the teacher and her students were able to evaluate instructional multimedia designs in their classroom activities, differences and communalties of students' and researchers' evaluations, and the gender differences found in instructional multimedia designs

Keywords: Evaluation, Inquiry, Multimedia, Project-based Learning, Technology

1. INTRODUCTION

Reform efforts have proposed project-based learning in the science classroom to promote student inquiry (American Association for the Advancement of Science, 1989; National Research Council, 1996). According to Blumenfeld et al. (1991) project-based approaches engage students in "relatively long-term, problem focused, and meaningful units of instruction" (p. 371). In science education, project-based learning approaches have provided students with authentic and meaningful contexts such as researching the water quality of a local river (Krajcik et al., 2000) or using original weather data to explain atmospheric phenomena (Songer, 1996). In other projects, students have used multiple historical data sources and argumentation templates to explain evolutionary theory (Reiser et al., 2001) or used discussion forums to debate scientific issues (Linn et al., 1998). A central feature of most science project-based approaches is that students' learning outcomes

Yasmin B. Kafai, University of California, Los Angeles, Department of Education, 2331 Moore Hall 951521, Los Angeles, CA 90095-1521

R. Pintó and D. Couso (eds.), Contributions from Science Education Research, 435–449.

are often represented in a final collaborative artifact such as a multimedia presentation, database, instructional software or a virtual model.

The quality of content integration in these student-generated projects has been a general issue (Carver et al., 1992). To evaluate the science in students' multimedia projects different approaches have been proposed. Hay et al. (1994) counted the length, number of links, and multimedia integration in documents. Kafai et al. (1998, 2004) identified different types of document units and developed design-differentiated scores that accounted for individual contributions in collaborative science projects. Further analysis pointed out that those students designing process-oriented instructional designs benefited most in terms of science learning. Orion et al. (2000) scored multiple science content and design features to assess the level of explanation provided by students in their multimedia presentations. Their findings included low explanation levels in students' multimedia presentations as well as significant amount of time students spend on decorative aspects of multimedia designs. Students themselves felt they did not benefit from having designed a multimedia presentation after having produced a research report. In all of these approaches teachers or researchers, not students, conduct these difficult multifaceted evaluations.

The following paper presents and discusses a project-based learning approach, called learning science by design (LSBD) (Kafai, 2005, 2006) in which teams of elementary students were asked to create and implement instructional software to teach younger students in their school about science. This 12 week-long project was integrated within the regular fourth and fifth grade science curriculum on human physiology. Each student working in a team developed a multimedia instructional design to provide an answer to his or her research question. In our analyses, we wanted to address the issue of assessing student-generated multimedia artifacts. Many project-based approaches ask students to generate artifacts such as communal databases (Scardamalia et al., 1994) or presentations (Hay et al., 1994; Orion et al., 2000; Spitulnik et al., 1998). In the LSBD project students were asked to generate instructional multimedia software to teach younger students, i.e., third graders, in their school about science. The intended audience of their software is not only the teacher and other peers but also third graders. These third grades actually visit the classroom twice to review project progress and provide feedback. Audience consideration offers an additional lens through which to assess multimedia artifacts. We were also interested in how not just the teacher and researchers but also how the students were able to assess such aspects and other features important to science inquiry in their products. For that reason, we asked students to use a rubric developed with a previous LSBD class to assess and review their fellow students' software products. We also developed a rubric, for researcher use, based on model by Scardamalia et al. (1992) that assessed students' attempt of system level science explanations. Our interest here was to contrast and compare student-generated reviews with those conducted by the teacher and researchers. In our discussion we address issues around how to assess the integration of science in multimedia projects and how to develop various assessments that support the students as well as the teachers in integrating science knowledge and skills.

2. RESEARCH CONTEXT

2.1 Study Participants

The participants in this study were 34 elementary students from a fourth/fifth grade combination classroom in a K-6 West Los Angeles school. There were 19 fourth graders (10 girls and 9 boys) and 15 fifth graders (6 girls and 9 boys) in an ethnically diverse class. All students had previous computer experience, mostly in using Internet applications for searching and software applications such as word-processing in other classes. The classroom teacher and researchers created nine single-gender teams composed of fourth and fifth graders. Each team had its own iMac station, with two remaining stations designated for Internet research. All the computers contained Microworlds™ software, which students used to program their science software. The front of the room was the focal point of the classroom with a rug for class gatherings and a white board for instruction.

2.2 Classroom Project

The LSBD lasted 12 weeks spread across four months with school vacation and statewide testing taking the other time. Each student had the same task to create instructional software for younger students at their school that answered the software developer's personal question about human physiology. Each student team was given a planning board consisting of three form-board panels to place a project calendar with important dates such as project milestones and school breaks to help them plan their project development. (See Marshall & Kafai, 1998, for a discussion on the changing functions of the planning board over the course of a software design project.) At the beginning of the project, the students/developers chose a question and refined it under the guidance of their teammates, teacher and researchers. Then the students researched multiple sources to find the answer to their question and developed simulation ideas for their software. Integrated throughout the project were different assessment activities such as team presentations, third grade visits and showcases to facilitate planning, software and content development. These were designed to help students plan their projects and hold them accountable to their teacher, fellow students and the third grade audience. Student teams also held two planning meetings, 10 to 15 minutes long, to prepare third grade visits (week 3) and review their questions and instructional designs (week 8). Once the project started, on average students spent 60 to 75 minutes per day on science and software design-related activities. It is important to note that the project curriculum was dynamic and non-structured; most activities were scheduled based on a perceived need by the teacher and/or the researchers to discuss particular topics. However, it was based on both National and California science standards. (See Galas, 1997–1998 for a more complete description of a similar curriculum.) Science instruction, from traditional lecture to a virtual eye dissection, was interwoven within the software design project.

2.3 Data Collection and Analysis

All classroom sessions were videotaped and parts of discussions were transcribed. All software projects were collected at the end of the project. Students' collaborative software projects were evaluated in two ways: (1) Peer evaluations and (2) Researcher evaluations. In teams, the students reviewed and rated their peers' software projects using a five-point Likert scale for each of four parameters: content accuracy, amount of content, understandability of science content for third graders, and software as a "simulation". Teams also wrote comments on any of the parameters. Scores were compiled for each student, team averages computed and comments transcribed and coded.

The research team designed a five-point coding scheme to assess the scientific level of explanation found within the final software product. It was based on a similar scheme in Scardamalia et al. (1992) developed for assessing students' quality of science explanations in text but not in multimedia format. Within our scheme we took into account that the students' software simulations were multi-faceted and integrated many information resources (i.e., web information, parent's input, text and library books). In our coding, we also addressed whether the resulting instructional software design included text, animations, and/or audio. The codes, ranging from one to five, were assigned in the following way: 1 = No answer; 2 = Listed facts with no integration; 3 = Information presented is more complex and includes some attempt at integration; 4 = System presentation of information is present in its beginning stages and information is presented in the student's own words; and 5 = Outstanding case which answers the question in an integrated fashion in the student's own words. The following examples illustrate the application of the level of explanation code (see Figures 1a, b, c, and d).

Madeline selected as her research questions "What is happening in your body during an asthma attack?" and designed the following instructional software screens. Her instructional design was rated a five because she successfully combined different media elements and provided an explanation of the process and terms in her own words (see Figure 1a).

An example for a rating of four can be found in Ian's final project. His question was, "How does the HIV virus attack the T4 lymphocytes?" In Ian's instructional design the upper circle is the HIV virus and the lower circle is a T4 Lymphocyte. When the attack button is pushed the two circles come together, allowing for replication of the virus' genetic material. While Ian's design combined different media elements and illustrated process, its wording was copied from information pages and thus not understandable for the intended third grade audience (see Figure 1b).

Ernest's project received a three according to our rating scheme. His software is a list of facts with some attempt at integration. His instructional design is composed of four screens. Page 1 asks, "Why does lead affect the body?" Page 2 is of a baby in then when a user presses the button it places a father with the audio stating, "Baby come here, baby come here" then added to the screen is the text, "The baby just ate lead so he can't hear his father" (see Figure 1c).

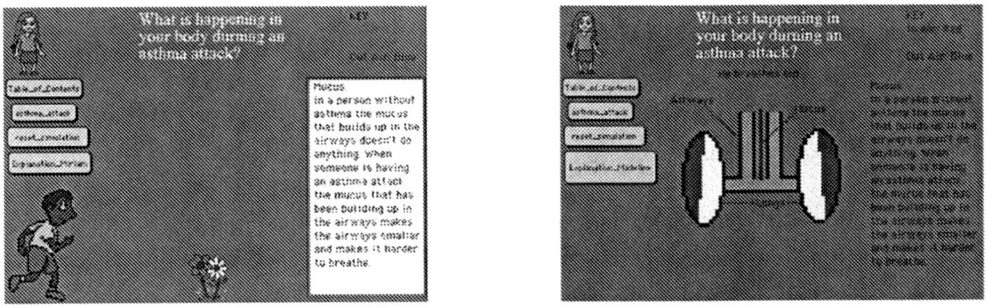

Figure 1a. Madeline's instructional software screens

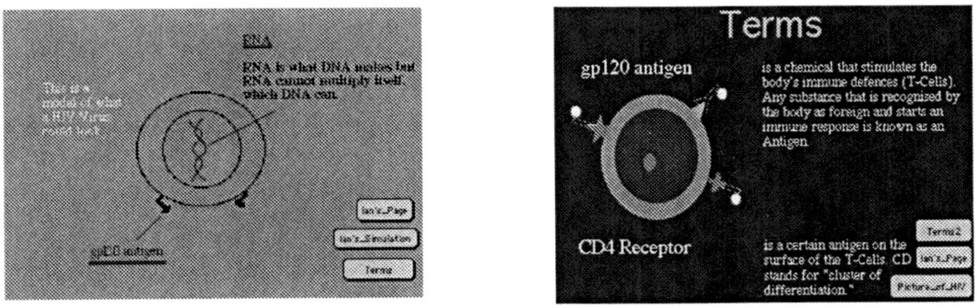

Figure 1b. Ian's instructional software screens

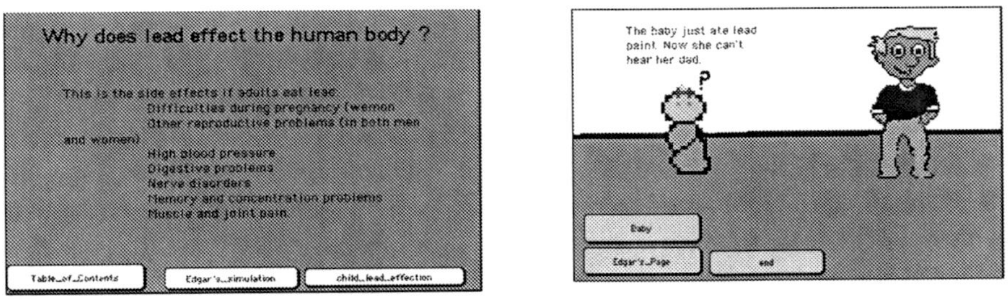

Figure 1c. Ernest's simulation software screens

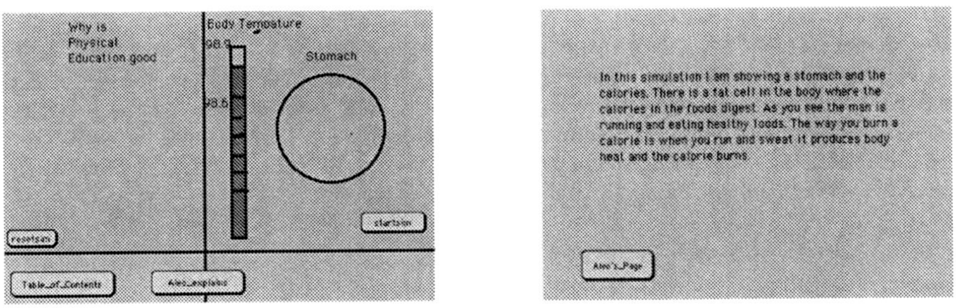

Figure 1d. Alec's simulation software screens

Finally, an example of a project rated a two is Alec's project on physical education. Alec's final question was "Why is physical education good for the body?" His simulation is not complete. He was seen working on his project at the very last possible minute during the final project showcase. None of the buttons work in the project and there are problems with navigation within the software (see Figure 1d).

Two independent raters coded all of the students' multimedia projects. Rater disputes on particular projects were then resolved through discussion in order to obtain final codes for analysis.

3. RESULTS

Our results section is divided into two parts. The first part examines three different assessment activities – third grader visits, team presentations, and final software simulation showcase – that were part of the LSBD project classroom. In the second part, we examine students' and researchers' reviews of final instructional projects in more detail.

3.1 Classroom Assessments

3.1.1 User evaluations Throughout the LSBD class, teams had multiple opportunities to review and evaluate each others' projects. In Week 3, students met with the third graders, the software users, for the first time. A key feature of these user visits is that they provide opportunities for student designers to engage in explanations of their instructional science designs. At the time of this first user testing, students had not been able to program much of their software. Consequently, the software designers take the third graders on a virtual tour of their future software. For example, Alyssa describes to George and Natalie, the third graders visiting, her instructional design and her current and future screen ideas. When George, a third grader points to the screen and states: "This, I don't really, I don't really understand this." Alyssa elaborates on her prospective design:

> Well, cause I don't have a question page, but we're supposed to have, well, what it's supposed to do is that, um, see, she's hot and so when you're hot, why the temperature goes up, that's what it's showing on the thermometer, and then it goes back to the regular temperature and you, like, start to sweat, that's why she had those little ugly blue spots on her. And another word for sweat is perspire and then on the next page it says 'it sure is cold out here,' what happens is that . . . And so what happens is that she gets really cold so she gets goose bumps and that's why the hand shows up there. Press start. She gets cold and then that goes down and then that goes up, she gets goose bumps and then the hair, some of the goose bump hair thingies it comes up from her body and then the blood vessels come in closer until they keep warm.

Alyssa explains her screen; she refers to it as her page, and the science concept she is researching. She attempts to make the content understandable for the third graders when she says, "another word for sweat is perspire." During this third grade visit, Alyssa has the chance to connect to one of the project's multiple audiences.

After the third graders left the classroom, the students wrote down their reflections about what they heard from the third graders and what they have to do. Alyssa wrote, "They liked my page, but it was not very understandable. ... I need to make an explanation page and I also need to do more programming." In the following classroom discussion, students from different teams presented various issues to the whole class. A common complaint mentioned by the designers was that the third grader did not understand their instructional designs but didn't feel comfortable saying so. Kelly points out that "Well, the 3rd grader only said one thing over and over, it was good. I could tell by her body language that she could not understand half of the content and only spoke up after much coaxing and said she didn't understand." Other students such as Justin commented on the third graders' focus on software design features rather than the science content:

It's kind of also the idea that the 3rd graders only, not really, maybe understanding it but not giving much feedback. Because our group, at least both of our 3rd graders, they both kind of paid more attention to like there was an unnecessary turtle on the page by accident, and when we cut it they kind of got obsessed with like the cutting idea ... And the sound effects were like the drinking of the arsenic bottle. [...] They kind have kept on imitating the gulp, gulp, gulp sound and they didn't really comprehend the actual what was happening and we kept trying to explain.

The teacher reminds students that they often do the same when looking at other's software and urges them to consider this in their software planning. A week later, she organizes team presentations to help students elicit feedback from their peers.

3.1.2 Team presentations Each team presented their software on a large screen monitor to their peers gathered on the carpet at the front of the room. An example of a Madeline's presentation is seen below. Like Alyssa during the first third grade visit, Madeline explains the science content and asks for feedback:

My question is – what is going on in your body during an asthma attack? And I don't have a simulation yet, but I'm gonna try to explain it to you. Um, you have sinuses that are up here and you have like mucus and the mucus gradually drips down and it's in your airways, but it doesn't really do anything. And like, what I'm gonna have is somebody walking along and he smells a flower that he's allergic to. And that triggers something in his body which causes [his] airways to swell up. And with the mucus it becomes very small. And so when you breathe in, your lungs, they get all the air in, but you can't get all of the air out. And what you do when you breathe in, you breathe in good and bad stuff. And when you breathe out you breathe out – you get rid of the stuff you don't need. And what happens when you get an asthma attack is that you can't get rid of all the stuff you can't need which causes you to cough.

Due to time constraints, students wrote their feedback as opposed to telling the students directly. The feedback along with the third graders' comments were typed and given to each team. These statements reflect the "critical standards" (Erickson & Lehrer, 1998) for evaluating software developing within the classroom. Examples of the feedback given to Madeline are shown below. They range from comments on the simulation to asking questions about content: "Good simulation idea! Good info!" or "Good start...what triggers the airway to swell?," "Great info.

Good idea. Start working, it will be good.", "I liked your simulation idea.," "Nice explanation Nice ideas. Nice info." "You should get together with Kelly because she has the same question."

During the following weeks, students used the feedback from the third graders and their peers to continue to develop their software. These previous activities gave participants a chance to learn how to explain and refine the answers to their research questions all the while developing science content to make it more accessible and understandable. Especially important is the fact that students had multiple opportunities to explain and justify their software design decisions. Each team presentation and follow-up third grade visit provided chances for student reflection.

3.1.3 Final showcase In the "final showcase" students addressed their peers and focused on the final software products. In teams, they looked for integration of audience consideration, science content and software development that answered a student's question. We can see in Table 1, the 4th/5th graders' comments about their peers' software, which ranged from praise to process, to a focus on the science content and how it was represented. We then list student scores received for each of the case study team members. We added all individual scores for a team and calculated the average which resulted in 17.8 for team 9. The final column list the researchers' scoring of the final projects for team 9.

TABLE 1. Peer Evaluation Comments for Team 9 (Week 12)

Student	Student comments	Student score	Researcher score
Alyssa	Good Simulation Learned a lot Very good and explanitory [sic] Needs work on nothing, good info	14.6	15
Cindy	Good simulation/ project This simulation was very easy to understand and to start it. Showed more than tell	15.8	20
Madeline	Good simulation On the verge of being too much text Very understandible [sic]. Simulation showed a lot of information Good representation Check spelling	14.3	20
Tracy	Good simulation Learned a lot Good project. It was understandible [sic]. VERY GOOD. We liked how you showed the arteries clogging up. Couldn't understand recording. Showed not told! Excellent work.	16.8	16
Average		15.4	17.8

3.2 Student and Researchers Final Software Evaluations

All students reviewed each others' final multimedia projects on a five-point scale addressing the accuracy and amount of science information, the understandability of the science content, and the representation of the content. The average scores for other teams can be seen in Table 2. The average of individual students' review scores was 12.16 (STD 3.32) with a range of ratings between 6.1 and 17.8. There were no differences between fourth and fifth graders' final multimedia reviews. We found, however, that all boys' scored less on average than girls teams with the exception of one girl team (Team 6 $X = 10.2$). The average score for boys was 10.53 (STD 2.83) whereas the girls received an average of 13.99 (STD 2.89), a significant difference ($z = 1.64$; $p < 0.00$). The last column lists the researcher level of explanation score, which we will describe in further detail in the next Section. The students also wrote comments about their peers' software. Student comments ranged from process noting the science content and how it was represented such as, "Simulation showed a lot of information. Good representation," to comments about personal learning. For example one team noted, "Very good. It showed a lot. We never knew what mucus was for." These comments reflected the critical standards found within the classroom. One such standard, 'showing not telling,' was mentioned often throughout the comments.

All the students' final multimedia projects were reviewed by the research team using the a five-point level of explanation score based on a scheme developed to assess writing quality of science explanations by Scardamalia et al. (1992). The average individual score was 12.91 (STD 4.81) with a range of ratings between 4 and 20. Here again, we did not find any differences in the project reviews between fourth and fifth graders. We found that girls' teams (M 14.69; STD 4.99) on average scored higher than boys' teams (M 11.33; STD 4.17); this difference was also significant ($z = 1.64$; $p < 0.00$). A comparison between the student- and researcher-generated review scores showed a significant, positive correlation ($R = 0.85$; $P < 0.00$).

TABLE 2. Peer Evaluation and Level of Explanation Team Scores of Final Projects

Team	Gender B = Boy G = Girl	Peer evaluation score	Level of explanation score[1]
1	B	11.7	12
3	B	12.2	12.4
5	B	7.7	7.0
7	B	10.1	11.0
8	B	11.7	13.9
2	G	13.9	16
4	G	16.6	18
6	G	10.2	7.0
9	G	15.4	17.8

[1] Original scale 1-5. Multiplied original score by 4 to put on same scale as Peer Evaluation.

4. DISCUSSION

In our analyses we focused not only on how different assessment activities could become an integral part of students' science inquiry environment but also on how students assessed their own learning products in terms of science inquiry compared to reviews conducted by the researchers. In the following Sections we discuss the process through which students were able to assess their instructional multimedia designs within their classroom activities, differences and commonalties of students' and researchers' reviews, and the gender differences found in instructional multimedia designs.

4.1 Assessment Integrated into Classroom Activities

One goal of our research was to document and understand how assessment of science multimedia projects could become an integral part of classroom activities. We found in our analyses of selected classroom activities such as the third grade usability sessions, team presentations and final showcase that students had ample opportunities to review and reflect on their multimedia projects. Two aspects were key in this effort: (1) multiple audiences and (2) iterative development. We found that each audience afforded a particular science focus. For example, the third grade visits provided opportunities to explain and expand science explanations and improve interface design features. During the team presentations in front of the entire class, the teacher focused on research questions and the comprehensiveness of the students' answers whereas during the showcases she evaluated the integration of science in the software designs. It is important to note that while each assessment had a particular audience, these were not exclusive to the session. In fact, all audiences were present during all sessions to a greater or lesser extent. For example, the teacher was guiding whole class discussions after third grade visits; at this time, she was helping students to articulate what good feedback would mean for them and helping them to prepare the next assessment session.

We also found that the iterative development of assessment throughout the design project allowed students to continuously revisit and revise their designs. Students had multiple opportunities to practice their own assessment skills and to develop and apply critical standards. The third graders came at different time points to review and provide feedback on software development. The team presentations were also scheduled at the beginning and at the end of the project to provide feedback. At each time point, the feedback focused on different, yet related aspects. While the emphasis during the first team presentation session was on getting students committed to a research question and the research needed to provide an answer, the second team presentation session focused on finishing the research and instructional multimedia designs. In a similar vein, the first third grade visits provided the designers with an opportunity to give expanded explanations on what they intended to design, as most designs were not implemented on the computer yet but available on paper or in the designers' minds. The second third grade visit provided a nearly finished product, thus the focus moved to check for needed explanations and interface feature improvements in the available instructional designs. The final showcase provided yet another opportunity to review the instructional multimedia designs as complete products rather than in parts, sketches or verbal explanations.

We see each assessment activity as complementary to each other and it is the combination of all of them, which helped students as a community with their teacher to develop the critical standards to assess their software designs. Most importantly, these multimedia reviews were an integral part of on-going classroom routines and not a separate activity placed at the end of the project (Orion et al., 2000). A further point concerns the combined learning of the software design tool and the science content. We have argued elsewhere (Kafai & Ching, 2001) that in fact the learning of the software design tool and the science content can be mutually beneficial to each other. As students move beyond a focus on representing more than just facts to explaining processes and systems, they invariably need to learn more programming skills to achieve such dynamic representations. Conversely, as students see other students develop and implement such dynamic representations in their instructional science designs, they might reexamine their own instructional designs and shift to explanations of processes. The critical standard of 'showing, not telling' became a guiding principle in classroom discussions and students' reviews.

It should be noted that all the students had participated in a similar LSBD project in the fall and thus were familiar with the software design tool. We know from previous research (Ching & Kafai, forthcoming) that observed software design teams in their first design project in the fall that teams spend considerable time apprenticing the new students in the classroom into different tool features. What we have seen then in our analyses is the benefit of students being familiar with the software design tool which could explain students' increased focus on content production and lessened interest in exploring different technology features.

4.2 Assessment of Students' Multimedia Science Projects

In our study we compared and contrasted two different approaches: a researcher-generated level of science explanation score that assessed the integration of science content with multimedia features with a student-generated score that was composed out of four different aspects: understandability, science content representation and quantity and software simulation quality. With our researcher-generated level of science explanation score we built on previous work by Scardamalia et al. (1992). It was our goal to develop one score rather than several that a teacher could use to assess the quality of science explanations found in the kind of instructional multimedia projects created by students in the LSBD project. Our score integrated various aspects such as appropriateness of wording, inclusion of multimedia elements, and accuracy in comparison to the student-generated reviews that provided four different questions as guides. We found that the assessment of students' final multimedia projects was not always an easy one, partially because the choice of research questions or available information differed from student to student. For example, Ian's research question about AIDS was a difficult one and the information available on the Internet, in textbooks, and at home was not easily converted and worded for an elementary audience, not to mention third graders. Evaluating each student's final project also gave the researchers an appreciation for projects that were not highlighted in the day-to-day workings of the classroom. For example, Cindy's project was given the highest score on the researcher's scale and never talked about in the classroom.

The student-generated review process grew out of work with a previous LSBD class. Here we discussed with students and teachers the kind of questions and the format that could be used to review their final multimedia designs. We found that guiding questions such as "How accurate was the information presented?" "How much science information was there in the software?" or "How understandable was the information presented?" allowed students to assess different features of their peer's software designs. While comments were not always consistent, in conjunction with the ratings they provided a more complete assessment. We found that the first question concerning the accuracy of content might be a hard one to answer even for a knowledgeable fifth grader, and thus we suggest replacing it with "How much new information did you learn? And a follow up question: "How can you check if this information is correct?" Our further analysis indicated that students' assessments of their peer's projects did not veer substantially from those conducted by researchers. This is a promising finding given that we do not know of other approaches in which students, and not teachers or researchers, review multimedia science projects. The format of 'guiding questions' could serve as a model for student-directed reviews of other multimedia science projects. We have found this approach of providing guiding questions to be successful in other project-based approaches with elementary students as well (Kafai & Gilliand-Swetland, 2001).

What we learned from both researcher and student reviews of final multimedia products is that either approach can provide sensible feedback to students and the teacher. From our previous discussion it is clear that this evaluation is not a standalone process but part of a larger integrative assessment effort which started with third grade visits and team presentations and continued in daily interactions within teams, other students and the teacher. The development and application of critical standards to review research questions and answers was an on-going process that this particular classroom community harnessed over time in successive projects. This was most evident in fifth graders' contributions to classroom discussions and in their guidance of team sessions but practiced by all students.

4.3 Absence and Presence of Differences in Multimedia Project Assessments

Lastly, we wanted to address two other findings from our research: the absence of differences between fourth and fifth graders and the presence of gender differences in final multimedia projects scores. We did not find differences in the level of science explanation or in peer-directed evaluations of instructional designs created by fourth and fifth graders. This finding might seem perplexing at first, yet we need to realize that even for fifth graders research questions were new, just as they were for the fourth graders. One would expect that fifth graders would be more agile in handling the multiple project demands such as sharing computer resources, conducting research, coordinating group work, and programming. While fifth graders were more familiar with the software design tool and the project activities in general, they were also called upon more to help their younger peers finish their software designs. Fifth graders thus did not spend more time on their instructional designs; instead they finished them earlier and helped other students, often teammates but sometimes also students from other teams. It is also possible that we reached a limit on how sophisticated multimedia designs can become with the programming knowledge and content understanding most fifth graders possessed.

Another surprising finding was that all, with the exception of one, girls teams received better scores for their multimedia projects than the boys teams. Other science assessments conducted at the beginning and the end of the project did not reveal any significant gender differences; neither did tests of students' programming proficiency. In this LSBD project, we had single-gender teams work on developing instructional multimedia designs. At the moment, we do not have a single explanation for this finding other than gender differences in collaborative apprenticeship interactions that were observed in previous design projects (Ching & Kafai, forthcoming). In this study, which preceded the current project by a year, Ching analyzed apprenticeship interactions within mixed-gender software design teams. She found that girls adopted more collaborative forms of assisted guidance towards the more inexperienced students in their teams whereas the boys tended to opt for more directed forms. In other words, girls involved younger students from the beginning in more software design activities whereas boys tended to take over software activities at the expense of providing inexperienced students with the opportunity to practice. It could well be that such forms of collaborations have a longer lasting impact on the quality of final software designs. What this points out is that the integration of assessment with learning has to happen on multiple levels within the classroom community, with in student teams, classroom discussions and teacher interactions to be successful. Another possible explanation is that the teaching or instructional design format of the multimedia software was something more girls than boys are familiar or comfortable with in school. It is also possible that the instructional design format provided a better context to learn programming or software design skills. Rather than learning such skills for its own sake, in the LSBD project, the purpose of learning programming commands and constructs served the larger goal of creating a functional application as found in commercial software.

ACKNOWLEDGEMENTS

An Early CAREER Award from the National Science Foundation to the first author (REC-9632695) supported this research. This paper is based on a previous presentation of preliminary results at the annual meeting of the American Educational Research Association in 1999 in New Orleans, LA. We would like to thank the teacher, Cathleen Galas, and her science students at UCLA's Seeds University Elementary School for their participation.

REFERENCES

American Association for the Advancement of Science. (1989).*Science for all Americans*. New York: Oxford University Press.

Blumenfeld, P., Soloway, E., Marx, R., Krajcik, J., Guzdial, M., & Palincsar, A. (1991). Motivating project-based learning: Sustaining the doing, supporting the learning. *Educational Psychologist, 26*, 369–398.

Carver, S. M., Lehrer, R., Connell, T., & Erickson, J. (1992). Learning by hypermedia design: Issues of assessment and implementation. *Educational Psychologist, 27*(3), 385–404.

Ching, C. C., & Kafai, Y. B. (forthcoming). Peer Pedagogy: Student collaboration and reflection in learning through design. *Teachers College Record.*

Erickson, J., & Lehrer, R. (1998). The evolution of critical standards as students design hypermedia documents. *The Journal of the Learning Sciences, 7*(3 & 4), 351–386.

Galas, C. (1997–1998). From presentation to programming: Designing something different, not the same thing differently. *Learning and Leading With Technology, 25(*4), 18–21.

Hay, K., Guzdial, M., Jackson, S., Boyle, R., & Soloway, E. (1994). Students as multimedia composers. *Computers & Education, 23*, 301–317.

Kafai, Y. B. (2005). The classroom as "living laboratory": Design-based research for understanding, comparing, and evaluating learning science through design. *Educational Technology, 45*(1), 28–34.

Kafai, Y. B. (2006). Constructionism.In K. Sawyer (Ed.), *Cambridge handbook of the learning sciences* (pp. 35–46). Cambridge University Press.

Kafai, Y. B., & Ching, C. C. (2001) Affordances of collaborative software design planning for elementary students' science talk. *Journal of the Learning Sciences, 10*(3), 323–363.

Kafai, Y. B., & Gilliand-Swetland, A. (2001). The integration of historical sources into elementary science education. *Science Education, 85*, 341–379.

Kafai, Y. B., Ching, C. C., & Marshall, S. (1998). Children as designers of educational multimedia software. *Computers & Education, 29*, 117–126.

Kafai, Y. B., Ching, C. C., & Marshall, S. (2004). Learning affordances of collaborative multimedia design. In M. Rabinowitz, F. C. Blumberg, & H. Everson (Eds.), *The impact of media and technology on instruction*. Mahwah, NJ: Lawrence Erlbaum Associates.

Krajcik, J., Blumenfeld, P., Marx, R., & Soloway, E. (2000). Instructional, curricular and technological supports for inquiry in science classrooms. In J. Minstrel, & E. H. V. Zee (Eds.), *Inquiring into inquiry learning and teaching in science* (pp. 283–315). Mahwah, NJ: Lawrence Erlbaum Publishers.

Linn, M., Bell, P., & Hsi, S. (1998). Using the Internet to enhance student understanding of science: The knowledge integration environment. *Interactive Learning Environments, 6*(1–2), 4–38.

Marshall, S. K., & Kafai, Y. B. (1998). Children's development of planning tools for managing complex software design projects. In A. Bruckman, M. Guzdial, J. Kolodner, & A. Ram (Eds.), *Proceedings of international conference of the learning sciences* (pp. 202–208). Charlottesville, VA: AACE.

National Research Council. (1996). *National science education standards*. Washington, DC: National Academy Press.

Orion, N., Dobowski, Y., & Dodick, J. (2000). The educational potential of multimedia authoring as a part of the earth science curriculum–A case study. *Journal of Research in Science Teaching, 37*(10), 1121–1153.

Reiser, B. J., Tabak, I., Sandoval, W. A., Smith, B. K., Steinmuller, F., & Leone, T. J. (2001). BGuILE: Strategic and conceptual scaffolds for scientific inquiry in biology classrooms. In S. M. Carver, & D. Klahr (Eds.), *Cognition and instruction: Twenty-five years of progress* (pp. 263–305). Mahwah, NJ: Lawrence Erlbaum.

Scardamalia, M., Bereiter, C., Brett, C., Burtis, P. J., Calhoun, & Lea, N. (1992). Educational applications of a networkedcommunal database. *Interactive Learning Environments, 2*(1), 45–71.

Scardamalia, M., Bereiter, C., & Lamon, M. (1994). The CSILE project: Trying to bring the classroom into World 3. In K. McGilly (Ed.), *Classroom lessons: Integrating cognitive theory and classroom practice* (pp. 201–228). Cambridge, MA: MIT Press.

Songer, N. (1996). Exploring learning opportunities in coordinated network-enhanced classrooms: A case of kids as global scientists. *The Journal of the Learning Sciences, 5*(4), 297–327.

Spitulnik, M. W., Stratford, S., Krajcik, J., & Soloway, E. (1998). Using technology to support students' artifact constructions in science. In B. J. Fraser, & K. G. Tobin (Eds.), *International handbook of science education* (pp. 363–382). Dordrecht: Kluwer Academic Publishers.

THORSTEN BELL, SASCHA SCHANZE, WOLFGANG GRÄBER, JAMES D. SLOTTA, DORIS JORDE, H. B. BERG, T. STRØMME, ANJA NEUMANN, SIGMAR-OLAF TERGAN AND ROBERT H. EVANS

34. TECHNOLOGY-ENHANCED COLLABORATIVE INQUIRY LEARNING: FOUR APPROACHES UNDER COMMON ASPECTS

Abstract: Collaborative inquiry is seen as a promising, but also a demanding way of learning. Over the past years several computer learning environments have been designed to support learners in doing inquiry. Students may investigate complex problems from everyday life or socio-scientific issues using simulations, datasets, media-enriched content information, text editors, mapping tools, graphical modelling etc. This paper describes and compares four different approaches to collaborative inquiry, developed in the projects ParIS, WISE, Viten, and Co-Lab. Commonalities and differences in their models of inquiry learning are highlighted. Building on and synthesizing research results from these projects the paper draws conclusions on issues relevant for designing collaborative inquiry learning environments, like guidance and freedom, knowledge construction, and integration of external knowledge

Keywords: Collaborative inquiry learning, Knowledge construction, Knowledge integration, Scaffolding, Technology-based learning environments

1. INTRODUCTION

Inquiry learning is a dynamic approach that engages the learner in activities like asking questions, making discoveries, designing and performing experiments, making predictions or drawing conclusions from data. Such inquiry processes are seen as fruitful and important in science education as they support knowledge and skill acquisition in a learner-centred way. However, while many studies discuss the complexity of inquiry learning, little evidence has been provided of its effectiveness in comparison to more traditional learning methods. Indeed, learners may encounter substantial challenges in the course of inquiry learning (de Jong and van Joolingen 1998). Moreover, teachers are also seriously challenged by the process of integrating inquiry into their practice (Krajcik et al. 1994).

In recent years many technology-based learning environments have been developed, taking learner difficulties into account and scaffolding their activities by providing strategies, tools and materials (Quintana et al. 2004). Integrating such technologies into the classroom, however, is not a straightforward process, and does

Thorsten Bell, Leibniz-Institute for Science Education (IPN), University of Kiel, Olshausenstr. 62, D-24098 Kiel, Germany

R. Pintó and D. Couso (eds.), Contributions from Science Education Research, 451–463.

not guarantee success. Technology innovations must be designed in accordance with a model that accounts for their role within the various phases of inquiry learning. This paper presents four such approaches, each of which provides inquiry projects within authentic contexts, as well as support for learners as they are involved with contemporary socio-scientific issues. Each of these approaches emphasises different aspects of collaborative inquiry learning, and each has developed distinct tools to support its emphasis. First, ParIS (Partnership Industry and School; www.projekt-paris.uni-kiel.de/paris) is a resource-based learning approach that engages students in self-regulated processes with everyday questions using the Web as a central learning resource. Second, WISE (Web-based Inquiry Science Environment; http://wise.berkeley.edu) is an internet-based platform where students collaboratively work on inquiry projects, making use of evidence from the Web. In a related effort, third, the Norwegian project Viten.no (www.viten.no) also offers Web-based inquiry projects as an instructional tool for teaching about socio-scientific controversies in science. Fourth, Co-Lab (Collaborative Laboratories for Europe; www.co-lab.nl) has a focus on conducting experiments and on constructing graphic models that explain empirical data. This contribution discusses these various approaches, emphasising their differences in terms of inquiry learning, process regulation, knowledge construction and knowledge integration. The paper highlights advantages of the different approaches and draws attention to places where the projects could learn from each other.

2. THEORETICAL BACKGROUND

In collaborative inquiry learning, student groups engage in a number of activities that serve to investigate a research question or to solve a problem (Quintana et al. 2004; Tabak 2004). Capturing the students' attention is a prerequisite for sustained self-guided inquiry. This calls for realistic and relevant problems that are typically embedded within complex "theme fields" offering multiple perspectives and levels of description, as well as different ways of addressing the problem. Many different approaches to collaborative inquiry learning have been investigated, including ParIS (Neumann et al. 2005), WISE (Slotta 2002), Viten (Mork and Jorde 2004), and Co-Lab (van Joolingen et al. 2005). Apart from their differences with respect to their models of inquiry (detailed in 3.1) these approaches can also be seen to possess many features in common as follows. Previous attempts to formulate a "process schema" of inquiry (e.g., de Jong et al. 2002; Löhner et al. 2005) included processes of orientation, hypothesizing, experimenting, and conclusion. Inquiry starts with an orientation where students are confronted with new phenomena and research questions arise. While hypothesizing, learners build and express their understanding of the problem and formulate testable statements; another term for this phase could be knowledge construction. Hypotheses may be tested in experiments – if experiments are possible – but may also be tested for consistency with the student's own knowledge and other information; therefore, a more general term may be knowledge checking. Knowledge construction and knowledge checking are interrelated processes that have to proceed iteratively, i.e. in learning cycles, until the results are satisfying. Inquiry can be said to end with the student drawing, documenting and presenting conclusions that explain the phenomena and address the research question. While the four inquiry approaches discussed in this contribution may all be described in these general terms, they each offer their own interpretation of this schema, emphasising different aspects, levels of problem description, and student activities (see below). The process schema of inquiry discussed above presents the learner with several generic difficulties as reported by de Jong and van Joolingen (1998): Students have difficulties in generating and adapting hypotheses, in designing experiments, in interpreting data and in regulating their inquiry. Teachers are likewise challenged with respect to finding authentic problems, effectively scaffolding students, creating a learning community, and integrating cognitive tools into their teaching (Krajcik et al. 1994). This interpretation of inquiry as problem solving within complex, information-rich, multi-faceted theme fields suggested several core issues to be addressed within this paper. Research insights from the four approaches are presented with respect to the following aspects: (1) Pre-structuring and self-regulation of the inquiry process (see 3.2), (2) Building complex, coherent knowledge (see 3.3), and (3) Incorporating existing external knowledge into the learners' own understanding (see 3.4). Aspect (1) addresses the delicate balance between effectively supporting students in their learning processes and leaving them enough freedom (de Jong and van Joolingen 1998) so they can build ownership of the task, can have their own experiences, and particularly have the joy of finding results themselves. Self-regulation is especially important to achieve higher learning goals like strategic skills, learning how to inquire, and knowledge about the nature of science. Aspect (2) addresses the challenge of building useful knowledge within complex contexts or problems where many knowledge elements, including new information, must be integrated into a complex mental model that supports the solution of a problem as well as its justification. Ideally, the model's parts should be consistent with each other and with other bodies of related knowledge and evidence. Aspect (3) is concerned with the process of checking one's understanding against existing, especially theoretical knowledge that can be collected from many sources (e.g. from teachers or peers, articles, textbooks, the Internet, etc.).

This paper brings together design perspectives (Section 3.1) and research results (Sections 3.2–3.4) from the ParIS, WISE, Viten, and Co-Lab projects, providing hints on how to support learners in some of the main problems they encounter within inquiry. Some of the insights can be synthesized into general conclusions, while others remain specific to the individual projects.

3. FOUR APPROACHES RECONSIDERED UNDER COMMON ASPECTS

3.1 Foci of Collaborative Inquiry Learning

The four approaches ParIS, WISE, Viten and Co-Lab all support inquiry in complex authentic socio-scientific and/or personally relevant issues. Inquiry learning focuses on letting learners act like scientists to realize the problems of the issue and take part in finding solutions for the problems. In these processes, the learners have to construct their own research plan and the teacher or the learning environment gives freedom to the learners to do their own research. All approaches have inquiry parts where students work in a self-regulated manner (see below). The results of these parts depend on the students' work. The goal is not to lead the students to one specific solution because there often is not ONE solution. Orienting, knowledge constructing, knowledge checking and concluding processes can be found in each approach. The approaches differ in the emphasis that is put on the respective processes and in the level of freedom in different processes.

The ParIS project is a resource-based learning approach that has been designed to tackle the problem of competence in self-regulated learning in the field of scientific literacy. Conventional lessons are supplemented by phases of project-based work engaging students in self-regulated coping with everyday questions using the Internet as a central learning resource. With mind mapping, a visual-spatial learning strategy has been implemented for fostering the management of knowledge and resources (McCagg and Dansereau 1991), promoting the development of cognitive and meta-cognitive competencies. In small groups the learners collaboratively evaluate information, integrate it into their map and present the results to the whole learning group. In order to support effective learning, the resource-based learning scenario has been embedded in a 'cognitive apprenticeship' approach (Collins et al. 1989; see 3.2).

In WISE and Viten, inquiry projects employ socio-scientific issues like whether genetically modified foods are safe to eat, or whether global climate change is caused by humans (Jorde et al. 2005; Slotta 2005). An important AIM of these Web-based inquiry environments is to help students understand different perspectives for regarding such complex problems, as well as to make connections between the issues and the relevant scientific knowledge (Linn and Hsi 2000; Slotta 2002). WISE and Viten provide well designed curricular projects and powerful support environments, engaging students in a variety of activities (e.g., animations, note taking, textual materials, etc.). Within each inquiry project, knowledge acquisition processes include orientation, knowledge constructing and knowledge checking phases. In subsequent steps within the project, students apply the acquired information by comparing, contrasting, critiquing, sorting out and reconceptualising their own scientific ideas, incorporating new information, evaluating alternative accounts, and connecting everyday and scientific ideas. These information application processes are scaffolded by the Web-based learning environment, as well as through social interactions within the classroom, such as problem sharing, debating, arguing, critiquing, valuing or discussing. A good example of how such Web-based inquiry environments can support socio-scientific inquiry is seen in the international collaboration study between the U.S.A. and Norway (Slotta 2005). In each country, students completed an inquiry project concerned with wolf management using WISE or Viten, respectively. Afterwards, using an online discussion forum, students from the two countries joined a single debate about two key aspects of the issue: global interdependence and biodiversity in forest ecosystems. With more than 3000 wolves in Minnesota (U.S.A.) and only 10 in Norway (on average), American students were able to contribute important insights to the Norwegians about biodiversity. In return, the Norwegian students proved to be more savvy about issues concerning global interdependence, and could help their American peers understand why the preservation of a habitat in one part of the world was important to people living far away in another. Thus, participation in this curriculum allowed opportunities to students that would be inaccessible if their exchanges had been limited to peers within their own country. The American students were observed to benefit from the discussion about global interdependence, as they showed particularly improved results in this topic in the post-test. This study illustrates the promise of bringing in new ideas and new perspectives on a topic that could not be achieved in a local discussion group.

The Co-Lab learning environment also supports on-line communication in a collaborative way. Topics like water management or greenhouse effect are of global interest and have different aspects in different regions (e.g. water shortage in southern countries vs. flood problems in other regions). Collaborating in international groups might be fruitful for discussing these different aspects. The focus of inquiry in Co-Lab is, however, more on understanding quantitative phenomena, experimenting, and constructing a model that explains the results of the experiments, reproducing the empirical data and making appropriate projections. Comparisons of model outcomes and empirical data lead to a report of the results or to new questions and further inquiry processes. Though they also have general background information about the topic, the learners' main task is to express their own knowledge about the phenomena.

3.2 Guidance and Freedom

The issue of freedom and guidance in inquiry is handled in different ways by the four approaches under consideration, but some convergence also emerges. As described in 3.1 the focus of inquiry projects within WISE and Viten is to engage students in reasoning, critiquing or debating relevant science problems. WISE and Viten projects mainly consist of the two processes outlined in 3.1: Knowledge acquisition, where the learning environment suggests a sequence of learning steps by offering information and scaffolds knowledge expression; and knowledge application: In order to construct their own understanding, students must integrate the collected information and chunks of knowledge, and are supported by the learning environment e.g., in writing an online article or preparing for a debate in a mock TV show. This integration is the somewhat unstructured inquiry process employed within WISE and Viten; there are no right or wrong answers, per se, nor any explicit solutions to problems.

The ParIS conception makes use of the 'cognitive apprenticeship' approach (Collins et al. 1989) that has foci on modelling, scaffolding and reflecting designed for fostering self-regulated learning and the transfer of knowledge into practice. These basic features of cognitive apprenticeship are translated into the global structure of the ParIS learning process. After introducing and discussing the topic's relevance, the teacher's activities in the "Modelling Phase" (working on a complex issue as an expert problem solver) serve as a model for students. In the following phase of "Group Work" where students can choose from a set of topics, the teacher acts as a coach. The students go through cycles of collecting and elaborating information. In such resource-based learning scenarios students often suffer from cognitive overload and need help through scaffolding and cognitive tools. In order to foster learning, the teacher introduces self-regulated strategies (by modelling), which the students are expected to apply, such as collecting, documenting the origin of resources, evaluating the adequacy, and organizing ideas and data according to some logical pattern. This whole process is supported by a mind mapping tool ("MindManager Smart", see Section 3.3).

Co-Lab projects, e.g., the Greenhouse effect project, have the following design (cf. Bosler et al. 2004): After an introduction into the topic, students enter the computer system and first work on a base module (graphically modelling the basic structure of a radiation balance), then on a main module (modelling the radiation balance of the earth), and in the end on an optional module of free inquiry (e.g., adding CO_2 production by car traffic). Each module consists of a sequence of levels, each initiated with a new phenomenon (simulation, dataset or remote experiment) motivating the students and initiating an extension of the students' model. Within a level students have to find their own learning pathway, while guidance and advice is given by the teacher acting as a coach and by Co-Lab's process coordinator tool. The process coordinator is a tree structure of learning goals, sub-goals and hints, provided by the designer (respectively the teacher) in advance which may be visited and edited by the learners. Fading guidance is realized across the levels and modules (the process coordinator gives less advice at higher levels).

The Co-Lab conception of inquiry learning was tested in eight school courses in Germany at the upper secondary level. Co-Lab units typically consisted of 15–20 lessons. At that time, the full Co-Lab environment was unavailable; i.e. the process coordinator tool could not yet be tested, but students could experiment with a simulation and construct graphical models. Theoretical background information was provided on paper. Some of the courses were visited, and field notes were taken. The study yielded the following insights on how to embed inquiry in the classroom (cf. Krajcik et al. 1994): In order to enable more effective student learning, in the beginning of the project most teachers added a small Co-Lab "modelling course" by demonstrating the modelling and having the students model a simple phenomenon presented as a real experiment (e.g., cooling coffee pot). Thus, the sequence of fading guidance was modified into: modelling demonstration; simple modelling task for students (strongly supported); base module (scaffolded), main module (quite open), and eventually a free inquiry module (done by good students). In this sequence of fading support, about two thirds of the student groups achieved good results, i.e. running models that captured the main structures of the greenhouse issue; the rest needed stronger support from the teacher the entire time. It remains to be tested whether larger parts of the scaffolding can be provided by the process coordinator tool. Teacher involvement in the Co-Lab study led to the emergence of an interesting parallel between approaches: As in the ParIS conception, a teacher demonstration phase ("modelling" in terms of cognitive apprenticeship) was added. Some differences remain: Co-Lab has a module conception with a smoother fading of guidance; the same effect can be achieved in ParIS, as well as in WISE and Viten, through teacher action that is flexibly adapted to student needs.

3.3 Knowledge Construction

In inquiry learning, the complexity of realistic issues normally has to be reduced drastically in order to make them manageable for students. The first reduction is introduced by the designers of an inquiry project under an educational perspective. A further reduction is done by the students who select relevant information and organize their own (mental) model that should be able to cover a limited area of the problem field. But still the remaining content model will be complex, at least from the students' perspective. This is even a precondition for meaningful inquiry. The four approaches use different methods to support students in handling content complexity. One possibility realized in ParIS is to provide the students in the modelling phase with strategies for processing complex information (see 3.2). Another way is to enable students to express their knowledge in external representations. WISE and Viten make use of electronic notebooks in their projects; students are prompted to enter answers to questions or express their opinion in a written format. They may manoeuvre back and forth in the program, and they also have the opportunity to edit their texts in the notebook when they have found new aspects or have changed their mind (iterative approach). In Viten, teachers have access to the electronic notebooks and use this option to monitor student progress and to provide feedback on student work. Studies have shown the effectiveness of this tool giving examples in which students improved their answers on their own when revisiting them, and other examples in which students reviewed their answers after electronic teacher comment (Jorde et al. 2005). Teacher scaffolding is an important tool provided with the Viten software as teachers can review student results shortly after the work is done and influence the ongoing learning process. The monitored learning process serves as a preparation for the concluding task, e.g. to write an online article or to take part in a debate on the issue under investigation.

ParIS deploys visual-spatial representations (mind maps) to be constructed by the students. The self-regulated student must be able to apply optimal learning strategies (Simons 1993). Learners carry out the problem solving processes iteratively imitating the expert's behaviour until they have reached their goals which leads to a cyclic structure at the local level. The students acquire the learning content actively, articulate and reflect upon their results, discuss them in groups and explore new information from different resources. For the most part, the students insert their findings and their observations and interpretations of the experiments with own words into a digital mind mapping tool (Buzan 1974). Mind mapping is suggested to support self-regulated learning and problem solving (Jonassen et al. 1993). In order to externalise the individual problem space, the students create a personal map, step by step, after each step reflecting their own and the group's project work and problem solution. The result is a complete map that summarizes the student's work and considerations.

Research data concerning the acceptance of the instructional approach of ParIS relate to both the 'cognitive apprenticeship' approach as a dominant instructional strategy and the use of mind mapping as a visual-spatial strategy for fostering self-regulated, resource-based learning. According to the teachers' statements, the implementation of the 'cognitive apprenticeship' in ParIS was very successful in class. The design approach could be integrated into the lesson well and could be executed within the time frame. The students rated the 'cognitive apprenticeship' type of lesson as interesting and indicated that independent work would be more possible than in traditional lessons with a more teacher-centred style. The analysis of the questionnaire as well as interview data from studies with different student populations (grades 8 to 13) revealed that the teachers and the students described the digital mapping as "well-suited for the classroom". They particularly liked the possibility of entering links, attaching notes and making clean, quick changes according to one's own wishes. Two thirds of the students continued to use the program at school after the end of the study.

While learning with Co-Lab, students integrate knowledge they gain step by step into a graphic model (stock-and-flow modelling). The model should be runnable and reproduce experimental data that can be obtained from simulations, datasets or remote experiments. Running capability and reproduction of data also indicate some degree of coherence, viz. internal coherence and coherence with respect to experiments, but not necessarily with respect to existing theory (see 3.4). In iterative processes of experimentation and model construction (= inquiry learning cycles going through Co-Lab levels) the student models become more and more complex. At each level there are new phenomena and experiments to be included into the predictive range of the model, thought to render new motivation.

The Co-Lab classroom study outlined in 3.2 showed: Co-Lab's level structure really stimulates most students to enter iterative processes of model extension, though in many groups fewer cycles were executed than planned by the designers. Still, in about one third of the groups, clear motivation by the teacher was needed to enter new cycles, but the majority found the task sufficiently motivating. During model construction, peer discussion within and among learning groups as well as teacher feedback turned out to be a crucial factor for success. Nearly all student groups could achieve running (i.e. coherent) models that represented a part of the problem. However, the level of complexity was very different in that talented students could achieve enormous complexity. The method allows for solutions at very different levels of expertise. The graphic models also strongly supported student presentations: Using their models they could show their knowledge as an integrated whole.

3.4 Knowledge Integration

In 3.1 it was pointed out that using, comparing, evaluating information and integrating different ideas into learners' own knowledge space is a main focus of inquiry learning in most of the approaches' projects. ParIS as a resource-based learning approach predominantly focuses on the integration of external knowledge. When the students are confronted with a complex problem from everyday life, they are normally missing content knowledge as well as adequate problem solving strategies with which to resolve it. ParIS teachers provide these missing parts through modelling an expert's behaviour on problem solution. After defining the inquiry question, one has to activate available pre-knowledge as a base for needs assessment. The most important part in ParIS is the search for new information and its evaluation and integration into the individual knowledge structure. In addition to the modelling, the scaffolding and coaching by the teachers the students are supported through the mind mapping tool. This tool helps to systematically externalize content and concept knowledge, to relate pre-knowledge with new information, to reflect newly developed structures and to communicate and present results. WISE and Viten promote valuing and evaluating processes through communicating activities like discussing or debating. In Co-Lab, background reading information on the topics' underlying theories is provided, but the learning environment does not force the learner to read it. The classroom study described in 3.2 showed that students tended not to read too much of the information. Novices often built models with their own ad-hoc terms and concepts and not with concepts from standard theory. The use of their own terms does not have to be prevented from the beginning because it can make it easier for the students to express their own knowledge. But the students need to learn to look for evidence of their theories and that at a certain point communicating their ideas is easier using common terms. Therefore, there is a need for the Co-Lab environment to provide instruction that encourages the learners to use background information in every new modelling step. A technical solution of the Co-Lab software is the process coordinator that scaffolds the whole learning process. Here specific advice could stimulate the students to use the background materials repeatedly. This possibility has to be tested. Other solutions rely on the teacher. The social interaction parts between groups like in Viten or WISE are examples of the good practice to use background information. In these processes the students have the opportunity to recognize that there are other ideas that could or should be considered.

All in all, the experiences show that guidance through the learning environment can help the learners with studying background material. All learning environments provide technical solutions to support the processes. In Co-Lab the functionality of the process coordinator as an instructional tool is promising, but needs testing.

4. DISCUSSION AND CONCLUSIONS

Bringing together expertise from the ParIS, WISE, Viten, and Co-Lab projects, some issues related to characterizing collaborative inquiry learning and to some of its difficulties as well as possible remedies are revealed. In all four projects, knowledge construction and checking are closely interrelated and proceed in iterative attempts. In three projects (WISE, Viten and Co-Lab), complex, contemporary problems, especially socio-scientific issues, were described as motivating for learning groups and particularly suitable for collaborative inquiry (Bosler et al. 2004; Jorde et al. 2005; Slotta 2005). Typically, these projects leave enough freedom for self-guided investigation. They do not prescribe just one correct solution, but instead, foster debate and argument within as well as among learning groups.

The four approaches to inquiry have different perspectives to problems. Co-Lab emphasises quantitative experimenting and modelling of phenomena, while ParIS focuses on the collection and organization of rich information. WISE and Viten aim at socio-scientific discussion and debate. All four inquiry approaches tend to provide stronger guidance for processes that are not of central interest and leave greater freedom for student construction in core processes. In central construction activities peer discussion becomes particularly important in order to promote a deeper level processing of information: In Co-Lab, mainly dynamic models are discussed by the students, whereas in ParIS it's knowledge structures in the form of mind maps about which they converse. In WISE and Viten the arguments, opinions, and decisions are mainly focused in a socio-scientific context. In the latter case, international collaboration becomes especially interesting due to different perspectives across cultures.

An analysis of these four projects reveals that one main means used to make inquiry accessible and fruitful for students is to provide an appropriate degree of guidance. The cognitive apprenticeship approach used in ParIS is a good example in that it suggests that an expert such as the teacher should demonstrate the treatment of a problem, before students start to apply the strategies and techniques on their own. So, for example, strong support is often necessary in the student work phase, particularly in the beginning, as seen in ParIS. Similarly, in Co-Lab, teacher involvement was high in the beginning of each module sequence such as when the teacher performed a demonstration, but then diminished from help with simple student tasks to more difficult ones with open problems and less teacher assistance.

Another challenge in doing inquiry revealed by these projects is the appropriate incorporation of external knowledge. This is a key issue not only in student inquiry, but also in expert studies, e.g. in the social sciences. Student inquiry is more consistent with this model if a phase within the whole learning process is explicitly targeted at the collection and organization of information, as is done in ParIS, WISE and Viten. If information has to be collected iteratively, e.g. parallel to modelling efforts as in Co-Lab, prompting this activity has to be repeated. On the other hand, whenever possible, students need to learn to perform inquiry activities independently. In Co-Lab this conception did not work well with novices (see 3.4). Here, it remains to be tested whether support through a process coordinator tool is a good compromise between guidance and freedom. Analysis of these four technology-based inquiry projects reveals solutions to another issue: Solving problems based on a rich collection of information needs support for organizing and representing complex knowledge structures. Information technology has developed several tools for constructing complex bodies of knowledge collaboratively, e.g., electronic notebooks, whiteboards for free sketching, mind mapping tools (in ParIS), concept mapping tools (in WISE) and editors for graphic models (in Co-Lab). Using such tools students construct an artefact that serves to explicate ideas ("communication through the artefact") and also as a forum of cognitive conflict. Synthesising the reported experiences we suggest that students can obtain very good construction results under the following two conditions: (a) The knowledge structure (the student "model") should be built iteratively; in this building process, knowledge construction and checking promote one another. The inquiry project has to be structured in a way that continuous work on the complex model is possible and that new motivation is created for every step of model extension (see Co-Lab in 3.2). (b) The construction tool should allow for repeated feedback during the building process. In collaborative inquiry, first feedback comes from peers, within or among learning groups. Moreover, for most learners feedback from the teacher is necessary. It may be given by direct coaching in the classroom, but in some tools can also be delivered online, synchronous to student work or during work breaks (see Viten in 3.3). Construction tools use different formats of representation. Of course, the format has to fit the purpose as exemplified by the approaches under discussion: The textual format is appropriate for tasks aiming at articles, presentations, discussions and debates with well-formulated statements and arguments. The graphic format has the advantage of easily providing an overview and of enabling parallel (fast) processing of information. Graphic tools (model editors, mapping tools) seem to be advantageous when representing complex structures of items and relations or constructing runnable models. We hypothesise that a combination of tools with different formats could bring specific benefit: First descriptions and hypotheses are usually expressed verbally or written down; complex constructions on the basis of many pieces of knowledge could benefit from graphic representation; and drawing conclusions, discussing and presenting them again needs the textual format.

5. OUTLOOK

Based on evidence from the four research projects WISE, Viten, ParIS, and Co-Lab, some general conclusions on design and support of collaborative inquiry learning can be made. Syntheses of approaches seem to make sense with respect to beneficially structuring learning environments and to combining different tools that support students' construction of knowledge. Other common elements of collaborative inquiry learning from these projects are that (1) Student products are seen more and more as useful learning objects: Students could communicate and/or combine their products. (ParIS and Co-Lab) (2) Collaboration across school and country borders is an interesting feature of modern technology; beneficial arrangements are possible as shown in Viten and WISE. (3) From the teacher's perspective, the issue of time management and of implementing inquiry sequences into the classroom is of great importance. Very different models are possible and were used in WISE, Viten, ParIS and Co-Lab inquiry projects. (4) More generally speaking, the role of teachers and of schools in dissemination of inquiry has to be considered.

Since these four projects have both unique and common solutions to the problems of learning via inquiry, there may well be some benefit from international co-operation in the development of inquiry projects. It remains a task for the future to disseminate new insights into the respective projects. Networks for research and development, like the EU project CIEL (Collaborative Inquiry and Experiential Learning) and NetCoIL, the Scientific Network for Collaborative Inquiry Learning of partners from Europe and America, can play a strong role in bringing research and development cultures together to enhance inquiry learning.

REFERENCES

Bosler, U., Bell, T. & van Gastel, L. (2004). Co-Lab: Supporting teacher development of inquiry learning. (In D. Zandvliet (Ed.), *NARST 2004 Conference, Conference Proceedings*. Vancouver (Canada): NARST, CD ROM)

Buzan, T. (1974). *Use Both Sides of Your Brain*. (New York: E.P. Dutton)

Collins, A., Brown, J.S. & Newman, S.E. (1989). 'Cognitive Apprenticeship': Teaching the crafts of reading, writing, and mathematics. (In L.B. Resnick (Ed.), *Knowing, Learning, and Instruction: Essays in Honor of Robert Glaser* (pp. 453–494). Hillsdale, NJ: Erlbaum & Associates)

De Jong, T. & van Joolingen, W.R. (1998). Scientific discovery learning with computer simulations of conceptual domains. *Review of Educational Research, 68(2)*, 179–201

De Jong, T., van Joolingen, W.R., Lazonder, A., Ootes, S., Savelsbergh, E.R. & Wilhelm, P. (2002). *Co-Lab Specifications. Part 1 Theoretical background (Technical Report)*. (Enschede, The Netherlands: University of Twente)

Jonassen, D.H., Beissner, K. & Yacci, M. (Eds.) (1993). *Structural Knowledge. Techniques for Representing, Conveying, and Acquiring Structural Knowledge*. (Hillsdale, NJ: Lawrence Erlbaum Associates)

Jorde, D., Breiteig Berg, H. & Strømme, T. (2005). The Challenges of using ICT to promote collaborative inquiry learning – the case of Viten.no. (In R. Pintó and D. Couso (Eds.), *Proceedings of the Fifth International ESERA Conference on Contributions of Research to Enhancing Students' Interest in Learning Science* (pp. 1509–1512). Barcelona: Spain)

Krajcik, J., Blumenfeld, P., Marx, R. & Soloway, E. (1994). A collaborative model for helping middle grade teachers learn project-based instruction. *The Elementary School Journal, 94(5)*, 483–497

Linn, M.C. & Hsi, S. (2000). *Computers, Teachers, Peers: Science Learning Partners*. (Mahwah: Erlbaum)

Löhner, S., van Joolingen, W.R., Savelsbergh, E.R. & van Hout-Wolters, B. (2005). Students' reasoning during modeling in an inquiry learning environment. *Computers in Human Behavior, 21(3)*, 441–461

McCagg, E.C. & Dansereau, D.F. (1991). A convergent paradigm for examining knowledge mapping as a learning strategy. *Journal of Educational Research, 84(6)*, 317–324

Mork, S.M. & Jorde, D. (2004). We know they love computers, but do they learn science? A study about the use of information technology and controversy in science instruction. *Themes in Education, 5(1)*, 69–100

Neumann, A., Gräber, W. & Tergan, S.-O. (2005). ParIS – Visualizing Ideas and Information in a Resource-Based Learning Scenario. (In S.-O. Tergan and T. Keller (Eds.), *Knowledge and Information Visualization*, LNCS 3426 (pp. 244–269). Berlin: Springer)

Quintana, C., Reiser, B.J., Davis, E.A., Krajcik, J., Fretz, E., Duncan, R.G., Kyza, E., Edelson, D., Soloway, E. (2004). A scaffolding design framework for software to support science inquiry. *The Journal of the Learning Sciences, 13(3)*, 337–386

Simons, P.R.-J. (1993). Constructive learning: The role of the learner. (In Th. M. Duffy, J. Lowyck and D.H. Jonassen (Eds.), *Designing Environments for Constructive Learning* (pp. 291–313). Berlin: Springer)

Slotta, J.D. (2002). Designing the Web-based Inquiry Science Environment. In S. Hooper (Ed.), *Educational Technology, 42(5)*, 15–20

Slotta, J.D. (2005). Learning from our peers in international exchanges: When is worth doing, and how can we help it succeed? (In R. Pintó and D. Couso (Eds.), *Proceedings of the Fifth International ESERA Conference on Contributions of Research to Enhancing Students' Interest in Learning Science* (pp. 1505–1508). Barcelona: Spain)

Tabak, I. (2004). Synergy: A complement to emerging patterns of distributed scaffolding. *The Journal of the Learning Sciences, 13(3)*, 305–335

Van Joolingen, W.R., de Jong, T., Lazonder, A.W., Savelsbergh, E.R. & Manlove, S. (2005). Co-Lab: Research and development of an online learning environment for collaborative scientific discovery learning. *Computers in Human Behavior, 21(4)*, 671–688

35. VISUALIZING THE QUANTUM ATOM

Abstract: Students from the advanced high school classes up to the last university years have difficulties in grasping the notions involved in the description of the Quantum Atomic Model (QAM). Many researchers suggest that using Information and Communication Technologies (ICT) for visualizing the QAM would improve students' understanding. An empirical study was conducted with 20 first-year students of the Department of Primary Education. They interacted with two Internet-based software packages, considered as representative 3D visualizations of the QAM. For the analysis of our results, a qualitative approach was taken by using SOLO (Structure of the Observed Learning Outcomes) taxonomy. Our results indicated that software packages concerning the 3D representations of QAM do not help students to understand scientific concepts and the atomic shape. The article proposes the use of Virtual Reality Technologies for the creation of atomic visualizations based on scientific data that support conceptual change

Keywords: 3D visualization, Information and Communication Technologies, Quantum Atomic Model, SOLO taxonomy

1. INTRODUCTION

It is generally accepted that Quantum Mechanics has a mysterious and exciting flavour because it deviates markedly from intuitive expectations and the 'normalcy' of the classical world. Yet, this strange theory is remarkably successful at describing the behaviour of real physical systems.

The structure of matter is a fundamental topic in science education from primary school up to the university level. It is well documented that students at all levels have conceptual difficulties in understanding the concepts associated with the particle nature of matter and mostly with the atomic models (Harrison and Treagust, 2000; Taber, 2003).

The Quantum Atomic Model (QAM) covers a part of the upper secondary and tertiary curriculum in many countries. As learning about the QAM involves a fundamental reconceptualization in many areas (Thacker et al., 2002), students from the advanced high school classes up to the last university years have difficulties in grasping the main notions. Students' misconceptions concern concepts such as the charge cloud, the probability of electron localization, indeterminacy, spin (Johnston et al., 1998; Budde et al., 2002a, b; Cassinelo and Gallego, 2005), as well as fundamental principles like the uncertainty principle (Müller and Wiesner, 1999; Budde et al., 2002a) or the law of quantum measurement (Johnston et al., 1998;

Dr. Tassos A. Mikropoulos, Department of Primary of Education, University of Ioannina, Greece

R. Pintó and D. Couso (eds.), Contributions from Science Education Research, 465–475.

Budde et al., 2002a) which are far from their intuition. It is also found that mechanical atomic models are dominating among secondary school and university students. These models are mostly taught in high school classes, they are more plausible, and their optical representations are concrete and comprehensible, so students recall them easily and even draw them without difficulty, since among other natural learning styles the visual–spatial is a prominent one (Barnea and Dori, 2000). Thus, many researchers propose visualization tools and suggest that using Information and Communication Technologies (ICT) for visualizing the QAM would improve students' understanding for three main reasons:

- It is complicated to create comprehensive three dimensional (3D) visual representations of the QAM without the advantage of ICT (Byrne, 1996; Barnea and Dori, 1999, 2000; Dori and Barak, 2001; Cataloglou and Robinett, 2002; Trindade et al., 2002; Barak, 2007).
- As the QAM is covered by a highly mathematical formalism and there is not yet consensus about how it might be taught less abstractly (Johnston et al., 1998), it can be described in a qualitative way taking advantage of ICT. Using the visualizations, students will be able to assimilate more abstract mathematical methods later in their careers (Byrne, 1996; Tuvi and Nachmias, 2001; Cataloglou and Robinett, 2002).
- The use of visualizations is an appropriate method for probing student understanding as many of them are not able to visualize the QAM in 3D space and to rotate the 3D model in their mind. Also, students lack depth perception and have limited sense of perspective when all they have seen are two dimensional (2D) representations or mathematical models (Hurwitz et al., 1999; Cataloglou and Robinett, 2002).

By the term 'visualization' we mean the optical hermeneutic transformation of scientific data and not the artistic rendering of a phenomenon. Visualization is the optical representation of information as the result of a simulation based on a scientific model.

The software packages referring to visualizations of the QAM which have been used as educational tools have the following characteristics: they are 2D or 3D visualizations based on simulations according to Quantum Atomic Theory and provide a certain level of interactivity. By rotating, for example, the 3D representations, students may become familiar with the shapes in space.

However,

- it is not often clarified if the wave function ψ, the probability density ψ^2, or the electron density is represented through the 3D visualizations;
- the use of different colours for all these representations may create misconceptions;
- the nucleus position is not referred in relation to different representations;
- the difference between the hydrogen wave functions (orbitals) and the wave functions describing polyelectronic atoms is not explicitly mentioned;
- it is not made clear under which conditions the hydrogen orbitals, except 1s, describe the atom state.

Since quite a few reports describe the results of using these visualizations as educational tools, we investigate if they help students to understand the basic concepts involved in the construction of the QAM and create mental models in coherence with scientific knowledge.

The axes of our study are to investigate the following:

- How students conceive the 3D representations
- If students are able to connect the visualizations of wave functions with the basic principles of Quantum Theory
- Which are students' mental models for the atom after interacting with the visualizations
- If students prefer 3D representations and for what reason.

Here, a part of our research is presented concerning the visualization of the QAM using ICT. More specifically, the goal of our study is to propose dynamic 3D visualizations of the atom giving an integrated picture, avoiding misconceptions coming from the classical atomic models, as well as from the piecemeal presentation of ψ and ψ^2 or the lack of their comprehensive visualization.

2. METHODS AND SAMPLE

Since 1999 in upper secondary school the Greek chemistry courses have included the QAM, the atomic structure according to this model, as well as the explanation of covalent bonding based on valence bond theory without the use of any mathematical formalism. At the same grade, physics courses have included in chronological order all the atomic models which had been constructed before the development of Quantum Mechanics.

An empirical study was conducted in March 2004. Twenty first-year students of the Department of Primary Education, University of Ioannina, participated. These 17-year-old students have been taught the topics under study during their last year in high school, as described above.

Students interacted with two Internet-based software packages, considered as representative 3D visualizations of the QAM. First, it was 'Orbitron' that presents visualizations of atomic orbitals from 1s up to 7g. It includes representations of the wave function ψ, its radial distribution, electron density, and electron density with dots (Winter, 2002). Secondly, students used 'visualization of atomic orbitals using VRML' (Blauch, 2001). This involves representations of ψ radial distribution, electron density, and isosurface plots.

We chose these two packages as the most appropriate comparing with others found in the Web because by using them it is made clear which wave function is represented each time, and it is clarified if hydrogen orbitals or orbitals of a polyelectronic atom are represented and electron density plots and isosurface plots are included for every wave function.

After the interaction with the visualizations referred to hydrogen 1s, 2px, 2py, 2pz orbitals and electron density with dots (eψ2), each student participated in a semi-structured interview. The most important sets of questions are presented below:

1. What these images represent? What kind of information do we get? Do they concern a certain chemical element? (we refer to 1s, 2px, 2py, 2pz orbitals visualizations). Describe how you conceive the notion of 'orbital'.
2. What these images represent? What kind of information do we get? Do they concern a certain chemical element? (we refer to 1s, 2px, 2py, 2pz electronic density plots with dots, which we call charge cloud as well). Describe how you conceive the notion of charge cloud.
3. Is it possible any of these images to represent an atom, if it were possible to see it somehow? Where it would be the nucleus?
4. According to uncertainty principle it is not possible to determine with accuracy the position and velocity of a particle. Why does it happen? Is there any relation between this principle and the images you have seen?
5. Do you prefer these 3D orbital representations than the 2D ones which are shown in your textbook?

Each one of the above sets consists of more than one question. This is because our aim was to make an in-depth analysis of the students' mental representations concerning the topic under study. For the analysis of our results, a qualitative approach was taken since we are interested in the exploration of cognitive content of students' answers by making an in-depth analysis of them. In the qualitative outlook, it is assumed that students learn cumulatively, interpreting and incorporating new material with what they already know, their understanding progressively changing as they learn (Biggs, 1994). Since knowledge is described with qualitative terms that focus on the cognitive content of students' answers, we use SOLO (Structure of the Observed Learning Outcomes) taxonomy for the data analysis (Biggs and Collis, 1982).

The SOLO taxonomy is an important application of cognitive theory to modern education. The SOLO taxonomy has evolved steadily since its initial formulation and now stands as a detailed model that can be used to explore and help in the interpretation of cognitive development in a range of learning areas. In so doing, it provides insights into the way understanding develops. This taxonomy provides five different levels of understanding, from prestructural to the deepest level, known as extended abstract. The reason we have chosen SOLO taxonomy is because we wanted to classify the students' answers into hierarchical levels and compare them with the scientific acceptable knowledge that is the extended abstract fifth level.

The SOLO taxonomy classifies students' understanding in the following five hierarchical levels.

Prestructural: Student is destructed or misled by irrelevant aspects. He avoids or repeats the question, makes an irrelevant association. In a transitional stage, he uses inadequately a relevant datum.

Unistructural: Student focuses on the relevant domain and works with a single aspect. He selects one relevant datum and closes on that. In a transitional stage, he selects two relevant but inconsistent data.

Multistructural: Student provides correct material with discrete, separate pieces of information that may be combined to provide a composite picture. He selects two or more relevant data, uses them inconsistently, and reaches an alternative conclusion. In a transitional stage, he recognizes inconsistencies but cannot resolve them.

Relational: Student offers an integrating understanding of the information. His answer has a coherent structure and meaning. He uses most or all relevant data, integrate them with a relating concept, and reaches a right conclusion. In a transitional stage, he tries to generalize his conclusion without success.

Extended abstract: Student provides abstract principles or hypotheses that show the specific example as just one of many possible results.

The 20 students' answers have been examined and analysed towards the following components for being evaluated and classified to a certain level:

- The recognition of characteristics for describing an atomic model according to a scientific theory.
- The correlation of the above elements in order to be consistent with the basic principles of the scientific theory according to which the atomic model is constructed.
- The deduction of a consistent conclusion for the representation of an atom according to the atomic model described above.

Transitional responses have been also detected that carry more information than usual in the level the student is emerging from, but he is not reaching at the next SOLO level.

3. DATA ANALYSIS AND RESULTS

Among the students' answers we have detected the first four SOLO levels and two transitional ones showing their development of understanding, after their interaction with the visualizations (Table 1).

More specifically, students' performance to each question is presented below.

Question set 1

Most of students' answers (17) reached up to the third SOLO level. They did not realize that the representations which they have seen were graphs of hydrogen ψ wave function at the ground state (1s orbital) or at the first excited state (2p orbitals).

TABLE 1. Students' SOLO Levels ($N = 20$)

Solo levels	Question 1	Question 2	Question 3	Question 4
1. Prestructural	2	6	16	18
2. Unistructural	5	6	3	1
2→3 Transitional	5	1	0	0
3. Multistructural	5	3	0	1
3→4 Transitional	3	3	0	0
4. Relational	0	1	1	0

Answer example of the 2–3 transitional level:

x, y, z are the orbitals. 1s, 2p, 2s are subshells. The colour intensity shows the number of electrons. The orbitals' shapes are the same for all chemical elements. I don't remember what an orbital exactly is. Is it the space where the electrons move? No....

The student referred more than one characteristic concerning the orbital, but they are not all correct.

Five students, whose answers are categorized to the third level, recognized the graphs as orbitals, but they were not able to combine the information given by the picture with that they had learned about wave functions in order to differentiate the two atomic orbitals. They simply referred Schrödinger equation without connecting its solutions with the concept of the orbital.

Answer example of the multistructural level 3:

They are orbitals... There are one 1s orbital and three 2p orbitals. There is the Schrödinger equation... There is a probability to find out the electron somewhere. The orbitals are always the same for all the chemical elements.

The student referred different characteristics concerning the atomic orbitals, most of them correct, but she did not even try to combine them.

Three students whose answers reached the 3rd to 4th transitional level tried to explain the different schemes of atomic orbitals 1s and 2p, the correlation between orbitals and Schrödinger equation or the difference between the orbitals for hydrogen and polyelectronic atoms without success.

Answer example of the 3–4 transitional level:

I can see the 1s and 2p orbitals for hydrogen; they must be different for polyelectronic atoms... I remember that Schrödinger equation gives us some information about orbitals, I can't remember which one. We have one 1s orbital but three 2p orbitals because for 1s: $n=1$ and $l=0$ (one value) and for 2p: $n=2$ and $l=-1, 0, 1$ (three values). I don't know why they have different representations.

The student referred different characteristics which are all correct. He also tried to relate them but he did not arrive to a correct conclusion about the information given by the 3D representations with which he had interacted before.

Question set 2

The students had great difficulty in describing what a charge cloud is, because it is not explicitly expressed in the Chemistry textbook. They confused the charge cloud with that of the orbital, as they have seen that both have the same contour. Consequently, 6 of the students are at the prestructural level and 16 did not attain further than the third level. There was only one answer at the relational level.

Answer example of the prestructural level 1:

They are orbitals. All these dots... I don't know, we can only measure the force from the nucleus to hydrogen.

The student gave an irrelevant answer as he confused the concepts 'orbital' and 'charge cloud'.

Answer example of the relational level 4:

'They are charge clouds; the densest area of the cloud is where we have the bigger probability of finding the electron. But there is very small possibility for the electron to be found too far from the nucleus, that's why there are quite a few dots there'.

The student used all the relevant data correctly and reached to the correct conclusions, but she was not able to generalize so as to arrive at the formulation of the uncertainty principle.

Question set 3

Almost all of the students did not know neither if the orbitals' graphs represent the atom, nor the position of the nucleus. Only one gave an answer which is classified at the fourth (relational) level. This may have happened because the notions of ψ, ψ^2, or electron density in their textbooks are not related with the shape of an atom at the ground or excited state.

Answer example of the unistructural level 2:

'These shapes must have some relation with the atom's shape, but I don't know exactly...the nucleus is at the middle of them'.

The student referred only one relevant data. Except that she was not able to say something about the atom's shape.

Answer example of the relational level 4:

'In general the electron orbitals in the outer shells must determine the scheme of an atomic model. So the hydrogen atom must be spherical as the charge cloud which is determined by the 1s orbital. The nucleus is at the centre of the sphere'.

In this answer the information given by the optical representation was related with student's knowledge for arriving at a right conclusion about the atomic model. Therefore the student had conceived that all these 3D images concern the atomic models but he could not generalize about the scheme of an atomic model in the ground or in the excited state.

Question set 4

All the students seemed not to have assimilated the uncertainty principle. Only one student's answer categorized to the third level, while 19 answers considered being at the two first levels.

Answer example of the prestructural level 1:

'We can't measure the position and the velocity of an electron exactly'.

The student repeated the uncertainty principle without giving any other explanation.

Answer example of the multistructural level 3:

'It is applied to the electrons, because they move very fast, so a very precise instrument is required'.

The student related this principle with the microcosm, but she was not able to go deeper and face the electrons as quantum and not classical objects.

Concerning the fifth question set, 16 students preferred the 3D representations than the 2D shown in their textbooks. They experienced the different 3D graphs concerning the structure of charge density and they understood better their shape in space. They also asked for more interactivity that might help them to comprehend the atom shape.

According to our findings, most of the students did not overcome the third level of SOLO taxonomy, while no one reached the fifth. They were able to refer some of the QAM characteristics, but they could not correlate them with the principles of Quantum Mechanics in order to come to a conclusion about the shape of the hydrogen atom and the properties of its electron.

4. CONCLUSIONS AND IMPLICATIONS

The aim of our study was to explore student's mental representations about the QAM. In particular, we tried to discover if the use of 3D graphics concerning atomic orbitals help them to describe the QAM and grasp the main notions involved in coherence with most important principles. The results of our study revealed that the students do not conceive:

- The concept of the orbital as the wave function ψ (or probability function of presence), which describes the state of an electron in an atom and not as 'a region in space inside which there is a given probability for the electron to be located'.
- The information given by the wave function ψ, the probability density ψ^2, and the charge density concerning the atom state they describe.
- How the charge density is correlated with the notion of the charge cloud.
- The differences between the state function ψ for describing the hydrogen atom and the polyelectronic atoms in the ground state.
- How the energy level determines the shape of the charge cloud in space for the hydrogen atom and the polyelectronic ones.
- That the uncertainty principle is an inherent attribute in microcosm; it is the result of the wave–particle duality and the unavoidable interaction between the observed quantum object and the observing instrument.
- How the uncertainty principle is related with the charge cloud notion and the non-deterministic nature of microcosm according to Quantum Mechanics.
- Which is the shape of an atom in the ground state.
- How the shape of an atom is changed if this atom interacts with a photon and it 'moves' to an excited state.

Besides, the students did not mention the electron's wave nature as they supposed electrons to be particles that move very fast. So, in agreement with our results Johnston and his colleagues (1998), Petri and Niedderer (2001), Olsen (2002), Tsaparlis and Papafotis (2002) have found that students at the last class of high school or at university level describe electrons as classical particles.

Summarizing, our results indicated that the students confused the characteristics of planetary and QAM models even after having interacted with the visualizations. They also could not distinguish the two atomic models and figure out their limitations. These findings are in coherence with those of other researchers (Unal and Zollman, 2000; Petri and Niedderer, 2001; Olsen, 2002; Dimopoulos and Kalkanis, 2005). It seems that 3D visualizations concerning the QAM do not improve students' understanding. The didactic transformation together with the specific learning activities are those that count in combination with them.

On the other hand, it is accepted that it is not possible to combine the graphs of radial and angular wave functions in order to create the representation of an orbital and the corresponding charge density in space (Murell et al., 1985). In other words, it is not possible to visualize the atomic model of a chemical element to support students for the creation of mental images consistent with scientific knowledge and to overcome the difficulties that come from the Bohr model.

In order to overcome these difficulties, we suggest the use of ICT for creating dynamic 3D visualizations of atoms according to Quantum Mechanics principles, following Margel's approach stating that "it is generally accepted that active visualization-based learning can improve understanding and retention, but at the same time interpretation of visual experience highly depends on existing knowledge" (Margel et al., 2004). For the creation of a 3D environment concerning the visualization of the QAM, we have to take into consideration the following remarks:

- to integrate the visualization in a proper educational environment so as to conduct students to comprehend the abstract notions and construct their personal knowledge;
- to explicitly describe the concepts involved in the visualization (e.g. charge density), for students not to keep their misconceptions or construct new ones.

Our intention is to create visualizations of the charge cloud in order to represent the atomic model of hydrogen and picture it in different states.

More specifically, our visualizations will have the following characteristics:

- Give a sense of the 3D space and the spatial distribution of charge clouds, using specific peripheral devices such as stereoscopic glasses.
- Give the possibility to students to freely navigate outside and inside the atom.
- Give the possibility to students to interact and change energy states.
- Give the possibility to students to comprehend the electron's properties.

We believe that the above attributes are implemented using Virtual Reality (VR) technologies and we work to build educational dynamic virtual environments based on scientific data. Virtual Reality has the ability to give substance to abstract concepts and to visualize situations which cannot be seen otherwise and moreover to immerse the student within them (Trindade et al., 2002). So, it seems that VR is a powerful tool for visualizing complex data for helping students to create mental representations that better approach the scientific models. Educational virtual environments visualizing both the microcosm such as the plant cell and megacosm such as our solar system have been developed and evaluated by our group, giving positive learning outcomes (Bakas and Mikropoulos, 2003; Mikropoulos et al., 2003).

REFERENCES

Bakas, C. H. & Mikropoulos, T. A. (2003). Design of virtual environments for the comprehension of planetary phenomena based on students' ideas. International Journal of Science Education, 25, 949–967

Barak, M. (2007). Transition from traditional to ICT-enhanced learning environments in undergraduate chemistry courses. Computers & Education, 48, 30–43

Barnea, N. & Dori, Y. (1999). High-school chemistry students' performance and gender differences in a computerized molecular modeling learning environment. Journal of Science Education and Technology, 8, 257–271

Barnea, N. & Dori, Y. (2000). Computerized molecular modelling – the new technology for enhancing model perception among chemistry educators and learners. Chemistry Education: Research and Practice in Europe, 1, 109–120. Retrieved September 20, 2005, from URL:http://www.uoi.gr/cerp/

Biggs, J. (1994). Student learning research and theory – where do we currently stand? (In G. Gibbs (Eds.), Improving Student Learning – Theory and Practice (pp. 1–19). Oxford: Oxford Centre for staff development)

Biggs, J. B. & Collis, K. F. (1982). Evaluating the Quality of Learning. The SOLO Taxonomy. (New York: Academic Press)

Blauch, D. (2001). Retrieved March 10, 2004, from www.chm.davidson.edu/ChemistryApplets/AtomicOrbitals/

Budde, M., Niedderer, H., Scott, P. A. & Leach, J. (2002a). 'Electronium': A quantum atomic teaching model. Physics Education, 37, 197–203

Budde, M., Niedderer, H., Scott, P. A. & Leach, J. (2002b). The quantum atomic model 'Electronium': A successful teaching tool. Physics Education, 37, 204–210

Byrne, C. (1996). Water on tap: The use of virtual reality as an educational tool. Dissertation, University of Washington. Retrieved March 25, 2000, from http://www.hitl.washington.edu/projects/learning_center/ chemistry/

Cassinelo, A. & Gallego, A. (2005). The quantum mechanical picture of the world. American Journal of Physics, 73, 273–281

Cataloglou, E. & Robinett, R. W. (2002). Testing the development of student conceptual and visualization understanding in quantum mechanics through the undergraduate career. American Journal of Physics, 70, 238–251

Dimopoulos, V. & Kalkanis, G. (2005). Simulating quantum states of the atom of hydrogen: A simulation program for non-physics major's students. (In R. Pintó & D. Courso (Eds.), Proceedings of the Fifth International ESERA Conference on Contribution of Research to Enhancing Students' in Learning Science (pp. 548–552). Barcelona: Spain)

Dori, J. Y. & Barak, M. (2001). Virtual and physical molecular modeling: Fostering model perception and spatial understanding. Educational Technology & Society, 4, 61–74

Harrison, A. & Treagust, D. (2000). Learning about atoms, molecules and chemical Bonds: A case study of multiple-model use in Grade 11 Chemistry. Science Education, 84, 352–381

Hurwitz, C., Abegg, G. & Garik, P. (1999). How computer simulation affect high school students' reasoning in quantum chemistry. Paper presented at the National Association for research in science teaching (NARST), Retrieved October 23, 2000, from URL: http://www.phys.ksu.edu/perg/papers/narst/

Johnston, D., Crawford, K. & Fletcher, P. (1998). Students' difficulties in learning quantum mechanics. International Journal of Science Education, 20, 427–446

Margel, H., Eylon, B. & Scherz, Z. (2004). 'We actually saw atoms with our own eyes'. Conceptions and convictions in using the scanning tunneling microscope in junior high school. Journal of Chemical Education, 81, 558–566

Mikropoulos, T. A., Katsikis, A., Nikolou, E. & Tsakalis, P. (2003). Virtual environments in biology teaching. Journal of Biological Education, 37, 176–181

Murell, J. N., Kettle, S. F. A. & Tedder, J. N. (2nd Ed.) (1985). The Chemical Bond. (John Wiley & Sons Ltd.) West Sussex, England

Müller, R. & Wiesner, H. (1999). Students Conceptions of Quantum Physics. (Paper presented at the National Association for research in science teaching (NARST), Retrieved October 23, 2000, from URL: http://www.phys.ksu.edu/perg/papers/narst/

Olsen, R. (2002). Introducing quantum mechanics in the upper secondary school: a study in Norway. International Journal of Science Education, 24, 565–574

Petri, J. & Niedderer, H. (2001). Atomic physics in upper secondary school – Layers of conceptions in individual Cognitive Structure. (In D. Psillos, P. Kariotoglou, V. Tselfes, G. Bisdikian, G. Fassoulopoulos, E. Hatzikraniotis & M. Kallery (Eds.), Science Education Research in the Knowledge Based Society, ESERA 2001 (pp. 579–581). Art of Texts, Thessaloniki)

Taber, K. S. (2003). Mediating mental models of metals: Acknowledging the priority of the learner's prior learning. Science Education, 87, 732–758

Thacker, B. A., Leff, H. S. & Jackson, D. (2002). An introduction to the theme issue. American Journal of Physics, 70, 199

Trindade, J., Fiolhais, C. & Almeida, L. (2002). Science learning in virtual environments: A descriptive study. British Journal of Educational Technology, 33, 471–488

Tsaparlis, G. & Papafotis, G. (2002). Quantum chemical concepts: are they suitable for secondary students? Chemistry Education: Research and Practice in Europe, 3, 129–144, Retrieved January 6, 2005, from URL:http://www.uoi.gr/cerp/

Tuvi, I. & Nachmias, R. (2001). Current state of web sites in science education – Focus on atomic structure. Journal of Science Education and Technology, 10, 293–303

Unal, R. & Zollman, D. (2000). Students' Description of an Atom: A phenomenographic analysis, Retrieved April 8, 2002, from http://www.phys.ksu.edu/perg/papers/vqm/AtomModels.PDF

Winter, M. (2002). Orbitron. Retrieved March 19, 2004, from www.shef.ac.uk/chemistry/orbitron/

36. EVALUATION OF THE HYPERMEDIA LEARNING ENVIRONMENT "PHYSICS FOR MEDICAL STUDENTS" WITHIN TWO DIFFERENT SETTINGS

Abstract: The hypermedia learning environment "Physics for Medical Students" was developed at the University of Düsseldorf. Several settings for the implementation of this hypermedia learning environment in the context of physics education of medical students were designed and evaluated

In this paper we report on two studies that deal with the comparison of the hypermedia learning environment to a labwork session concerning the learning efficiency. The comparison was made with regard to the major objectives for physics education of medical students

Keywords: Evaluation, Hypermedia learning environment, Labwork, Medical students, Multimedia, Physics education

1. INTRODUCTION

In Germany, physics education of medical students is usually implemented in the first academic year and comprises a lecture and a labwork course. In general, the labwork course allows to deal with the content more intensely and actively which is why it has become the focus of our studies.

At the University of Düsseldorf, two learning environments for this target group were designed and evaluated: a labwork course (Theyßen et al. 2002; Theyßen 2005), and in the sequel, a hypermedia learning environment (HML) (Theyßen 2002).

1.1 Theory-Based Development of the Learning Environments

Several investigations on physics labwork (e.g. Toothacker 1983; Welzel et al. 1998; Psillos & Niedderer 2002; Neumann 2005) show that lecturers and learners attribute numerous objectives to labwork courses. However, these objectives are not always reached satisfactorily. As a consequence, Séré (2002) suggests the design of "targeted labwork". This means that

- a careful choice must be made among the various possible objectives and aims,
- this choice is to be made according to research outcomes and the target group, and
- the chosen objectives will be made transparent to the students.

Heike Theyßen, University of Dortmund /Institute of Physics Education, Otto-Hahn-Straße4, 44221 Dortmund, Germany

R. Pintó and D. Couso (eds.), Contributions from Science Education Research, 477–487.

The methodical framework for the development of the labwork course was the model of educational reconstruction (Kattmann et al. 1998). The development was based on scientific clarification of objectives and topics on the one hand and an investigation of the students' learning processes on the other hand. Research outcomes from both areas were equally considered in the formulation of criteria which the learning environment should meet. These criteria were the guidelines for the construction of the labwork course and the benchmarks for its evaluation. All in all, the development was an iterative process of research, construction of instruction and evaluation (Theyßen et al. 2002; Theyßen 2005).

The subsequent development of the HML was a continuation of this process. It was based on the same research outcomes, the results of the iterative optimisation of the labwork course, and findings on learning with new media (e.g. Gerdes 1997). The main reasons for this continuative development are given in the following sections. They are based on hypotheses concerning the learning efficiency of new media as well as on practical considerations concerning the transfer to other universities.

As an important step towards targeted labwork, surveys among lecturers and physicians were used to identify the topics and the major objectives of a physics labwork course for medical students. This group was regarded as the experts for the demands students have to cope with during their further studies and professional life. The surveys were supplemented by interviews and analyses of textbooks that are used in the subsequent university courses. The main results are (Theyssen 2005):

- The topics should be closely linked to medical studies.
- The major objectives for this target group are that
  - students are learning physics related content that is relevant for their further studies and professional life,
  - students are enabled to link theory to practice, especially to apply the physics related content to medical problems.

Concerning the selection of topics, these results are consistent with those of Ucke (1977). He states that the topics have to be determined by their relevance with regard to the further studies and the subsequent professional life of the students.

The major objectives are quite different from those identified for labwork courses in science education within the EU-study "Labwork in Science Education" (Welzel et al. 1998), e.g. "learn experimental skills" and "get to know the methods of scientific thinking". The third, frequently mentioned objective "link theory to practice" is represented here in a form that is specific for the target group.

Obviously the special target group demands for priorities in physics education that are different from the priorities scientists in general attribute to physics labwork. The acquirement of experimental skills is of secondary importance.

These objectives and the results of investigations on students' learning conditions formed the guideline for the development of the labwork course (Theyßen 2005) and the subsequent development of the HML. Due to the same developmental guideline, both learning environments share the following characteristic features:

- The physics related content is highly relevant for medical studies.
- To make the relevance obvious for the students, the physics related content is embedded in a medical context. For example, the basics of geometric optics are treated in the context of the function of the human eye, and the physics of gases are connected with the process of inspiration and expiration.
- Students are guided by a step-by-step description of the labwork tasks. The tasks start off with the careful observation of phenomena and continue with qualitative as well as quantitative investigations of the properties of physical values and the relations between them.
- As a consequence, there is no need for students to acquire the theoretical background in advance. The abstract description of physics related content (e.g. the use of formulas) is always based on their experiences with the concrete phenomena.
- Numerous questions and tasks are embedded in the labguide and the HML respectively. They demand the verbalisation of observations, the interpretation of data and the application of physics related content to medicine related problems.

However, the two learning environments differ in the media supplied for the experimental tasks, and in the restrictions these media impose on the time structure of learning. Both aspects are explicated right below.

In the HML, the experimental setups of the labwork session are represented by Interactive Screen Experiments (ISEs) (Kirstein 2001). ISEs are digital videos of real experiments. The important additional feature of an ISE compared to a normal videotape is that the user can manipulate several parameters of the experimental setup. This manipulation is performed by means of the computer mouse and directly within the representation of the experimental setup. Compared to the real experiment, the number of parameters that can be manipulated is usually restricted and can be chosen according to the educational objectives. Therefore, ISEs can focus the students' attention on selected features of the experiment and facilitate the repetition of experiments. The acquirement of experimental skills is hardly possible.

In the labwork course, each session is divided into two phases: During the experimental phase, students deal with experimental tasks described in the labguide through which they gather a lot of experimental experience. Students work in teams of two, a tutor is present and (as a consequence) the time schedule and location for the occupation with the content are predetermined. During the postprocessing phase, students acquire the theoretical background. All necessary information and advices concerning its relations with the experimental tasks are included in the labguide. Since the postprocessing is usually done at home, students have no access to the experiments. Repeating observations and measurements is not possible.

In the HML, experiment and theory are closely and contemporary interwoven. The theory is integrated whenever it is deemed useful for the learning process, usually immediately after the observations and investigations based on ISEs. There is no artificial interruption between an experimental phase and a post-processing phase. Since the HML is available on the internet, students have access to the content at all times and from almost everywhere. Repeating experiments and corresponding observations is always possible.

1.2 Settings for the HML

Since the HML is at students' disposal on the internet, it can be used within various settings, i.e. under various conditions for the students' occupation with its content. Several authors emphasise that the learning efficiency of new media highly depends on the setting in which the media are applied (e.g. Clark 1994; Kerres 2001). Thus for the evaluation of the HML, the setting has to be clearly defined. In the studies that are presented here, two very different settings with almost complementary boundary conditions have been evaluated. In both settings, students deal with a predetermined section of the HML. For both studies the physics of gases and respiration were chosen exemplarily.

The first setting is very similar to labwork. Like during a labwork session students have to work in teams of two. The time and location for the occupation with the content are predetermined. Thus, the time schedule and the possibility of repetitions are restricted due to the limited total time. However, the experimental phase is not separated from the postprocessing phase as in the labwork session. Up to 10 teams work in the same room and are supported by a tutor who is present in the room. Due to the presence of a tutor, the HML itself offers restricted support only. Advices to find the correct solution or answer are available for most of the tasks and questions, but no complete example solutions are offered.

The second setting allows for maximum individuality. First of all students are free to choose the media they work with: labguide and experiments or HML with ISEs. In case they choose the HML, they may choose the social form (e.g. teamwork or individual work), time and location for their occupation with the content. Thus the time schedule can be arranged very individually. Repetitions are possible and the total time for the occupation with the content is up to the students. Tutors can be consulted only via email or phone. To compensate for this, complete example solutions are available in the HML in addition to the advices.

1.3 Idea of the Studies and Research Questions

With the labwork session and the HML in the two different settings, we have three learning environments. They are very similar in content and tasks, but differ in the media and the restrictions to the time structure of learning. The main idea of the studies is to compare the HML in each setting with the labwork session concerning the learning outcome. The concept of learning outcome used in these studies refers to the major objectives identified for physics education of medial students. Accordingly, the research questions focus on the students' physics-related knowledge and their ability to apply this knowledge in physics- and medicine-related contexts:

Can students

(Q1) ... identify a physics-related statement (definitions and correlations) as right or wrong?

(Q2) ... verbalise physics-related statements on request?
(Q3) ... use these statements to deal with physics-related and medicine-related tasks?

2. METHODS AND SAMPLE

2.1 Methods of Inquiry

Four different tests were developed and applied to answer the research questions.

"Basics" concerns basic mathematical knowledge that the students need to cope with the tasks in the labguide and in the HML respectively. The items refer to the basics of fractions, the calculation of percentage, the interpretation of diagrams, the rule of proportion and the use of values and units. It consists of 13 multiple choice items and two open questions. About 70% of the items are published items of TIMSS II and TIMSS III.

The other three parts concern the physics and medicine related content of the treatment (the labwork session and the section of the HML respectively).

"Rating" refers to research question Q1. It consists of 50 statements that have to be rated as true or false. The statements are generated by means of content analysis of the labguide and the HML. All statements are part of both learning environments. They cover largely the content of the treatment and have quite similar levels of demands. Exemplary statements taken from the test are: "If you keep the temperature of a certain amount of gas constant and enlarge the pressure, the volume shows a proportional change." Or: "During inspiration the inhaled air is moistened to a relative humidity of 50%."

Half of the statements are true. Thus the statistical probability of a correct rating is 0.5. For each correct rating one point is assigned, so the maximum score that can be reached in this test is 50 points. Students are requested to rate each statement and to guess in case of doubt. If a statement has not been rated at all, 0.5 points are assigned corresponding to the probability of guessing the correct answer.

In a pilot study the test was applied with 64 statements. An item analysis based on the test results was used to optimise the test and reduce the number of items.

"Verbalisation" refers to research question Q2. It consists of 3 open questions demanding the verbalisation of physics and medicine related content. An exemplary question is: "How are pressure, volume and temperature associated with each other for a certain amount of gas? If you give an equation, specify the variables."

"Application" refers to research question Q3. It consists of 7 open questions which demand the application of this content in a medicine related content. An exemplary question is: "What is the ratio of the volume before and after, if a certain amount of gas is heated from 20°C to 37°C (pressure is constant)?"

For both tests with open questions, the assessment was based on a detailed model solution for each task. A maximum number of points was attributed to each important step of the model solution. This was negotiated among two experts. Students' solutions were assessed according to this scheme. The maximum score in "Verbalisation" is 6 points, the maximum score in "Application" is 21 points.

2.2 First Study: Labwork versus HML in the First Setting

The study was performed with 56 medical students in spring of 2003, just before the start of the summer term. All participants were volunteers. The treatment itself was part of the labwork course and the participants of the study had attended it prior to the study. They received a small financial compensation for the higher expenditure of time that was caused by additional tests, especially post-tests. The topic "physics of gases and respiration" was chosen exemplarily for the study (see Hüther & Theyßen, 2005).

All students took part in a pre-test comprising the tests "Basics" and "Rating". According to the results of the pre-test, the students were organised into two groups with almost the same medial score and standard deviation. According to the results of a pilot study, the medial test scores in "Verbalisation" and "Application" were less than 5% of the maximum score. As their results were too low to be used for the formation of the groups, those tests were only applied in the post-test.

The group "Labwork" (27 students) took part in the labwork session and used the labguide for the post-processing of the content. Both phases were separated by a break and during the postprocessing phase, students had no access to the experiments. The group "Hypermedia" (29 students) worked out the same content by means of the HML. The HML was applied in the first setting described above that is very similar to labwork. Both groups spent the same amount of time on the content. Students worked together in pairs and were supposed to interact. Tutors were present in both groups so that support was readily available.

After the treatment, both groups performed the same post-test comprising the parts "Rating", "Verbalisation", and "Application".

2.3 Second Study: Labwork versus HML in the Second Setting

The study was performed with 275 medical students during the summer term of 2003. All medical students of that year (except the participants of the first study) took part. The topic "physics of gases and respiration" was the same as in the first study. Compared to the first study, the number of tests had to be reduced, since treatment and tests had to be integrated into the regular time schedule of the labwork course. The post-test was restricted to the test "Rating" only because the additional expenditure of time had to be minimised.

As in the first study, all students took part in a pre-test comprising the tests "Basics" and "Rating" in order to measure their previous knowledge. It is a characteristic feature of the setting that the students have the free choice among the two learning environments. Thus a formation of the groups according to the pre-test scores was not possible in this study.

225 students decided to take part in the labwork session and do the postprocessing with the labguide without access to the experiments (group "Labwork"). 50 students decided to work out the same content by means of the HML (group "Hypermedia"). According to the second setting described above, they could choose the time, location and social form for their occupation with the content.

The disproportionate number of students who chose the labwork session is attributed to their familiarity with the labwork sessions. The choice between the

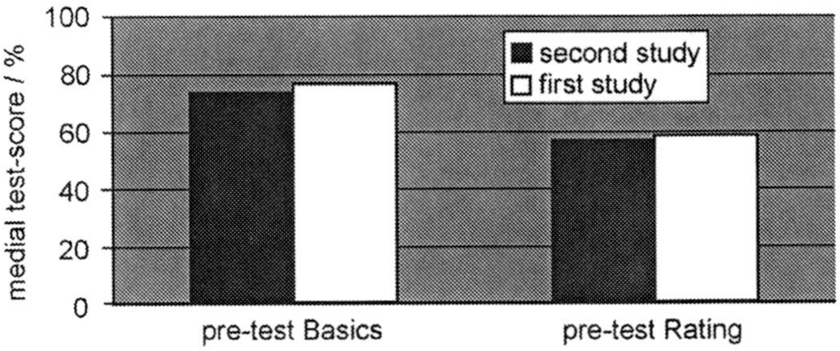

Figure 1. Comparison of the two studies concerning the previous knowledge of the participants

learning environments is usually made before the students get to know the HML whereas they have experienced labwork during several other sessions.

After the treatment both groups performed the same post-test comprising only the part "Rating".

3. DATA ANALYSIS AND RESULTS

Since all students had performed the same pre-test, the participants of both studies can be compared. Figure 1 shows the medial scores for the participants of the first and the second study and for both parts of the pre-test. All results are given in percent of the maximum score.

The average previous knowledge is the same for the participants of both studies. This applies for basic mathematics as well as for the physics related content of the treatment. Consequently, the participants of the first study can be regarded as a representative sample of all medical students in Düsseldorf in 2003.

With regard to the content of the test, the medial scores of the part "Basics" are quite low. Nevertheless, their basic mathematical knowledge was sufficient for the students to cope with the tasks during the treatment.

For the test "Rating" the scores are quite high. This can be attributed to the fact that the probability of guessing the correct answer is 50%. Thus a medial score of about 60% indicates a very small previous knowledge.

The results of the two studies will now be discussed in detail.

3.1 First Study: Labwork versus HML in the First Setting

Figure 2 shows the medial test scores for the groups "Hypermedia" and "Labwork" in the first study. Again the results are given in percent of the maximum score. The error bars indicate the standard deviation. The statistical spread is obviously much larger for the open questions than for the other tests. This can be attributed to the high probability of guessing the correct answers in the tests "Basics" (multiple-choice items) and "Rating" (decision true or false).

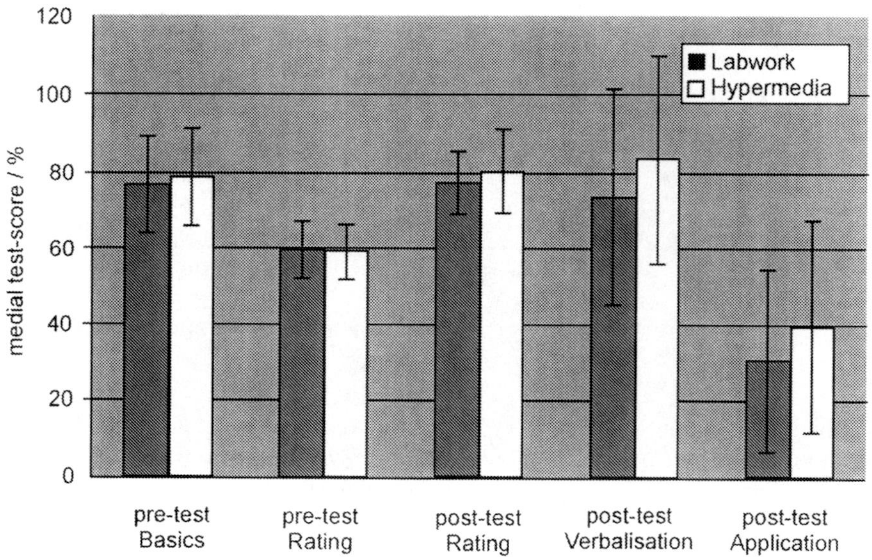

Figure 2. Results of pre- and post-tests for the first study (Hüther & Theyßen, 2005)

On the left hand side the diagram shows the results of the two parts of the pre-test. According to the formation of the groups, the medial score and the uncertainty of measurement are almost the same for both groups.

On the right hand side the diagram shows the results of the three parts of the post-test. There is a slight tendency to a better test performance of the group "Hypermedia", especially concerning the verbalisation and application of the content (Q2 and Q3). A statistical analysis reveals however that none of the differences between the groups are significant. The part "Rating" was performed in pre- and post-test so that the learning increase concerning the rating of given statements could be calculated. This increase is highly significant within both groups.

3.2 Second Study: Labwork versus HML in the Second Setting

Figure 3 shows the medial test scores for the groups "Hypermedia" and "Labwork". Again the results are given in percent of the maximum score, the error bars indicate the standard deviation.

On the left hand side of the diagram the results of the two parts of the pre-test are compared with each other. Obviously the free choice of the learning environment yielded two groups with the same average previous knowledge. In other words, the HML was not favoured by students with either higher or lower previous knowledge regarding both, the basic mathematical knowledge and the content related knowledge.

Concerning the test "Rating", the standard deviation is almost identical for both groups and for the pre- and post-test. It is of the same order of magnitude as in

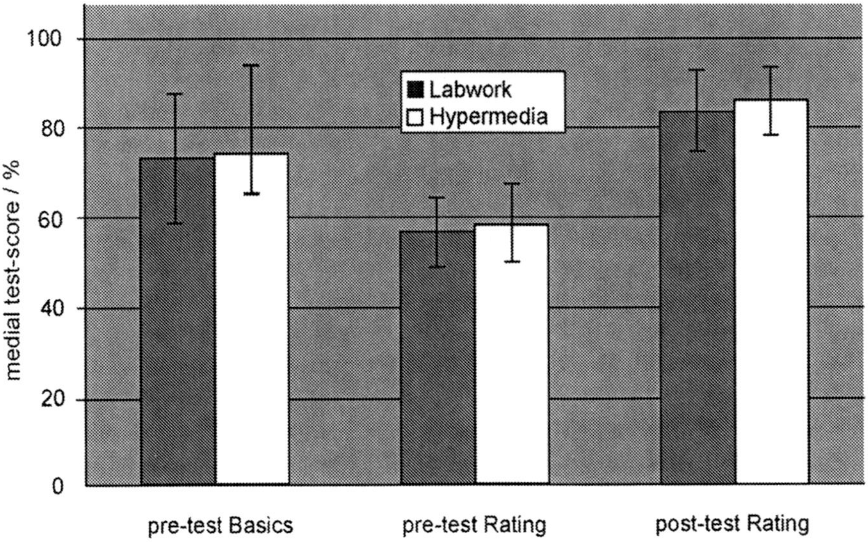

Figure 3. Results of pre- and post-tests for the second study

the first study. Concerning the pre-test "Basics", the standard deviation is distinctly higher, especially in the group "Hypermedia".

On the right hand side the diagram in Figure 3 shows the results of the post-test "Rating". Again, there is no significant difference between the groups, but as in the first study, the established learning increase between pre- and post-test is highly significant for both groups. The same applies if the evaluation is restricted to those students with higher or lower previous knowledge, and equally for male or female students in both groups.

4. CONCLUSIONS

A new labwork course and a hypermedia learning environment (HML) for physics education of medical students were designed and evaluated with regard to the learning efficiency. The development was targeted at the realisation of the major objectives that were identified for this special target group: to learn relevant physics related content and to link theory to practice. According to former studies (Theyßen et al. 2005; Theyßen, 2005), several improvements in physics education of medical students have been achieved with the implementation of the new labwork course. However, due to financial limits, its implementation at other universities often failed.

The subsequently developed HML has followed the same developmental guideline and is very similar in content and structure. It is implemented in two settings that differ in the external restrictions to the learning process (e.g. time schedule and social form), and in the intensity of support offered by tutors.

Two studies were carried out in order to compare the learning efficiency of the HML in each setting to that of the labwork course. Both studies were restricted to the same topic: physics of gases and respiration. For both studies, a pre-post-test design was chosen and the learning efficiency was taken as the students' ability...

- ... to identify physics related statements (definitions and correlations) as right or wrong,
- ... to verbalise physics-related statements on request and
- ... to use these statements to deal with physics-related and medicine-related tasks.

Both studies yield that there are no significant differences between the labwork session and the HML concerning the acquirement of the abilities listed above. A (not significant) trend in the post-test-data indicates a slight advantage of the use of the HML over the labwork course in the setting similar to labwork. This refers particularly to the ability to verbalise and apply physics related content. The learning increase between pre- and post-test is highly significant for each treatment.

Independently from the setting, the HML allows for the same (significant) learning increase as the labwork session. A hypothesis for further studies could be that in the first setting, the support given by the tutors compensated for the restrictions on the time structure of learning, whereas in the second setting, the individual time structure compensated for the restricted support.

From these results we conclude that the implementation of the HML is a suitable alternative to the implementation of the labwork course. In contrast to the labwork course with its complex and expensive experimental setups, the HML can easily be transferred to other universities. The results reported here have already encouraged lecturers at several universities to include the HML as a regular part of their labwork courses for medical students. Single labwork sessions (not labwork courses as such) have been replaced completely by the HML.

Further studies with larger samples of students or focussing on other objectives of labwork should follow to complete these results and to increase the acceptance of HML among lecturers and students.

ACKNOWLEDGEMENTS

We are grateful for various kinds of support:

The studies we presented here were planned and realised in close cooperation with Dieter Schumacher (University of Düsseldorf) and Elke Sumfleth (University of Duisburg-Essen).

The studies were supported by the German Research Foundation and the federal state of North Rhine-Westphalia.

REFERENCES

Clark, R. E. (1994). Media will never influence learning. *Educational Technology Research and Development, 42(2)*, 21–29

Gerdes, H. (1997). *Lernen mit Text und Hypertext.* (Lengerich: Pabst)

Hüther, M. & Theyßen, H. (2005). Vergleichende Unteruchungen zur Lernwirksamkeit einer hypermedialen Lernumgebung und eines Praktikums. Zeitschrift für Didaktik der Naturwissenschaften 11, 117–129

Kattmann, U., Duit, R. & Gropengießer, H. (1998). The Model of Educational Reconstruction. Bringing together issues of scientific clarification and students' conceptions. (In H. Bayrhuber & F. Brinkman (Eds.), *What – Why – How? Research in Didactics of Biology* (pp. 273–282). Kiel: IPN.)

Kerres, M. (2001). *Multimediale und telemediale Lernumgebungen – Konzeption und Entwicklung.* (München: Oldenbourg)

Kirstein, J. (2001). The educational value of "Interactive Screen Experiments" – a new representation of experiments with multimedia technology. (In D. Psillos, P. Kariotoglou, V. Tselfes, G. Biskian, G. Fassoulopoulos, E. Hatzikraniotis & M. Kallery (Eds.), *Proceedings of the Third International Conference on Science Education Research in the Knowledge Based Society, II* (pp. 468–470). Thessaloniki: Art of Text)

Neumann, K. (2005). Didaktische Rekonstruktion eines physikalischen Praktikums für Physiker. (In H. Niedderer & H. Fischler (Eds.), *Studien zum Physiklernen* (Band 38). Berlin: Logos.)

Psillos, D. & Niedderer, H. (Eds.) (2002). *Teaching and Learning in the Science Laboratory.* (Dordrecht: Kluver)

Séré, M. (2002). Towards targeted labwork. (In D. Psillos & H. Niedderer (Eds.), *Teaching and Learning in the Science Laboratory* (pp. 255–259). Dordrecht: Kluver.)

Theyßen, H. (2002). Physik für Mediziner – Didaktisches Konzept und inhaltliche Umsetzung. (In I. Siebert (Ed.), *Schriften der Universitäts – und Landesbibliothek Düsseldorf, Band 34* (pp. 7–22). Düsseldorf: ULBD.)

Theyßen, H., Aufschnaiter, S. V. & Schumacher, D. (2002). Development and evaluation of a laboratory course in physics for medical students. (In D. Psillos & H. Niedderer (Eds.), *Teaching and Learning in the Science Laboratory* (pp. 91–104). Dordrecht: Kluver.)

Theyßen, H. (2005). Didaktische Rekonstruktion eines Physikpraktikums für Medizinstudierende. *Zeitschrift für Didaktik der Naturwissenschaften, 11*, 57–27

Toothacker, W. S. (1983). A Critical Look at Introductory Laboratory Instruction. *American Journal of Physics, 51*, 516–520

Ucke, C. (1977). *Didaktik naturwissenschaftlicher Praktika: Beispiel Physik für Mediziner.* (Hamburg: AHD)

Welzel, M., Haller, K., Bandiera, M., Hammelev, D., Kouramas, P., Niedderer, H., Paulsen, A., Robinault, K. & Aufschnaiter, S. V. (1998). Ziele, die Lehrende mit experimentellen Arbeiten in der naturwissenschaftlichen Ausbildung verbinden – Ergebnisse einer europäischen Umfrage. *Zeitschrift für Didaktik der Naturwissenschaften, 4(1)*, 29–44

THORID RABE AND HELMUT F. MIKELSKIS

37. THE ROLE OF LANGUAGE IN LEARNING PHYSICS WITH COMPUTER-BASED MULTIMEDIA

Abstract: This chapter presents results of a laboratory study, in which students of grade 12 learned physics with a computer-based program on diffraction of light by single slit ($N = 80$). Text coherence plus text-picture-references (text surface design) and the instruction for self-explanations were investigated in their influence on knowledge acquisition in physics in an empirical 2x2 design. A knowledge test on optics was implemented in pre-, post- and follow-up testing. Qualitative evaluation of students' conceptions on diffraction and interference of light complements the quantitative analysis of the knowledge test. Results show a significant impact of self-explanation activity on students' knowledge in the follow-up test that was conducted 6–8 weeks after the intervention. Significant effects of text surface design did not emerge. Misconceptions and inadequate use of models of light can be observed at all points in time though the knowledge gains are, in general, very high given the short time of intervention (one session of about 60 minutes)

Keywords: Misconceptions, Multimedia, Optics, Self-explanation, Text surface design, Text coherence

1. INTRODUCTION AND THEORETICAL FRAMEWORK

The role of written and spoken language in learning physics has been discussed for a long time (since the 1970s at the latest). Students report difficulties in understanding their teacher, reading textbooks and in their use of technical language. Language has therefore been described as a possible barrier to learning physics (Wellington and Osborne, 2001). This is critical, because language is both: a mean to an end (since it is used to communicate knowledge) and an end in itself (since the appropriate use of language is an important outcome of physics instruction) (Yore, 2003).

The importance of computer-based physics instruction increases in formal learning settings (school) as well as in informal settings. Computer-based instruction seems promising wherever individualized learning is implemented. Research in science education is intended to find out what conditions foster computer-based learning. Among other multimedia principles (Mayer, 2001), research points out the influence of text design and in particular of personalized messages (Moreno and Mayer, 2000).

University of Potsdam, Institute of Physics, Didaktik der Physik, Am Neuen Palais 10, 14469 Potsdam, Germany

R. Pintó and D. Couso (eds.), Contributions from Science Education Research, 489–502.

What is the learner's role in the processes of reading and learning? Constructivist approaches in research on physics education, cognitive theories on text comprehension and multimedia learning share common ground: The learner must process the information actively and has to construct or modify a coherent mental representation of the learning content. Without such learner activities learning cannot take place. In the following, we distinguish technically between "comprehension" as an appropriate, but only current change of mental structures, and "learning" as a lasting change of knowledge.

During the process of knowledge construction, inferences are made among parts of new information and new information is connected to prior knowledge (Schnotz and Bannert, 2003). The cognitive processing takes place in working memory which is limited in capacity. Three types of cognitive load can be distinguished: *Intrinsic cognitive load* is evoked by the learning content itself. Didactical reconstruction (Duit et al., 2005) can be used to find an appropriate level for learners. The instructional design can cause high *extraneous cognitive load* that constrains knowledge construction. The *germane load* is generated by cognitive and metacognitive activities and should be on a high level (Brünken et al., 2005). Learners with little prior knowledge profit from guidance of attention in order to foster coherence formation, because they have difficulties to identify relevant information and process it.

From research on text comprehension we learn that understanding and processing of texts happens as an interplay of features of the text surface, the text content and the learner himself. Concerning the design of text surface, criteria on readability and understandability have been reported. However, they have not been implemented in multimedia as evaluation has shown (Starauschek and Rabe, 2004). In the context of multimedia learning the following two attributes of the text surface seem promising to reduce the extraneous cognitive load.

Local text coherence (in some theoretical approaches also called "cohesion") describes the connectivity of a text on the level of subsequent sentences (Rickheit and Schade, 2000). The number of recurrences of substantives in subsequent sentences can be chosen as indicator of local text coherence. The following example demonstrates recurrance: "A high number of recurrences indicates a high local *textcoherence*. For novices a text with high local *textcoherence* is easier to process" (recurrence of the substantive *textcoherence*). In contrast the recurrence misses in the following changed version of the same sentences: "A high number of recurrences indicates a high local *textcoherence*. For novices a text with this *text feature* is easier to process" (recurrence is replaced by synonym). (The recurrence of nouns is highlighted in the screenshot in Figure 1, too.)

A second feature of the text surface is the amount of *references to the associated pictures* given in the text. The rationale is that interrelations in content of pictorial and verbal information are highlighted. By this the extrinsic cognitive load is reduced and coherence formation is fostered. To connect pictorial and verbal information seems important in particular in the context of multimedia learning (see Mayer, 2001).

Local text coherence and text-picture-references might support text and picture comprehension of physics beginners who are not familiar with language and content of physics. An influence on learning as a sustainable change of knowledge and on the capability to transfer knowledge cannot be expected.

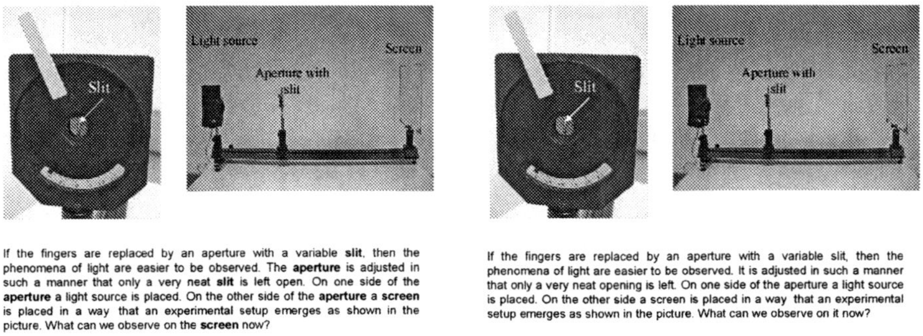

Figure 1. Screenshots from the computer program: Coherent and incoherent textversions

A problem with computer-based learning settings is the false belief of some learners that no effort is necessary for comprehension and learning. To address this problem one can enhance the germane cognitive load as a kind of internal support for the learner The instruction to self-explain seems auspicious to activate learners cognitively. According to de Leeuw and Chi (2003) a reader's efforts to explain a text to themselves go along with monitoring one's understanding, which also leads to a deeper engagement in knowledge construction in order to overcome self-diagnosed problems of understanding. Self-explanation is an intentional strategy that requires conscious thought and effort. Therefore it might not only lead to better comprehension of content, but also help to create sustainable changes of the mental structure. Research has shown a self-explanation effect in ways of a correlation between the number of spontaneous self-explanations and the results of a problem-solving test. Furthermore prompted self-explanations lead to a deeper understanding of texts as well (Chi et al., 1994). Research on self-explanation in the context of learning physics is just at the beginning.

Learners' characteristics influence learning processes strongly. In physics first of all prior domain knowledge influences learning and students' conceptions (Vosniadou, 1999). Spatial abilities are fundamental in learning with multimedia (Mayer, 2001) and verbal abilities are relevant in learning with texts. Additionally students' cognitive and metacognitive learning strategies of students might interact with the instruction to self-explain that aims at stronger use of such strategies (Lind et al., 2005).

2. RESEARCH QUESTIONS AND HYPOTHESIS

1. How do local textcoherence and text-picture-references in combination influence the knowledge construction with a computer program on optics?
2. How does the instruction to self-explain during learning with a computer program on optics influence knowledge construction?
3. What (mis-)conceptions of students in optics can be observed before and after instruction?

3. METHOD AND SAMPLE

A computer program on diffraction of light by single slit was designed, that brings up the issue of diffraction and interference at an introductory level. It aims at students in grade 12 with little prior knowledge on wave optics. In principle it could be used as part of a more extensive learning unit on wave optics, in which students study individually to some extent (the development of such a unit is not part of the presented study).

Findings on students' conceptions on diffraction and interference of light (Ambrose et al., 1999) as well as research on learning about models in physics (Mikelskis-Seifert and Leisner, 2005) were considered. Initially students are confronted with phenomena of diffraction of light, which gives reason to discuss assumptions and limitations of the ray model and the wave model of light. Diffraction is described qualitatively with the wave model of light and the principle of Huygens. Afterwards the structure of the interference pattern is explained with superposition of waves, and its topology is discussed referring to width of slit, wavelength and optical retardation as crucial factors.

Two textversions of the computer program were generated only differing in the text surface design – not in content. Both versions fulfil likewise general demands on comprehensibility, indicated by the same amount of personalized messages, similar ratings of readability, same amount of passive phrases and of technical terms. The textversions differ clearly in two criteria: In the "incoherent" textversion only 16% of the sentences are connected by recurrence of nouns and only two references to pictures are made. In the "coherent" textversion the local textcoherence is at 68% and 40 text-picture references are given. Thus we have two textversions: (a) version 1: low local textcoherence, few text-picture-references; (b) version 2: high local textcoherence, numerous text-picture-references. It can be seen from the screenshots from the program in Figure 1 that the two versions do not differ in content at all and provide identical information.

An experimental study was conducted in a 2x2-design with textversion and instruction as independent variables. The two dimensions of the variable textversion have been described above. The instruction to self-explain was given to two of the experimental groups. Knowledge on diffraction and interference of light is the dependent variable. It was assessed at three points in time – pre-test, post-test and a follow-up test.

The sessions in which the students learned individually with thecomputer program were organized in a laboratory setting. The students prompted to self-explain worked in the presence of an instructor who explained the task in the beginning. Students were asked to read the text and speak out aloud during self-explaining, since the sessions were recorded. Students were given guiding questions as a support for self-explanation such as "What information is new to you? How does it connect to your prior knowledge? Which questions come up to you?" Students, who were not instructed to self-explain, worked individually as well, but in the presence of others and an instructor.

TABLE 1. Distribution of the sample

	Textversion 1		Textversion 2	
	Female	Male	Female	Male
No self-explanation	**21**		**18**	
	4	17	4	14
With self-explanation	**20**		**21**	
	7	13	7	14

In total, 106 students participated in the study voluntarily. All of them attended basic or intensive physics courses in grade 12 at the German Gymnasium (mean age: 17.8 years; 34 female students). The students were distributed to the four experimental groups randomly (see Table 1). For organizational reasons it was not possible to parallelize the groups with regard to prior knowledge or other characteristics of the students. Out of all students, only 80 participated in the follow-up test due to organizational reasons. Procedure and instruments of the investigation are presented in Table 2.

An important part of the investigation was a test on knowledge in optics and in particular on diffraction of light. This test was employed identically at all points in time. It was developed with respect to the content structure of the computer program and anticipated misconceptions of students in particular in the field of models of

TABLE 2. Procedure and Instruments of the Investigation

Pre-test ($n = 106$) (2–3 weeks before treatment)	• **Cognitive ability test** on spatial and verbal abilities (Heller and Perleth, 2000) • **Questionnaire on** – Self-assessed computer experience – Self-assessed experience of self-competency in physics (Wild et al., 1997) – cognitive and metacognitive learning strategies (5 scales) (Wild et al., 1997) • **Knowledge test on optics**
Treatment ($n = 106$)	Individual work with the computer program in one of the four experimental groups (approx. 60 minutes)
Post-test ($n = 106$) (immediately after treatment)	• **Questionnaire** on usability of the computer program • **Knowledge test on optics**
Follow-up test ($n = 80$) (about 8 weeks after treatment)	• **Knowledge test on optics**

light. The overall 26 items in the knowledge test cover tasks on retention as well as on transfer of knowledge. For retention students had to remember information that is directly given in the program. For transfer they had to apply such knowledge or information to new or modified situations like diffraction at a grid, double slit or differently shaped openings. Additionally two items referred to knowledge on refraction of light.

4. DATA ANALYSIS AND RESULTS

For descriptive analysis of variables and analysis of variance, SPSS (Version 13.0) was used. The evaluation of control variables shows no significant differences between the experimental groups. In the subtests on verbal and spatial abilities the sample reaches values slightly below the age-standardized values (sample: $mean_{verbal}$=7.8, SD=2.44, $mean_{spatial}$=15.76, SD=3.69; standardized: $mean_{verbal}$=10.03, SD=3.00, $mean_{spatial}$=17.06, SD=3.53). The scale on experience of competency in physics has a central tendency as the mean for the group is 3 (SD=0.80) on a scale from 1 to 5. The values for the five scales on cognitive and metacognitive learning strategies group around 3 as well. Asked about their interest in physics in general, students rated themselves on a scale from 1 to 5 as rather interested (mean=3.55, SD=1.16).

The evaluation of the questionnaire on computer experience reveals that more than half of the students (54%) have been using learning software sporadically only and 20% never used it before. Asked if they would favour an implementation of such computer programs in physics courses, they answer ambiguously with “partly”.

The results of the questionnaire on the usability in general suggest that all treatment groups find the computer program convenient for learning. The amount of text and pictures and their combination as well as the amount of information per page are judged as appropriate. The students characterize the language as very understandable (mean=4.03, SD=0.99 on a 5-ary scale) and no significant differences concerning textversions occur. Female students (mean=3.65, SD=0.88) judge the program as more challenging than male students (male students: mean=3.21, SD=0.69; p=0.014). Another interesting detail emerges in students’ answers to the question if they liked to be addressed directly in the texts of the computer program. Female students like this significantly better (mean=4.52, SD=1.09) than the male students (mean=3.96, SD=1.11, p=0.018). This goes along with studies on students’ interests in physics showing that female students dislike the impersonal style widespread in physics (Hoffmann et al., 1998).

For statistical evaluation of the knowledge tests, sumscores were counted for each time of assessment. Each item was given a maximum of points, attainable if it was answered completely and correctly. Partly correct answers obtained a smaller number of points. For items with open answer format, that is for verbal and graphical tasks, categorization systems were worked out by using content analysis (Mayring, 2003). The coding of the students’ answers according to these categories by two independent raters yielded a sufficient level of inter-coder-reliability. Three items had to be excluded from further analysis, because of too high item difficulties. A maximum sumscore of 45 points was then achievable in the knowledge test. In the following analysis only those students (N=80) are included for whom the data sets are complete.

The scores for the whole sample are shown in Figure 2 (all values for the sumscores show normal distribution within the sample according to a Kolmogorow–Smirnov test).

The results of the pre-test (mean score=9.6) indicate that the students have very little and at the same time heterogeneous (SD=5.6) prior knowledge on diffraction and interference of light. They know little about the wave model of light and try to apply assumptions of the ray model to explain diffraction phenomena. These are often explained by refraction of light or interpreted as shadow formation (e.g. a fine grid causes a shadow pattern). Thus the students in our study can be considered as novices in our special content area. According to the school curriculum they have been taught about both models of light in lower grades (grade 10 and lower). But experience shows that most of the students are only familiar with the ray model of light and with the phenomena reflection and refraction of light. In order to control heterogeneous prior knowledge of the students, the knowledge test was already employed in the pre-test.

Compared to the prior knowledge the sumscore in the post-test is significantly higher (T-Test: $p < 0.001$), students show much better understanding of the content area directly after they learned with the program. That they did not reach even higher scores can be explained with the short time of intervention (about 60 minutes learning time). Additionally the variance within the sample is very high (post-test: SD=8.1).

The score of the follow-up test is significantly higher than the one of the pre-test, too ($p < 0.001$). This result is interpreted as a sign for learning and a permanent change of the mental structures of the students. On the other hand the mean score of the follow-up test reveals a typical trend in students knowledge as it is lower than the post-test score immediately after the intervention (students had no additional

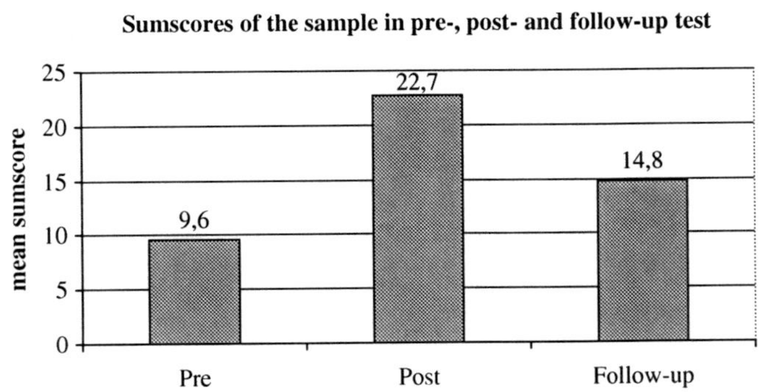

Figure 2. Sumcores of the sample

lessons on optics during the whole investigation). Students fall back into using the ray model of light instead of applying the more appropriate wave model of light. Some can still explain how diffraction patterns emerge in general, but fail to predict or explain their structure in detail.

How do text surface design and instruction to self-explain influence the learning outcome? Analysis of variance (General Linear Model with repeated measurements) reveals that the text surface design does not explain any differences between the experimental groups ($F(1{,}76)=0.590$, $p=0.445$). In contrast to our expectations, local text coherence and text-picture-references do not foster comprehension of physical content (post-test) nor learning as a permanent change of knowledge structure (follow-up test).

The variable instruction to self-explain shows a highly significant main effect ($F(1{,}76)=7.3$, $p=0.009$, $\varepsilon=0.31$ middle effect) and a significant interaction with time. The influence of self-explaining depends on the point in time: The scores of the "non-explainers" and the "self-explainers" only differ in the follow-up test (t-Test $p<0.001$, see Figure 3). Students who have learned with self-explanation retain more knowledge on diffraction and interference of light compared to the students who did not get a special instruction. Only in the long term it becomes obvious that self-explaining fosters a deeper processing of the content.

The effects for the variable instruction are the same if only those items are included into analysis that directly refer to the content of the computer program. This means concerning retention of content the students only profit in the long term.

Analysis of items, that demand a transfer of the physical content, that was learned with the computer program, shows a tendency that self-explanations additionally have an effect in the post-test (ANOVA: $F(3{,}76)=2.22$, $p=0.093$). Students who were prompted to self-explain seem to process the information in a way that enables them to use this knowledge for explanation of new situations and phenomena in unknown situations, for example diffraction and interference of light by a double slit, by a grid or diffraction of water waves by a barrier.

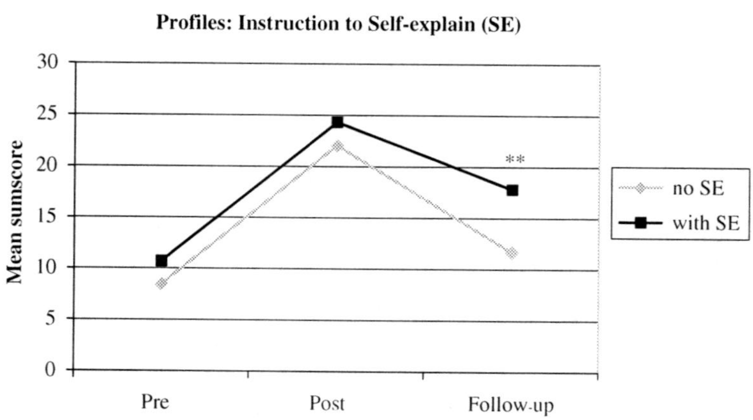

Figure 3. Profiles: Instruction to self-explain

The effect size for self-explanation increases if spatial abilities and self-assessed competency in physics are included in an analysis of covariance (ε=0.41: large effect).

Qualitative analysis of single items allows to gain deeper insight into students' understanding of diffraction and interference of light.

Figure 4 displays answer frequencies for a multiple choice item (Item 5; students had the chance to choose more than one answer).

In pre-test, half of the students chose the answer (b) saying that the left part of the interference pattern disappears. One can assume that these students mix up diffraction phenomena and shadow formation, because they do not realize that an even thinner slit will cause diffraction in the same manner as before, only that the pattern becomes wider. These students might use assumptions of the ray model as "the way of the light is blocked and because of that a shadow is formed". This is how shadow formation is explained in lower grades in school physics and students are familiar with this explanation. Even after the instruction, in which the diffraction pattern was discussed, 20% of the students persist in this concept of shadow formation. Though the correct answer (d) is chosen by more than 50% in the post-test, the distractors still seem reasonable to a lot of students. The answer frequencies for the follow-up test reveal how difficult it is to change students' conceptions permanently. The correct answer is chosen less often and the wrong alternatives gain higher frequencies. Students seem to "fall back" into their old conceptions, which does not surprise given the very short time of intervention.

Item 5

A diffraction pattern on a screen is generated by a vertical single slit. What can one observe if the left half of the slit is covered?

a) the maxima on the right side of the screen disappear
b) the maxima on the left side of the screen disappear
c) the maxima move further together
d) the maxima move further apart
e) nothing changes

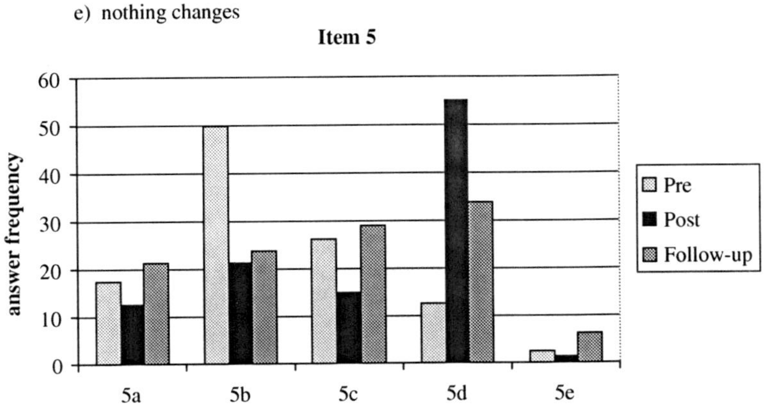

Figure 4. Evaluation of multiple choice Item 5

As an example for a task with a verbal answer format, Item 9b will be discussed. Students are asked to explain the following situation: "Light passes a fine grid and causes an observable coloured pattern on a screen behind it." The answers were categorized according to the procedure of qualitative content analysis (Mayring, 2003) (Table 3).

The task requires that students transfer their knowledge on wave model of light and diffraction by a single slit to a new situation, in which the slit is replaced with a grid. The answers in categories A–C reveal that these students are more or less able to perform this transfer. Some of them only focus on diffraction and interference respectively. In contrast the answers in categories D–F suggest that students apply the ray model of light, that is inadequate in this context, or they mix up assumptions from both models of light. In the pre-test category D with answers referring to refraction is found most frequently, followed by category E referring to shadow formation (see Figure 5; students who did not answer the question are excluded from this analysis).

These categories indicate that students try to use their prior knowledge on ray optics to explain the phenomenon. In post-test almost half (49%) of the answers belong to one of the categories A–C mentioning diffraction and/or interference as an explanation. The frequency of category B ("diffraction") in post- and follow-up test indicates that students' attention was focused more on diffraction of light than on interference during learning. A reason might be that diffraction was explained qualitatively in the first part of the computer program whilst interference was discussed towards the end in a more technical way.

TABLE 3. Categories of item 9b

Item 9b: Categories and examples of students' answers		
	Description of category	*Example: students' quotations*
A	Diffraction and interference of waves of light are named as reasons for the pattern	"An interference pattern is generated, because the light waves are diffracted by the grid and they overlap behind it."
B	Only diffraction as reason for the pattern	"Light falls on the grid and it is diffracted unequally and different colours can be observed."
C	Only interference as reason for the pattern	"Because of the optical retardation waves overlap and at certain places only certain colours remain."
D	Refraction as reason	"Light is refracted and separated into its colours"
E	Grid function as a "wavelength filter"/ causes shadow formation	"The colours of light have different wavelengths. If the amplitude of the waves is too large, they do not fit through the grid. Because of that only some colours pass the grid and one observes a coloured pattern." "Rays of light cannot pass some of the bars of the grid"
F	Light is decomposed/ Wavelength of light is changed	"The light is split into smaller rays and by that into different wavelengths"

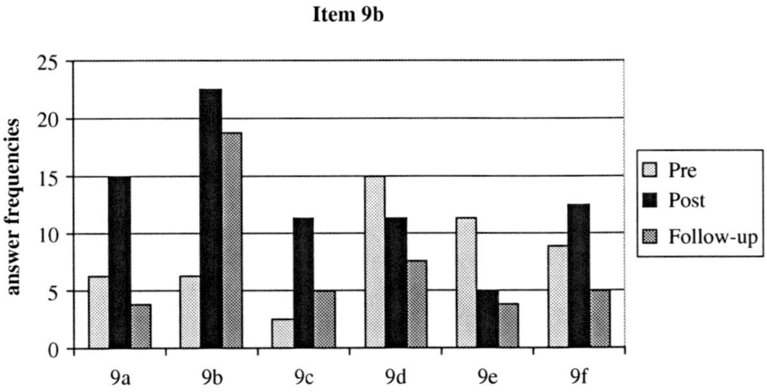

Figure 5. Evaluation of verbal task Item 9b

Students' answers in the knowledge test reveal that they often mix up assumptions of ray model and wave model of light. They come up with ideas that can be described as "hybridizations" or "hybrid models". Figure 6 shows two sketches of a student that he or she draws as an answer to the following question: "Please draw the interference pattern of light (consisting of one wavelength) as it appears on a screen if (a) a single slit is used; (b) a square opening is used; (c) a circular opening is used" (Item 4).

The student sketches the structure of the interference patterns as it would be caused by a single slit. He or she does not realize that the changed shape of the opening causes additional interference maxima. Instead of giving the whole pattern a new structure he or she changes the form of the maxima: If a square opening is used, the maxima have a square shape too and correspondingly the maxima have a circular shape if a circular opening is used. The pattern becomes a mixture of an interference pattern – it consists of maxima and minima – and of a shadow pattern, produced by the shape of the opening – the maxima adopt the shape of the opening.

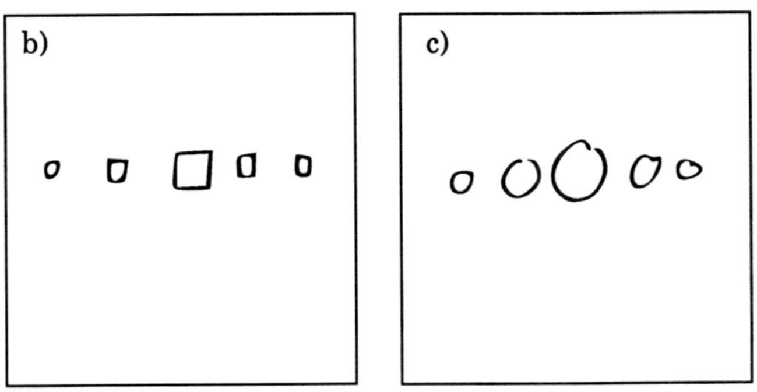

Figure 6. Student's answers to Item 4

5. CONCLUSIONS

The results of our investigation show that the computer-based tutorial on diffraction of light fulfils basic standards of usability and is adaptable for learning physics in school in respect to understandability and relevance of content. Students are able to learn about diffraction on an introductory level using the program. As the program covers only a small part of wave optics it should be implemented into a curriculum that is sensitive towards an explicit learning about models in the domain of optics.

Although our assumptions suggested that a text surface with high local textcoherence and text-picture-references should facilitate text comprehension of novices in the domain of physics, no effect is observable in our data. This might be traced back to two reasons. First the text of the computer program fulfils standard criteria of understandability. These might be helpful enough to cover up effects of our variable. The second and more plausible reason is the age and with it the prior experience of the students. Students in grade 12 might have learned to process specialized texts. They have become used to the language of textbooks in physics and are trained to understand difficult text.

Strong effects of the variable self-explanation are not found directly after the intervention. Results of the post-test only show a tendency for an effect on transfer. This contradicts our expectation that the prompted activity of the learners would lead to better comprehension immediately.

A difference in the depth of processing only manifests itself weeks after the intervention in the follow-up test. Students who were asked to self-explain are able to remember much more information from the learning sequence and to transfer it. According to the model of comprehension and learning by Schnotz (2002; Schnotz and Bannert, 2003), students' mental structures have changed in a sustainable way and this enables them to reconstruct their knowledge more adequately.

How is this to be explained? In general, the instruction to self-explain seems to trigger a higher level of cognitive activity and thereby a stronger germane load. Students' attention is focused on the subject that has to be learned. Additionally students might consult their prior knowledge more frequently and build up stronger connections between newly processed information and prior knowledge (Cote et al., 1998). This leads to a stable change of their mental structures.

While self-explaining the learners can realize inconsistencies within their prior knowledge or in their comprehension of new information. Such monitoring of one's own comprehension might be crucial especially in learning physics. Here students' (mis-) conceptions have proven to be resistant to change. Often they are even combined with the scientific conception. This was observable in our study in the qualitative analysis of items. However, the overall evaluation of the knowledge test suggests that students, who have self-explained, fall back into misconceptions less frequently than others. They might have used the self-explanation to "self-repair" their conceptions (Chi et al., 1994; McNamara, 2004).

Self-explanations can also be interpreted in terms of reading or learning strategies (Lind et al., 2005). Self-explanations include elaboration strategies, which foster integration of new knowledge into the existing knowledge, as well as reduction and organization strategies. These refer to the extraction of relevant information and their organization in order to facilitate the processing.

The qualitative analysis of students' answers shows the overall increase of scientific understanding. Nevertheless they reveal some typical misconceptions. The most typical ones are the inadequate application of the ray model of light and the hybridization of ray model and wave model of light. Students do not consider the model character of these descriptions and therefore are not aware of their limited explanatory power concerning physical phenomena. This leads us to the conclusion already mentioned above that students should learn optics in a model-sensitive approach. An explicit learning about models could help them to overcome the observed difficulties in handling the models of light.

Concerns that self-explanation without feedback could strengthen misconceptions of the learners (Cote et al., 1998) can be disapproved as far as we look at the whole sample of our study. Whether individual learners were mislead by their own explanation cannot be ruled out, and single case studies could shed light on this question in future.

In conclusion, our results suggest that an activation of learners in ways of "making them talk about physics and their learning process" helps them to learn. Therefore tasks demanding verbalization and use of language should be implemented more often in general and especially in computer-based physics instruction. Especially in instructional settings that demand autonomous learning of students it seems promising to create tasks that induce activities similar to self-explanations. Studies investigating, for example, pairwise explaining instruction would lead further in this direction.

REFERENCES

Ambrose, B. S., Shaffer, P. S., Steinberg, R. N., & McDermott, L. C. (1999). An investigation of student understanding of single-slit diffraction and double-slit interference. *Am. J. Phys, 67*(2), 146–155.

Brünken, R., Seufert, T., & Zander, S. (2005). Förderung der Kohärenzbildung beim Lernen mit multiplen Repräsentationen. *Zeitschrift für Pädagogische Psychologie, 19*(1/2), 61–75.

Chi, M. T. H., deLeeuw, N., Chiu, M.-H., & Lavancher, C. (1994). Eliciting self-explanations improves understanding. *Cognitive Science, 18*, 439–477.

Cote, N., Goldman, S. R., & Saul, E. U. (1998). Students making sense of informational text: Relations between processing and representation. *Discourse Processes, 25*(1), 1–53.

deLeeuw, N., & Chi, M. T. H. (2003). Self-explanation: Enriching a situation model or repairing a domain model? In G. Sinatra & P.Pintrich (Eds.), *Intentional Conceptual Change* (pp. 55–78). Mahwah, NJ.: Erlbaum.

Duit, R., Gropengießer, H., & Kattmann, U. (2005). Towards science education research that is relevant for improving practice: The model of educational reconstruction. In H. E. Fischer (Ed.), *Developing Standards in Research on Science Education* (pp. 1–9). London: Taylor & Francis Group.

Heller, K. A., & Perleth, C. (2000). *KFT 4-12+R. Kognitiver Fähigkeitstest für 4. bis 12. Klassen. Revision*. Göttingen: Beltz.

Hoffmann, L., Häußler, P., & Lehrke, M. (1998). *Die IPN-Interessenstudie Physik*. Kiel: IPN.

Lind, G., Friege, G., & Sandmann, A. (2005). Selbsterklären und Vorwissen. *Empirische Pädagogik, 19*(1), 1–27.

Mayer, R. E. (2001). *Multimedia Learning*. Cambridge: University Press.

Mayring, P. (2003). *Qualitative Inhaltsanalyse. Grundlagen und Techniken.* Weinheim und Basel: Beltz Verlag.

McNamara, D. S. (2004). SERT: Self-Explanation Reading Training. *Discourse Processes, 38*(1), 1–30.

Mikelskis-Seifert, S., & Leisner, A. (2005). Investigation of effects and stability in teaching model competence. In K. Boersma,M. Goedhart, O. deJong & H. Eijkelhof (Eds.), *Research and the Quality of Science Education* (pp. 337–351). Dordrecht: Springer.

Moreno, R., & Mayer, R. E. (2000). Engaging students in active learning: The case for personalized multimedia messages. *Journal of Educational Psychology, 92*(4), 724–733.

Rickheit, G., & Schade, U. (2000). Kohärenz und Kohäsion. In K. Brinker, G. Antos & U. A. (Eds.), *Text und Gesprächslinguistik. Ein internationales Handbuch zeitgenössischer Forschung.* Berlin, New York: de Gruyter.

Schnotz, W. (2002). Towards an integrated view of learning from text and visual displays. *Educational Psychology Review, 14*(1), 101–120.

Schnotz, W., & Bannert, M. (2003). Construction and interference in learning from multiple representation. *Learning and Instruction, 13*, 141–156.

Starauschek, E., & Rabe, T. (2004). Multimedialernen im naturwissenschaftlichen Unterricht: Bemerkungen zu kognitionspsychologischen Lerneffekten multimedialer Wissensrepräsentation beim Physiklernen. In A. Pitton (Ed.), *Zur Didaktik der Physik und Chemie. Vorträge der GDCP-Tagung in Berlin 2003* (pp. 114–116).

Vosniadou, S. (1999). Conceptual change research: State of the art and future directions. In W. Schnotz, S. Vosniadou & M. Carretero (Eds.), *New Perspectives on Conceptual Change* (pp. 3–13). Amsterdam U.A.: Elsevier.

Wellington, J., & Osborne, J. (2001). *Language and literacy in science education.* Buckingham, Philadelphia: Open University Press.

Wild, K. A., Schreyer, I., Lewalter, D., & Krapp, A. (1997). *Dokumentation und Analyse der Erhebungsverfahren in der Intensivstudie Berufsschulunterricht. Berichte aus dem DFG-Projekt "Bedingungen und Auswirkungen berufsspezifischer Lernmotivation". Nr.3* (No. 3): Universität München.

Yore, L. D. (2003). Examining the literacy component of science literacy: 25 years of language arts and science research. *International Journal of Science Education, 25*(6), 689–725.

AUTHOR INDEX

SUBJECT INDEX

Printed in the United States
89650LV00001B/70-96/A